Springer Tracts in Mechanical Engineering

About this Series

Springer Tracts in Mechanical Engineering (STME) publishes the latest developments in Mechanical Engineering - quickly, informally and with high quality. The intent is to cover all the main branches of mechanical engineering, both theoretical and applied, including:

- Engineering Design
- Machinery and Machine Elements
- Mechanical structures and stress analysis
- Automotive Engineering
- Engine Technology
- Aerospace Technology and Astronautics
- Nanotechnology and Microengineering
- Control, Robotics, Mechatronics
- MEMS
- Theoretical and Applied Mechanics
- Dynamical Systems, Control
- Fluids mechanics
- Engineering Thermodynamics, Heat and Mass Transfer
- Manufacturing
- Precision engineering, Instrumentation, Measurement
- Materials Engineering
- Tribology and surface technology

Within the scopes of the series are monographs, professional books or graduate textbooks, edited volumes as well as outstanding PhD theses and books purposely devoted to support education in mechanical engineering at graduate and post-graduate levels.

Indexed by SCOPUS and Springerlink.

To submit a proposal or request further information, please contact: Dr. Leontina Di Cecco Leontina.dicecco@springer.com or Li Shen Li.shen@springer.com.

Please check our Lecture Notes in Mechanical Engineering at http://www.springer.com/series/11236 if you are interested in conference proceedings. To submit a proposal, please contact Leontina.dicecco@springer.com and Li.shen@springer.com.

More information about this series at http://www.springer.com/series/11693

Falah Alobaid

Numerical Simulation for Next Generation Thermal Power Plants

Springer

Falah Alobaid
Forschungsgruppenleiter des Inst. für
Energiesysteme und Energietechnik (EST)
Technische Universität Darmstadt
Darmstadt
Germany

ISSN 2195-9862 ISSN 2195-9870 (electronic)
Springer Tracts in Mechanical Engineering
ISBN 978-3-030-09437-9 ISBN 978-3-319-76234-0 (eBook)
https://doi.org/10.1007/978-3-319-76234-0

Printed on acid-free paper

This Springer imprint is published by the registered company Springer International Publishing AG part of Springer Nature
The registered company address is: Gewerbestrasse 11, 6330 Cham, Switzerland

Static is simple, but the universe is dynamic

Falah Alobaid

Foreword

The conventional design of thermal power plants mainly focuses on high process efficiency, while market requirements more and more target operation flexibility due to the increased penetration of renewable energy sources. Five major approaches to maintain the supply security of future electricity systems are distinguishable in general:

(1) Expansion of high-voltage transmission infrastructure: The adequate addition of high-voltage transmission infrastructure in combination with smart power electronics is essential to accommodate the growing renewable capacity and ensures security of supply. By increasing the number of interconnections between major load centres, the transmission capacity and robustness of the electricity grid can be greatly improved. For sufficient flexibility of the overall system, the grid expansion must be complemented by additional measures. Environmental and political challenges are common causes for the delayed expansion of grid infrastructure.

(2) Enrolment of demand response: Demand response is the ability of end users to reduce load according to price signals or dispatch rules. This measure can be applied to mitigate or counteract short-term grid imbalances.

(3) Modification to system operations: Improved forecasting decreases the uncertainty in net load, resulting in reduced utilisation of expensive peaking capacity and a cost-effective modification of system operations. Both (2) and (3) are support measures under development with the potential to effectively reduce costs associated with the transformation of the energy system.

(4) Large-scale energy storage (also called grid-energy storage): These systems can be charged or discharged flexibly to allow for high shares of renewable electricity feed-in. Grid-energy storage is, in theory, ideally suited to balance intermittent power supply and demand. Pumped hydroelectric storage is commonly used, but it is limited to suitable geographical conditions. Other technologies in various stages of development include battery energy storage, flywheel energy storage, compressed air energy storage, power-to-gas and thermal energy storage. Despite recent advances in battery storage driven by the

automotive industry, the specific costs of storage systems still limit their use to small-scale applications and there is currently no economically viable storage technology available for the required capacities (TWh rather than MWh).

(5) Flexibility improvement of existing infrastructure of electricity supply (dispatchable power generation): This approach represents the most economic option for increasing system flexibility. In many countries without abundant natural resources suitable for large hydro or geothermal energy, power generation is mainly based on thermal power plants and will continue to do so in the foreseeable future, considering plant lifetime of up to 40 years. The operating flexibility of thermal power plants is limited by technical constraints such as ramp rates and minimum load limit. Existing power plants can be retrofitted with optimised components and control circuits to mitigate these constraints and to meet enhanced flexibility requirements. Highly dispatchable generating units such as combined-cycle power plants and gas engines are also available to replace outdated plants.

This book is built around the approach of the dispatchable generating units and offers significant benefits to students, engineers, researchers and industrial experts. Mathematical models including one-dimensional dynamic process simulation and three-dimensional computational fluid dynamics (CFD) are developed and validated with the operation data from real plants. The book can be divided into three main parts:

- In the first part, the mathematical background for the one-dimensional dynamic process modelling of thermal power plants is shown. The thermal hydraulic models such as the mixture flow model and the two-fluid flow models are explained in detail. Furthermore, the process, automation and electrical components required for the dynamic simulation of energy systems are described.
- The second part presents the mathematical background for the three-dimensional CFD simulation of gas-solid flows such as quasi-single-phase, two-fluid and single-particle methods. The focus is on the discrete element method, including the forces and the moments of force acting on particles. Here, the determination of inter-phase values (volumetric void fraction, momentum and heat transfers) is carried out using a new procedure known as the offset method. Furthermore, the particle-grid method that allows the refinement of the grid resolution independent of the particle size is introduced.
- The last part shows relevant results on dynamic process simulations for different technologies of combined-cycle power plant, pulverised coal-fired power plant, concentrated solar power plant and municipal waste incinerator. Furthermore, the application of CFD models to different energy systems is shown as well, e.g. 1 MWth combustion chamber, 1 MWth fluidized bed, 60 MWth waste incinerator, large-scale lignite and hard coal-fired power plants. Measurements obtained from real energy systems are applied to validate the process and CFD simulation models.

My gratitude is pointed to the author, who achieved a great contribution to the field of theoretical modelling and practical application of energy systems. The extensive review for this book summarises state-of-the-art modelling techniques ranging from 1D steady-state and dynamic approaches to 3D-CFD reactor models with high reference to industry applications. The author can be sure that there will be many grateful readers benefiting from the detailed summary on mathematical formulations for the one-dimensional dynamic process simulations and the three-dimensional computational fluid dynamics models, not only because the modelling gained particular importance nowadays, but also of the comprehensively way of writing. Once again, I would like to thank the author for his considerable efforts in writing this book.

Darmstadt, Germany
March 2018

Bernd Epple
Technische Universität Darmstadt

Preface

This book provides researchers, developers and practitioners with a major contribution to mathematical models' developments (including one-dimensional dynamic process simulation and three-dimensional computational fluid dynamics (CFD)) and their applications to various energy systems (e.g. combined-cycle power, pulverised coal-fired power, concentrated solar power and fluidized bed systems).

Process, automation and electrical components as well as the mathematical background for process modelling, i.e. the mixture flow model and the two-fluid flow models (four-equation, five-equation, six-equation and seven-equation flow model), are explained in detail. CFD numerical methods for the modelling of gas–solid flows (quasi-single-phase, two-fluid and single-particle methods) are presented, too. Special attention is given to the discrete element method, including the forces and the moments of force acting on particles. Relevant studies that describe the application of dynamic process simulations and CFD models to different energy systems are shown. Measurements obtained from real thermal power plants are used to validate the process and CFD simulation models.

The book entitled "Numerical Simulation for Next Generation Thermal Power Plants" is based on my habilitation thesis and represents the results obtained during my time at the Department of Energy Systems and Technology at the Technische Universität Darmstadt.

I would like to express my gratitude to

Prof. Dr.-Ing. Bernd Epple (Technische Universität Darmstadt),

Prof. Dr.-Ing. Johannes Janicka (Technische Universität Darmstadt),

Prof. Dr.-Ing. Hartmut Spliethoff (Technische Universität München),

Prof. Dr.-Ing. Jürgen Karl (Friedrich-Alexander-Universität Erlangen-Nürnberg) and Prof. Dr.-Ing. Timo Hyppänen (Technische Universität Lappeenranta),

and all people, who directly or indirectly supported me to accomplish this book.

I wish also to thank Dr.-Ing. Ralf Starkloff, Dr.-Ing. Nicolas Mertens, Dr.-Ing. Wisam Abed Kattea AL-Maliki, Dr.-Ing. Ralf Postler, Alexander Stroh, M.Sc., and Jan May, M.Sc., for their scientific contributions to this book. Special thanks to my friends Jochen Ströhle, Daniel Binzer, Vitali Kez, Nabil Baraki and Mohamad Aladib.

Exceptionally, I wish to express my gratitude to my family for supporting me during the long time period of my education.

Darmstadt, Germany
March 2018

Falah Alobaid

Contents

Nomenclature

Latin Symbols

A	Surface, vessel cross-section area, pre-exponential factor [m^2], [m^2], [1/s]
a	Translational acceleration [m/s^2]
B	Magnetic flux density [N/A m]
b	Angular acceleration [rad/s^2]
C	Coefficient, correction factor [–], [–]
c	Concentration [kmol/m^3]
c_p	Specific heat capacity [J/kg K]
D	Surface diffusion coefficient, volume diffusion coefficient, diameter [m^2/s], [m^3/s], [m]
d	Displacement, diameter [m], [m]
E	Young's modulus, electric field intensity, energy, rate of entrainment [N/m^2], [N/A s], [J], [–]
e	Restitution coefficient [–]
F	Force [N]
f	Force density [N/m^3]
G	Shear modulus, particle acceleration [N/m^2], [m/s^2]
g	Standard gravity [m/s^2]
H	Pump head [m]
h	Enthalpy, height, rotation unit vector [kJ/kg], [m], [–]
$\mathbf{I}$	Unit matrix [–]
I	Moment of inertia, current [kg m^2], [A]
J	Impulse vector, superficial velocity [kg m/s], [m/s]
K	Stodola coefficient, heat transfer coefficient [–], [W/m^2 K]
k	Stiffness coefficient, reaction rate constant [N/m], [1/s]
k_R	Rotational stiffness coefficient [s]
l	Length [m]

M	Moment of force, molar mass [N m], [kg/kmol]
m	Mass [kg]
$\dot{m}$	Mass flow rate [kg/s]
N	Number of particle [–]
n	Number of north magnetic poles [A m]
P	Power [W]
p	Static pressure [N/m^2]
Q	Heat flow, capacity [J], [$A\ s$]
q	Charge, rate of heat flow [A s], [J/s]
$\mathbf{R}$	Rotation matrix [–]
R	Radius, rate of stratification, resistance [m], [–], [Ω]
r	Radius, position vector, reaction rate [m], [m], [kmol/m^3 s]
S	Source term, saturation level [different], [–]
s	Distance, unit vector [m], [–]
$\mathbf{T}$	Stress tensor for Newtonian fluids [N/m^2]
T	Torque, temperature, computing time [N m], [K], [s]
t	Time [s]
U	Electric potential, internal energy, voltage [kg m^2/A s^3], [J], [V]
u	Translational velocity [m/s]
V	Volume, velocity [m^3], [m/s]
$\dot{V}$	Volume flow rate [m^3/s]
W	Weight [kg]
w	Angular velocity [rad/s]
X	Mass fraction, molar fraction [kg/kg$_{mix}$], [mol/mol$_{mix}$]
x	Steam concentration [kmol/m^3]
x, y, z	Cartesian coordinates [m]

Greek Symbols

α	Heat transfer coefficient, collision angle [W/K m^2], [rad]
α_{UDS}	Blending factor [–]
β	Resistance coefficient, restitution coefficient, diffusion parameter [kg/s m^3], [–], [–]
γ	Surface tension, isentropic exponent [N/m], [–]
δ	Penetration depth, diameter [m], [m]
e	Restitution coefficient [–]
ε	Volumetric void fraction (porosity), emissivity [–], [–]
Γ	Mass transfer [kg/s m^3]
η	Damping coefficient, rolling damping constant, polytropic expansion efficiency, efficiency [N s/m], [kg m^2/s], [–], [–]
θ	Contact angle, inclination angle [rad], [rad]
λ	Thermal conductivity, bulk viscosity [W/K m], [kg/m s]

μ	Dynamic viscosity, friction coefficient, relative permeability [kg/m s], [–], [–]
μ_r	Rolling friction coefficient [m]
ξ	Fill angle [rad]
ω	Angular velocity [rad/s]
ρ	Density [kg/m^3]
σ	Surface tension [N/m]
τ	Viscous stress, stress tensor [N/m^2], [N/m^2]
ν	Poisson's number, stoichiometric coefficient, specific volume [–], [–], [m^3/kg]
ϕ	Physical value [different]
φ	Angular displacement, rotation angle, internal friction angle, surface charge [rad], [rad], [rad], [A s/m^2]
Θ	Granular temperature [m^2/s^2]
χ	Void fraction [–]
Ω	Computational domain [–]
$\partial\Omega$	Grid surface [m^2]
Π	Pressure ratio [–]

Constants

A	Hamaker constant [kg m^2/s^2]
a	Lattice constant [m]
c	Speed of sound, speed of light [m/s], [m/s]
h	Planck's constant [kg m^2/s]
k_B	Boltzmann's constant [kg m^2/s^2 K]
R	Universal gas constant [N m/kmol K]
ε_0	Absolute permittivity (electric constant) [A^2 s^4/kg m^3]
ε_{rel}	Relative permittivity (dielectric constant) [–]
λ	London constant [kg m^8/s^2]
σ	Stefan–Boltzmann constant [kg/s^3 K^4]
hw	Lifshitz-van der Waals constant [kg m^2/s^2]

Dimensionless Numbers

Kn	Knudsen number
Pe	Peclet number
Pr	Prandtl number
Re	Reynolds number
Sc	Schmidt number
Stk	Stokes number

Subscripts and Indices

acc	Accumulation
adh	Adhesion
ads	Adsorption
air	Air
ann	Annular
aper	Aperture
attemp	Attemperator
ave	Average
Bac	Backward
Bas	Basset
bc	Boundary condition
Beet	Beetstra
bin	Binary
bio	Biomass
blow	Blow-down
Boud	Boudouard
bub	Bubble
buo	Buoyancy
CV	Control volume
calc	Calcination
cap	Capillary
carb	Carbonation
cav	Cavity
cha	Char
char	Characteristic
clean	Cleanliness
col	Collision
cool	Cooling
com	Compressor
con	Contact
cont	Control
conv	Convection
cpl	Close packing limit
cri	Critical
D	Dimensional, derivative
DC	Device control
dam	Damping
den	Denominator
der	Derivation
dif	Diffusion
dis	Displaced
down	Downcomer

dra	Drag
dro	Droplet
dru	Drum
dust	Dust
dyn	Dynamic
Erg	Ergun
e	East
ela	Elastic
ele	Electric
elst	Electrostatic
$end - los$	End-loss
eq	Equilibrium
equ	Equivalent
ext	External
FF	Feedforward
Fos	Foscolo
f	Fluid or function
$f \rightarrow P$	Fluid to particle
fg	Flue gas
fil	Filter
fin	Final
flu	Fluid
for	Forward
fri	Frictional
fuel	Fuel
fw	Feedwater
Gar	Garside
Gid	Gidaspow
g, gas	Gas
gra	Gravitation, grain
Hertz	Hertz's theory
Hooke	Hooke's theory
h	Hydraulic, heat
het	Heterogeneous
hom	Homogeneous
hor	Horizontal
I	Integral
IAM	Incidence angle modifier
i	Component, notation, particle index
ilme	Ilmenite
in	Inner
inl	Inlet
int	Internal, interfacial, inter-stage, integration
isen	Isentropic
iwal	Particle i-wall

j	Component, notation, particle index
Koch/Hill	Koch and Hill
k	Component, notation
kin	Kinetic
l, liq	Liquid
lift	Lift
lin	Linear
loc	Located
load	Load
low	Lower
Mags	Magnus
m	Mixture
mag	Magnetic
mas	Mass
max	Maximum
mech	Mechanical
mem	Memory
min	Minimum
NC	Non-condensable
n	Normal, north
nec	Neck
nom	Nominal
num	Numerator
OC	Oxygen carrier
opt	Optical
out	Outer, outlet
ox	Oxidation
P	Particle, cell centre, proportional
$P \rightarrow f$	Particle to fluid
PP	Particle–particle
Pwal	Particle-wall
par	Parcel
pre	Pressure
pu	Pump
pyr	Pyrolysis
R	Rotation
r	Rolling
rad	Radiometric, radiation, radial
rc	Raw coal
ref	Reference, refractive
rel	Relative
relax	Relaxation
res	Resultant
riser	Riser
S	Shear flow

s	Solid, surface, south
Saf	Saffman
Sya	Syamlal and O'Brien
sat	Saturation
sca	Scattered
sep	Separation
sg	Search grid
sgc	Search-grid cell
shad	Shadowing
sin	Sintering
slide	Slide
sol	Solid bridge
sorb	Sorbent
ss	Solid–solid interaction
st	Steam
sta	Stack
static	Static
sto	Storage
stop	Stop
sub	Sub-cooling
sup	Superficial, superheating
sur	Surround
T	T-transition
t	Tangential
ter	Terminal
the	Theoretical
tot	Total
tra	Transition
track	Tracking
tur	Turbulent
turb	Turbine
u	Momentum
up	Upper
V	Volume
vac	Vacuum
val	Valve
van	Van der Waals
ver	Vertical
ves	Vessel
vir	Virtual
vm	Virtual mass
Wen&Yu	Wen and Yu
w	West, wall
waf	Water- and ash-free
wal	Wall

wat	Water
–	Average
^	Non-dimensional
0	Initial, standard, reference
(0)	Before the collision, before the contact

Chemical Symbols

Ar	Argon
C	Carbon
CaO	Calcium oxide
$CaCO_3$	Calcium carbonate
CH_4	Methane
C_2H_4	Ethane
C_2H_6	Ethane
C_3H_8	Propane
CO	Carbon monoxide
CO_2	Carbon dioxide
COS	Carbonyl sulphide
Fe_2O_3	Ferric oxide
$FeTiO_3$	Ilmenite
H_2	Hydrogen
H_2O	Water
H_2S	Hydrogen sulphide
HCl	Hydrogen chloride
N_2	Nitrogen
NO	Nitric oxide
NO_2	Nitrogen dioxide
NO_x	Nitrogen oxides
O_2	Oxygen
TiO_2	Titanium dioxide
SO	Sulphur monoxide
SO_2	Sulphur dioxide
SO_x	Sulphur monoxide/dioxide/trioxide

Abbreviations

ABB	Asea Brown Boveri (multinational corporation)
AC	Alternating current
AF	Auxiliary fan
ANSYS	Commercial CFD software
APH	Air preheater
ASME	American Society of Mechanical Engineers

ASPEN	Commercial process simulation programme
APROS	Commercial process simulation programme
BARRACUDA	Commercial CFD software
BDS	Backward difference scheme
BFP	Boiler feedwater pump
BMCR	Boiler maximum continuous rating
BTL	Biomass to liquid
BUS	Burn-out supply
CAES	Compressed air storage system
CCL	Calcium carbonate-looping
CCPP	Combined-cycle power plants
CCS	Carbon capture and storage
CCV	Circulation control valve
CDS	Central differencing scheme
CFB	Circulating fluidized bed
CFD	Computational fluid dynamics
CIRC	Circulation
CLC	Chemical-looping combustion
CP	Circulation pump
CPFD	Computational particle fluid dynamics
CPH	Condensate preheating
CPU	Central processing unit
CSP	Concentrated solar power
CVFEM	Control volume based finite element method
D	Derivative controller
1D	One-dimensional
2D	Two-dimensional
3D	Three-dimensional
DB	Dead band value
DBS	In-house process simulation programme
DC	Direct current
DEM	Discrete element method, diffuse element method
DEMEST	In-house CFD/DEM code
DFGD	Dry flue gas desulfurization
DNI	Direct normal irradiance
DNS	Direct numerical simulation
DO	Discrete ordinates
DOOSAN	Multinational corporation
DTRM	Discrete transfer radiation model
DYMOLA	Commercial process simulation programme
EBSILON	Commercial process simulation programme
ECO	Economiser
EDC	Eddy dissipation concept
EDM	Eddy dissipation model
EFG	Element-free Galerkin method

EMMS	Energy minimisation multiscale
EST	Energy systems and technology department
ESP	Electrostatic precipitator
EVA	Evaporator
FD	Finite difference method
FDS	Forward difference scheme
FE	Finite element method, feedforward controller
FG	Flue gas
FGC	Flue gas condenser
FTT	Fast Fourier transform
FUNC	Function
FV	Finite volume method
GE	General Electric (multinational corporation)
GGH	Gas–gas heat exchanger
GIMP	Raster graphics editor (programme)
GT	Gas turbine
HCE	Heat collection element
HFO	Heavy fuel oil
HP	High pressure
HPBPCV	High-pressure bypass control valve
HPMSCV	High-pressure main steam control valve
HRA	Heat recovery area
HRSG	Heat recovery steam generator
HTF	Heat transfer fluid
HV	Hysteresis value
I	Integral controller
IGCC	Integrated gasification combined cycle
IDF	Induced draft fan
IP	Intermediate pressure
IPBPCV	Intermediate-pressure bypass control valve
IPMSCV	Intermediate-pressure main steam control valve
ISCC	Integrated solar combined-cycle power plant
JModelica	In-house process simulation programme
KTDG	Kinetic theory of dense gases
KTGF	Kinetic theory of granular fluids
LBM	Lattice Boltzmann method
LCOE	Levelised costs of energy
LES	Large eddy simulation
LFO	Light fuel oil
LHV	Lower heating value
LP	Low pressure
LPBPCV	Low-pressure bypass control valve
LPMSCV	Low-pressure main steam control valve
LV	Limit value
LU	Lower upper

MATLAB	Commercial mathematical software
MCFC	Molten carbonate fuel cell
MD	Molecular dynamics approach
MLPG	Meshless Local Petrov-Galerkin method
MODELICA	Multidomain modelling language
MPI	Message passing interface
MPICH2	Software for message passing interface
MP-PIC	Multiphase particle-in-cell
MSW	Municipal solid waste
NG	Natural gas
NPSH	Net positive suction head
ODE	Ordinary differential equation
OpenFOAM	Open-source CFD software
P	Proportional controller
P1	Radiation model
PAFC	Phosphoric acid fuel cell
PBE	Population balance equation
PD	Proportional-derivative controller
PDE	Partial differential equation
PEM	Proton exchange membrane fuel cell
PH	Preheater
PI	Proportional–integral controller
PIC	Particle-in-cell method
PID	Proportional-integral-derivative controller
PISO	Pressure-implicit with splitting of operators
PPSD	Commercial process simulation programme
POC	Post-oxidation chamber
PSIC	Particle source in cell method
PSD	Particle size distribution
PUM	Partition of unity method
PV	Photovoltaic cells
QUICK	Quadratic upwind interpolation for convective kinematic
RANS	Reynolds-averaged Navier–Stokes method
RDF	Refuse derived fuel
RELAP	Commercial process simulation programme
REF	Recirculation fan
RF	Rotary feeder
RH	Reheater
RKPM	Reproducing kernel particle method
RMS	Root mean square
SC	Steam cycle, screw conveyor
SCA	Solar collector assemble
SCE	Solar collector element
SCM	Shrinking core model
S-CO_2	Supercritical carbon dioxide

SCR	Selective catalytic reduction
Select	Selector
SFB	Stationary fluidized bed
SH	Superheater
SIMPLE	Semi-implicit method for pressure-linked equations
SIMPLEC	Semi-implicit method for pressure-linked equation consistent
SimulationX	Commercial process simulation programme
SIMULINK	Commercial mathematical software
SM	Spectral method
SNCR	Selective non-catalytic reduction
SOFC	Solid oxide fuel cell
SOR	Successive over-relaxation
SPH	Smoothed particle hydrodynamic method
ST	Steam turbine
TCP/IP	Transmission control protocol/Internet protocol
TGA	Thermogravimetric analysis
TGF	Transport gas fan
TOC	Total organic carbon
UDS	Upwind differencing scheme
VM	Vortex method
WFGD	Wet flue gas desulfurization
WS	Water/steam

Chapter 1
Introduction

The increased expansion of intermittent electricity generation in the energy grid, notable wind power and photovoltaics can lead to a seemingly paradox situation of negative electricity prices at times of low demand and/or high renewable electricity output. The main reason here is the inflexibility of dispatchable electricity generations, e.g. coal and nuclear power that continue generating power despite the negative price signal from the electricity market in order to avoid a cost-intensive unit shutdown. Several potential solutions to maintain security of supply and to improve flexibility of the future electricity system can be considered such as expansion of high-voltage transmission infrastructure, enrolment of demand response, modification to system operations, large-scale energy storage and flexibly dispatchable power generation (Adams et al. 2010). Each of these solutions differs in its potential impact, technological maturity and economic viability. The author is convinced that the future electricity system will comprise all of these concepts, to varying degrees and with the possible integration of value-adding processes beyond electricity (e.g. power-to-fuels).

In many countries without abundant natural resources suitable for the use of large hydro or geothermal energy, power generation is mainly based on thermal power plants and will continue to do so in the foreseeable future, considering plant lifetimes of up to 40 years (Walter and Epple 2016). The operating flexibility of thermal power plants is limited by technical constraints such as ramp rates and minimum load limit. Existing thermal power plants can be retrofitted with optimised components and control circuits to mitigate these constraints and thus to meet the enhanced flexibility requirements. Highly dispatchable generating units such as combined-cycle power plants are also available to replace outdated power plants. In addition to the integration of large-scale energy storage systems in the existing thermal power plants the utilisation of low-rank solid fuels in fluidized beds and the co-firing of biomass in existing coal-fired power plants are promising technologies for a secure, environmental and sustainable energy supply.

Basically, it is distinguished between conventional power plants that use fossil fuels such as natural gas, petroleum or solid fuels (coal, biomass, municipal waste,

F. Alobaid, *Numerical Simulation for Next Generation Thermal Power Plants*,
Springer Tracts in Mechanical Engineering,
https://doi.org/10.1007/978-3-319-76234-0_1

refuse derived fuel or even dried sewage sludge) and nuclear plants that use nuclear fuels such as uranium-235. In conventional thermal power plants, the fossil fuel is combusted, producing a high temperature heat flow. Part of this flow is converted into mechanical energy, driving a steam turbine coupled with an electrical generator and the rest is dissipated from the system into the atmosphere as residual heat (Spliethoff 2010). A thermal power plant ideally runs at steady-state design load, but it is also required to operate on, so-called off-design load conditions due to the fluctuations of supply and demand as well as the increased penetration of renewable energy sources. In Europe, these requirements translate into new operating challenges that can be divided in three categories (Alobaid et al. 2017). Firstly, higher load gradients for both positive and negative load changes are required. Furthermore, the dynamic of shutdown/start-up procedures should be optimised in response to a sudden load change in the grid. Secondly, the potential operating range of the thermal power plant has to be re-evaluated in accordance with the technical minimal load limit. Lower minimal load limit potentially reduces the number of shutdown/start-up procedures and can directly impact on the lifetime consumption of thermally stressed components. Thirdly, high efficiency at part load is relevant, since the thermal power plants that were originally operated almost continuously at nominal load should now run in load-following operation. Here, a thermo-economic optimisation at different base loads and off-design load conditions is necessary.

Mathematical models contribute to a better understanding of the process and play an important role for increasing the power plant efficiency (Flynn 2003). Generally, design and optimisation of energy systems start with steady-state modelling. Here, it is assumed that the power plant operates continuously at its design base load. The steady-state models do not require control structures and are mathematically based on mass, momentum, species and energy balances. Using steady-state simulation models, analyses of the thermodynamic properties of working fluid, mass and energy flows as well as process efficiency can be conducted for a series of operating points. However, steady-state simulation models do not provide any information about transient conditions. The relevant next step is therefore the process analysis with a dynamic simulation model. The latter presents an effective tool for assessing the control strategies, capabilities and the limitations when the operating system is close to the critical points. Dynamic simulation is preferred for the proposal stage of a thermal power plant project, e.g. to check whether or not the load changes according to specific customer requirements are feasible without unacceptable lifetime consumption in thick-walled components. It is also a cost-efficient approach to support unit commissioning and regular operation by estimating component lifetime and directing maintenance. However, investigation into the dynamic performance of thermal power plants requires detailed information of process and its control structures. The inherent complexity of the governing differential conservation equations and the numerical solution methods leads to highly complicated software with long development periods.

For the transient modelling of thermal power plants, different process components such as pipe, heat exchanger, drum and pump etc. are required. In addition to process components, a thermal power plant includes several automation and electrical systems. The accurate description of control structures is essential in order to obtain a meaningful dynamic response. The consideration of electrical components in the dynamic simulation is required to calculate the electrical power consumption at base loads and make sure that automation and process components get the needed electric power during transients. Furthermore, the effect of possible failures in electrical network on the system can be evaluated.

In thermal power plants, two-phase flows are present in the water/steam evaporator circuits. As this flow type is complicated and diverse, a number of two-phase models with various levels of complexity were proposed in literature. There are typically two classes of models for describing the two-phase flows, namely mixture flow model and two-fluid models (Yadigaroglu and Hewitt 2017). In the first category, the two-phase flow is considered as a mixture and treated as a single-phase flow with fairly complex thermodynamic properties. The gas and liquid phases are assumed to be in thermodynamic equilibrium with equal velocity, pressure and temperature. The second category of two-phase flow models (four-equation, five-equation, six-equation and seven-equation flow model) treats each phase separately and accordingly two sets of conservation equations for both phases are formulated. In addition to the conservation equations, adequate constitutive equations and experimental correlations are required. These can be tables for thermodynamic and transport properties or relations for heat transfer coefficients. The four-equation or five-equation version of the two-fluid models supposes mechanical and thermal equilibrium, but not chemical equilibrium. The classic six-equation flow model is a single pressure model. Here, both phases are in mechanical equilibrium, but not in chemical and thermal equilibrium. The seven-equation version of the two-fluid models treats each phase separately and does not assume, by contrast to the six-equation flow model, pressure equilibrium between the phases.

The dynamic simulation model can be used efficiently to investigate the design and control of the entire power plant during transients, but does not completely resolve the flow. Computational fluid dynamics (CFD), by contrast, offers a detailed insight into the flow pattern, providing qualitative and in many cases quantitative predictions. Generally, the numerical methods of CFD can be applied to individual components of thermal power plants, e.g. combustion chambers or heat exchangers. In the last decades, the CFD simulation of multi-phase flows, especially the gas-solid flow in terms of gasification and combustion of solid fuels, has attracted attention in the scientific literature. The modelling, the characterisation and the advanced understanding of this flow present now a major challenge since its behaviour is still unknown and difficult to predict in many applications. Basically, there are different approaches for the representation of gas-solid flows, namely quasi-single-phase, two-fluid and single-particle method (Alobaid 2013). The quasi-single-phase model assumes that a gas-solid flow is homogeneous and accordingly the conservation equations for mass, momentum and energy of the single-phase flow are used to model the physical values of the mixture. In the two-fluid method, also known as Euler-Euler model, each phase is regarded as a continuum and is mathematically calculated by solving the conservation equations, considering interaction terms between both phases. The single-particle method, also known as Euler-Lagrange model, considers the solid as disperse phase, i.e. for each particle a set of transport equations is solved, while the gas phase is modelled using modified conservation equations of the single-phase flow. In the single-particle method, the particle-particle and particle-wall collisions can be stochastically modelled or deterministically detected. For the deterministic collision detections, two models are used for the collision treatment: the hard sphere model and the soft sphere model. In the latter, also known as discrete element method (DEM), the particles can overlap each other or penetrate into the wall, resulting in a contact force that depends on the penetration depth and can be modelled using a spring-damper-slider system. In case of stochastic collision models, only the hard sphere model can be applied. In this model, single binary collisions are considered as instantaneous processes, which restrict the application of this model to dilute granular systems. The properties of particles after the collision are related to the properties of particles before the collision through momentum and energy balances.

1.1 State of the Art

Dynamic process simulation models, based on differential equation systems and numerical solution procedures, can be considered as a virtual representation of all essential systems of the power plant. The complexity of these simulation models entails challenges that reflect on the amount of research and application. The state of the art technologies for the dynamic simulation of conventional power plants and the main outcomes of the literature research are summarised in a recent comprehensive review paper (Alobaid et al. 2017) and here briefly described as follows:

- Combined-cycle power plant: According to World Energy Outlook (in 2015), the gas-fired power generation contributed to 23% global electricity supply, dominated by combined-cycle power plants (CCPP). The modern concept of CCPP is mainly driven by increasing performance of the gas turbine. Today, a 1 + 1 arrangement of gas turbine and steam turbine in combination with a triple-pressure reheat heat recovery steam generator (HRSG) is state of the art, reaching a nominal process efficiency of up to 61%. Since the gas turbine is an inherently flexible component, studies in the literature are largely focused on the dynamic response of the water/steam bottoming cycle, in particular the heat recovery steam generator. Detailed modelling and calculation of start-up as well as dynamic optimisation under thermal stress restraints transients are conducted [among others: (Angerer et al. 2017; Benato et al. 2015a, b; Casella and Pretolani 2006; Faille and Davelaar 2009; Hack et al. 2012; Kim et al. 2000; Li et al. 2017a; Romero et al. 2017; Rossi et al. 2017; Sindareh-Esfahani et al. 2014, 2017; Walter and Linzer 2005; Wan et al. 2016)]. However in most studies, no information about the accuracy level or model deviations is available.

- Pulverised coal-fired power plant: Coal-fired power plants play a major role for the total share of the worldwide electricity generation. The specific contribution varies from country to country and depends on several factors including local resources, political framework and access to world market. The main components of a coal-fired power plant are steam generator, mills, feedwater preheaters, steam turbine and flue gas cleaning devices. Modern coal-fired power plants use single reheat as well as several low-pressure and high-pressure feedwater preheaters, resulting in a thermal process efficiency of up to 46%. Although all coal-fired power plants are operating on the same concept, each one is individually engineered, which in turn leads to different dynamics. Here, the load change and start-up behaviour by increasing flexibility requirements are in the focus of interest. In the literature, the dynamic process models of pulverised coal-fired power plants are limited to subsections of the power plant. There is a few number of full-scale dynamic models available, but the validity of these models is constrained by several boundary conditions [among others: (Bhambare et al. 2007; Chen et al. 2017; Hauschke and Leithner 2009; Hübel et al. 2017; Jin et al. 2014; Neuman et al. 2002; Oko and Wang 2014; Richter et al. 2015; Schuhbauer et al. 2014; Zehtner et al. 2008; Zindler et al. 2008)]. Furthermore, the lack of available data for validation of the developed models is a major problem.
- Municipal waste incineration: In many industrialised countries, incineration is the preferred choice for the large-scale disposal of municipal solid waste. Advantages of this technology include decreasing of original waste volume as well as producing stable and odourless residue. The heat flow released during the waste combustion (lower heating value of approximately 10 MJ/kg) is recovered by a water/steam circuit, reaching an electrical gross efficiency of up to 30%. Modern municipal waste incineration is equipped with complex air pollution control devices, removing the solid residues, nitrogen oxides and sulphur oxides from the flue gas. Although the municipal waste incinerator has been applied commercially for many decades, the dynamic behaviour has not yet been investigated in the literature.
- Concentrated solar power plant: This technology uses mirrors, lenses or a combination of both to concentrate solar rays on a receiver to heat up a fluid, which then directly or indirectly runs a thermodynamic process to generate electricity. Currently, levelised costs of energy (LCOE) are relatively high. However, recent technical improvements show potential for efficiency increase and cost reductions. In 2013, the total installed capacity of the concentrated solar power (CSP) amounted to 3.6 GW. Due to the inherently dynamic operation of CSPs, a significant number of studies is available in the literature, focussing on the transient behaviour of the solar field. Some other studies investigate the dynamic behaviour of sub-systems such as the thermal energy storage, considering stable power output and improvement of capacity factor [among others: (Diendorfer et al. 2014; El Hefni and Soler 2015; Falchetta and Rossi 2014; García et al. 2011; Hakkarainen et al. 2017; Larraín et al. 2010; Liu et al. 2015; Luo et al. 2015; Manenti and Ravaghi-Ardebili 2013; Nelson et al. 2018;

Österholma and Pålssonb 2014; Powell and Edgar 2012; Russo 2014; Stuetzle et al. 2004; Wagner and Wittmann 2014)]. However, most studies use a simplified steady-state model rather than a detailed dynamic model of the power block.

Several engineering problems involve reactive gas-solid flows, where the particles of a wide range of shapes and sizes undergo time dependent turbulent motion, e.g. fluidized beds, gasification or combustion of solid fuels as well as CO_2 capture methods such as chemical-looping and carbonate-looping combustion. Here, different physical and chemical processes at diverse length and time scales are involved, including mass, momentum and heat transfers between gas and solid phases. The literature research on existing CFD models, applied to energy systems, shows the following:

- CFD models, based on quasi-single-phase method, are frequently applied to thermal power plants (e.g. pulverised coal-fired power, combined-cycle power and incinerator), making the thermochemical conversion of solid fuel more effective, economically efficient and environmentally friendly [among others: (Backreedy et al. 2005; Bhuiyan and Naser 2015, 2016; Cléirigh and Smith 2014; Costa et al. 2016; Díez et al. 2005; Galindo-García et al. 2014; Le Bris et al. 2007; Lin and Ma 2012; Liu et al. 2016; Park et al. 2013; Sheng et al. 2004; Shin et al. 2012; Xia et al. 2014; Zha et al. 2017)]. Here, different kinetic models have been suggested, various numerical methods were investigated and a set of parameters was introduced, summarised in (Bhutta et al. 2012; Chen et al. 2012; Ravelli et al. 2008; Stopford 2002; Walter and Epple 2016).
- CFD models, based on two-fluid and single-particle method, are mainly applied to simulate the hydrodynamic behaviour of gas-solid flows in cold-flow small-scale facilities [among others: (Altantzis et al. 2015; Bizhaem and Tabrizi 2017; Gryczka et al. 2009; Herzog et al. 2012; Loha et al. 2014; Lv et al. 2014; Farid et al. 2017; Huilin et al. 2005; Liu et al. 2017; Sahu et al. 2018; Vashisth et al. 2015; Wang et al. 2015)]. Although few works used the two-fluid model and the single-particle method combined with a stochastic collision model for the simulation of large-scale systems, but the thermochemical reaction mechanism is not considered in most cases.
- More sophisticated CFD models are required to describe the thermochemical conversion of solid fuels, e.g. coal and biomass. The complexity of the process has many reasons such as the inhomogeneous composition of raw material and the large number of chemical components that appear as intermediates or end products. Furthermore, the heat and mass transfer rates are highly depended on the particle geometry. Relative big particles correspond to large temperature gradients and own longer fluid residence times (Di Blasi 1998). The particle shape has also certain effects on the pyrolysis process. Compared to the spherical particles, the ellipsoidal particles with the same volume need more time for completing the pyrolysis. While the released volatile components and gases outweigh for spherical particles during the pyrolysis process, the particles with an ellipsoid shape are dominated by the rest of pyrolysis products (e.g. tar)

(Babu and Chaurasia 2004). In addition, the correct consideration of particle-particle/wall collisions has an influence on flow, gasification and combustion models and consequently qualitative on the accuracy of obtained results.

1.2 Objectives

This book provides researchers, developers and practitioners with a major contribution to mathematical models' developments and their applications to various energy systems. The objectives of this work are as follows:

- The dynamic behaviour of combined-cycle power plants and pulverised coal-fired power plants during load changes, shutdown/start-up procedures and malfunctions is investigated. Here, detailed dynamic models of existing power plants (e.g. Prai Power Station and Heilbronn Power Station) are developed, including all components and all required automation processes with minimum boundary conditions, based on the real design specifications, construction drawings and control schemes. The numerical results obtained are compared with operation data obtained from real power plants. Such detailed and validated models are not available in the literature. In addition, the dynamic behaviour of an oxyfuel pulverised coal-fired power plant and the influence of fast gas turbine start-ups on the dynamic behaviour of a supplementary-fired HRSG, an once-through HRSG with sub-critical and super-critical steam properties are presented.
- A full-scale dynamic model of Andasol II parabolic trough power plant that includes all components of the solar field, the two-tank thermal storage system, the power block and all required automation processes is developed. This detailed modelling of the whole parabolic trough power plant is of high relevance in order to analyse the dynamic interaction of sub-systems (solar field, thermal energy storage, power block) and the system-level plant behaviour with high accuracy. In contrast to combined-cycle power, pulverised coal-fired power and parabolic trough power, no dynamic simulation of a waste incineration power plant exists to the author's knowledge. In this work and for the first time, dynamic simulation results for the 60 MW_{th} waste incineration power plant, built in Finland, are shown.
- In the literature, CFD studies that are dealing with the numerical simulation of energy systems using two-fluid method and Euler-Lagrange/multiphase particle-in-cell (MP-PIC) model are rarely presented. In this work, several thermochemical reaction models for the combustion of solid fuels, based on quasi-single-phase, two-fluid and Euler-Lagrange/MP-PIC models, are developed and validated towards measurements obtained from entrained-flow reactor, 1 MW_{th} combustion chamber, 1 MW_{th} carbonate-looping pilot plant, 1 MW_{th} chemical-looping pilot plant, 60 MW_{th} waste incineration and large-scale coal-firing power plants.

- Since the Euler-Lagrange/discrete element method (DEM) has been successfully employed by Tsuji et al. (1992, 1993) to simulate the hydrodynamic behaviour of dense gas-solid flows in a laboratory scale fluidized bed, significant advances have been accomplished in this field. From 1995, Kuipers and his team engage in the numerical simulations of cold flow fluidized bed using the DEM model, where relevant contributions were published by (Deen et al. 2007; Li et al. 2017b; Link 2006; Sutkar et al. 2013, 2016). Further recent works from other researchers are for example (Ayeni et al. 2016; Kildashti et al. 2018; Zhang et al. 2017; Zhong et al. 2016; Zhu et al. 2007). However, the consideration of thermochemical reaction mechanism is still in the early stages. The challenges for the energy systems simulations as actual state of research are to combine the gas flow and the homogenous reactions with the particle motions and the heterogeneous reactions. Therefore, a numerical simulation programme (DEMEST), based on the deterministic Euler-Lagrange/DEM model, for the simulation of reactive gas-solid flows is developed. The particle-particle/wall and gas-solid interactions are considered by tracking all individual particles. For this purpose, the discrete element method is applied and further developed. The gas-solid interaction is performed using a new procedure, known as the offset-method. The proposed method is highly precise in determination of the inter-phase values (volumetric void fraction, momentum and heat transfers). An additional grid, so-called particle-grid, in which the physical values of solid phase is computed, is introduced. The suggested procedure allows the variation of the fluid grid resolution independent of the particle size. The collision detection between particle-particle/wall is achieved with the aid of the particle search-grid method. The use of the particle search-grid method enhances the efficiency of collision detection between collision partners. The developed Euler-Lagrange/DEM model is qualitative and quantitative compared with the measurements as well as with two-fluid and Euler-Lagrange/MP-PIC models. Furthermore, kinetic models for drying, pyrolysis, char-oxidation and gas phase reactions as well as the associated heat and mass transfer processes are implemented in the extended DEMEST programme. In combination with the simulation of the gas-solid flow, it is possible to calculate the local concentrations of solid and gaseous substances, the local gas and particle temperatures, the local heat release and the heat transfer rates. Three series of warm simulations are carried out, realising the coupling of the Euler-Lagrange/DEM model with the thermochemical reaction mechanism.
- The efficiency and accuracy of the Euler-Lagrange/DEM model is investigated in detail. Parameter studies are carried out, in which stiffness coefficient and fluid time step are varied for different particle numbers and diameters. The obtained results are compared with the measurements in order to derive the optimum parameters for DEM simulations. The results suggest that the application of higher stiffness coefficients (more than 10^5 N/m) improves the simulation accuracy slightly, but the average computing time increases exponentially. For time intervals larger than five milliseconds, the results show that the average computation time is independent of applied fluid time step,

while the simulation accuracy decreases extremely by increasing the size of fluid time step. The use of fluid time steps smaller than five milliseconds leads to negligible improvements in the simulation accuracy, but to exponential rise in the average computing time.

- In the DEM model, the number of simulated particles is limited to the availability of computational resources. For the simulation of large-scale industrial systems, few real particles can be modelled as a representative particle (parcel) with same material properties as the particles in the system. The parcel approach or coarse grain particle model is further developed to simulate a 3D cold-flow fluidized bed test rig, carbonator and calciner reactors of the 1 MW_{th} carbonate-looping pilot plant. For the validation of the coarse grain particle model, the measured pressure, the solid velocity and the volumetric void fraction of solid at different heights along the 3D cold-flow and warm-flow fluidized bed reactors are used.

1.3 Structure

In Chap. 2, the mathematical background for the process modelling of energy systems is presented. Here, different thermal hydraulic models, i.e. the mixture flow model and the two-fluid flow models (four-equation, five-equation, six-equation and seven-equation flow model) are explained. Furthermore, process, automation and electrical components required for the dynamic simulation of thermal power plants are described and supported by model examples.

In Chap. 3, CFD numerical methods for the modelling of gas-solid flows are described, including quasi-single-phase, two-fluid and single-particle methods. Advantages of these models are discussed in detail. Special attention is given here to the discrete element method, including the forces and the moments of force acting on particles.

In Chap. 4, relevant studies that describe the application of dynamic process simulation to specific thermal power plant technologies such as combined-cycle power plant, pulverised coal-fired power plant, concentrated solar power plant and municipal waste incineration are presented. Furthermore, the application of CFD models to different energy systems is shown as well. Measurements obtained from real energy systems are used to validate the process and CFD simulation models.

In Chap. 5, relevant conclusions with regard to future research in the field of dynamic process and CFD simulations for thermal power plants are given.

References

Adams J, O'Malley M, Hanson K (2010) Flexibility requirements and potential metrics for variable generation: implications for system planning studies. NERC, Princeton, NJ

Alobaid F (2013) 3D modelling and simulation of reactive fluidized beds for conversion of biomass with discrete element method. tuprints, Germany

Alobaid F, Mertens N, Starkloff R, Lanz T, Heinze C, Epple B (2017) Progress in dynamic simulation of thermal power plants. Prog Energy Combust Sci 59:79–162
Altantzis C, Bates R, Ghoniem A (2015) 3D Eulerian modeling of thin rectangular gas–solid fluidized beds: estimation of the specularity coefficient and its effects on bubbling dynamics and circulation times. Powder Technol 270:256–270
Angerer M, Kahlert S, Spliethoff H (2017) Transient simulation and fatigue evaluation of fast gas turbine startups and shutdowns in a combined cycle plant with an innovative thermal buffer storage. Energy 130:246–257
Ayeni O, Wu C, Nandakumar K, Joshi J (2016) Development and validation of a new drag law using mechanical energy balance approach for DEM–CFD simulation of gas–solid fluidized bed. Chem Eng J 302:395–405
Babu B, Chaurasia A (2004) Dominant design variables in pyrolysis of biomass particles of different geometries in thermally thick regime. Chem Eng Sci 59:611–622
Backreedy R, Fletcher L, Jones J, Ma L, Pourkashanian M, Williams A (2005) Co-firing pulverised coal and biomass: a modeling approach. Proc Combust Inst 30:2955–2964
Benato A, Bracco S, Stoppato A, Mirandola A (2015a) Dynamic simulation of combined cycle power plant cycling in the electricity market. Energy Convers Manag 107:76–85
Benato A, Stoppato A, Mirandola A (2015b) Dynamic behaviour analysis of a three pressure level heat recovery steam generator during transient operation. Energy 90:1595–1605
Bhambare K, Mitra SK, Gaitonde U (2007) Modeling of a coal-fired natural circulation boiler. J Energy Res Technol 129:159–167
Bhuiyan AA, Naser J (2015) CFD modelling of co-firing of biomass with coal under oxy-fuel combustion in a large scale power plant. Fuel 159:150–168
Bhuiyan AA, Naser J (2016) Modelling of slag deposition and flow characteristics of coal combustion under oxy-firing condition in a 550 MW tangentially fired furnace. Appl Therm Eng 106:221–235
Bhutta MMA, Hayat N, Bashir MH, Khan AR, Ahmad KN, Khan S (2012) CFD applications in various heat exchangers design: a review. Appl Therm Eng 32:1–12
Bizhaem HK, Tabrizi HB (2017) Investigating effect of pulsed flow on hydrodynamics of gas-solid fluidized bed using two-fluid model simulation and experiment. Powder Technol 311:328–340
Casella F, Pretolani F (2006) Fast start-up of a combined-cycle power plant: a simulation study with Modelica. In: Proceedings 5th international modelica conference, pp 3–10
Chen L, Yong SZ, Ghoniem AF (2012) Oxy-fuel combustion of pulverized coal: characterization, fundamentals, stabilization and CFD modeling. Prog Energy Combust Sci 38:156–214
Chen C, Zhou Z, Bollas GM (2017) Dynamic modeling, simulation and optimization of a subcritical steam power plant. Part I: plant model and regulatory control. Energy Convers Manag 145:324–334
Cléirigh CTÓ, Smith WJ (2014) Can CFD accurately predict the heat-transfer and pressure-drop performance of finned-tube bundles? Appl Therm Eng 73:681–690
Costa M, Massarotti N, Mauro A, Arpino F, Rocco V (2016) CFD modelling of a RDF incineration plant. Appl Therm Eng 101:710–719
Deen N, Van Sint Annaland M, Van der Hoef M, Kuipers J (2007) Review of discrete particle modeling of fluidized beds. Chem Eng Sci 62:28–44
Di Blasi C (1998) Physico-chemical processes occurring inside a degrading two-dimensional anisotropic porous medium. Int J Heat Mass Transf 41:4139–4150
Diendorfer C, Haider M, Lauermann M (2014) Performance analysis of offshore solar power plants. Energy Procedia 49:2462–2471
Díez LI, Cortés C, Campo A (2005) Modelling of pulverized coal boilers: review and validation of on-line simulation techniques. Appl Therm Eng 25:1516–1533
El Hefni B, Soler R (2015) Dynamic multi-configuration model of a 145 MWe concentrated solar power plant with the ThermoSysPro library (tower receiver, molten salt storage and steam generator). Energy Procedia 69:1249–1258
Faille D, Davelaar F (2009) Model based start-up optimization of a combined cycle power plant. Power Plants Power Sys Control, 197–202

Falchetta M, Rossi A (2014) Dynamic simulation of the operation of a molten salt parabolic trough plant, comprising draining procedures. Energy Procedia 49:1328–1339

Farid MM, Jeong HJ, Kim KH, Lee J, Kim D, Hwang J (2017) Numerical investigation of particle transport hydrodynamics and coal combustion in an industrial-scale circulating fluidized bed combustor: effects of coal feeder positions and coal feeding rates. Fuel 192:187–200

Flynn D (2003) Thermal power plant simulation and control. IET

Galindo-García IF, Vázquez-Barragán AK, Rossano-Román M, (2014) CFD simulations of heat recovery steam generators including tube banks. In: ASME 2014 power conference, Baltimore, MD

García IL, Álvarez JL, Blanco D (2011) Performance model for parabolic trough solar thermal power plants with thermal storage: comparison to operating plant data. Sol Energy 85:2443–2460

Gryczka O, Heinrich S, Deen N, van Sint Annaland M, Kuipers J, Jacob M, Mörl L (2009) Characterization and CFD-modeling of the hydrodynamics of a prismatic spouted bed apparatus. Chem Eng Sci 64:3352–3375

Hack H, Fan Z, Seltzer A, Alvarez J (2012) Advanced methods of HRSG design for life cycle optimization under fast startups. In: POWERGEN international, pp 11–13

Hakkarainen E, Sihvonen T, Lappalainen J (2017) Dynamic modelling and simulation of CSP plant based on supercritical carbon dioxide closed Brayton cycle. In: AIP conference proceedings. AIP Publishing, p 070004

Hauschke A, Leithner R (2009) Dynamic simulation of fouling and optimization of sootblowing intervals in a hard coal fired power plant. In: Energy, environment, ecosystems and sustainable development (EEESD'09). Athens, pp 301–304

Herzog N, Schreiber M, Egbers C, Krautz HJ (2012) A comparative study of different CFD-codes for numerical simulation of gas–solid fluidized bed hydrodynamics. Comput Chem Eng 39:41–46

Hübel M, Meinke S, Andrén MT, Wedding C, Nocke J, Gierow C, Hassel E, Funkquist J (2017) Modelling and simulation of a coal-fired power plant for start-up optimisation. Appl Energy 208:319–331

Huilin L, Shuyan W, Yunhua Z, Yang L, Gidaspow D, Ding J (2005) Prediction of particle motion in a two-dimensional bubbling fluidized bed using discrete hard-sphere model. Chem Eng Sci 60:3217–3231

Jin B, Zhao H, Zheng C (2014) Dynamic simulation for mode switching strategy in a conceptual 600 MWe oxy-combustion pulverized-coal-fired boiler. Fuel 137:135–144

Kildashti K, Dong K, Samali B, Zheng Q, Yu A (2018) Evaluation of contact force models for discrete modelling of ellipsoidal particles. Chem Eng Sci 177:1–17

Kim TS, Lee DK, Ro ST (2000) Analysis of thermal stress evolution in the steam drum during start-up of a heat recovery steam generator. Appl Therm Eng 20:977–992

Larraín T, Escobar R, Vergara J (2010) Performance model to assist solar thermal power plant siting in northern Chile based on backup fuel consumption. Renew Energy 35:1632–1643

Le Bris T, Cadavid F, Caillat S, Pietrzyk S, Blondin J, Baudoin B (2007) Coal combustion modelling of large power plant, for NOx abatement. Fuel 86:2213–2220

Li D, Hu Y, He W, Wang J (2017a) Dynamic modelling and simulation of a combined-cycle power plant integration with thermal energy storage. In: Automation and computing (ICAC), 2017 23rd international conference on. IEEE, pp 1–6

Li Z, Janssen T, Buist K, Deen N, van Sint Annaland M, Kuipers J (2017b) Experimental and simulation study of heat transfer in fluidized beds with heat production. Chem Eng J 317:242–257

Lin H, Ma X (2012) Simulation of co-incineration of sewage sludge with municipal solid waste in a grate furnace incinerator. Waste Manag 32:561–567

Link JM (2006) Development and validation of a discrete particle model of a spout-fluid bed granulator. PrintPartners Ipskamp

Liu S, Faille D, Fouquet M, El-Hefni B, Wang Y, Zhang J, Wang Z, Chen G, Soler R (2015) Dynamic simulation of a 1 MWe CSP tower plant with two-level thermal storage implemented with control system. Energy Procedia 69:1335–1343

Liu Y, Fan W, Li Y (2016) Numerical investigation of air-staged combustion emphasizing char gasification and gas temperature deviation in a large-scale, tangentially fired pulverized-coal boiler. Appl Energy 177:323–334

Liu H, Li J, Wang Q (2017) Simulation of gas–solid flow characteristics in a circulating fluidized bed based on a computational particle fluid dynamics model. Powder Technol 321:132–142

Loha C, Chattopadhyay H, Chatterjee PK (2014) Effect of coefficient of restitution in Euler-Euler CFD simulation of fluidized-bed hydrodynamics. Particuology 15:170–177

Luo N, Yu G, Hou H, Yang Y (2015) Dynamic modeling and simulation of parabolic trough solar system. Energy Procedia 69:1344–1348

Lv X, Li H, Zhu Q (2014) Simulation of gas–solid flow in 2D/3D bubbling fluidized beds by combining the two-fluid model with structure-based drag model. Chem Eng J 236:149–157

Manenti F, Ravaghi-Ardebili Z (2013) Dynamic simulation of concentrating solar power plant and two-tanks direct thermal energy storage. Energy 55:89–97

Nelson J, Johnson NG, Doron P, Stechel EB (2018) Thermodynamic modeling of solarized microturbine for combined heat and power applications. Appl Energy 212:592–606

Neuman P, Pokorny M, Varcop L, Weiglhofer W, Javed A (2002) Engineering and operator training simulator of coal-fired steam boiler. In: Proc 10th Int Conference MATLAB

Oko E, Wang M (2014) Dynamic modelling, validation and analysis of coal-fired subcritical power plant. Fuel 135:292–300

Österholma R, Pålssonb J (2014) Dynamic modelling of a parabolic trough solar power plant. In: Proceedings of the 10th international modelica conference, Lund, Sweden

Park HY, Baek SH, Kim YJ, Kim TH, Kang DS, Kim DW (2013) Numerical and experimental investigations on the gas temperature deviation in a large scale, advanced low NOx, tangentially fired pulverized coal boiler. Fuel 104:641–646

Powell KM, Edgar TF (2012) Modeling and control of a solar thermal power plant with thermal energy storage. Chem Eng Sci 71:138–145

Ravelli S, Perdichizzi A, Barigozzi G (2008) Description, applications and numerical modelling of bubbling fluidized bed combustion in waste-to-energy plants. Prog Energy Combust Sci 34:224–253

Richter M, Möllenbruck F, Starinsk A, Oeljeklaus G, Görner K (2015) Flexibilization of coal-fired power plants by dynamic simulation. In: Proceedings of the 11th international modelica conference, pp 715–723. Linköping University Electronic Press, Versailles, France, Sept 21–23, 2015

Romero A, Chacartegui R, Becerra JA, Carvalho M, Millar DL (2017) Analysis of the start-up and variable load operation of a combined cycle power plant for off-grid mines. Int J Glob Warming 13:330–352

Rossi I, Sorce A, Traverso A (2017) Gas turbine combined cycle start-up and stress evaluation: a simplified dynamic approach. Appl Energy 190:880–890

Russo V (2014) CSP plant thermal-hydraulic simulation. Energy Procedia 49:1533–1542

Sahu AK, Raghavan V, Prasad B (2018) Temperature effects on hydrodynamics of dense gas-solid flows: application to bubbling fluidized bed reactors. Int J Therm Sci 124:387–398

Schuhbauer C, Angerer M, Spliethoff H, Kluger F, Tschaffon H (2014) Coupled simulation of a tangentially hard coal fired 700 ° C boiler. Fuel 122:149–163

Sheng C, Moghtaderi B, Gupta R, Wall TF (2004) A computational fluid dynamics based study of the combustion characteristics of coal blends in pulverised coal-fired furnace. Fuel 83: 1543–1552

Shin H, Kim D, Ahn H, Choi S, Myoung G (2012) Investigation of the flow pattern in a complex inlet duct of a heat recovery steam generator. Energy Power 2:1–8

Sindareh-Esfahani P, Habibi-Siyahposh E, Saffar-Avval M, Ghaffari A, Bakhtiari-Nejad F (2014) Cold start-up condition model for heat recovery steam generators. Appl Therm Eng 65: 502–512

Sindareh-Esfahani P, Tabatabaei SS, Pieper JK (2017) Model predictive control of a heat recovery steam generator during cold start-up operation using piecewise linear models. Appl Therm Eng 119:516–529

Spliethoff H (2010) Power generation from solid fuels. Springer, Berlin

Stopford PJ (2002) Recent applications of CFD modelling in the power generation and combustion industries. Appl Math Model 26:351–374

Stuetzle T, Blair N, Mitchell JW, Beckman WA (2004) Automatic control of a 30 MWe SEGS VI parabolic trough plant. Sol Energy 76:187–193

Sutkar VS, Deen NG, Kuipers J (2013) Spout fluidized beds: recent advances in experimental and numerical studies. Chem Eng Sci 86:124–136

Sutkar VS, Deen NG, Patil AV, Salikov V, Antonyuk S, Heinrich S, Kuipers J (2016) CFD–DEM model for coupled heat and mass transfer in a spout fluidized bed with liquid injection. Chem Eng J 288:185–197

Tsuji Y, Tanaka T, Ishida T (1992) Lagrangian numerical simulation of plug flow of cohesionless particles in a horizontal pipe. Powder Technol 71:239–250

Tsuji Y, Kawaguchi T, Tanaka T (1993) Discrete particle simulation of two-dimensional fluidized bed. Powder Technol 77:79–87

Vashisth S, Motlagh AA, Tebianian S, Salcudean M, Grace J (2015) Comparison of numerical approaches to model FCC particles in gas–solid bubbling fluidized bed. Chem Eng Sci 134:269–286

Wagner PH, Wittmann M (2014) Influence of different operation strategies on transient solar thermal power plant simulation models with molten salt as heat transfer fluid. Energy Procedia 49:1652–1663

Walter H, Epple B (2016) Numerical simulation of power plants and firing systems. Springer, Berlin

Walter H, Linzer W (2005) Numerical simulation of a three stage natural circulation heat recovery steam generator. na

Wan A, Gu F, Jin J, Gu X, Ji Y (2016) Modeling and optimization of shutdown process of combined cycle gas turbine under limited residual natural gas. Appl Therm Eng 101:337–349

Wang T, He Y, Kim DR (2015) Granular temperature and rotational characteristic analysis of a gas–solid bubbling fluidized bed under different gravities using discrete hard sphere model. Powder Technol 271:35–48

Xia Z, Li J, Wu T, Chen C, Zhang X (2014) CFD simulation of MSW combustion and SNCR in a commercial incinerator. Waste Manag 34:1609–1618

Yadigaroglu G, Hewitt GF (2017) Introduction to multiphase flow: basic concepts, applications and modelling. Springer, Berlin

Zehtner W, Spliethoff H, Woyke W (2008) Analysis and optimisation of operation of modern hard coal-fired power plants through simulation. VGB PowerTech 88:28–32

Zha Q, Li D, Wang CA, Che D (2017) Numerical evaluation of heat transfer and NOx emissions under deep-air-staging conditions within a 600 MW e tangentially fired pulverized-coal boiler. Appl Therm Eng 116:170–181

Zhang H, Zhou Z, Yu A, Kim S-Y, Jung S-K (2017) Discrete particle simulation of solid flow in a melter-gasifier in smelting reduction process. Powder Technol 314:641–648

Zhong W, Yu A, Liu X, Tong Z, Zhang H (2016) DEM/CFD-DEM modelling of non-spherical particulate systems: theoretical developments and applications. Powder Technol 302:108–152

Zhu H, Zhou Z, Yang R, Yu A (2007) Discrete particle simulation of particulate systems: theoretical developments. Chem Eng Sci 62:3378–3396

Zindler H, Walter H, Hauschke A, Leithner R (2008) Dynamic simulation of a 800 MW_{el} hard coal one-through supercritical power plant to fulfill the great britain grid code. In: 6th IASME/WSEAS international conference on heat transfer, thermal engineering and environment, Rhodes, Greece, pp 184–192

Chapter 2
Process Simulation

Conventional thermal power plants are traditionally responsible for compensation of daily and seasonal load variations. The increased penetration of renewable energy sources in the generation of electrical power recently raises technical and economic challenges for the operation of these plants due to the uncertainty of supply and demand. Existing thermal power plants have to be retrofitted with optimised components and control circuits to improve their operation mode concerning the load change times as well as the rate of shutdown and start-up. In addition to the experimental works, mathematical models contribute to a better understanding of the process and can play an important role for increasing the power plant flexibility. The design and optimisation of thermal power plants start with steady-state modelling, where analyses of the thermodynamic properties of working fluid, mass and energy flows as well as process efficiency can be obtained for a series of operating points. The investigation into the dynamic performance of thermal power plants requires besides the unsteady solution of conservation equations, the implementation of dynamic boundary conditions, the control structures and their associated components. The governing differential equations and the required numerical solver make the dynamic simulation codes very complex tools with long development periods (Alobaid et al. 2017).

The first process simulation programme of thermal power plants dated back to 1960s. A technique was developed, in which a water/steam circuit was solved numerically. The circuit was constructed using individual components that connected through lines. The main components of the water/steam circuit as well as the connection lines were systematised with codes, so that the entire circuit can be defined as a sequence of numbers and interpreted by the computer. This representation (selecting from the library and building component by component) has mainly been maintained in most simulation programmes since. Nowadays, process simulation programmes combine graphical user interface with detailed models for flow, thermodynamic, mass and heat transfer. They predict the performance of energy systems with high accuracy and enable rapid assessments of:

F. Alobaid, *Numerical Simulation for Next Generation Thermal Power Plants*,
Springer Tracts in Mechanical Engineering,
https://doi.org/10.1007/978-3-319-76234-0_2

- new power plant design
- power plant optimisation and efficiency improvement
- process modifications and retrofitting of existing power plants
- power plant security and safety
- operating behaviour at base loads, off-design, start-up and shutdown
- operating behaviour during malfunctions

The physical and mathematical backgrounds of these programmes show various degrees of complexity, depending on the application case:

In steady-state simulation, the time derivations disappear from the conservation equations. In case of zero-dimensional modelling, the local discretisation is not considered. The modelling of thermal power plant components such as heat exchanger, pump, condenser, turbine, etc., results in an algebraic system of equations with inputs and output parameters of components (pressure, enthalpy, mass flow and concentration). In case of one-dimensional modelling, the power plant components are discretised between the inlet and the outlet along the flow in finite objects, so-called numerical grid. Here, an algebraic system of equations at each discrete location is obtained. In case of two or three-dimensional modelling, additional local discretisation along the coordinates is required, resulting in more detailed calculation of the thermal power plant components.

In dynamic simulation, the time derivations must be taken into consideration. In case of zero-dimensional modelling, the components are modelled without local discretisation. For one-dimensional modelling, the transient behaviour and local discretisation along the flow are computed. In case of two or three-dimensional modelling, a further discretisation along the local coordinates is required.

In quasi steady-state simulation, the time derivation of certain components is not relevant and can be neglected in the conservation equations, which in return simplifies the system of equations significantly.

The knowledge of the steady-state data of components of a process is sufficient in most of practical engineering applications. Here, input and output parameters of the power plant components such as boiler, preheater, turbine etc. are computed. Furthermore, design calculations at different loads can also be conducted with such steady-state simulation models. The dynamic simulation enables more detailed data of the transient behaviour of the entire thermal power plant with its related control structures. Several in-house developed codes and commercial software programmes for steady-state and dynamic process simulation of thermal power plants are available, e.g. EBSILON Professional, DBS, APROS, ASPEN PLUS DYNAMICS, FLOWNEX and Power Plant Simulator & Designer (PPSD). Some programmes provide specialised component libraries for steady-state and time-dependent simulation of energy systems, including simple cycle plants, combined-cycle as well as combined heat-and-power plants. In others, missing components have to be modelled by user such as MATLAB SIMULINK that offers a generic programming tool. Using the non-proprietary object-oriented, equation-based language (MODELICA), complex physical systems, containing mechanical and control components can be modelled. Based on MODELICA, different non-commercial or commercially simulation environments are also available, for example DYMOLA, JModelica.org and SimulationX.

Employing the previous steady-state and dynamic programmes, different kind of energy systems can be simulated with the following energy sources: hard coal, lignite, oil or natural gas, municipal solid waste, biomass, geothermal energy source, solar energy source, nuclear heat source etc.; with various combustion systems: grate firing, fluidized bed (fixed, steady-state or circulating), pulverised coal firing with different types of mills (e.g. ball and tube mills, bowl mill, beater-wheel mills), oil or gas firing (e.g. gas turbine) etc.; with various operation of evaporators: natural circulation, forced circulation, once-through boiler with and without superimposed circulation; with various pressure operation: fixed-pressure, sliding-pressure, modified sliding pressure; with various boiler construction: box or cube boiler, single-pass boiler, two-pass boiler, multi-pass boiler.

2.1 Thermal Hydraulic Models

The state of the art in modelling the thermal hydraulics of thermal power plants differs in the fundamental physical models. Here, various formulations have been suggested and different numerical methods for compressible and incompressible flows were investigated. The complexity of the process is explained by the occurrence of several two-phase flow regimes, heat conduction in solid structures, heat transfer between fluid and solid structures, heat and mass transfer between gas and liquid. Different approaches are proposed in the literature in order to model the two-phase flow in a thermal power plant such as mixture flow model and two-fluid models, including four-equation, five-equation, six-equation and even seven-equation flow model. In two-fluid models, two sets of conservation equations are formulated, governing the mass, momentum and energy balance for each phase. This formulation presents considerable difficulty by reason of mathematical complexity and the uncertainty in modelling the interaction between phases at the inter-phase boundary. Generally, the interaction terms cannot be derived from fundamental physical laws and are based on empirical assumptions in most cases. Solving the resulting differential equations requires higher computational effort and entails parameters that may cause numerical instability, especially due to improper selection of interaction terms. The difficulties associated with the two-fluid models can be significantly reduced by formulating the two-phase flow in terms of the mixture flow model. Here, three characteristic fluid variables are computed, including local pressure, total mass flux and temperature or enthalpy, represented by three conservation equations (mass, momentum and energy) of mixture. Due to its simplicity and applicability to a wide range of two-phase flows, this model is of considerable relevance since the response of the total mixture and not of each constituent phase is often sufficient. The two-fluid model becomes, however, more appropriate for special applications since it offers the possibility to include non-equilibrium thermodynamic situations into the formulation. Furthermore, the two-fluid model describes the phase boundaries in an easier way. This problem can clearly be observed in the mixture flow model that uses many closure models, resulting in approximate solutions and accordingly accuracy restrictions for certain applications.

The steady-state and dynamic behaviour of thermal power plants can also be described using the lumped parameter model, also known as the lumped element model. Here, the description of the physical system is simplified by means of discrete control volumes (zones or lumps) that are connected each other using thermal resistors and capacitors with the assumption of small temperature differences inside each lump. The equivalent thermal network consists of thermal resistance, thermal capacitances and power losses inside the system. The lumped parameter model can be applied to electrical and mechanical systems, heat transfer processes and thermo-hydraulic analysis of conventional thermal power plants. The application of the lumped parameter model is suitable, when a simplified formulation of the transient behaviour of the process is required. The advantage of such an approach lies in the reduced computational cost, but it does not offer the same accuracy of more complex numerical methods (mixture and two-fluid models). Further information can be found for example in Bergman et al. (2011).

In the following sections, the mixture flow model and the different two-fluid models are explained in detail as in Alobaid et al. (2017), Walter and Epple (2016).

2.1.1 Mixture Flow Model

The one-dimensional mixture flow model (also known as homogeneous or three-equation flow model) assumes thermodynamic equilibrium between phases. The mixture flow model is represented by three-partial differential equations for mass, momentum and energy that describe the steady-state and dynamic behaviour of the characteristic variables. For single-phase flow components (e.g. superheater, reheater, economiser and steam turbine), the three characteristic fluid variables are the local pressure, the total mass flux and the fluid temperature or the fluid enthalpy for sub-cooled water or superheated steam. In case of two-phase flow components (e.g. evaporator and condenser), the three variables are complemented by the void fraction. The void fraction can be computed by adding a fourth additive constitutive equation. The latter is a drift-flux correlation that describes an adequate relation between different two-phase parameters, e.g. a relation between the steam quality or the steam mass flux and the steam void fraction. The drift-flux closure laws that allow the use of slip relation between the phases are based on theoretical, empirical or semi-empirical approaches. Several studies have been carried out, resulting in numerous void fraction experiments and different drift-flux correlations, which show significant deviations. Bhagwat and Ghajar (2012) evaluated the correlations available in the literature and observed that the tested correlations, although predicting the void fraction with desired accuracy at a certain point, were inaccurate for a broad range of operating conditions. Furthermore, the correlations that predict the void fraction accurately for vertical pipe orientation fail in case of inclined pipe orientations. According to Bhagwat and Ghajar (2014), the recommended correlations by the above mentioned studies lose their accuracy at higher pressure, large pipe diameter and for fluids with higher dynamic viscosity than water. In conclusion, there is no closure relation that reliably predicts the void fraction for a suitable range of flow patterns, void fractions, diameters, orientations of pipe and particularly fluid properties.

The local pressure, the total mass flux and the enthalpy is described by three conservation equations of the mixture. The mass conservation equation is expressed by:

$$\frac{\partial \rho}{\partial t} + \frac{\partial (\rho u)}{\partial z} = S. \tag{2.1}$$

The momentum conservation equation is written as:

$$\frac{\partial (\rho u)}{\partial t} + \frac{\partial (\rho u^2)}{\partial z} = -\frac{\partial p}{\partial z} + F_{gra} + F_{wal} + f(val + form + pu). \tag{2.2}$$

The energy conservation equation is given as:

$$\frac{\partial (\rho h_0)}{\partial t} + \frac{\partial (\rho u h_0)}{\partial z} = \frac{\partial p}{\partial t} + q_{wal}. \tag{2.3}$$

In these equations, F_{gra} is the gravitational acceleration force per volume, F_{wal} and q_{wal} represent the friction force per volume and the heat flow through walls per volume. The symbols ρ and u refer to density and longitudinal velocity of fluid, respectively. The function f considers the pressure losses due to valve and form frictions as well as the hydrostatic pressure differences and the pressure force of a pump. The total enthalpy h_0 is the static enthalpy including the kinetic energy of the flow. In the mass equation, the source term S can contain additional mass flows into the system, or vice versa. The pressure derivative term appears in the energy equation due to the fact that the total enthalpy is used instead of the internal energy U.

$$\frac{\partial U}{\partial t} = \frac{\partial h_0}{\partial t} - \frac{\partial p}{\partial t}. \tag{2.4}$$

In thick-walled tubes, e.g. drum and feedwater storage tank, the lower part is pure water and the upper part is pure steam. Here, the composition of the outflowing fluid from the tank is determined by the water level and the connected branch height. Generally, the number of connected branches to the tank is not limited. The branch inlet height must, however, be in the height range of the tank. When the water level is below the branch height, the flow consists of steam, while it consists of water if the water level is above the branch height. In between, there is a transition region, where the composition of the leaving fluid is gradually changing from water to steam.

In case of two-phase systems, the fundamental physical laws of mixture flow model are the conservation equations of mass, momentum and energy with the movement between the two phases using drift-flux correlations. In this model, the four characteristic fluid variables are the local pressure, the total mass flux, the enthalpy and the void fraction. The continuity equation of mixture is expressed by:

$$\frac{\partial(\rho_m)}{\partial t} + \frac{\partial(\rho_m u_m)}{\partial z} = S_m. \tag{2.5}$$

The symbols ρ_m, u_m and S_m represent the density, the fluid velocity and the injection or leakage of the mixture, respectively.

The momentum equation of mixture by neglecting the effect of surface tension can be written as:

$$\begin{aligned} &\frac{\partial(\rho_m u_m)}{\partial t} + \frac{\partial(\rho_m u_m^2)}{\partial z} + \frac{\partial}{\partial z}\left(\frac{\chi \rho_{gas} \rho_{liq}}{(1-\chi)\rho_m} V_{gas,j}^2\right) \\ &= -\frac{\partial p}{\partial z} + F_{gra} + F_{wal} + f(val + form + pu). \end{aligned} \tag{2.6}$$

Here, the symbol χ is the void fraction of gas phase and ρ_{liq} denote the density of gas and liquid phases, respectively. The symbol $V_{gas,j}$ is the drift velocity of gas phase with respect to the volumetric centre of mixture.

The energy equation of mixture is defined as:

$$\frac{\partial(\rho_m h_{0,m})}{\partial t} + \frac{\partial(\rho_m u_m h_{0,m})}{\partial z} + \frac{\partial}{\partial z}\left(\frac{\chi \rho_{gas} \rho_{liq}}{\rho_m} V_{gas,j}\left(h_{0,gas} - h_{0,liq}\right)\right) = \frac{\partial p}{\partial t} + q_{wal}. \tag{2.7}$$

Zuber and Findlay (1965) proposed the following relation for $V_{gas,j}$:

$$V_{gas,j} = u_{gas} - C_0 j, \tag{2.8}$$

where j is the superficial velocity and C_0 represents the distribution parameter:

$$C_0 = \frac{<\chi j>}{<\chi><j>}. \tag{2.9}$$

The expression within the angle brackets ($< \; >$) indicates the cross-sectional averaged flow properties. Equation (2.8) can be written as:

$$V_{gas,j} = \frac{j_{gas}}{\chi} - \left(j_{gas} - j_{liq}\right). \tag{2.10}$$

The superficial velocity of gas or liquid is an artificial velocity, determined by assuming that only a given phase is flowing in a certain cross-sectional area. It can be determined as:

$$j_k = \frac{\dot{V}_k}{A}. \tag{2.11}$$

Here, the subscript k is either gas or liquid, $\dot{V}_k$ is the volume flow rate of the phase k and A is the cross-sectional area. Using the drift-flux correlation, the drift velocity and void fraction can be defined, which in turn can be used to obtain the velocity of the gas phase u_{gas} and the liquid phase u_{liq} as:

$$u_{gas} = u_m + \frac{\rho_{liq}}{\rho_m} V_{gas,j}, \tag{2.12}$$

$$u_{liq} = u_m - \frac{\chi}{1-\chi} \frac{\rho_{gas}}{\rho_m} V_{gas,j}, \tag{2.13}$$

with the mixture density ρ_m:

$$\rho_m = \chi \rho_{gas} + (1-\chi)\rho_{liq}. \tag{2.14}$$

In conclusion, the formulation of the mixture flow model is based on the mixture balance equations. In two-phase flow regions, the relative motion between phases is considered by a kinematic constitutive equation. The constitutive equations that define the relative motion between phases are generally derived by taking into account interfacial geometry, body-force field, shear stresses and interfacial momentum transfer. Due to its simplicity and applicability to a wide range of two-phase flow, the mixture flow model with drift-flux correlations is often used when the response of the total mixture and not of each constituent phase is required. Therefore, this flow model can be found in most of process simulation programmes such as APROS, ASPEN PLUS DYNAMICS, FLOWNEX and PPSD. It should, however, be distinguished between the drift-flux models, where it is assumed that the two-phase flow can be expressed by the momentum equation of mixture complemented by a kinematic equation specifying the relative motion between phases and the two-fluid models (see the following section), where the velocity of each phase is solved separately.

2.1.2 Two-Fluid Model

The two-fluid models, also known as heterogeneous or Euler-Euler flow model, formulate separate conservation equations of mass, momentum and energy for gas and liquid phases. The two-fluid models describe the two-phase flow more accurately than the mixture flow model, but the definition of the interaction terms between phases is difficult. This is due to the fact that these interaction terms cannot be determined from physical laws and are generally obtained from experiments under several artificial assumptions. Here, careful study of the interfacial constitutive equations is required since the improper selections of these terms may result in numerical instability. Due to the increased number of differential equations and closure relations, the two-phase fluid models, in contrast to mixture flow model, are related to higher computational cost and are suitable for thermodynamic non-equilibrium applications. Accordingly, the complex and time-consuming two-fluid models are only implemented in few process simulation programmes such as APROS and RELAP.

The two-fluid models are characterised by two sets of conservation equations. Therefore, the velocities and temperatures of each phase are independent, as opposed to the mixture flow model with drift-flux correlations, where there is only one momentum and one energy equation for mixture. The four-equation or five-equation version of the two-fluid models assumes mechanical and thermal equilibrium, but not chemical equilibrium. In the six-equation flow model, the phases are in mechanical equilibrium (they are at the same pressure), but not in chemical and thermal equilibrium. The seven-equation version of the two-fluid models allows the phases to be completely in non-equilibrium state. In the latter, each phase has its own pressure, own velocity and temperature. According to Stuhmiller (1977), the seven-equation flow model avoids the non-hyperbolicity of the six-equation flow model, which can lead to ill-posed Cauchy problems. Mathematically, a hyperbolic partial differential equation of order n has a well-posed initial value problem for the first $n - 1$ derivatives, i.e. the Cauchy problem is solved locally for certain initial data of any non-characteristic hypersurface (Ransom and Hicks 1984). In all two-fluid models (excepting the seven-equation model), the void fraction is generally determined by the interfacial drag. The latter is calculated based on drift-flux correlations for the distribution parameter and void-weighted area-averaged drift velocity.

Non-condensable gases may affect the operation safety of thermal power plants. For example, dissolved oxygen and carbon dioxide in the feedwater leads to serious corrosion damages in the steam generator components and therefore should be removed using a deaerator system, for example. Generally, the non-condensable gases can be nitrogen, oxygen, helium or hydrogen that are mixed with steam phase or dissolved in liquid phase with the possibility to be transferred from one phase to another (Hänninen and Ahtinen 2009). The modelling of the non-condensable gases can be considered in the formulation of the two-fluid models and has already taken into account the process simulation programmes that are used in safety analysis work.

In the following sections, different versions of the two-fluid models, including four-equation, five-equation, six-equation and seven-equation flow model are described in detail.

2.1.2.1 Four-Equation Flow Model

The four-equation model assumes instantaneous phase change and equilibrium of two phases. This version of the two-fluid models contains one mixture mass equation, one mixture energy equation and two momentum equations for gas and liquid phases. The phase velocities are independent from each other, in contrast to the mixture flow model with drift-flux correlations that uses one momentum equation for mixture. The four conservation equations, describing the physical properties of pressure and temperature of mixture as well as the gas and liquid velocities are expressed by:

$$\frac{\partial(\chi\rho_{gas} + (1-\chi)\rho_{liq})}{\partial t} + \frac{\partial(\chi\rho_{gas}u_{gas} + (1-\chi)\rho_{liq}u_{liq})}{\partial z} = S. \tag{2.15}$$

$$\begin{aligned}&\frac{\partial(\chi\rho_{gas}u_{gas})}{\partial t} + \frac{\partial(\chi\rho_{gas}u_{gas}^2)}{\partial z}\\ &= F_{gra,gas} + F_{wal,gas} + F_{liq,gas} - \chi\left(\frac{\partial p}{\partial z} + \frac{\partial p_{ext}}{\partial z}\right).\end{aligned} \tag{2.16}$$

$$\begin{aligned}&\frac{\partial(1-\chi)\rho_{liq}u_{liq}}{\partial t} + \frac{\partial(1-\chi)\rho_{liq}u_{liq}^2}{\partial z}\\ &= F_{gra,liq} + F_{wal,liq} + F_{gas,liq} - (1-\chi)\left(\frac{\partial p}{\partial z} + \frac{\partial p_{ext}}{\partial z}\right).\end{aligned} \tag{2.17}$$

$$\begin{aligned}&\frac{\partial}{\partial t}\left[\chi\rho_{gas}h_{0,gas} + (1-\chi)\rho_{liq}h_{0,liq}\right] + \frac{\partial}{\partial z}\left[\chi\rho_{gas}u_{gas}h_{0,gas} + (1-\chi)\rho_{liq}u_{liq}h_{0,liq}\right]\\ &= \frac{\partial p}{\partial t} + q_{wal}.\end{aligned} \tag{2.18}$$

In these equations, the subscripts *gas* and *liq* refer to gas and liquid phases. The forces F_{gra}, F_{wal} and F_{ik} account for gravitation, wall friction and interfacial friction per volume, respectively. The subscript *ext* refers to an external pressure source/sink such as pump and losses due to valve and component-specific friction. The symbol q_{wal} denotes the wall heat flow per volume.

2.1.2.2 Five-Equation Flow Model

In the five-equation model, mass balances for each phase are required instead of one mixture mass in case of the four-equation flow model. Here, the five partial differential equations of two-phase flow can be formulated either with thermal equilibrium or with drift-flux correlation.

The first version of the five-equation model assumes mechanical and thermal equilibrium between phases (pressure and temperature are kept equal), but the phases are generally not in chemical equilibrium. The mass and momentum conservation equations are solved for gas and liquid separately, while the energy equation is only solved for gas-liquid mixture.

The mass balance equation by neglecting the diffusion term in the Eulerian form is written for gas phase as:

$$\frac{\partial(\chi\rho_{gas})}{\partial t} + \frac{\partial(\chi\rho_{gas}u_{gas})}{\partial z} = S_{gas} + \Gamma_{liq,gas}, \tag{2.19}$$

and for liquid phase as:

$$\frac{\partial(1-\chi)\rho_{liq}}{\partial t} + \frac{\partial((1-\chi)\rho_{liq}u_{liq})}{\partial z} = S_{liq} + \Gamma_{gas,liq}. \tag{2.20}$$

The terms S_{gas} and S_{liq} represent the injection and leakage of gas and liquid phase. The interfacial mass transfer of gas $\Gamma_{liq,gas}$ describes the evaporation and the interfacial mass transfer of liquid $\Gamma_{gas,liq}$ considers condensation. The sum of interfacial mass transfer of liquid and gas is zero:

$$\Gamma_{gas,liq} + \Gamma_{liq,gas} = 0. \tag{2.21}$$

The gas momentum balance equation is defined by:

$$\begin{aligned} &\frac{\partial(\chi\rho_{gas}u_{gas})}{\partial t} + \frac{\partial(\chi\rho_{gas}u_{gas}^2)}{\partial z} \\ &= F_{gra,gas} + F_{wal,gas} + F_{liq,gas} - \chi\left(\frac{\partial p}{\partial z} + \frac{\partial p_{ext}}{\partial z}\right). \end{aligned} \tag{2.22}$$

The liquid momentum balance equation is expressed by:

$$\begin{aligned} &\frac{\partial((1-\chi)\rho_{liq}u_{liq})}{\partial t} + \frac{\partial((1-\chi)\rho_{liq}u_{liq}^2)}{\partial z} \\ &= F_{gra,liq} + F_{wal,liq} + F_{gas,liq} - (1-\chi)\left(\frac{\partial p}{\partial z} + \frac{\partial p_{ext}}{\partial z}\right). \end{aligned} \tag{2.23}$$

In the momentum equations, the term inside the time derivative expresses the temporal change of gas or liquid flow rate. The term inside the space derivative is equal to the changes of flow momentum along the integration axis z. The last term on the right-hand side expresses the influence of pressure on the momentum conservation due to the axial pressure distribution and the external pressure forces such as pump or losses due to valve and form friction.

The energy balance equation for gas-liquid mixture is written as:

$$\frac{\partial}{\partial t}\left[\chi \rho_{gas} h_{0,gas} + (1-\chi)\rho_{liq} h_{0,liq}\right] + \frac{\partial}{\partial z}\left[\chi \rho_{gas} u_{gas} h_{0,gas} + (1-\chi)\rho_{liq} u_{liq} h_{0,liq}\right] \\ = \frac{\partial p}{\partial t} + q_{wal}. \tag{2.24}$$

In the second version of the five-equation model, the phase velocities are coupled by a functional relation (drift-flux correlation). Here, the mass and energy conservation equations are solved for gas and liquid separately, while the momentum equation is only solved for gas-liquid mixture. The summation of gas and liquid momentum Eqs. (2.22) and (2.23) results in the momentum conservation equation of mixture:

$$\frac{\partial \rho_m u_m}{\partial t} + \frac{\partial \rho_m u_m^2}{\partial z} = F_{gra} + F_{wal} - \left(\frac{\partial p}{\partial z} + \frac{\partial p_{ext}}{\partial z}\right), \tag{2.25}$$

where ρ_m is the mixture density and u_m is the mixture velocity. The drift-flux model describes the superficial velocity of the gas u_{gas} as a function of superficial mixture velocity j, void fraction χ, drift-flux velocity $V_{gas,j}$ and distribution parameter C_0.

For gas phase, the energy balance equation is expressed by:

$$\frac{\partial(\chi \rho_{gas} h_{0,gas})}{\partial t} + \frac{\partial(\chi \rho_{gas} u_{gas} h_{0,gas})}{\partial z} = \chi \frac{\partial p}{\partial t} + \Gamma h_{0,liq,gas} + q_{wal,gas} + q_{liq,gas}. \tag{2.26}$$

For liquid phase, the energy balance equation is formulated as follows:

$$\frac{\partial}{\partial t}\left[(1-\chi)(\rho_{liq} h_{0,liq})\right] + \frac{\partial}{\partial z}\left[(1-\chi)(\rho_{liq} u_{liq} h_{0,liq})\right] \\ = (1-\chi)\frac{\partial p}{\partial t} - \Gamma h_{0,gas,liq} + q_{wal,liq} + q_{gas,liq}. \tag{2.27}$$

The heat flows per volume on the right-hand side of the energy equations are defined separately for the heat transfer from wall to gas $q_{wal,gas}$, wall to liquid $q_{wal,liq}$ and between phases $q_{liq,gas}$ and $q_{gas,liq}$.

2.1.2.3 Six-Equation Flow Model

The six-equation flow model, in contrast to the four-equation model and the five-equation model, has attracted more attention in the scientific literature. This flow model allows chemical and thermal non-equilibrium (velocity and temperature disequilibrium between phases), but assumes mechanical equilibrium (the phases are at the same pressure). The six-equation flow model is suitable for water/steam mixture with high mass and heat transfer between phases. It is characterised by separate conservation equations of mass, momentum and energy for gas and liquid phases. However, the formulation of two sets of conservation equations presents considerable difficulty by reason of mathematical complexity and modelling uncertainty of interaction terms between phases. The six-equation model is therefore more prone to numerical instability, compared to the mixture flow model.

This section describes the six-equation flow model, as implemented in the advanced process simulation software (APROS) (Hänninen 2009). The solution of the six-equation flow model is based on the one-dimensional six partial differential equations, from which the pressure, the void fraction, the phase velocities and the phase enthalpies are solved. If the two-phase flow also includes non-condensable gases, additional conservation equations, describing the behaviour of the non-condensable gases are required.

The mass conservation equation for phase k is written as:

$$\frac{\partial(\chi_k \rho_k)}{\partial t} + \frac{\partial(\chi_k \rho_k u_k)}{\partial z} = \Gamma_{ik}. \tag{2.28}$$

The momentum conservation equation for phase k is formulated as:

$$\begin{aligned} &\frac{\partial(\chi_k \rho_k u_k)}{\partial t} + \frac{\partial(\chi_k \rho_k u_k^2)}{\partial z} \\ &= -\chi_k \frac{\partial p}{\partial z} + \Gamma_{ik} u_{ik} + F_{gra,k} + F_{wal,k} + F_{ik} + f_k(val + form + pu). \end{aligned} \tag{2.29}$$

The energy conservation equation for phase k is expressed by:

$$\frac{\partial(\chi_k \rho_k h_{0,k})}{\partial t} + \frac{\partial(\chi_k \rho_k u_k h_{0,k})}{\partial z} = \chi_k \frac{\partial p}{\partial t} + \Gamma_{ik} h_{0,ik} + q_{ik} + q_{wal,k} + F_{ik} u_{ik}. \tag{2.30}$$

The subscript k refers to l = liquid or g = gas. The subscript ik describes the interface between two phases and the subscript wal, k denotes the interface between one phase and the wall. The term Γ is the mass exchange rate between phases. The function f_k considers the effects of valves, pumps and friction on the flow. The terms F and q represent the average friction force per volume and the heat flow per volume, respectively. In the energy equation, the symbol h_0 is the total enthalpy including the kinetic energy. In the six-equation flow model, the wall friction $F_{wal,k}$, the interfacial friction F_{ik}, the interfacial heat flow q_{ik} and wall heat flow $q_{wal,k}$ are modelled by means of empirical correlations.

The gravitation force per volume is determined using the following relation:

$$F_{gra,k} = \chi_k \rho_k g \cos \theta, \tag{2.31}$$

where θ is the inclination angle and g denotes the standard gravity. The mass exchange rate (interfacial mass transfer) Γ is obtained by forming the energy balance for the phase boundary as follows:

$$\Gamma_{ik} = -\Gamma_{ki} = -\frac{q_{i,liq} + q_{i,gas} - q_{wal,i}}{h_{gas,sat} - h_{liq,sat}}. \tag{2.32}$$

The symbols $h_{gas,sat}$ and $h_{liq,sat}$ represent the saturation enthalpies of gas and liquid. The interfacial heat flow is calculated separately for liquid and gas phases as:

$$q_{i,gas} = -K_{i,gas}(h_{gas} - h_{gas,sat}), \tag{2.33}$$

$$q_{i,liq} = -K_{i,liq}(h_{liq} - h_{liq,sat}). \tag{2.34}$$

Here, the terms h_{gas} and h_{liq} are the static enthalpies of gas and liquid, respectively. For evaporation and condensation, separate heat transfer correlations are required. The interfacial heat transfer coefficients $K_{i,gas}$ and $K_{i,liq}$ depend strongly on both phase flow velocities and void fraction. The interfacial heat transfer coefficient of gas is calculated with the Lee-Ryley correlation, if $h_{gas} < h_{gas,sat}$ (vapour is sub-cooled, condensation) (Lee and Ryley 1968):

$$K_{i,gas} = \frac{6(1-\chi)\lambda_{gas}\left(2 + 0.74\mathrm{Re}_{dro}^{1/2}\,\mathrm{Pr}_{gas}^{1/3}\right)}{\delta_{dro}^2 c_{p,gas}}, \tag{2.35}$$

with the Reynolds number of droplet flow Re_{dro}, the Prandtl number of gas phase Pr_{gas}, the thermal conductivity of gas phase λ_{gas} and the heat capacity of gas phase $c_{p,gas}$. The droplet diameter δ_{dro} that describes the interface size of two phases has a considerable influence on the calculation of the interfacial heat transfer coefficient as well as the interfacial friction force. It is determined as:

$$\delta_{dro} = \min\left(\frac{8\sigma}{\rho_{gas}\Delta u_{ik}^2}; 1.73\sqrt{\frac{\sigma}{g(\rho_{liq} - \rho_{gas})}}\right). \tag{2.36}$$

The term Δu_{ik} represents the relative velocity between phases, σ is the surface tension, ρ_{gas} and ρ_{liq} denote the density of gas and liquid, respectively. The droplet Reynolds number is calculated as follows:

$$\mathrm{Re}_{dro} = \frac{\rho_{gas}\Delta u_{ik}\delta_{dro}}{\mu_{gas}}. \tag{2.37}$$

Here, the symbol μ_{gas} is the dynamic gas viscosity. If $h_{gas} \geq h_{gas,sat}$ (evaporation), the interfacial heat transfer coefficient of gas is reduced by multiplying Eq. (2.35) by the variable a:

$$a = \frac{1}{1 + \frac{\left(T_{gas} - T_{gas,sat}\right)}{1000}}, \tag{2.38}$$

resulting in the following relation:

$$K_{i,gas} = \frac{6000(1-\chi)\lambda_{gas}\left(2 + 0.74\mathrm{Re}_{dro}^{1/2}\,\mathrm{Pr}_{gas}^{1/3}\right)}{\delta_{dro}^2 c_{p,gas}\left(\left(T_{gas} - T_{gas,sat}\right) + 1000\right)}. \tag{2.39}$$

The interfacial heat transfer coefficient of liquid phase $K_{i,liq}$ is calculated during vaporisation $\left(h_{liq} > h_{liq,sat}\right)$ as follows:

$$K_{i,liq} = \frac{1.2 \cdot 10^{-8} \rho_{liq}^2 u_{liq}^2 \cdot \exp(4.5\chi)}{\mu_{liq}\,\mathrm{Pr}_{liq}}. \tag{2.40}$$

Here, the term u_{liq} represents the liquid phase velocity, Pr_{liq} and μ_{liq} are the Prandtl number and the dynamic viscosity of liquid phase. During condensation $\left(h_{liq} \leq h_{liq,sat}\right)$, the interfacial heat transfer coefficient of liquid increases and is calculated in droplet flow according to Shah correlation (Shah 1979):

$$\begin{aligned} K_{i,liq} = & \frac{0.092\mathrm{Re}_{dro}^{0.8}\,\mathrm{Pr}_{liq}^{0.4}\,\lambda_{liq}}{D_H^2 c_{p,liq}}\left[(1-x)^{0.8} + 3.8x^{0.79}(1-x)^{0.04}\left(\frac{p_{cri}}{p}\right)^{0.38}\right] \\ & + E\frac{6(1-\chi)\lambda_{liq}}{\delta_{dro}^2 c_{p,liq}}. \end{aligned} \tag{2.41}$$

The symbol D_H is the hydraulic diameter of the flow channel, Pr_{liq}, λ_{liq} and $c_{p,liq}$ are the Prandtl number, the thermal conductivity and the heat capacity of liquid phase, respectively. The terms E and x denote the rate of entrainment and the steam concentration in the flow. The critical pressure of the steam/water mixture p_{cri} is equal to 22.06 MPa.

The wall heat flow $q_{wal.k}$ is determined depending on the heat transfer zone. Generally, three heat transfer zones are distinguished: a wetted wall, a dry wall and a transition zone between the wetted wall and the dry wall. When the wall temperature is lower than the saturation temperature of liquid, the single-phase flow is assumed to be in contact with the wall. During this stage, the heat flux rises with increasing wall temperature. If the heat flux has exceeded the critical heat flux, the wall starts drying out and accordingly the heat transfer decreases sharply. The transition zone ranges from the critical heat flux temperature to the minimum film boiling temperature. Above this temperature, the dry wall zone stars. Here, only the gas phase touches the wall and the heat flux begins to increase again. The critical heat flux and the minimum film boiling temperature are used for the selection of the heat transfer zone.

The heat flux on the wetted wall can be defined according to several correlations, e.g. the Dittus-Boelter correlation for forced convection and the Thom correlation for nucleate boiling (Groeneveld and Snoek 1986). The nucleate boiling starts when the wall temperature exceeds the saturation temperature of liquid. The heat flow is expressed as follows (Hänninen 2009):

$$q_{wal,liq} = \begin{cases} \underbrace{\frac{\lambda_{liq}}{D_H^2}\left(0.023\mathrm{Re}_{liq}^{0.8}\mathrm{Pr}_{liq}^{0.4}\right)\left(T_{wal} - T_{liq}\right)}_{A} & case\ T_{wal} \le T_{liq,sat} \\ A + 1971 \cdot \exp\left[\left(2.3 \cdot 10^{-7}p\right)\left(T_{wal} - T_{liq}\right)^2\right] & case\ T_{wal} > T_{liq,sat} \end{cases} . \quad (2.42)$$

If the wall temperature is higher than the minimum film boiling temperature, the heat transfer correlations of the dry wall are used. The heat flow between the dry wall and the gas phase is calculated according to Bestion as follows (Bestion 1990):

$$q_{wal,gas} = \max\left(K^1_{wal,gas}; K^2_{wal,gas}; K^3_{wal,gas}; K^4_{wal,gas}\right)\left(h_{wal} - h_{gas}\right), \quad (2.43)$$

with the Berenson coefficient for pool boiling $K^1_{wal,gas}$ (Berenson 1961). The heat transfer coefficients $K^2_{wal,gas}$ and $K^3_{wal,gas}$ describe laminar and turbulent forced convections, while $K^4_{wal,gas}$ is the heat transfer coefficient of natural convection.

In the transition zone between the wetted wall and the dry wall, the heat flow is interpolated between the heat flow of the dry zone and the critical heat flow. The latter is calculated using the Zuber-Griffith correlation for lower mass flow density and the Biasi correlation for higher mass flow density (Groeneveld and Snoek 1986).

The friction force between the single-phase (gas or liquid) and the wall of the flow channel is computed using the following relation:

$$F_{wal,k} = \frac{-2f_{wal,k}\rho_k u_k |u_k|}{D_H}. \quad (2.44)$$

The phase friction coefficients are calculated employing Blasius correlations and laminar formula. The phase friction coefficients are then determined depending on the void fraction for gas phase and for liquid phase as:

$$f_{wal,gas} = \max\left[\frac{16}{\mathrm{Re}_{gas}}\chi^5; 0.079\mathrm{Re}_{gas}^{-0.25}\chi^5\right], \quad (2.45)$$

$$f_{wal,liq} = \max\left[\frac{16}{\mathrm{Re}_{liq}}\left(1-\chi^5\right); 0.079\mathrm{Re}_{liq}^{-0.25}\left(1-\chi^5\right)\right]. \quad (2.46)$$

The interfacial friction force F_{ik} (the friction between liquid and gas phases) is highly reliant on the flow regime. Generally, it is distinguished between stratified and non-stratified flows, including bubbly, annular and droplet flow regimes. Based on the void fraction χ and the rate of entrainment E, the interfacial friction force is formulated as:

$$F_{ik} = (1 - E)\left[([1 - \chi])F_{ik,bub} + \chi F_{ik,ann}\right] + EF_{ik,dro}. \tag{2.47}$$

The interfacial friction of bubbly flow $F_{ik,bub}$ is determined according to (Bestion 1990):

$$F_{ik,bub} = \left(\frac{29\rho_{gas}}{\delta_{bub}} + \frac{f_\delta F_{\mu,liq}^{0.25}\rho_{liq}}{D_H}\right)\chi(1-\chi)^3 \Delta u_{ik}|\Delta u_{ik}|, \tag{2.48}$$

where Δu_{ik} represents the relative velocity between phases. The bubble diameter δ_{bub} can be estimated using the following relation:

$$\delta_{bub} = \left(\frac{1}{D_H^2} + \frac{(\rho_{liq} - \rho_{gas})}{f^2\sigma} g\right)^{-0.5}, \tag{2.49}$$

with the variable f that depends on the void fraction:

$$f = \begin{cases} 1.3 + 15.7\chi^3(256 - 768\chi) & \text{case } \chi < 0.25 \\ 17 & \text{case } \chi \geq 0.25 \end{cases}, \tag{2.50}$$

and the variable f_δ that depends on the bubble diameter and the hydraulic diameter of the flow channel:

$$f_\delta = 2.81 + 34\left(\frac{\delta_{bub}}{D_H}\right)^5 \left(6 - \frac{5\delta_{bub}}{D_H}\right). \tag{2.51}$$

The viscosity factor $F_{\mu,liq}$ of liquid phase is expressed as follows:

$$F_{\mu,liq} = \mu_{liq}\left[\frac{g(\rho_{liq} - \rho_{gas})}{\rho_{liq}^2\sigma^3}\right]^{0.25}. \tag{2.52}$$

In the annular flow regime, the interfacial friction force per volume is determined according to Wallis relation (1969):

$$F_{ik,ann} = \frac{0.01[1 + 75(1 - \chi)]\rho_{gas}\Delta u_{ik}|\Delta u_{ik}|}{D_H}. \tag{2.53}$$

In the droplet flow regime, the interfacial friction force per volume is calculated as a function of the droplet diameter δ_{dro} (Bestion 1990):

$$F_{ik,dro} = \frac{0.75(1-\chi) f_{dro} \rho_{gas} \Delta u_{ik} |\Delta u_{ik}|}{\delta_{dro}}, \tag{2.54}$$

with the friction coefficient f_{dro} of the droplet flow:

$$f_{dro} = \frac{24}{\mathrm{Re}_{dro}} + \frac{3.6}{\mathrm{Re}_{dro}^{0.313}} + \frac{0.42}{1+4.25 \cdot 10^4 \cdot \mathrm{Re}_{dro}^{-1.16}}. \tag{2.55}$$

For specific cases, non-condensable gases can be modelled and therefore additional equations may be required. This can be a non-condensable gas in steam phase or a dissolved component in liquid phase. Assuming that a non-condensable gas and the steam form a homogeneous mixture with same temperature and velocity, then one additional partial differential equation for the non-condensable gas density is required:

$$\frac{\partial(\chi \rho_{NC})}{\partial t} + \frac{\partial(\chi \rho_{NC} u_{gas})}{\partial z} = S_{NC}. \tag{2.56}$$

The subscript NC refers to the non-condensable gas. The term S_{NC} describes formation/reduction of the non-condensable gas in the steam.

The dissolved gas in liquid phase is modelled by a mass transport equation of the dissolved gas as:

$$\frac{\partial}{\partial t}\left[(1-\chi)\rho_{liq} X_{NC}\right] + \frac{\partial}{\partial z}\left[(1-\chi)\rho_{liq} u_{liq} X_{NC,dis}\right] = S_{NC}. \tag{2.57}$$

Here, the symbol $X_{NC,dis}$ is the molar fraction of the dissolved gas.

It should be mentioned here that the massive existence of the non-condensable gas in the steam phase or dissolved gas in the liquid phase must also be considered in the conservation equations of gas and liquid phases. For example, the total gas enthalpy can be defined as:

$$h_{0,gas} = X_{NC} h_{NC} + (1 - X_{NC}) h_{0,gas}. \tag{2.58}$$

Furthermore, the total gas density is calculated as a sum of the steam partial density and the non-condensable gas partial density. Similar to gas density, the total pressure is the sum of the partial pressures. The heat and mass transfers between wall and gas, wall and liquid, gas and liquid must also include additional terms in order to consider the influence of non-condensable gases on these terms.

2.1.2.4 Seven-Equation Flow Model

The six-equation flow model accounts for significant non-equilibrium of the phases, but still assumes pressure equilibrium. The seven-equation model (also known as two-pressure flow model) allows the phases to be completely independent of each other. It solves the fluid dynamic interface problems and the two-phase flow system simultaneously, resulting in individual pressure, velocity, temperature and chemical potential for both phases. The model has two sets of mass, momentum and energy conservation equations as well as one volume fraction evolution equation that describes how the fluid composition changes with time. Although the seven-equation flow model is different from the six-equation flow model, many of the closure models in the six-equation flow model can also be used. The closure laws are determined from the flow parameters and their derivatives, coupled with empirical correlations describing all regimes of a two-phase flow (Berry et al. 2010).

In the process simulation of thermal power plants, it is more common to apply the two-fluid model with pressure equilibrium. The system of the partial differential equations related to the pressure equilibrium suffers, however, from improper mathematical properties. The eigenvalues of the Jacobian matrix are not always real and may assume complex values, which in turn lead to an ill-posed Cauchy problem. The seven-equation model systematically allows seven real eigenvalues and is shown to have a well-posed basis of eigenvectors, particularly in the context of compressible two-phase flows. In comparison to models based on pressure equilibrium assumption, the unconditionally hyperbolic property makes the two-pressure seven-equation model very attractive. Recently, this flow model has gained interest for modelling of a wide range of applications, including non-equilibrium dispersive two-fluid flow, free-surface two-fluid flow under the influence of gravity, boiling and flashing of superheated liquid as well as the bubble collapse in light-water reactor (Berry et al. 2013). However, the seven-equation model also suffers from several difficulties such as the model complexity and the presence of neoconservative products, i.e. the model cannot be equivalently recast in full conservative form. These neoconservative products naturally disappear, when the model corresponds to two decoupled gas dynamic systems and the void fraction is locally constant in space.

This section describes the seven-equation model, as implemented in the RELAP software (Berry et al. 2014). The solution of the seven-equation flow model is based on the one-dimensional seven partial differential equations, from which the void fraction, the phase pressures, the phase velocities and the phase enthalpies are solved.

The volume fraction evolution is expressed by:

$$\frac{\partial(\chi A)}{\partial t} + \frac{\partial(\chi u_{\text{int}} A)}{\partial z} = \underbrace{A\frac{\left(p_{gas} - p_{liq}\right)}{\eta}}_{\textit{Pressure relaxation rate}} + \underbrace{\frac{\Gamma A_{\text{int}} A}{\rho_{\text{int}}}}_{\textit{Interfacial mass transfer term}} \quad . \tag{2.59}$$

The mass balance equations for gas and liquid phases are formulated as:

$$\frac{\partial(\chi\rho_{gas}A)}{\partial t}+\frac{\partial(\chi\rho_{gas}u_{gas}A)}{\partial z}=\underbrace{\Gamma A_{\text{int}}A}_{\text{Interfacial mass transfer}}, \tag{2.60}$$

$$\frac{\partial((1-\chi)\rho_{liq}A)}{\partial t}+\frac{\partial((1-\chi)\rho_{liq}u_{liq}A)}{\partial z}=-\Gamma A_{\text{int}}A. \tag{2.61}$$

The momentum balance equations for gas and liquid phases are written:

$$\begin{aligned}\frac{\partial(\chi\rho_{gas}u_{gas}A)}{\partial t}+\frac{\partial\left(\left(\rho_{gas}u_{gas}^2+p_{gas}\right)\chi A\right)}{\partial z}=&\underbrace{p_{\text{int}}A\frac{\partial\chi}{\partial z}}_{\text{Volume fraction term}}+\underbrace{\chi p_{gas}\frac{\partial A}{\partial z}}_{\text{Pressure term}}+\underbrace{A\beta\left(u_{liq}-u_{gas}\right)}_{\text{Velocity relaxation rate}}\\&+\underbrace{\Gamma A_{\text{int}}Au_{\text{int}}}_{\text{Interfacial mass transfer}}\underbrace{-f_{wal,gas}\rho_{gas}\chi\left(u_{gas}-u_{wal}\right)^2(\pi A)^{0.5}}_{\text{Wall drag}}\\&\underbrace{-\frac{1}{2}f_{\text{int}}\rho_{gas}\left(u_{gas}-u_{\text{int}}\right)^2A_{\text{int}}^{0.5}}_{\text{Interfacial viscous drag}}+\underbrace{\chi\rho_{gas}gA}_{\text{Gravitational term}},\end{aligned} \tag{2.62}$$

$$\begin{aligned}\frac{\partial((1-\chi)\rho_{liq}u_{liq}A)}{\partial t}+\frac{\partial\left(\left(\rho_{liq}u_{liq}^2+p_{liq}\right)(1-\chi)A\right)}{\partial z}=&\,p_{\text{int}}A\frac{\partial(1-\chi)}{\partial z}+(1-\chi)p_{liq}\frac{\partial A}{\partial z}\\&+A\beta\left(u_{gas}-u_{liq}\right)-\Gamma A_{\text{int}}Au_{\text{int}}-f_{wal,liq}\rho_{liq}(1-\chi)\left(u_{liq}-u_{wal}\right)^2(\pi A)^{0.5}\\&-\frac{1}{2}f_{\text{int}}\rho_{liq}\left(u_{liq}-u_{\text{int}}\right)^2A_{\text{int}}^{0.5}+(1-\chi)\rho_{liq}gA.\end{aligned} \tag{2.63}$$

The energy equations are as follows:

$$\begin{aligned}\frac{\partial(\chi\rho_{gas}h_{0,gas}A)}{\partial t}+\frac{\partial((\rho_{gas}h_{0,gas}+p_{gas})\chi u_{gas}A)}{\partial z}=&\underbrace{p_{\text{int}}u_{\text{int}}A\frac{\partial\chi}{\partial z}}_{\text{Volume fraction term}}\underbrace{-\frac{\bar{p}_{\text{int}}}{\eta}A\left(p_{gas}-p_{liq}\right)}_{\text{Pressure relaxation term}}\\&+\underbrace{\bar{u}_{\text{int}}\beta A\left(u_{liq}-u_{gas}\right)}_{\text{Velocity relaxation term}}\underbrace{-\Gamma A_{\text{int}}A\left(\frac{p_{\text{int}}}{\rho_{\text{int}}}-h_{0,gas,\text{int}}\right)}_{\text{Interfacial mass heat transfer}}+\underbrace{K_{\text{int},gas}\left(T_{\text{int}}-T_{gas}\right)A_{\text{int}}^{0.5}}_{\text{Interfacial heat transfer}}\\&+\underbrace{\chi K_{wal,gas}\left(T_{wal}-T_{gas}\right)\left[4\pi A+\left(\frac{\partial A}{\partial x}\right)^2\right]^{0.5}}_{\text{Wall heat transfer}}+\underbrace{\chi\rho_{gas}gu_{gas}A}_{\text{Gravitational term}},\end{aligned} \tag{2.64}$$

$$\frac{\partial((1-\chi)\rho_{liq}h_{0,liq}A)}{\partial t}+\frac{\partial((\rho_{liq}h_{0,liq}+p_{liq})(1-\chi)u_{liq}A)}{\partial z}=p_{\text{int}}u_{\text{int}}A\frac{\partial(1-\chi)}{\partial z}$$
$$-\frac{\bar{p}_{\text{int}}}{\eta}A(p_{liq}-p_{gas})+\bar{u}_{\text{int}}\beta A(u_{gas}-u_{liq})-\Gamma A_{\text{int}}A\left(\frac{p_{\text{int}}}{\rho_{\text{int}}}-h_{0,liq,\text{int}}\right)$$
$$+K_{\text{int},liq}(T_{\text{int}}-T_{liq})A_{\text{int}}^{0.5}+(1-\chi)K_{wal,liq}(T_{wal}-T_{liq})\left[4\pi A+\left(\frac{\partial A}{\partial x}\right)^2\right]^{0.5}$$
$$+(1-\chi)\rho_{liq}gu_{liq}A. \tag{2.65}$$

Most of the two-phase nomenclatures used in these sets of equations have already been mentioned. Here, the interfacial mass transfer Γ has the unit of $[\text{kg}/\text{m}^5\text{s}]$, while in other two-fluid models has the unit of $[\text{kg}/\text{m}^3\text{s}]$. The symbols A and A_{int} represent the flow cross-sectional area and the specific interfacial area between phases (gas and liquid) per unit volume. The terms p_{gas} and p_{liq} denote the gas pressure and the liquid pressure. The symbol β describes the inter-phase momentum transfer coefficient (resistance coefficient). The terms p_{int} and u_{int} represent the interfacial pressure and the interfacial velocity that exerted on the surface of a two-phase control volume, where the volume fraction gradients exist. These interfacial variables are expressed using the following equations:

$$u_{\text{int}}=\bar{u}_{\text{int}}+\text{sgn}\left(\frac{\partial(1-\chi)}{\partial z}\right)\left(\frac{p_{gas}-p_{liq}}{Z_{gas}+Z_{liq}}\right), \tag{2.66}$$

$$p_{\text{int}}=\bar{p}_{\text{int}}+\frac{Z_{gas}Z_{liq}}{Z_{gas}+Z_{liq}}\text{sgn}\left(\frac{\partial(1-\chi)}{\partial z}\right)(u_{gas}-u_{liq}), \tag{2.67}$$

with the phasic relation:

$$Z_k=\rho_k c_k. \tag{2.68}$$

Here, k denotes liquid or gas and c_k is the phasic sound speed. The average values of interfacial pressure and velocity can be defined as follows:

$$\bar{u}_{\text{int}}=\frac{Z_{gas}u_{gas}+Z_{liq}u_{liq}}{Z_{gas}+Z_{liq}}, \tag{2.69}$$

$$\bar{p}_{\text{int}}=\frac{Z_{gas}p_{liq}+Z_{liq}p_{gas}}{Z_{gas}+Z_{liq}}. \tag{2.70}$$

Similar to the six-equation flow model, the non-condensable gases can be included in the framework of the seven-equation flow model. The non-condensable gases can be a part of the gas phase or dissolved in the liquid phase. In this case, additional partial differential equations are expressed for non-condensable gases in the gas phase (e.g. nitrogen, helium and hydrogen in the steam) and for dissolved components in the liquid phase (e.g. nitrogen, helium, hydrogen and even boron dissolved in water).

2.2 Process Components

In thermal power plants, flows of various reactants are injected into the combustion chamber, where the reactions take place between oxidising agents (generally air or oxygen) and fossil fuels such as natural gas, oil or coal. The heat released is transferred through radiation and convection to water/steam side. Here, the working fluid (generally water) flows with aid of boiler feedwater pumps through cylindrical hollow bodies with a circular profile and an uniform wall thickness. The cylindrical hollow components used in energy systems can be divided into thin-walled tube and thick-walled tube, based on their wall thicknesses. The tubes of the economisers, evaporator and superheater as well as the connecting pipes and valves belong to the thin-walled tubes, while other components such as drum and header belong to thick-walled tubes. Finally, the superheated steam expands in a steam turbine that is connected to an electrical generator, converting the mechanical energy in electrical energy.

The process components required for the modelling of thermal power plants and energy systems include point, thin-walled tube (e.g. pipe, valve, heat exchanger), thick-walled tube (e.g. header, drum), turbomachines (e.g. turbine, pump, compressor) etc. These are explained below according to APROS and ASPEN PLUS DYNAMICS programmes as well as (Epple et al. 2012; Alobaid et al. 2017).

2.2.1 *Connection Point*

A point component (no real process component as a counterpart) is used to connect different kinds of process components together. It may have at least one inlet flow and one outlet flow. The following mass and energy balances can be expressed:

$$\sum_{i=1}^{i=k} \dot{m}_{inlet,i} = \sum_{i=1}^{i=j} \dot{m}_{outlet,i}, \tag{2.71}$$

$$\sum_{i=1}^{i=k} \dot{m}_{inlet,i} h_{inlet,i} = \sum_{i=1}^{i=j} \dot{m}_{outlet,i} h_{outlet,i}. \tag{2.72}$$

The subscripts k and j represents the number of inlet and outlet flows, respectively. The pressure of all outlet flows involved is assumed to be equal. The point model can be applied to different working fluids such as water/steam, flue gas, organic working fluids, etc. In a water/steam circuit, the fluid is water, when the void fraction is zero. If the void fraction is equal to unity, the fluid is steam. The fluid is a mixture of liquid water and steam, when the void fraction is between 0 and 1. For other gaseous and liquid fluids such as air, flue gas, natural gas or oil, the void fraction is set to 1 (gases) or 0 (liquids). In case of mixing two fluids with different compositions such as natural gas and air, the materials balances of the individual substances must be computed and the mixture enthalpies calculated accordingly.

2.2.2 Thin-Walled Tube

Thin-walled tubes are used as a representation of several different structures in thermal power plants such as pipes and heat exchangers. The thin-walled tubes incorporate a heat transfer model between wall and fluid, heat storage and a pressure loss of the flow. The knowledge of the wall temperature is of relevance in order to define the material and thermal stresses. If the maximum surface temperature is exceeded, damages to the component can be occurred. Furthermore, the heat storage in the wall or the stored heat release from the wall should be considered, especially during the transient processes. This is due the fact that the heat storage/release can affect the state of the streaming working fluid in the component significantly. For reasons of simplification, the wall temperature of thin-walled tubes can be modelled with a constant temperature in the radial direction. Berndt (1984) calculated the wall temperature of thin-walled tubes, formulating the wall energy balance under the assumption of infinite thermal conductivity in the radial direction and negligible thermal conductivity in axial and tangential directions as follows:

$$A_{wal}\Delta x\frac{\partial}{\partial t}\left(\rho_{wal}c_{p,wal}T_{wal}\right) = \alpha A_{in}\left(T_{flu} - T_{wal}\right) + \dot{Q}. \tag{2.73}$$

Here, the symbol A_{wal} is the cross-section surface of the wall, ρ_{wal} and $c_{P,wal}$ represents the density and the heat capacity of the wall material, $\dot{Q}$ is the heat flow, α denotes the heat transfer coefficient, A_{in} is the inner surface area of the wall, T_{flu} and T_{wal} are the temperature of the working fluid and the wall, respectively. In order to reduce the computational cost, it is assumed that the physical characteristics of the wall material such as density and heat capacity are not subjected to major change during the transient.

The thin-walled tubes used in thermal power plants, explained below, include pipe, valve, heat exchanger and attemperator.

2.2.2.1 Pipe

The gaseous and liquid fluids have to be transported through pipes that cause pressure losses. In order to overcome these losses and to generate a constant mass flow rate, pumps are generally required. In some applications (evaporator with natural circulation operation), the density differences at the inlet and the outlet of the evaporator system give rise to a constant mass flow without having a pump. Numerically, the pipe component that is defined between two connection points is used to calculate the fluid flow. Here, the shape and dimensions of the pipes should be specified. When it is necessary, the heat storage into the pipe material or the heat release out from the pipe can be taken into account.

The pressure drop of the flow in the pipe is expressed as follows:

$$\Delta p_{pip} = f_k \frac{l}{D} \frac{\rho_k u_k^2}{2}. \tag{2.74}$$

The symbol f_k represents the friction coefficient, ρ_k and u_k are the density and the velocity of the fluid, l and D denote the length and the diameter of the pipe, respectively.

2.2.2.2 Valve

In thermal power plants, the main task of a valve is to achieve desired fluid flow rates. In the process simulation, the valve model is considered as a flow resistance. The pressure drop over the valve is computed as a sum of the pressure drop due to friction coefficient f_k and the pressure drop due to the flow resistance coefficient of the valve f_{val}:

$$\Delta p = \left(f_{val} + f_k \frac{l}{D} \right) \frac{\rho_k u_k^2}{2}. \tag{2.75}$$

The valve can be controlled using automation components and can also be connected to electrical components to simulate its behaviour during the loss of electricity. There are different kinds of valves used in thermal power plants such as control valve, shut off valve, check valve and safety valve.

Control valve: The flow resistance of a control valve is calculated as a function of the valve position that is a non-dimensional value and varies between 0 and 1. The valve is fully closed when the valve position is 0 and is fully opened when the valve position is 1. The control valve is specified by a function, so-called characteristic curve, describing the correlation between the mass flow rate and the valve position. Generally, the characteristic curve can be linear (mass flow rate increases linearly with valve travel) or equal percentage (mass flow rate increases exponentially with valve travel). If it is necessary, more accurate curves or different kind of curve types such as parabolic, hyperbolic or square root can also be applied. The function of the control valve is adjusted by the automation system, which gives the setpoint to the valve position. The valve model receives the new setpoint and computes how much the valve should move during the time step taking into account the driving time of the valve. The driving time of the valve is defined as the time required by the valve actuator to open the valve from the fully closed position to the fully opened position or vice versa. Generally, large valves in thermal power plants use hydraulic pressure or electricity, allowing typical driving times between 10 and 30 s. The electrical system sends signal and provides electricity to the valve actuator. The valve model can also incorporate loss of electrical energy supply. In this case, there are three possible modes of valve operation: to open, to close or to stay at the same position.

Shutoff valve: The shutoff valves include different types such as butterfly valve, flap valve and conical seat valve. The characteristic curve of shutoff valves is usually unknown, but the driving time and the flow resistance coefficient at a fully open valve are required. However, there are generic characteristic curves of the flow resistance coefficient as a function of the valve position for each valve type. The binary signals TRUE and FALSE control the shutoff valve. If the signal is TRUE, the valve starts opening and the valve begins to close when the signal is FALSE. The shutoff valve opens/closes during a time step taking into account the driving time of the valve. In break down cases (electrical power is lost), the shutoff valve either stays in its current position, starts closing or opening depending on the valve configuration.

Check valve: The check valve, also known as non-return valve or one-way valve, allows the fluid to flow in only one direction. It is either fully open or fully closed. If the pressure difference over the valve is positive, i.e. the pressure at the valve inlet is higher than the pressure at the valve outlet, the valve is open. If the pressure difference over the valve is negative, the check valve is close. The flow velocity is computed as a function of mass flow rate, fluid density and check valve diameter. The pressure drop can be calculated with aid of the characteristic curve that describes the pressure drop as a function of the flow velocity.

Safety valve: The safety valve has the function of increasing the safety of thermal power plants by limiting the pressure in pressurised vessels. Here, the fluid is automatically discharged from a drum or other components, when the pressure exceeds the specified limit. The characteristic curve of a safety valve defines the mass flow rate as a function of the pressure instead of the valve position in case of the control valve.

2.2.2.3 Heat Exchanger

In thermal power plants, heat exchangers perform the heat transfer between one or more fluids. Based on the requirements, a variety of different heat exchangers has been developed. Basically, the heat exchangers are designed as regenerator or recuperator. In the regenerator heat exchanger, the heat transfer between fluids is carried out in two steps. In the first step, the heat flow is transferred to storage mass and in a second step the energy stored is emitted to the heat-absorbing fluid after a time delay. The charge/discharge cycle can be discontinuous or continuous. Example of regenerators is the thermal wheel, also known as a rotary air preheater heat exchanger. In recuperators, by contrast, the heat is transferred without delay between fluids through a solid wall. They represent the most commonly-used type of heat exchangers, including double-pipe heat exchangers, plate heat exchangers, shell and tube bundle heat exchangers etc. Depending on the flow configuration of the fluids involved in the heat exchange, it is distinguished among concurrent flow, countercurrent flow, cross-flow or a more complicated flow configuration. If phase-change occurs in any of the fluids in the heat exchanger, it is referred to the heat exchanger as a phase-change heat exchanger; otherwise the heat exchanger is a sensible heat exchanger. In thermal power plants, the heat exchangers used are listed as follows:

- Shell and tube bundle heat exchangers (economiser, evaporator, superheater and reheater): This type of heat exchanger transfers the heat from the flue gas to the water/steam side. It consists of numerous, equally long, heated tubes switched in parallel. The bundle of tubes is connected to one another by a header. The water, steam or water/steam mixture flow through the tubes, while the flue gas flows over the tubes, transferring the heat between two fluids. The shell and tube bundle heat exchangers can also be manufactured with finned heating surfaces, so-called finned tube heat exchanger, increasing the heat transfer between fluids significantly.
- Membrane wall heat exchanger (evaporator): This type of heat exchanger is installed on the combustion chamber walls of coal-fired power plants, so that the combustion of fuel with oxidizing agent takes place within the heat exchanger.
- Feedwater preheater: This type of heat exchanger heats the feedwater mass flow, before entering the steam generator using steam extractions form high and low-pressure turbines.
- Condenser: This type of heat exchanger condensates the turbine exhaust steam by means of cooling water or cooling air, depending on site conditions.
- Air and gas preheaters: This type of heat exchangers heats the air or the natural gas used for combustion using the heat of flue gas or process steam.

The transport equations required to model the heat exchanger used in thermal power plants are mass, momentum and energy balances for flue gas path and water/steam side. The mass and momentum balances for the tube wall are unnecessary, so that only the energy balance has to be solved, describing the heat transport from the flue gas to the tube wall and from the tube wall to the water/steam flow. In a tube bundle heat exchanger, the heat flows to the tube wall from the flue gas (through convection and thermal radiation) and from an optional flame-radiation zone of a combustion chamber. The heat is then transferred from the tube wall to the working fluid. Figure 2.1 illustrates the discretised structure of a counter-flow tube bundle heat exchanger as an example. The flue gas path and the water/steam tubes are discretised in equally-spaced control volumes with one calculation node in the centre and a calculation branch between two adjacent nodes. Here, each control volume consists of a horizontal pipe. In order to reduce the computational cost, each control volume may consist of many horizontal pipes. The flue gas and the water/steam sides are coupled by the heat flows through the tube walls. The tube walls can also be discretised as a radial structure of heat nodes and heat branches, which represent the mass and the thermal conductivity of the tube material.

2.2.2.4 Attemperator/Desuperheater

An attemperator, also known as injection cooler, limits the steam temperature to the setpoint value. In thermal power plants, attemperators are installed at multi-zone of

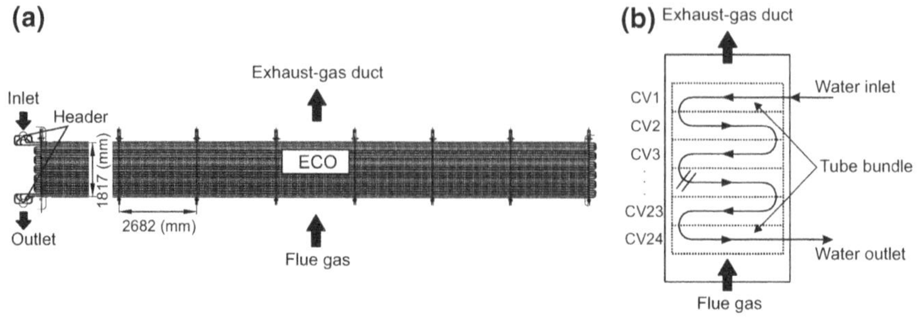

Fig. 2.1 Counter-flow heat exchanger of a low-pressure economiser of a HRSG: **a** real geometry (24 rows with 105 tubes per row) and **b** discretisation structure of the heat exchanger

superheater and reheater to control the temperature at the inlet of high and intermediate-pressure turbines. The attemperators use water directly from the boiler feedwater pumps. The injected mass flow rate is adjusted by an automation circuit that controls the valve position. While the attemperator controls the steam temperature, a desuperheater removes the superheating of the steam, reducing the steam temperature nearly to the range between 10 and 50 °C above the saturation temperature. Desuperheaters are generally found away from the boiler. For example, the bypass system of high-pressure turbine routes the high-pressure steam that is not admitted by the HP steam turbine into the cold reheater or condenser. Here, a desuperheater is installed that cools down the steam temperature to approximately 50 °C above the saturated steam temperature.

2.2.3 *Thick-Walled Tube*

The assumption of constant temperature in the radial direction for thin-walled tubes is acceptable for purposes of simplification. In the case of thick-walled tubes that are usually unheated by the flue gas such as drum and header, this assumption is non-permissible and can lead to incorrect appraisals, especially by calculating the heat storage/release in the walls. The temperature profile in a thick-walled tube can be determined by solving Fourier's differential equation of heat conduction for the whole wall thickness as:

$$\rho_{wal} c_{p,wal} \frac{\partial T_{wal}}{\partial t} = \frac{1}{r} \frac{\partial}{\partial r} \left(r \lambda_{wal} \frac{\partial T_{wal}}{\partial r} \right). \tag{2.76}$$

Here, the symbol λ_{wal} is the wall thermal conductivity. In order to simplify the problem, the wall can be divided into individual circular ring elements and the Fourier's differential equation is solved numerically to each segment. According to Walter (2001), this discretisation is achieved under the assumption that the thermal conduction in the axial direction is negligible and the temperature distribution is radially-symmetric.

In the thick-walled tubes, the rate of temperature change is usually required to prevent the thermal stresses and thus to prove operation safety during start-up procedures of boilers. Material stresses are calculated based on the recorded pressure and temperature history. The thermal stresses are proportional to the difference between integral average wall temperature and inner wall temperature:

$$\sigma_{th} = \frac{\alpha_T \beta_{lin} E_\sigma}{1 - \nu} \left(\bar{T}_{wal}(t) - T_{wal,in}(r,t) \right). \tag{2.77}$$

The linear thermal expansion coefficient β_{lin}, the modulus of elasticity E_σ (Young's modulus) and Poisson's number ν are material properties. The stress concentration factor α_T takes into account weakening of the cylinder wall due to connected tubes, i.e. this factor is strongly dependent on the geometry and weld joints at the T-transition piece. The average wall temperature of a thick-walled tube with a volume V can be calculated according to the following relationship:

$$\bar{T}_{wal} = \frac{1}{V} \int_V T_{wal} dV. \tag{2.78}$$

In case of a hollow cylinder, this relationship takes the form:

$$\bar{T}_{wal}(t) = \frac{2}{r_{out}^2 - r_{in}^2} \int_{r_{in}}^{r_{out}} T_{wal}(r,t) r dr. \tag{2.79}$$

Here, the symbols r_{in} and r_{out} are inner and outer radii of the hollow cylinder. From Eq. (2.79), it can be observed that the thermal stresses are a quadratic function of the wall thickness, which in turn rises with increasing the operation pressure.

During cold start-up procedure of the thermal power plant, the generated steam can condense on the still-cool wall surface of the drum. The heat transfer coefficient of the condensate is much higher than the heat transfer coefficients of the steam and the boiling water in the lower part of the drum. This leads to different heat fluxes along the wall of the drum and accordingly a heat transport in the tangential direction is developed. In order to consider the heat transport in radial and tangential direction, 2D calculations of the heat conduction in the drum wall is required. During hot and even warm start-ups, by contrast, small temperature differences between the upper and the lower part of the drum take place. Therefore, the heat transport of thick-walled hollow cylinders can be reduced to heat conduction in the radial direction.

In the following sections, the thick-walled tubes used in thermal power plants will be explained, including header, drum, separator and feedwater storage tank.

2.2.3.1 Header

A header is a large diameter pipe with a thick wall, connected to a large number of heating surface tubes with small diameters. The headers are arranged on the water/steam side at the inlet and the outlet of heat exchangers. The inlet header has the task to distribute water, steam or water/steam mixture as uniformly as possible on the tubes of the heat exchanger. The outlet header collects and homogenises the working fluid from the tubes of the heat exchanger and feeds it to interconnecting pipes. As described above, the thick wall of the header can store or release the heat from or to the working fluid during the transients, shut-down or start-up procedures. Furthermore, the temperature gradients generated in the header walls results in material and thermal stresses. Therefore, the headers used in thermal power plants are designed with stringent requirements for strength, corrosion and creep properties. The high mechanical properties of the headers are required on the one hand due to the operating under harsh conditions, e.g. high temperature and pressure, high rate of temperature and pressure changes, and on the other hand due to the large number of connected tubes that contribute to cross-sectional weakening.

2.2.3.2 Drum

A drum is a horizontal, cylindrical shaped tank with quite thick walls. In natural or force circulation thermal power plants, the drum represents the core element of evaporator system and has a variety of tasks. It is used as a link between downcomers and risers of the evaporator, supporting the circulation of the working fluid through evaporators. Furthermore, the drum achieves the separation of steam from water/steam mixture by force of gravity using the density difference between phases. The separation of the two-phase mixture can be improved, applying different types of separators using centrifugal force, e.g. cyclone separators or simply deflecting boxes. The saturated steam leaves the drum through a steam dryer (demister) that enhances the removal of liquid droplets entrained in the stream. However, the saturated steam may flow into the superheaters, including minimum water droplets. The water remains in the drum bottom and flows through the downcomers in direction of the evaporator with minimum steam bubbles as possible. Preheated feedwater is fed to drum through economisers with a certain sub-cooling value and mixed then with the saturated water in the drum. The content of salts in the preheated feedwater cannot leave with the saturated steam and remains in the evaporator system. Accordingly, a specific amount of the saturated water is constantly discharged from the drum to a blow-down tank, limiting the concentration of salts in the evaporation system. A schematic representation of the drum with all variables (feedwater and steam pipes, risers, downcomers and the blow-down pipe) is presented in Fig. 2.2.

The mathematical model that describes the physical processes in the drum, by contrast to the header model, are extremely complex. This is due the fact that the separation of the water/steam mixture takes place under highly-turbulent regime.

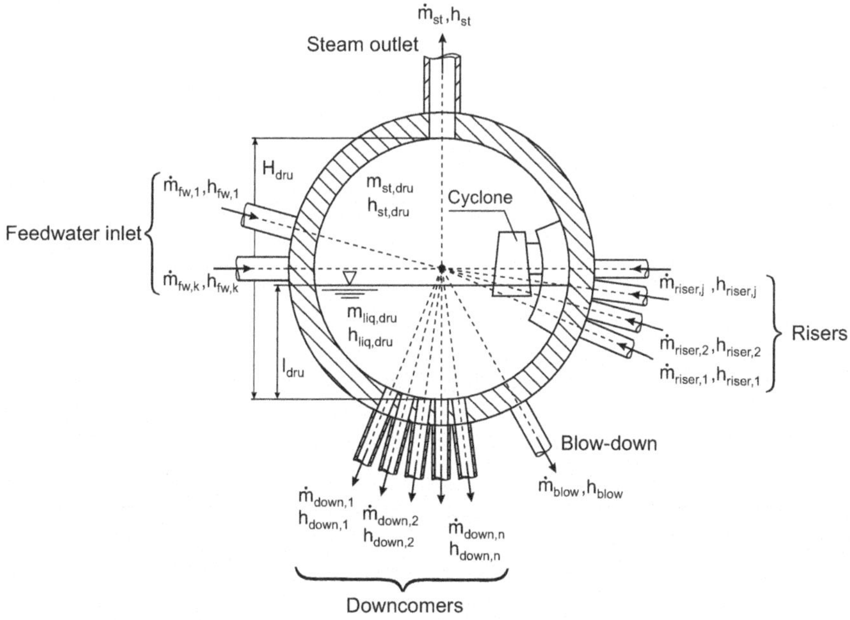

Fig. 2.2 Drum whit its connected tubes (Epple et al. 2012)

In the drum model, it can be assumed that the momentum transported into the drum with the inlet flows is completely dissipated, but it builds up again at the pipes, where the working fluid leaves the drum. For the geometric description of the drum, the volume V_{dru}, the height H_{dru} and the cross-sectional area A_{dru} are required. The mass of the water and the steam in the drum can be expressed by:

$$m_{liq,dru} = \rho_{liq,dru} A_{liq,dru} l_{dru}, \tag{2.80}$$

$$m_{st,dru} = \rho_{st,dru} A_{dru} (H_{dru} - l_{dru}). \tag{2.81}$$

Here, the symbol $A_{liq,dru}$ denotes to the cross-sectional area of the drum equipped by the water, $\rho_{liq,dru}$ and $\rho_{st,dru}$ are the densities of water and steam at the boiling point under the drum pressure. A feedwater control valve is used to regulate the water level in the drum l_{dru} by adjusting the feedwater mass flow rate entering into the drum.

According to mass and energy flows at the inlet and the outlet of the drum, the following balance equations of mass:

$$\frac{dm_{liq,dru}}{dt} + \frac{dm_{st,dru}}{dt} = \sum_{i=1}^{i=k} \dot{m}_{fw,i} + \sum_{i=1}^{i=j} \dot{m}_{riser,i} - \sum_{i=1}^{i=n} \dot{m}_{down,i} - \dot{m}_{blow} - \dot{m}_{st}, \tag{2.82}$$

and energy:

$$\frac{d(m_{liq,dru}h_{0,liq,dru})}{dt} + \frac{d(m_{st,dru}h_{0,st,dru})}{dt} = \sum_{i=1}^{i=k} \dot{m}_{fw,i}h_{0,fw,i} + \sum_{i=1}^{i=j} \dot{m}_{riser,i}h_{0,riser,i} - \sum_{i=1}^{i=n} \dot{m}_{down,i}h_{0,down,i} - \dot{m}_{blow}h_{0,blow} - \dot{m}_{st}h_{0,st} + V_{dru}\frac{dp_{dru}}{dt}, \quad (2.83)$$

can be applied. In the mass and energy balance equations, the subscripts k, j, n represent the number of feedwater pipes, the number of risers and the number of downcomers that are connected to the wall of the drum, respectively. The inlet mass flows are the feedwater mass flow rate from economisers $\dot{m}_{fw}$ and the water/steam mixture mass flow rate from riser $\dot{m}_{riser}$. The outlet mass flow rates are the water mass flow rate to downcomers $\dot{m}_{down}$ and the water mass flow rate to blow-down tank $\dot{m}_{blow}$ as well as the steam mass flow rate leaving at the drum apex $\dot{m}_{st}$. The pressure derivative term appears in the energy equation since the internal energy of water or steam is replaced with the corresponding total enthalpies h_0. The time derivative contained in the mass and energy balance equations can be approximated by using an appreciated numerical solution method.

2.2.3.3 Separator

A separator, also known as cyclone is a vertical, cylindrical shaped tank with a relative small diameter and high wall thickness. It is located at the evaporator outlet of once-through boilers, including the once-through boiler with part load recirculation and the once-through boiler with super-imposed recirculation. The once-through boilers can be operated at sub-critical and super-critical pressures, showing higher degree of operational flexibility. In the once-through boiler with super-imposed recirculation, the recirculation number amounts to around 1.3–1.7. Here, the separator is used to separate the water from water/steam mixture during operation. In the once-through boiler with part load recirculation, the working fluid is forced to flow through all heat exchangers in a single pass, including preheater, economiser, evaporator and superheater. Here, the separator is used to remove the water droplets within the steam flow and to separate the water from water/steam mixture during start-up and at low part loads. If the boiler is operated at super-critical pressure, no phase separation can take place. The working principle of separator is that the two-phase mixture enters tangentially into the separator, leading to a spiral motion of the gas phase (see Fig. 2.3). The water droplets in the stream are centrifuged to the separator wall, where they move downwards and are collected in the separator storage tank. The clean steam leaves then the upper part of the cyclone.

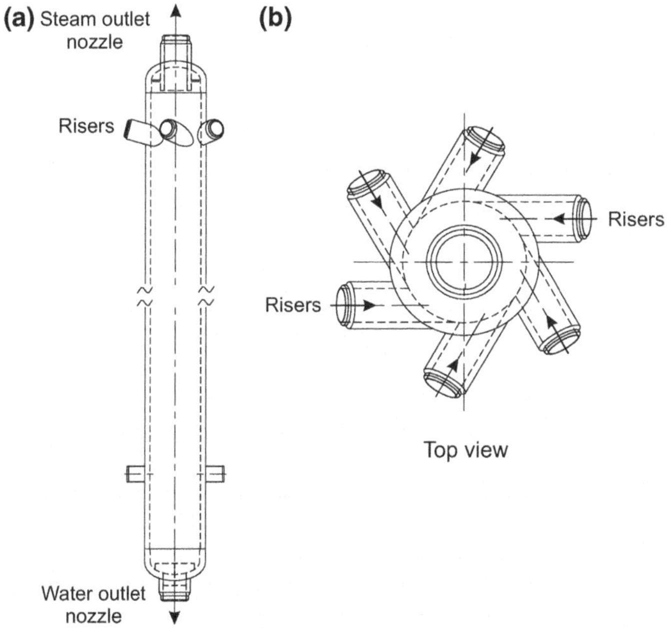

Fig. 2.3 Separator: **a** side view and **b** front view

Based on the drum model, the following applies for the mass balance:

$$\frac{dm_{liq,dru}}{dt} + \frac{dm_{st,dru}}{dt} = \sum_{i=1}^{i=k} \dot{m}_{header,i} - \dot{m}_{st} - \dot{m}_{liq}, \tag{2.84}$$

and for energy balance:

$$\begin{aligned} \frac{d\left(m_{liq,dru}h_{0,liq,dru}\right)}{dt} + \frac{d\left(m_{st,dru}h_{0,st,dru}\right)}{dt} = & \sum_{i=1}^{i=k} \dot{m}_{header,i}h_{0,header,i} - \dot{m}_{st}h_{0,st} \\ & - \dot{m}_{liq}h_{0,liq} + V_{dru}\frac{dp_{dru}}{dt}. \end{aligned} \tag{2.85}$$

Here, the subscript k denotes the number of the header pipes that are connected to the wall of the separator.

2.2.3.4 Feedwater Storage Tank

The feedwater that is pumped into the steam generator is supplied from a heated container with thick walls, known as a feedwater storage tank. The feedwater storage tank stores the feedwater in order to decouple the feedwater mass flow rate from the build-up of condensate. Generally, the storage capacity is designed to cover a few minutes of full load operation of the boiler. This stored energy can sometimes be used effectively in order to yield a rapid increase of the electrical power output (condensate throttling method). Here, the steam extraction valves to the low-pressure preheaters and the condensate valve to the feedwater tank is temporarily throttled or even closed very quickly. Accordingly, the extraction steam passes through the last turbine stages, resulting in a sudden increase in the electric power output (up to 5% within 30 s). This can last for several minutes depending on the volume of the feedwater storage tank. The condensate throttling method, although has no impact on the main steam pressure, has complicated dynamic behaviour since it results in variations of extraction steam flows as well as in changes in the water level of the feedwater storage tank.

In thermal power plants, the feedwater storage tank is equipped with a deaerator that removes the dissolved oxygen, carbon dioxide and other gases from the feedwater. Dissolved oxygen in feedwater attaches to metallic walls and forms oxides (rust). This in turn leads to serious corrosion damages in the steam generator components, including tanks, feedlines, pumps and heat exchangers. If carbon dioxide is also present, then it combines with water to form carbonic acid that causes additional corrosion. The feedwater storage tank and deaerator are modelled as hydraulic accumulators. In steady-state modelling, the mass balance of the storage is simplified to equalisation of the inflow and the outflow. In transient case, the mass balance equation of the storage can be expressed by:

$$\frac{dm_{sto}}{dt} = \dot{m}_{inl} - \dot{m}_{out}, \tag{2.86}$$

and the energy balance equation of the storage is written as:

$$\frac{d\left(m_{sto}h_{0,sto}\right)}{dt} = \dot{m}_{inl}h_{0,inl} - \dot{m}_{out}h_{0,out}. \tag{2.87}$$

Similar to the feedwater storage tank, a compressed air storage tank with a constant volume stores the air at higher pressure and uses it when necessary. Examples of using the compressed air storage system (CAES) are Huntorf and McIntosh power stations. At times of low power consumption, the surplus electricity in the power grid can be used to compress air into salt caverns. At times of high power consumption, the natural gas is burned with the compressed storage air, generating hot exhaust gas that is used for power generation. The gas turbine with compressed air energy storage can be started without additional energy from the power grid and reaches 100% of its nominal load in approximately 6 min. For the purpose of comparison, modern gas turbines without compressed air energy storage can be run up to full load within 20 min and up to half of the generated power is used to drive the compressor. The fast start-up of CAES power plants is of relevance as a standby power plant in case of electrical network failure and even to stabilise variances of fluctuating energy sources such as wind power or photovoltaics etc. For CAES, a large, well-explored and pressure-tight storage cavern is required, resulting in a limited number of suitable sites.

The compressed air storage tank is modelled as a storage component with a constant volume. Here, the transient mass and energy balance equations of the feedwater storage tank [Eqs. (2.86) and (2.87)] can be applied with a few modifications. The storage geometry of the tank should take into account a surface enlargement factor by the reason of highly fissured cavern walls.

2.2.4 *Turbomachines*

Turbomachines are a vital part of every energy system. They transfer energy between a rotor and a working fluid (gas or liquid). This energy transfer can take place from the rotor to the working fluid such as pump, compressor and fan or from the working fluid to the rotor such as steam and gas turbines. Modern turbomachines have smaller gaps between the rotor and the housing for minimising the leakage loss, achieving maximum efficiency.

The pressure of an incompressible fluid can be increased to a higher level using pumps, while the pressure of a compressible fluid is increased to a higher level using compressors. Fan and blower move large amounts of a compressible fluid with a low increase in the pressure. According to American Society of Mechanical Engineers (ASME), fans differ from blowers and compressors by the pressure ratio that can be achieved (up to 1.1 for fans, from 1.11 to 1.2 for blowers and more than 1.2 for compressors). In a steam turbine, the steam expands, resulting in mechanical work that is converted to electrical power by a generator. The gas turbine converts the thermal energy released from the combustion of natural gas or liquid fuels to mechanical energy. The resulting mechanical energy drives a generator, producing electrical power.

In the process simulation, turbomachine models are integrated into the water/steam side and can also incorporate the automation and the electrical system. This allows designer to evaluate their electric power consumption/generation during transients and to assess their behaviour during breakdown cases, e.g. when electricity is lost (i.e. blackout). Further information regarding the specific turbomachines is presented in the following sections.

2.2.4.1 Compressor

A compressor is a turbomachine device that increases the pressure and the enthalpy of a fluid. The pressure increase can be identified by the compressor pressure ratio that describes the ratio of the outlet pressure (discharge pressure) to the inlet pressure (suction pressure). The drive power required is usually supplied to the compressor shaft by an electric motor, a steam turbine or a gas turbine.

The compressors are designed for compressible low density flow (air or gas) and can be classified in two different types, namely intermittent flow or continuous flow. The intermittent flow compressors (positive displacement) include rotary compressor and reciprocating compressor, while the continuous flow compressors (rotor dynamic) include centrifugal compressor and axial compressor. The rotary compressor consists of two rotors within a casing. The compressing and displacement are influenced by the positive action of rotating elements. The reciprocating compressor increases the pressure by reducing the volume of the working fluid. This can be carried out by a piston within a cylinder as a compressing and displacing element, achieving high compression ratios. The centrifugal compressor converts angular momentum transferred by a set of rotating impeller blades (dynamic displacement) to the working fluid. The axial compressor uses arrays of aerofoils to compress the working fluid. The axial air compressor is one main component of the combined-cycle power plant. It compresses and supplies the fresh atmospheric air to the combustion chamber of the gas turbine.

Relevant input data for the compressor model include the dimensions of components, design point values for pressure and temperature before compressor, pressure ratio, rotation speed, characteristic curve and efficiency. Basically, the compressor model has one inlet flow and one outlet flow and therefore the mass balance is expressed by:

$$\dot{m}_{inl} = \dot{m}_{out}, \tag{2.88}$$

and the energy balance is defined as:

$$\dot{m}_{inl}h_{inl} = \dot{m}_{out}h_{out} + P. \tag{2.89}$$

The following equation can be used to calculate the required performance:

$$P = \dot{m}c_p(\bar{T})T_{inl}\Omega_{comp}. \tag{2.90}$$

Here, the symbol c_p is the air specific heat capacity, generally determined simply at an average temperature as:

$$\bar{T} = \frac{(T_{inl} + T_{out})}{2}. \tag{2.91}$$

The dimensionless ratio can be expressed using the following formula:

$$\Omega_{comp} = \Pi^{\frac{\gamma}{\eta_{pol,comp}}} - 1. \tag{2.92}$$

In Eq. (2.92), the term γ is the isentropic exponent. The optimum pressure ratio Π depends on the selected thermodynamic process and should provide the maximum efficiency of combined gas and steam processes. For classic Joule process, the design pressure ratio for modern gas turbines is approximately 18–20 and in case of reheat gas turbines up to 38. The polytropic compressor efficiency $\eta_{pol,comp}$ is defined as:

$$\eta_{pol,comp} = \gamma \frac{\ln \Pi}{\ln\left(\frac{\Pi^{\kappa}-1}{\eta_{isen,comp}}+1\right)}, \tag{2.93}$$

with the isentropic efficiency $\eta_{isen,comp}$:

$$\eta_{isen,comp} = \frac{h_{out,isen} - h_{inl}}{h_{out} - h_{inl}}. \tag{2.94}$$

The symbol h_{inl} is the state enthalpy at the inlet of the compressor, while $h_{out,isen}$ and h_{out} denote the state enthalpy at the compressor outlet for isentropic process and for actual process, respectively. The discharge temperature of the compressor can be obtained as follows:

$$T_{out} = \left(1 + \Omega_{comp}\right)T_{inl}. \tag{2.95}$$

2.2.4.2 Fan

A fan generates a pressure difference, allowing for a large mass flow rate of air or gas to flow against a resistance. The drive power required is supplied to the fan rotor from a rotating shaft (generally an electric motor). Basically, it is distinguished between centrifugal flow and axial flow fans. In the axial fan, air or gas flows axially along the fan shaft without any change in the flow direction. In the centrifugal fan, the air or gas changes its direction relative to the shaft (e.g. forward curved, backward curved or radial).

In thermal power plants, the fans used are forced-draft and induced-draft fans. The forced-draft fan is located at the inlet of the flue gas path, while the induced-fan is located at the outlet of the flue gas path. The forced-draft fan charges the fresh air into the combustion chamber via air preheaters. The induced-fan creates a certain amount of negative gauge pressure in the combustion chamber (the pressure is below atmospheric pressure) by sucking the flue gas out of the combustion chamber, forcing it to flow through the stack into the atmosphere. For this purpose, old thermal power plants use only the stack draft, i.e. the density difference between the hot flue gas in the combustion chamber and the cooler ambient air will generate a difference in pressure, forcing the flue gas to flow. Without the induced-draft fan, excess pressure may prevail in the combustion chamber, which should generally be prevented for safety reason due to the flue gas leakage. In the fan model, the input data such as geometry, dimension, characteristic curve, mechanical coupling should be specified. Furthermore, the automation and electrical systems for starting/stopping the motor and for controlling the inlet guide vanes can be modelled, if necessary.

2.2.4.3 Blower

A blower is used instead of the fan, when higher pressure ratio (approximately 1.11–1.2) is required. In the process simulation, the mathematical model of the fan can be used for the blower.

2.2.4.4 Pump

While the compressor is design for compressible and low density fluid such as air or gas, a pump is design for incompressible and high density fluid such water or oil. In the pump, the mechanical energy of the shaft is transformed into kinetic and potential energy of the flow. Each pump type has a characteristic curve that depends on its geometry and its rotation speed. At several locations in the thermal power plant, pumps are applied to transport the working fluid from one location to another. For example, the condensate pumps move the condensate water that accumulates in the condenser tank through the low-pressure preheaters into the feedwater storage tank. Then, the boiler feedwater pumps force the water to flow from the feedwater tank through high-pressure preheaters and the steam generator.

The pump model can include mechanical coupling, motor and busbar. The automation system controls the rotation speed of the pump either by controlling the frequency or directly by means of a mechanical coupling. The steady-state operation of the pump is computed as a function of the actual volumetric flow rate $\dot{V}_{liq}$ and the rotation speed w as follows:

$$H = \left(H_{max} - (H_{max} - H_{nom})\left(\frac{\dot{V}_{liq}}{\dot{V}_{liq,nom}}\right)^2\right)\left(\frac{w}{w_{nom}}\right)^2. \tag{2.96}$$

The initial values required for the pump model are the maximum pump head H_{max}, the nominal pump head H_{nom}, the nominal volumetric flow rate $\dot{V}_{nom}$ and the nominal rotation speed w_{nom}. The rotation speed of the pump can be controlled by automation and electrical systems. If the electricity is lost, the pump is coast-down according to the following relationship:

$$w = w_0\left(1 - \frac{\Delta t}{t_{stop}}\right). \tag{2.97}$$

Here, the symbol w_0 represents the rotation speed of the pump at old time step, Δt is the time step and t_{stop} denotes the coast-down time of the pump.

2.2.4.5 Steam Turbine

A steam turbine (ST) transforms the enthalpy of steam to mechanical energy. The steam turbine model calculates the enthalpy after the turbine module using the enthalpy before the turbine, the efficiency and the nominal values. The model may describe the steam turbine either with a single turbine stage or with many stages in order to consider the steam extractions. The mechanical power produced by the steam turbine is computed and transferred to the generator model that calculates the rotation speed of the steam turbine.

The pressure and enthalpy drops over the steam turbine are added as source terms in the momentum and energy equations. The pressure change in the turbine stage is specified at nominal load and computed using Stodola equation at the part loads (Cooke 1984). The pressure drop is a function of nominal mass flow rate $\dot{m}_{nom}$, nominal pressure at the turbine inlet $p_{inl,nom}$ and nominal pressure at the turbine outlet $p_{out,nom}$:

$$\dot{m} = \dot{m}_{nom} \frac{p_{inl}}{p_{inl,nom}} \sqrt{\frac{1 - \left(\frac{p_{out}}{p_{inl}}\right)^{\frac{n+1}{n}}}{1 - \left(\frac{p_{out,nom}}{p_{inl,nom}}\right)^{\frac{n+1}{n}}}}, \tag{2.98}$$

with:

$$n = \frac{\ln\left(\frac{p_{out}}{p_{inl}}\right)}{\ln\left(\frac{p_{out}}{p_{inl}}\right) - \ln\left(\frac{T_{out}}{T_{inl}}\right)}. \tag{2.99}$$

The variable $\dot{m}$ represents the mass flow rate, p_{inl} is the pressure before the steam turbine, p_{out} is the pressure after the steam turbine, T_{inl} and T_{out} denote the temperature at inlet and outlet of the steam turbine.

The enthalpy drop over a steam turbine is determined through the expansion efficiency. The basic expansion in the steam turbines is isentropic. Generally, the processes are not pure isentropic or isothermal but a function of the exponent. By giving different exponent values, all processes from isothermal to isentropic are covered. The expansion equation is expressed by:

$$\eta = \frac{1}{v} \frac{\partial h}{\partial p}. \tag{2.100}$$

If the specific volume v is solved from Eq. (2.100) and introduced in Johanson's equation:

$$pv = \frac{h}{\zeta}, \tag{2.101}$$

the following equation is then obtained:

$$\frac{\partial h}{h} = \frac{\eta}{\zeta} \frac{\partial p}{p}, \tag{2.102}$$

with:

$$\zeta = \frac{\gamma}{\gamma - 1}. \tag{2.103}$$

Here, the symbol γ is the isentropic exponent. By integrating the right side term of Eq. (2.102) from h_{out} to h_{inl} and the left side from p_{out} to p_{inl}, one can obtain:

$$\frac{h_{inl}}{h_{out}} = \left(\frac{p_{inl}}{p_{out}}\right)^{\frac{\eta}{\zeta}}. \tag{2.104}$$

Since the tabulated steam enthalpy is the water enthalpy plus the latent heat (evaporation) and the water enthalpy is set to 0 at $T = 0$ °C, the tabulated enthalpy cannot be applied. It has to be shifted with the reference enthalpy h_{ref}:

$$\frac{h_{inl} - h_{ref}}{h_{out} - h_{ref}} = \left(\frac{p_{inl}}{p_{out}}\right)^{\frac{\eta}{\zeta}}. \tag{2.105}$$

Equation (2.105) can be rewritten as:

$$h_{out} - h_{ref} = \left[\left(h_{inl} - h_{ref}\right)\left(\frac{p_{out}}{p_{inl}}\right)^{\frac{\eta}{\zeta}}\right]. \tag{2.106}$$

The specific enthalpy drop over a turbine section is determined using different models. When the fluid is steam, the specific enthalpy drop over a steam turbine is calculated as:

$$\Delta h_{turb,st} = \left(h_{inl,st} - h_{ref}\right) - \left(h_{out,st} - h_{ref}\right). \tag{2.107}$$

If the fluid contains water droplets, the specific enthalpy drop can be computed with the following formula:

$$\begin{aligned} \Delta h_{turb,dro} = {} & x\left[\left(h_{inl,st} - h_{ref}\right) - \left(h_{out,st} - h_{ref}\right)\right] \\ & + (1 - x)\left(h(p)_{inl,dro} - h(p)_{out,dro}\right). \end{aligned} \tag{2.108}$$

Introducing Eq. (2.106) in Eq. (2.107) and in Eq. (2.108), the following relations:

$$\Delta h_{turb,st} = \left(h_{inl,st} - h_{ref}\right) - \left(h_{inl,st} - h_{ref}\right)\left(\frac{p_{out}}{p_{inl}}\right)^{\frac{\eta}{\zeta}}, \tag{2.109}$$

$$\Delta h_{turb,dro} = x\left[\left(h_{inl,st} - h_{ref}\right) - \left(h_{inl,st} - h_{ref}\right)\left(\frac{p_{out}}{p_{inl}}\right)^{\frac{\eta}{\zeta}}\right] + (1 - x)\left(h(p)_{inl,dro} - h(p)_{out,dro}\right), \tag{2.110}$$

are obtained. Here, the subscripts *st* and *dro* refer to steam and water droplets, *inl* to the state before the turbine and *out* to the state after the turbine. The quantities x and η are the steam mass fraction and the polytropic expansion efficiency, respectively. The notation $h(p)$ means the saturated enthalpy at the mentioned pressure. The symbol ζ can be set equal to 4.27.

The mechanical power produced by the steam turbine can be calculated by knowing the mass flow rate through the steam turbine and the enthalpy drop:

$$P_{mech} = \sum_{i=1}^{i=k} \Delta h_{turb,i} \dot{m}_i. \tag{2.111}$$

Here, the subscript k denotes the number of the turbine stages and $\dot{m}_i$ is the mass flow rate that flows into the turbine stage i.

2.2.4.6 Gas Turbine

A gas turbine (GT) converts the flue gas enthalpy to mechanical energy, driving an electrical generator that generates electrical energy. The gas turbine represents the core element of a combined-cycle power plant (CCPP), in addition to compressor, combustion chamber and steam generator that generally known as heat recovery steam generator (HRSG). The latter is arranged downstream of the gas turbine and absorbs the waste heat of the GT to generate live steam. Typically, the CCPPs can be operated with different fuels, including nature gas, crude oil and biogas. They can extend their fuel range to cover biomass and coal through the application of integrated gasification combined-cycle (IGCC). The CCPPs deliver high power plant efficiency levels up to 58%. Also, combined-cycle power plants with efficiency levels greater than 60% are now running, for example Irsching 4 power station that is located in Irsching, Germany. Modern gas turbines can reach their nominal load within 20 min, while 70% of the nominal flue gas temperature and 60% of the nominal flue gas mass flow rate are already achieved approximately 7 min after the start (Walter and Hofmann 2011). Depending on whether a hot, warm, or cold start-up procedure is performed, the HRSGs reaches their nominal load 50, 90 or 170 min after the GT start (Alobaid et al. 2012).

By contrast to steam turbines, heavy-duty gas turbines with high inlet temperature require cooling of the first blade rows. If the cooling is not required, the mathematical model of the gas turbine is similar to the steam turbine. Here, the flue gas is used as a working fluid instead of the steam. In case of the air cooling, the entire gas turbine has to be divided into many stages. The cooling air of one stage is mixed with the combustion gases and can be considered into the mass balance equation of the turbine stage. Based on data from ABB Company, Wang and Leithner (1995) developed an empirical formula to calculate the cooling air mass flow required at nominal load:

$$\dot{m}_{cool,nom} = \dot{m}_{air}\left(3.1817 \cdot 10^{-4} T_{inl} - 0.2454\right). \tag{2.112}$$

The symbol $\dot{m}_{air}$ represents the total air mass flow rate and T_{inl} is the inlet temperature in degree Celsius. At part loads, the following equation for cooling air mass flow rate is proposed by Palmer and Erbes (1994):

$$\dot{m}_{cool,load} = \dot{m}_{cool,nom} \frac{p}{p_{nom}} \sqrt{\frac{T_{nom}}{T}}. \tag{2.113}$$

Here, the subscripts *nom* and *load* refer to nominal state and part load, respectively.

2.2.5 Additional Components

Besides the above mentioned process components, the detailed modelling of thermal power plants requires additional components such as mill, electrostatic precipitator and selective catalytic reduction unit. Furthermore, there is an increased attention on hybrid power plants that combine between different technologies to generate electrical power. Examples of hybrid systems are the integrated solar combined-cycle power plant (ISCC) and the hybrid fuel cell/gas turbine power plant.

2.2.5.1 Combustion Chamber

The combustion of fossil fuels takes place in a combustion chamber. The combustion chamber model has an oxidiser inlet (generally air or oxygen), a fuel inlet (such as natural gas, oil or coal) and a flue gas outlet. Compared to combustion process of liquid or gaseous fuels, the combustion of solid fuels (e.g. coal, biomass or municipal waste) is more complex, including three major mechanisms: drying, pyrolysis and oxidation. The flue gas formed in the combustion chamber of a steam generator or in a gas turbine consists mainly of O_2, N_2, CO_2, H_2O, SO_2 and Ar. The combustion calculation that gives the composition of the flue gas requires the type of fossil fuel, the fuel mass flow rate and the Air/fuel ratio. Determining the exact composition is of high relevance for computing the material properties of the flue gas. The mass flow balance in the combustion chamber can be expressed by:

$$\dot{m}_{out,fg} = \dot{m}_{inl,air} + \dot{m}_{inle,fuel}, \tag{2.114}$$

and accordingly the energy balance equation is written as:

$$\dot{m}_{out,fg} h_{out,fg} = \dot{m}_{inl,air} h_{inl,air} + \dot{m}_{inle,fuel} \mathrm{LHV}_{fuel}. \tag{2.115}$$

Here, the subscript *fg* refers to flue gas and the term LHV is the lower heating value. In addition to the mass and energy balances, a balance equation for each substance is solved. Usually, the combustion chamber is operated under atmospheric pressure like in pulverised coal-fired and incineration power plant. On the contrary, the combustion chamber of a gas turbine is pressurised.

2.2.5.2 Fluidized Bed

A fluidized bed is a bulk of solid particles located in a vertical vessel and the gas or the liquid streams form bottom via a porous plate or nozzles. The emerging forces from gas to particles cause a fluidization condition and offer an effective way for gas-solid interaction. The fluidized bed can be seen as a dense phase zone and a lean zone. In the dense zone, there is a higher concentration of solids near the air distributor plate. Here, it is distinguished between two phases, namely the emulsion phase (uniform mixing of gas and solid) and the bubble phase (only gas). The dense zone is followed with the lean zone, in which the concentration of solid decreases sharply as the flow moves upwards.

The fluidized bed as a firing system is characterised by several advantages.

- The continuous operation with the possibility of solid transport in and out of system
- High heat and mass transfer rates from gas to solid, resulting in uniform temperature gradients in the bed even with highly exothermic or endothermic reaction rates
- Very low NO_x and SO_2 emissions in addition to simple CO_2 capture
- The possible operation over a wide range of geometrical and mechanical properties of particles
- High solid mixing rate for polydisperse particles
- Simple construction and the suitability for large-scale operation
- The possibility of using different solid fuels such as coal, biomass or a mixture of both fuels
- The continuous ability of adding or withdrawing of catalysts

As a result of their competitive advantages, the fluidized bed systems show an increasing trend towards energy applications. This includes CO_2 capture in the thermal power plants by chemical or carbonate-looping process (Ströhle et al. 2014, 2015) as well as the solid fuels conversion, including gasification and combustion of coal, biomass, municipal solid waste, refuse derived fuel (RDF) and even fuel blends.

Different variables can influence the fluidization behaviour such as fluidization volume flow rate, particle size, static bed height and vessel geometry. The fluidization volume flow rate is an extensive variable due to the fact that it depends on the vessel cross-section area. By dividing the volume fluidization flow rate by the cross-section area of vessel, the fluidization volume flow rate per unit of cross-section area of vessel is obtained, which is also known as superficial fluidization velocity (intensive variable):

$$\left|\vec{u}_f^{sup}\right| = \frac{\dot{V}_f}{A_{ves}}. \tag{2.116}$$

In a fluidized bed with defined particle size, static bed height and vessel geometry, the fluidization regimes depend on the superficial fluidization velocity (see Fig. 2.4) as:

Fixed bed: At lower superficial fluidization velocities below the minimum fluidization velocity $\vec{u}_f^{min}$, the particles remain stationary packed since the gas flows through the interstitial space of particles. In this case, the bed behaves like a porous medium.

Particulate fluidization: If the superficial fluidization velocity increases, the force exerted by fluid on particles increases accordingly. At the point, where the gas velocity is between the minimum fluidization velocity and the minimum bubbling velocity, the drag force compensates the bed weight. In the particulate fluidization, the inter-particle distances between the particles become bigger and thus the bed expands, but bubbles are not formed. Here, the suspended particles in the gas characterise the bed behaviour.

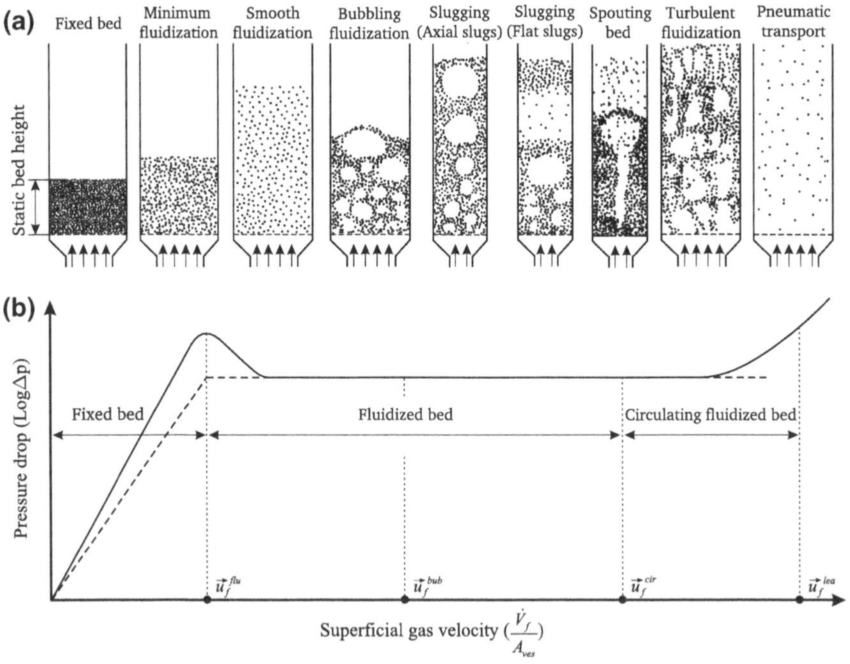

Fig. 2.4 Fluidization: **a** schematic representation of different flow regimes for gas-solid flow and **b** pressure drop versus superficial fluidization velocity (solid line: a real bed; dashed line: an ideal bed), extended from (Kunii and Levenspiel 1991)

Bubbling fluidization: If the superficial fluidization velocity exceeds the minimum bubbling velocity $\overrightarrow{u}_f^{bub}$, the bed becomes inhomogeneous and bubbles can be clearly distinguished. In this regime, the coalescence and breakup of bubbles may occur.

Slugging bed: A further increase in the superficial fluidization velocity enhances the rate of bubbles coalescence. As a result, cavities as large as the cross-sectional area of vessel can be formed. The slugging flow regime takes place, when the vessel diameter is small and/or the ratio of the vessel height to the vessel diameter is large.

Spouting bed: The gas flows through a small jet nozzle at higher superficial fluidization velocities. The particles used in this system are generally of group D (see Fig. 2.5). Here, the gas penetrates the entire bed, dividing the bed into two regions, namely dilute flow region and the annular flow region. In the dilute flow region, the particles move upward, while in the annular flow region the particles move downward to the core region.

Turbulent fluidization: If the superficial fluidization velocity exceeds the solid terminal velocity, the particles start moving faster and form streamers and clusters. The solid terminal velocity is the velocity that the particle reaches in free fall due to gravity. Compared to the slugging bed, the rate of bubbles breakup is enhanced, reducing the presence of large bubbles. Here, the overall structure of the bed transforms to be more homogeneous.

Pneumatic transport: At higher superficial fluidization velocity, the particles are completely entrained. The entire bed becomes as lean gas-solid suspension.

Besides the superficial fluidization velocity, boundary and operating conditions, the fluidization behaviour of the gas-solid flow depends also on the mechanical and physical properties of particles. (Geldart 1973) classified the particles into four groups based on their fluidization behaviour. Geldart identified the particles in terms of mean particle diameter and the difference between solid and fluid densities (see Fig. 2.5). The map was obtained empirically for air at ambient temperature and has been widely used in the research of the gas-solid flow.

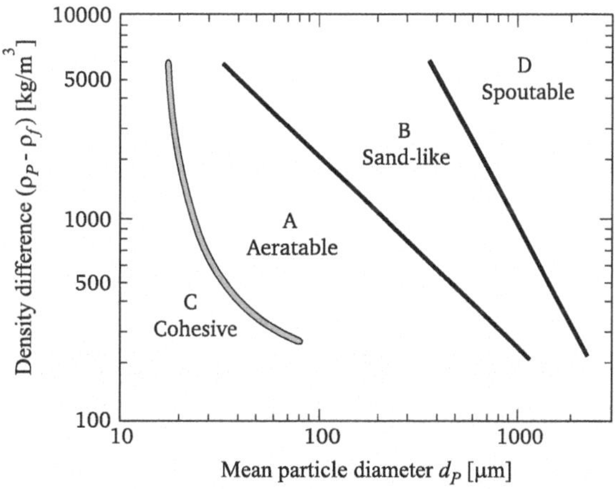

Fig. 2.5 Particle classifications after (Geldart 1973)

Group C: The particles are cohesive and have very small diameter $d_P \leq 20\,\mu\text{m}$. In fluidization for particles of group C, the short-range forces between the particles that result from interactions between particles (contact, van der Waals, electrostatic and capillary forces) dominate the aerodynamic forces. The particles tend to stick together and thus are difficult to fluidize. In case of fluidization, the bed expansion for particles of group C can be very high.

Group A: The particles are aeratable and have a typical diameter between $30 \leq d_P \leq 100\,\mu\text{m}$. In fluidization of particles in group A, the aerodynamic surface forces that result from the interactions between fluid and solid (drag, Basset, pressure, Saffman and Magnus forces) affect the fluidization behaviour more than the short-range forces. However, the short-range contact force may also play a significant role in dense systems. The minimum fluidization velocity is smaller than the minimum bubbling velocity. The bed expands homogeneously at fluid velocities between both minimum fluidization and bubbling velocities. In case of bubbling fluidization, a maximum stable bubble size can be reached.

Group B: The particles have a typical size range of $30 \leq d_P \leq 1000\,\mu\text{m}$. The minimum fluidization velocity and the minimum bubbling velocity have the same value. The bubble size increases with the bed height, but no maximum stable bubble size can be reached. In case of fluidization, the bed expansion for particles of group B is moderate.

Group D: The particles are coarse $d_P > 1$ mm and spoutable. In case of fluidization, the bed expansion is low. Compared to groups A, B and C, the mixing process for particles of group D is not effective.

In order to evaluate the bed behaviour during the fluidization, there are many identification numbers available such as absolute bed height, bed expansion ratio, bed fluctuation ratio and fluidization index (see Fig. 2.6), in addition to visual observation of solid distribution in the bed and pressure profile along the bed height. These identification numbers can be applied to any superficial fluidization velocity and thus they are suitable for all fluidization regimes. The absolute bed height can directly be defined by measuring the highest point in the bed, while the bed expands. The bed expansion ratio *ER* is determined as the ratio of average expanded bed height to the initial static bed height:

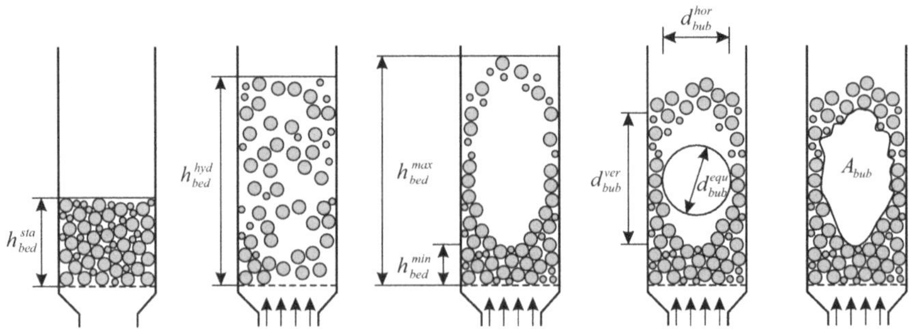

Fig. 2.6 Evaluation the bed behaviour during fluidization

$$ER = \frac{h_{bed}^{hyd}}{h_{bed}^{sta}}. \tag{2.117}$$

The bed fluctuation ratio *FR* is calculated as the ratio of maximum expanded bed height to minimum bed height within the bed fluctuates:

$$FR = \frac{h_{bed}^{max}}{h_{bed}^{min}}. \tag{2.118}$$

The fluidization index is a measure of the fluidization uniformity and can be determined as the ratio of bed pressure drop to bed material weight per unit area of cross-section of vessel:

$$FI = \frac{\Delta p_{bed}}{\left(W_{(bed)}/A_{ves}\right)}. \tag{2.119}$$

Here, it should be mentioned that the identification numbers depend on the superficial fluidization velocity, the particle size and shape, the static bed height and the density ratio between both phases.

For bubbling and slugging regimes, the equivalent bubble diameter can also be evaluated as an additional quantitative measurement of fluidization behaviour. Here, two approaches can be used to assess the equivalent bubble diameter d_{bub}^{equ} (Kuipers et al. 1991). In the first procedure, the equivalent bubble diameter is calculated based on the assumption that the bubble has an elliptical shape. The resulted equivalent bubble diameter is then defined as:

$$d_{bub}^{equ} = \sqrt{d_{bub}^{ver} d_{bub}^{hor}}, \tag{2.120}$$

where d_{bub}^{ver} and d_{bub}^{hor} represent the vertical and horizontal bubble diameters, respectively. In the second procedure, the equivalent bubble diameter is calculated from the measured bubble area as:

$$d_{bub}^{equ} = \sqrt{\frac{4}{\pi} A_{bub}}. \tag{2.121}$$

For modelling of the gas-solid flow in the fluidized bed, there are two different numerical approaches available (Euler-Euler and Euler-Lagrange models). In the Euler-Euler method, also known as two-fluid method, each phase is regarded as a continuum and is mathematically calculated by solving the balance equations. The Euler-Lagrange approach combines the continuum descriptions of fluid phase with the Lagrange representation of dispersed phase on the basis of Newton's transport equations. Although these numerical models described here are already implemented in 3D CFD simulation programmes such as ANSYS-FLUENT [among others publications (Alobaid 2015a; Gryczka et al. 2009)], OpenFOAM and CPFD-BARRACUDA [among others publications (Alobaid 2015b; Karimipour and Pugsley 2012; Stroh et al. 2016; Wu et al. 2009)], they are not suitable for 1D process simulation programmes used to model the entire thermal power plant system. Therefore, semi-empirical models such as Kunii and Levenspiel (Kunii and Levenspiel 1991) are preferred by reason of low computational cost. In this model, the fluidized bed is divided into two regions (dense and lean regions), interacting each other based on several assumptions. The dense region describes the lower part of the rector, where the volume fraction of solid is constant, while the volume fraction of solids in the lean region decreases exponentially with height.

2.2.5.3 Fuel Cell

A fuel cell converts the chemical energy stored in the fuel into electricity and forms water. The fuel cell has two electrodes, namely the anode and the cathode. Generally, hydrogen is the basic fuel, but fuel cells can also extend their fuel range to cover methane. There are four different types of fuel cells, including molten carbonate (MCFC), solid oxide (SOFC), phosphoric acid (PAFC) and proton exchange membrane (PEM). The selection of the fuel cell type depends mainly on the specific application; whether large or small scale, stationary or mobile.

The modelling of a fuel cell requires the consideration of chemical reactions and additional components involved. For example, the simulation model of the solid oxide fuel cell contains a natural gas reformer and a sulphur removal unit. The fuel cell model is connected on the one hand to the process components such as pipe or valve, supplying it with reactants and discharging products. On the other hand, it is connected to the electrical system, generating electricity. The voltage of the fuel cell is used as a boundary condition for the electrical system, while the electric current of the electrical network is transferred to the fuel cell model and applied as an input.

2.2.5.4 Weather

Among renewable energy sources, the solar energy offers a promising option for electricity generation in the countries with high solar radiation. The main drawback is only here the daily and monthly variation of the solar radiation. Solar technologies can be divided into concentrating solar power (CSP) and photovoltaic cells (PV). In the latter, the sunlight is directly converted into electrical energy using semiconducting materials. In the CSP, the sunlight of a large area is concentrated using mirrors or lenses onto absorber tubes, through which a heat transfer fluid (generally oil or water) passes. The thermal energy stored in the heat transfer fluid is used as a heat source for a power generation system. Here, it should not be confused with the concentrator photovoltaics that use concentrated sunlight to produce electricity via the photovoltaic effect. Concentrating solar power plants are showing increasing interest in field of research and application, mostly as parabolic trough collectors and solar tower collectors. This is due to the fact that the CSP technology can easily be coupled with thermal energy storage and with fossil fuel combustion system, increasing the plant availability, especially during low radiation periods.

The solar radiation module considers the variation of solar radiation at different periods of day. It calculates the solar position and both beam and diffusive irradiation on the horizontal surface according to input parameters such as geometric coordinates of the field, time, elevation from sea level and clear sky index. The data from solar radiation module can be used to determine the total amount of irradiation that is passed to different types of surfaces with various angles of inclination. The direct irradiance received per unit area by a surface normal to the sun is defined as direct normal irradiance (DNI). Using DNI values that obtained from the weather stations in the location, the absorbed solar energy by an absorber tube can be computed as follows:

$$S_{absorb} = (DNI)\cos(\theta)A_{aper}\eta_{opt}E_{track}f_{dust}f_{clean}f_{row,shad}f_{end-los}k_{IAM}, \qquad (2.122)$$

with the mirror aperture area A_{aper} and the inclination angle θ. The symbol η_{opt} represents the optical efficiency and E_{track} is a tracking error. The reduction in the absorption energy due to several factors can also be considered, including dust on the absorber glass cover f_{dust}, mirror cleanliness f_{clean}, row shadowing $f_{row,shad}$ and end-losses $f_{end-los}$ caused by spacing between solar collector elements, spacing between solar collector assembly as well as the non-zero incidence angle. When the angle of incidence increases, losses that can arise due to additional reflection and absorption by the glass envelope are corrected by the incidence angle modifier k_{IAM}.

2.2.5.5 Mill

A mill is a mechanical device that is used to break different types of solid material in small pieces by grinding, crushing or cutting. In pulverised coal-fired power plants, a pulveriser mill grinds the raw coal into a combustible dust. Here, the coal particles are dried by flue gas or hot air and milled to the fine size, so that a stable combustion and a complete burn-out of pulverised coal can be achieved in the combustion chamber. The coal properties such as volatile components, ash content, hardness, humidity and grinding fineness are the decisive factor for selecting the required mill. Basically, the coal pulverisers can be divided into three assembly groups: impact mills, gravity-force mills and external force mills. The impact mills such as hammer and beater wheel mills are classified as high speed mills. While the hammer mill is suitable for lignite and hard coal, the beater wheel mill is preferred for lignite, but can be used conditionally for hard coal. The gravity-force mills are low speed mills and well suited for wet and hard coal. Due to their construction, they are identified as tube mills. When considering the design of external force mills, different constructions are well-known such as bowl mill and roller mill. In such medium speed mills, the grinding bowl is rotated by a gearbox, while the grinding parts are pressed by external force either by springs or hydraulic cylinders against the grinding track. The external force mills are suitable for harder coals, showing different advantages like high grinding performance, low wear and low power consumption.

Although the dynamic of the mill has a significant influence on dynamic behaviour of the entire coal-fired power plant, the mill is generally modelled as a mixing point of coal and primary air for purpose of simplification. In this mixing point, the water content of the coal is evaporated under the assumption that no combustion will take place. If it is required, complex pulveriser models, e.g. (Fan and Rees 1994; Luo et al. 2011; Zhou et al. 2000) can also be applied to achieve accurate dynamic behaviour of pulverised coal-fired power plants.

2.2.5.6 Flue Gas Control

Particulate matter, nitrogen oxides (NO_x), sulphur oxides (SO_x) and carbon dioxide (CO_2) emissions emitted by combustion of fossil fuels contribute to global climate change and might present a hazard for health and environment. The energy systems in terms of conversion of fossil fuels as a major emission source are in the focus of attention. Thus, flue gas cleaning devices are considered as an essential part of modern thermal power plants.

Fossil fuels, except natural gas, contain non-combustibles that form the majority of the particulate in the flue gas. In case of coal combustion, ash and some amount of unburned carbon are carried by the flue gas from the furnace into the atmosphere. Particulate control equipment is therefore required to remove particulate, to keep the particulate from re-entering the flue gas and to discharge the collected material. The design and selection of particulate control equipment are related to firing system used (pulverised coal-fired or fluidized bed). Here, different types of particulate control are available such as electrostatic precipitators, fabric filters, mechanical collectors and venturi scrubbers.

Nitrogen oxides are formed during the combustion process depending on temperature, the supply of oxygen and nitrogen from both air and fossil fuels. The term NO_x denotes to cumulative emissions of nitric oxide (NO), nitrogen dioxide (NO_2) and other nitrogen-bearing species. In thermal power plants, there are three principal mechanisms of NO_x formation, namely thermal NO_x, fuel NO_x and prompt NO_x. A reduction in NO_x emissions can be achieved using primary and secondary techniques. The primary measures for NO_x aim to reduce both peak temperature and residence time at peak temperature, including fuel/air staging, over fire air, less excess air, flue gas recirculation and combustion optimisation. The secondary techniques are based on chemical reduction of NO_x such as selective catalytic reduction (SCR) and selective non-catalytic reduction (SNCR). The primary techniques, by contrast to secondary techniques, are much more cost-effective.

In addition to nitrogen oxides, the sulphur oxides contribute to the formation of acid rain and can stress the ecosystem on earth. Most sulphur emitted to the atmosphere oxidizes slowly to sulphur dioxide (SO_2) that is a reactive and acrid gas.

For SO_2 control, two strategies can be generally applied either the use of low sulphur coal or installing scrubbers. For the latter, several techniques are in service, while others are in the development or already in the pilot stage. Today, commercialised technologies are wet, semidry and dry processes. For thermal power plants, the technology of choice is the wet flue gas desulfurization (WFGD) scrubber and in case of lower sulphur oxide emissions is the dry flue gas desulfurization (DFGD).

In recent years, scientific researches confirm that human activities such as burning fossil fuels contribute to an increase in carbon dioxide concentration and accordingly to an increase in the average global temperature of earth. Lower CO_2 emissions can be achieved by increasing the use of renewable energy sources. However, the current and future growing demand for energy worldwide cannot completely be covered through the renewables. Besides the efficiency increase of thermal power plants, an additional way to reduce CO_2 emissions is the carbon capture and storage (CCS). Depending on the manner of CO_2 capture and the oxidation of fuel, it is distinguished between three CO_2 capture methods, namely pre-combustion, post-combustion and oxyfuel. In the pre-combustion approach, the carbon dioxide is separated before the combustion process. The coal is gasified in a first step at higher-pressure levels and the synthesis gas consists essentially of carbon monoxide and hydrogen. In a subsequent water-gas shift reaction, the carbon monoxide reacts under the supply of steam to carbon dioxide and additional hydrogen. The hydrogen produced can be combusted in a combined-cycle power plant, in which the only reaction product is water. This technology is called integrated gasification combined-cycle (IGCC) and pilot plants in realistic scale are already operated. The post-combustion process has the advantage that existing thermal power plants can already be retrofitted with CO_2 capture process. With a long life time of thermal power plants of up to 40 years, this is an important economic argument. The flue gas in the post-combustion process, which basically consists of nitrogen, oxygen, carbon dioxide and steam, is further treated. For this treatment, different concepts for the CO_2 capture are developed such as the chemical scrubbing of flue gas [among others: (Diao et al. 2004)] and the carbonate-looping process [among others: (Blamey et al. 2010; Ströhle et al. 2014)]. In the third method (the oxyfuel process), the coal is combusted with pure oxygen. A part of the flue gas is circulated into the combustion chamber in order to obtain lower adiabatic combustion temperature. After cleaning the flue gas from pollutants and separation of the steam by a condensation process, the flue gas consists of pure CO_2 that can be compressed for transport and storage. The oxyfuel process, although it is less complex from other CCS methods, is cost-intensive due to the provision of the pure oxygen using an air separation unit. Here, there is an enormous loss of overall efficiency of 8–14 percentage points, which leads to a significant increase in fuel consumption (Bouillon et al. 2009). By contrast, the combustion of solid fuels by means of the chemical-looping process (new oxyfuel method) enables a CO_2 capture with low energy input. Here, air and fuel are kept separate and the oxygen is transferred from the air to the fuel by use of an oxygen-carrier material (Adánez et al. 2018). Generally, particles containing a suitable metal oxide are used as oxygen-carriers, and these particles are moved between two coupled, circulated gas-solid fluidized beds (air and fuel reactors). While in the air reactor the oxygen-carrier is oxidized, the fossil fuel reacts in the fuel reactor with the oxygen-carrier material to carbon dioxide and steam. The reduced oxygen-carrier is then sent to the air reactor to be regenerated by the air. In an ideal case, the flue gas at the fuel reactor outlet consists of CO_2 and H_2O. The latter can be easily removed by a condensation process (Ströhle et al. 2015).

For the modelling of flue gas cleaning devices (dust removal, NO_x removal, SO_x removal and CO_2 capture), the standard library components of dynamic process simulation programmes are generally insufficient apart from a few exceptions (e.g. ASPEN PLUS DYNAMICS). Therefore, simple numerical models are used, in which the flue gas cleaning systems such as electrostatic precipitator, flue gas desulfurization and selective catalytic reduction unit are modelled as pressure drops and thermal masses. For CO_2 capture, by contrast, detailed numerical models can be found in the literature. However, these models are mostly based on equilibrium state (state-state) [among others publications (Li et al. 2016; Ohlemüller et al. 2015; Ströhle et al. 2014)], while dynamic process models are rarely used. The dynamic simulation of flue gas cleaning components should therefore be the subject of further research.

2.3 Automation Components

In dynamic simulation, the selection of suitable control structures is essential in order to accurately describe the behaviour of thermal power plants during transients. The automation components include measuring devices, analogue and binary modules, signal sources and controllers. The automation modules, described below, are explained according to process simulation software APROS and ASPEN PLUS DYNAMICS.

2.3.1 Measurement Modules

Measuring devices collect data on the physical properties and transmit them in analogue signals. The output signal of a measuring device can be used as an input signal for a control structure or for other purposes such as operation monitoring or data recording. Measuring values from energy systems include but are not limited to pressures, temperatures, mass flow rates and levels.

The pressure and temperature measurement modules can be applied to record the pressure and temperature from different process components such as points, tanks, headers, pipes, etc.

$$y^p_{out} = y^p_{inp}, \tag{2.123}$$

$$y^T_{out} = y^T_{inp}. \tag{2.124}$$

The flow measurement module can be used to measure the mass flow rate of pipes, channels and valves.

$$y^{\dot{m}}_{out} = y^{\dot{m}}_{inp}. \tag{2.125}$$

The level measurement module can be used to record the level from different components such as tanks and condensers.

$$y^L_{out} = y^L_{inp}. \tag{2.126}$$

It should be mentioned here that some measurements devices may cause a pressure drop in the flow. This pressure drop is typically not considered in measurement modules and can be considered separately, if required.

2.3.2 *Analogue Modules*

Analogue modules are applied to modify analogue signals. The input and output are always analogue signals. Additional binary input signal may be required for controlling tasks. The analogue modules can be divided in three groups, namely basic, static and dynamic modules. These are explained in detail in the following sections.

2.3.2.1 Basic Modules

Analog basic modules include adder, multiplier, divider, mean value, setpoint and signal splitter.

Adder: An adder module is applied to add or subtract signals $y_{inl,1}, y_{inl,2}$ and $y_{inl,i}$. The output signal y_{out} is calculated from the following equation:

$$y_{out} = \pm y_{inl,1} \pm y_{inl,2} \pm \ldots \pm y_{inl,i}. \tag{2.127}$$

Multiplier: A multiplier module can be used for multiplication of analogue signals $y_{inl,1}, y_{inl,2}$ and $y_{inl,i}$. The output signal is calculated according to the following relation:

$$y_{out} = y_{inl,1} * y_{inl,2} * \ldots * y_{inl,i}. \tag{2.128}$$

Amplifier: An amplifier is a special case of the multiplier module and used to amplify the input signal y_{inl} by a factor of k_P. The output signal is:

$$y_{out} = k_P * y_{inl}. \tag{2.129}$$

Divider: A divider is used to perfom a division of two analogue signals. The value of the output signal is calculated as follows:

$$y_{out} = \frac{y_{num}}{y_{den}}. \tag{2.130}$$

The denominator signal has to be non-zero. If the denominator of the divider module is zero, the division is not achieved and a warning massage is printed.

Mean value: A mean value module calculates the average value of analogue input signals. The output signal y_{out} is expressed by:

$$y_{out} = \frac{y_{inl,1} + y_{inl,2} + \ldots + y_{inl,i}}{N}. \tag{2.131}$$

The symbol N represents the number of input signals.

Setpoint: A setpoint is the target value of a variable. The setpoint module may have two operation modes: normal operation and tracking operation. In the normal operation, the output signal is constant. For example, the control system of the boiler aims to maintain the steam temperature at the inlet of the steam turbine at a constant temperature setpoint. In the tracking operation, the output signal is not constant and follows the value of the input signal either directly or with a given gradient.

Signal splitter: A signal splitter divides the same signal in two information flows. The input signal is y_{inl} and the outlet signals are $y_{out,1}$ and $y_{out,2}$.

$$\begin{aligned} y_{out,1} &= y_{inl}, \\ y_{out,2} &= y_{inl}. \end{aligned} \tag{2.132}$$

2.3.2.2 Static Modules

Analogue static modules can frequently be found in each control structure. The analogue static modules include delay, memory, switch, dead band, hysteresis, limiter, Max and Min selector, square root and polyline function. Some of the static modules such as dead band and limiter are a source of discontinuity, which in return may result in numerical instability during the simulation.

Delay: A delay module shifts the given value of an input signal by a time constant T. The input signal is y_{inl} and the outlet signal is y_{out}.

$$y_{out}(t) = \begin{cases} 0 & case\, t \leq T \\ y_{inl}(t-T) & case\, t > T \end{cases}. \tag{2.133}$$

Memory: A memory module is used as a memory for an analogue signal value. The module has an analogue input signal, an analogue output signal and a binary input signal. If the binary input signal has the value FALSE, the value of the output signal is equal to the value of the input signal. If the binary input value is changed to TRUE, the value of the output signal is fixed and remains independent of the input signal value changes.

Switch: A switch module can be considered as selector between two analogue signals. It has two input signals, an output signal and a binary input signal. The output signal follows the first input signal, if the binary input is TRUE and follows the second input signal, if the binary input is FALSE. This can be described mathematically with the help of the following equation:

$$y_{out} = \begin{cases} y_{inl,1} & case \quad y_{bin} = 0 \\ y_{inl,2} & case \quad y_{bin} = 1 \end{cases}. \tag{2.134}$$

Dead band: A dead band module has an analogue input signal and an analogue output signal. When the absolute value of the input signal is smaller than the defined dead band value (DB), the output signal is then equal to zero. If the input signal value is bigger than DB, the output signal is the input signal value minus the dead band value. If the input signal value is less than the negative dead band value (−DB), the output value is then the sum of the input signal value and the dead band value. Mathematically, these relations can be written according to the following equation:

$$y_{out} = \begin{cases} 0 & case \quad |y_{inl}| & < DB \\ y_{inl} - DB & case \quad y_{inl} & \geq DB \\ y_{inl} + DB & case \quad y_{inl} & < -DB \end{cases}. \tag{2.135}$$

Hysteresis: A hysteresis module is used to form a hysteresis effect to an analogue signal. The value of the output signal will follow the input signal, so long as the value of the input signal is changing to the same direction. If the direction changes, then the value of the output signal is held constant as long as the total change to that new direction is smaller than the defined hysteresis value.

Limiter: A limiter limits the signal y_{inl} by an upper limit L_{up} and a lower limit L_{low}. If the input signal value is between the given limits (within the predefined range), the output signal follows the input signal. This relation can be expressed mathematically as:

$$y_{out} = \begin{cases} y_{inl} & case \quad L_{low} \leq y_{inl} \leq L_{up} \\ L_{low} & case \quad y_{inl} < L_{low} \\ L_{up} & case \quad y_{inl} > L_{up} \end{cases}. \tag{2.136}$$

Max and Min: Max and Min modules are signal selectors. The value of the output signal is equal to the highest value of the input signals in case of a Max operator and is the smallest value of the input signals in case of a Min operator. The equation for maximum value selection for two input signals is:

$$y_{out} = \begin{cases} y_{inl,1} & case \quad y_{inl,1} \geq y_{inl,2} \\ y_{inl,2} & case \quad y_{inl,1} < y_{inl,2} \end{cases}. \tag{2.137}$$

and the equation for minimum value selection for two input signals is written as:

$$y_{out} = \begin{cases} y_{inl,1} & case \quad y_{inl,1} \leq y_{inl,2} \\ y_{inl,2} & case \quad y_{inl,1} > y_{inl,2} \end{cases}. \tag{2.138}$$

Square root: A square root module returns a square root of the input signal value. The output signal is expressed by:

$$y_{out} = \sqrt{y_{inl}}. \tag{2.139}$$

Function: A function module can be used to form a polyline (n-point cross line curve) between two variables. The value of the output signal is a function of the input signal and the given polyline. The polyline will be defined by a set of (x, y) coordinate pairs. If the input signal value is out of the range of the defined x coordinates, the value of the output signal is equal to the function value of the last x coordinate.

2.3.2.3 Dynamic Modules

Analogue dynamic modules include gradient, integrator, differentiator, derivator and filter.

Gradient: A gradient module limits the rate of variation for a variable. The rate of variation can be limited separately for increase and for decrease direction. The value of the output signal follows the changes of the input signal considering the defined gradient limits. The value of the output signal is the same as the input, if the change in the input signal is slower than the defined gradient or the input signal does not change.

Integrator and Differentiator: Integrator and differentiator modules are simple control components. The output value of the integrator module can be expressed by:

$$y_{out}(t) = \frac{1}{T_{int}} \int_0^t y_{inl}(\tau) d\tau. \tag{2.140}$$

The rate of change of the output value is determined according to the given integration time T_{int} and the value of the input signal. If the input signal is positive, the output value is increasing. If the value of the input signal is negative, the output value is decreasing. The output value remains constant, when the input value is zero. For example, if the integration time is set to 60 s and the input signal is 1.0, the output value will increase to 1.0 within one minute.

The differentiator module can be represented as first-order ordinary differential equation. The rate of change of the output value is calculated according to the given derivation time T_{der} and the value of the input signal y_{inl} as:

$$y_{out} = T_{der} \frac{dy_{inl}}{dt}. \tag{2.141}$$

Derivator: A derivator module defines the rapid changes of a variable. The output signal of the derivator depends on the derivation time and the derivation gain. The latter calculates the height of the impulse, while the derivation time determines the duration of the impulse. Using these parameters, it is possible to define the step response of the derivator module. If the value of the input signal does not change, the value of the output signal is zero. The output signal of this module is written as:

$$y_{out} = \frac{T_{der}}{(K_{der}T_{der} + 1)} \frac{dy_{inl}}{dt}. \tag{2.142}$$

Filter: A filter module performs a filtering operation on an analogue signal. The value of the output signal is determined according to the input signal value and the filter parameters. The simplest filter is a moving average filter. The only parameter in this case is the time constant of the filter T_{fil}.

2.3.3 Binary Modules

Binary modules, also known as Boolean logic elements, are required for the decision making in different kind of control circuits. In such modules, the input can be binary and/or analogue signals, while the output is always binary signal. Therefore, when expression is evaluated by binary modules, the output signal is ether zero or unity (expression with zero output means FALSE, expression with unity output means TRUE). The most relevant binary modules that used in the dynamic simulation of thermal power plants are explained below.

2.3.3.1 Basic Modules

Basic binary modules are AND, OR and NOT.

AND: The operator "AND" is applied to perform a logical operation AND for binary signals. The output signal is TRUE, when all input signals are in TRUE state. If one of the input signals is in FALSE state, the output signal is FALSE.

$$y_{out} = \begin{cases} 0 & case \quad y_{inl,1} = 0 \quad or \quad y_{inl,2} = 0 \\ 1 & case \quad y_{inl,1} = 1 \quad and \quad y_{inl,2} = 1 \end{cases}. \tag{2.143}$$

OR: The operator "OR" is used to perform a logical operation OR for binary signals. The output signal is TRUE, if anyone of the inputs signal is in TRUE state. The output signal is only in FALSE state, if all input signals are in FALSE state.

$$y_{out} = \begin{cases} 0 & case \quad y_{inl,1} = 0 \quad and \quad y_{inl,2} = 0 \\ 1 & case \quad y_{inl,1} = 1 \quad \text{or} \quad y_{inl,2} = 1 \end{cases}. \tag{2.144}$$

NOT: The operator "NOT" is used to invert the state of a binary signal. It has only one input signal and one output signal. The state of the output signal is:

$$y_{out} = \begin{cases} 0 & case \quad y_{inl} = 1 \\ 1 & case \quad y_{inl} = 0 \end{cases}. \tag{2.145}$$

2.3.3.2 Advanced Modules

Advanced binary modules include amongst others delay, switch, n/m selector, flip-flop, limit value checker, button, alarm and timer.

Delay: A delay module shifts a binary signal in time. It has one binary input signal and one binary output signal. The output signal will follow the input signal after a given delay time, which can be constant or varied by an analogue signal. Mathematically, the following equation can be formulated for the delay module:

$$y_{out}(t, \Delta T(t)) = y_{inl}(t - \Delta T(t)). \tag{2.146}$$

Switch: A switch module is a selector between two binary signals. It has two input binary signals, an output binary signal and an input binary control signal. The state of the output signal follows the state of either of the two input signals depending on the control binary signal.

$$y_{out} = \begin{cases} y_{inl,1} & case \quad y_{cont} = 0 \\ y_{inl,2} & case \quad y_{cont} = 1 \end{cases}. \tag{2.147}$$

n/m selector: A n/m selector is a logical selector module for binary signals. The symbol n is a given positive integer, while the symbol m denotes the number of input signals. The output signal is TRUE, when at least n numbers of input signals are in TRUE state. Otherwise, the output signal is FALSE.

Flip-flop: A flip-flop module is a bistable element, i.e. it has two stable states. This module that is applied in logic circuits as a memory element has two input signals (SET and RESET) and one output signal. The state of the output signal indicates, which of the input signals was most recently in TRUE state. The output signal can be expressed as follows:

- FALSE, if both input signals are FALSE.
- FALSE, if the RESET input signal is only in TRUE state.
- FALSE, if both inputs are in TRUE state and the RESET input signal is dominating.
- TRUE, if the SET input signal is in TRUE state.
- TRUE, if both inputs are in TRUE state and the RESET input signal is not dominating.

Limit value checker: A limit value checker module compares an analogue input value to a given limit value *LV*, resulting in the following output signal:

- FALSE, if the analogue input signal is bigger than the given limit value.
- TRUE, if the analogue input signal is smaller than the given limit value.

This relation can be expressed mathematically using the following equation:

$$y_{out} = \begin{cases} 0 & case \quad y_{inl} \geq LV \\ 1 & case \quad y_{inl} < LV \end{cases}. \tag{2.148}$$

The limit value checker module can also consider a hysteresis effect to an analogue signal. In this case, an additional hysteresis value *HV* has to be given. The output signal is:

- FALSE, if the analogue input signal is bigger than the given limit value plus the half of hysteresis value.
- TRUE, if the analogue input signal is smaller than the given limit value minus the half of hysteresis value.

The following equation is used to describe the limit value checker module:

$$y_{out} = \begin{cases} 0 & case \quad y_{inl} \geq LV + HV/2 \\ constant & case \quad LV - HV/2 \leq y_{inl} < LV + HV/2 \\ 1 & case \quad y_{inl} < LV - HV/2 \end{cases}. \tag{2.149}$$

Cold junction: A cold junction module is a special case of the limit value checker module. Here, the analogue signal is compared with another analogue signal, instead of the limit value in case of conventional limit value checker module. The output value is then:

- FALSE, if the first analogue input signal is bigger than the second analogue input signal.
- TRUE, if the first analogue input signal is smaller than the second analogue input signal.

- Accordingly, this relation can be expressed as follows:

$$y_{out} = \begin{cases} 0 & case \quad y_{inl,1} \geq y_{inl,2} \\ 1 & case \quad y_{inl,1} < y_{inl,2} \end{cases}. \tag{2.150}$$

Button: A button module has only an output signal. When the button is pushed, the output signal is in TRUE state. After a certain period of time defined by user, the output value returns to FALSE state.

Alarm: An alarm module provides alarms triggered by logical signals. If the input signal shows the desired value (TRUE or FALSE), the module sends an alarm message.

Timer: A timer module has an inlet and an output signal. When the inlet signal is set to TURE, the timer starts counting the time and compares it to a given time value. The output signal is in FALSE state, so long as the time value is not reached. When the time value is reached, the output signal is in TRUE state. The timer module can also be used as an analogue module. In this case, the time is given as analogue output value. Here, the timer can be reset via a RESET signal.

2.3.4 Signal Source Modules

Signal source modules can be applied to define transient setpoints or to generate binary or analogue signals. These include pulse generator, noise generator, sine or cosine wave generator, square wave generator and triangular wav generator. The mean value of the output signals for analogue signal generators is zero. In the signal source modules, the output signal is a function of time:

$$y_{out} = f(t). \tag{2.151}$$

Pulse generator: A pulse generator module generates binary pluses. The module has a binary input signal that is used to switch off the generator, an analogue input signal that is used to define a period of time and a binary output signal. When the input signal is FALSE, the output signal is always in FALSE state. If the input signal is changing from FALSE to TRUE, the module will generate pulses. Each pulse lasts for half of the defined period of time that can be constant or can vary according to an input analogue signal.

Noise generator: A noise generator module generates random signal with a constant power spectral density. The module has a binary input signal that is used to switch off the generator, an analogue input signal that is used to define a given variance and an analogue output signal. When the binary input is in FALSE state, the value of the output signal will be zero. If the generator is switched on, the module will generate normally distributed white noise of a given variance.

Sine and cosine wave generator: A sine wave generator module generates a sine wave. The amplitude and the time period of the sine wave can be constant or they can vary depending on analogue input signals. A cosine wave has a shape identical to that of a sine wave, with a phase-shift of $\pi/2$ radians.

Square wave generator: A square wave generator module generates square waves. It has one binary input, two analogue inputs and one analogue output. The amplitude and the time period of generated wave should be given. The output of the module is zero, if the binary input is in FALSE state. When the generator is switched on, the module will generate square waves according to the given amplitude and the time period that can be constant or varied following the analogue input signals.

Triangular wave generator: A triangular wave generator module generates triangular waves. The module can be switched on/off by a binary input signal. The amplitude and the time period of generated triangular waves can be defined by two analogue inputs.

2.3.5 Controller Modules

The well-known control schemes in process and energy systems are explained briefly below. These include feedback control, feedforward control, feedforward plus feedback control, cascade control and cascade plus feedforward control.

Feedback control scheme: A feedback control system represents the simplest form of closed loop control scheme. The error signal resulting from the comparison between the process output and the setpoint is used as a means for the controller.

Feedforward control scheme: A feedforward control replies to the moment when the disturbance occurs. The process variable adjustment is therefore based on the knowledge of the process using mathematical model and measurement disturbances without having to wait for a deviation in the process variable. Due to modelling errors and unmeasured disturbances, a flawless feedforward control is not possible and is generally used in combination with a feedback control.

Feedforward plus feedback control scheme: In this control system, the feedforward controller decreases or eliminates the effect of outer disturbances, while the feedback controller responds to setpoint variations. Compared to a feedback control system, this combined control technique can significantly give better response to a disturbance that can be measured, before it affects the process output.

Cascade control with and without feedforward control schemes: The cascade control consists of two loops: an inner loop and an outer loop. This control scheme leads to the benefits required, only when the inner loop has faster dynamics as compared to the outer loop (Song et al. 2002). The cascade control is frequently used within the control circuits of thermal power plants. A combination of cascade and feed-forward control is achieved, in which the primary and the secondary processes are controlled by a cascade control, while the disturbance rejection is obtained using a feedforward control.

Typical variants of controller used in these control schemes are P-controller, I-controller, PI-controller, PD-controller, and PID-controller. The output of the controller y_{out} is determined according to the control deviation of the controller and to possible feedforward input.

$$y_{out} = y_{out}^{P} + y_{out}^{I} + y_{out}^{D} + y_{out}^{FF}. \tag{2.152}$$

Here, the term y_{out}^{P} represents the output of the proportional (P) part of the controller, y_{out}^{I} is output of the integral (I) part of the controller, y_{out}^{D} denotes to the output of the derivative (D) part of the controller and y_{out}^{FF} is the output of the feedforward (FF) part of the controller.

The transient equation of a PID controller is expressed by:

$$y_{out} = K_P y_{inl} + \frac{K_P}{T_I} \int y_{inl} dt + K_P T_D \frac{dy_{inl}}{dt}. \tag{2.153}$$

The parameters K_P, T_I, T_D represent gain, integration time and derivation time of the controller, respectively. In order to obtain the PD or PI controller, $T_I \to \infty$ or $T_D = 0$ can be specified. Limited PID and PI controllers are generally applied to the control circuits of the thermal power plants. Here, the controller output is limited by a high value and a low value in order to remain in a physically meaningful range. For the determination of controller parameters, several tuning methods can be found in the literature such as stereotypical tuning methods (e.g. Ziegler Nichols, relay auto-tuning, pole placement and internal model control) and intelligent methods (e.g. fuzzy logic, genetic algorithms, artificial neutral networks and particle swarm optimization). Further information regarding the algorithms used in different controllers and their tuning methods can be found for example in Åström and Hägglund (1995), Bequette (2003) and Tan et al. (2012).

2.4 Electrical Components

In addition to process and automation components, a thermal power plant contains several electrical modules that are vital for the power plant operation. The consideration of electrical components in dynamic simulation of thermal power plants is of high relevance to calculate the electrical power consumption at base loads and to make sure that other components get the required electric power during transients. Furthermore, electrical components are used to evaluate the effect of possible failures in the electrical network on automation and process components and to study the plant behaviour at severe break down cases such as blackout. In the latter, the electricity supply is lost and accordingly all motor driven components trip. The passive system of the power plant should direct the power plant to a safe operation point, which can be assessed using dynamic process simulation models.

The electrical components used for the dynamic simulation of energy systems are node, line, switch, load, generator, transformer, direct current source, battery, inverter, alternating current/direct current converter and direct current/direct current converter. The node, switch, line and load components can be applied to both alternating current (AC) and direct current (DC) electrical circuits. The battery and DC/DC converter are direct current modules, while the inverter and AC/DC converter connect direct current and alternating current circuits. The generator and transformer can be used only in the alternating current system. The electrical modules described below are explained as implemented in the advanced process simulation software APROS.

2.4.1 Basic Modules

Basic electrical moulds include node, line, switch and load.

Electrical node: An electrical node (no real process component as a counterpart) has the main function to connect other electrical modules to each other. It can be applied for both alternating current and direct current electrical circuits.

Electrical line: An electrical line is a component for the modelling bulk transfer of electrical energy via the electrical transmission line. The module can be a one phase DC electrical line represented by a resistance or an AC electrical line with a pi equivalent circuit. Each electrical line has an input and an output connection node, defining the positive direction of line from the input node to the output node. The type of the electrical line is automatically defined according to the type of the connection nodes (both connection nodes must be either DC or AC). The electrical line has a built-in measurement for current, active power (electrical power consumption) and reactive power (imaginary power consumption). In direct current electrical circuits, the power is always active power in the steady-state. The reactive power may only exist during the transient cases. Transmission electrical lines become transmission networks, when interconnected with each other.

Electrical switch: An electrical switch is located between two electrical nodes and can be applied to direct or alternating current electrical circuits. The order of nodes is insignificant, but they have to be same type, either AC or DC nodes. The switch is controlled by a logical input. If the input signal is in TRUE state, the switch is closed. In this case, the switch is a kind of a transparent module type, i.e. the electrical nodes on both sides of the switch have the same voltage. If the input signal is in FALSE state, the switch is open and there is no connection through this module between nodes.

Electrical load: An electrical load module models the power consumption in an alternating current or direct current electrical circuits. The load is connected to one node either a DC node or an AC node. The load has a built-in measurement for current, active power and reactive power. The reactive power is zero, if the load is connected to a direct current node.

2.4.2 Current Sources Modules

The sources modules are power generation devices. It is distinguished between the alternating current generators (electrical generator) and the direct current sources (e.g. battery and photovoltaic).

Generator: A generator module is applied to model the production of electrical power in alternating current electrical circuits. The generator is connected to one node. The electrical power produced by the generator is the mechanical power multiplied by the efficiency. The generator has built-in regulators for frequency, voltage and power.

Battery: A battery is an electric device, converting the stored chemical energy into electrical energy. The battery module can be applied to store the energy of the one-phase direct current or to supply current to the direct current electrical circuits. The battery model is represented as a direct current source, variable initial resistor and energy charging level counter. The battery is connected to one DC node. The operation mode of the battery includes charge, discharge and stand-by. The operating characteristic during a battery discharge depends on the discharge capacity and the voltage.

The discharge capacity is a function of discharge current, ambient temperature and long term warehousing, while the voltage is a function of discharging current and state of energy storage. The discharge capacity Q can be described by Peukert exponential equation as follows:

$$Q(I) = CI^{1-n}. \tag{2.154}$$

The symbol I is the discharge current, n denotes the Peukert exponent with default value of 1.4 and C represents a factor that is calculated at the nominal load of the battery:

$$C = I^n t. \tag{2.155}$$

The discharge capacity of a battery decreases for reduced ambient temperature. For example, at −20 °C, the discharge capacity can be reduced to 60% compared to the discharge capacity at +20 °C. The capacity of a battery also decreases during the long term warehousing or stand-by without recharging. The discharge capacity decrease during a simulation time step Δt is calculated as:

$$\Delta Q = I^n I_{nom}^{1-n} \frac{\Delta t}{3600}. \tag{2.156}$$

The symbol I_{nom} is the nominal current. The voltage is determined with the following formula:

$$U = K_1 K_2 U_{nom}. \tag{2.157}$$

Here, U_{nom} represents the nominal voltage, the factors K_1 and K_2 are highly reliant on discharge current and on discharge time. The discharge current is determined by the electricity consumption of the direct current system. If the current is smaller than or equal to the nominal current, the capacity of the battery decreases during a simulation time step by the amount discharge current multiplied by the simulation time step. If the discharge current is bigger than the nominal current, the capacity is decreased with a larger amount.

The charging current can be expressed as a function of capacity as:

$$I = I_{nom} e^{\frac{Q - 0.6Q_{\max}}{0.5I_{nom}}}, \tag{2.158}$$

where $Q_{\max}$ is the maximum capacity of battery and I_{nom} represents the nominal charging current. The internal resistance of the battery can be changed to control the charging current. If direct current source in the simulated system cannot supply the required charging current, the actual current may be smaller than the value obtained from Eq. (2.158). At higher capacities, the charging current decreases exponentially.

The self-discharge process can be simulated by decreasing the capacity of the battery by a small amount during each time step. The self-discharge of a battery may be considered in all operation modes (discharge, charge and stand-by). The self-discharge time of the battery depends mainly on the ambient temperature. For example, at 40 °C, the capacity of the battery is assumed to change from the maximum capacity to zero in 9 months. The self-discharge time is 21 months at 30 °C, while it is 32 months at 20 °C.

Photovoltaic: A photovoltaic system, also known as a solar panel, is a module type for production of DC electrical power. The solar panel module is represented as DC current source with variable internal resistance and is connected to a weather component. The ideal DC source has constant internal resistance, while the internal resistance of the solar panel changes non-linearly with the solar radiation. The current and voltage generated by the solar panel module depend on the consumption of the electricity in the electrical circuit and the I-V curve of the solar panel.

2.4.3 DC and AC Modules

DC and AC modules required to simulate the electrical network include DC/AC inverter, AC/DC converter, AC/AC transformer and DC/DC converter.

DC/AC inverter: An inverter is an electrical device, converting the voltage from direct current to alternating current. It can be applied to convert direct current from solar panels, batteries or fuel cells to alternating current for the electrical grid. The DC/AC inverter is usually not suitable for inductive AC and sensitive electronic devices that can be damaged by poor waveforms. The inverter module has two connection nodes and simulates the transformation of DC electric power into three-phase AC electric power. The input electrical node is a DC node, while the output node is an AC type. The DC/AC inverter module is represented as one electrical line ending in a node with a high base admittance (nearly zero voltage) to the DC circuit, in addition to an AC three-phase generator for supplying power to the AC circuit. The connection between DC circuit and AC circuit is arranged in such a way that the AC voltage is reliant on the DC voltage and the AC apparent power is equal to the power losses of the DC side resistance. As a result, the AC load is transferred into the DC supply side. Generally, the DC current source is followed by the DC/AC inverter.

AC/DC converter: An AC/DC converter is an electrical device, converting the voltage from alternating current to direct current. Although this device can be found in several electrical circuits (e.g. charging batteries), it is characterised by disadvantages such as poor current overload capacity and expensive automatic regulators. The AC/DC converter module models the transformation of three phase AC electric power into DC electric power. The model is represented as one electrical line ending in a node with a high base admittance (nearly zero voltage) to the AC circuit and a direct current source that supplies the voltage or power to the DC node. The connection between AC and DC circuits is arranged to transfer the DC power loss into the AC supply side. In order to achieve that, the DC voltage should be a function of the AC voltage and the DC apparent power is equal to the power losses of the AC side resistance.

AC/AC transformer: An AC/AC transformer is an electrical device that transforms the voltage level between two AC circuits. The generated electricity with low level voltage must be transformed to higher voltage for efficient electrical power transmission. In this case, the electrical current decreases, which in turn results in a reduction in the ohmic losses and a reduction in cross-sectional area of the electrical lines. Increasing the voltage by a step-up transformer at the generating side of the power network decreases the capital cost and improves the voltage regulation of the system. With the help of step-down transformers at the receiving end, the high voltage is reduced to the desired level for distribution to the consumers. The AC/AC transformer consists basically of two or more coils that are electrically isolated from each other, but wrapped together around a closed magnetic iron circuit (core), allowing the electrical power to be transferred from one coil to other by induction. The transformer module is represented as one electrical line ending in a node with a high base admittance to the primary circuit and an alternating current generator for supplying the voltage or power to the secondary circuit (output node). The AC/AC transformer module has a primary and secondary connection node. Both connection nodes have to be alternating current nodes and the order of the nodes is important.

DC/DC converter: An DC/DC converter is an electrical device for voltage changes in DC electrical circuits. The DC/DC converter model is represented as one electrical line ending in a node with a high base admittance (nearly zero voltage) to the primary circuit and a direct current source for supplying the voltage or power to the secondary circuit (output node). The connection between primary and secondary circuits is arranged in such a way that the secondary voltage is dependent on the primary voltage and the secondary apparent power is equal to the power losses of the primary side resistance. Accordingly, the load of the secondary circuit (load side) is transferred into the primary circuit (supply side).

2.5 Conclusion

Most processes of thermal power plants are based on the generation of superheated steam at high-pressure, driving a steam turbine coupled with an electrical generator. In the evaporator systems of the power plants, the two-phase flows (i.e. water/steam flow regimes) are existing. Basically, there are two numerical approaches for describing the behaviour of these flows: the mixture flow model and two-fluid models. In the first approach, the water/steam flow is considered as a mixture and modelled as a single-phase flow with fairly complex thermodynamic properties. In the second approach, each phase is modelled as a separate fluid and accordingly two sets of conservation equations for gas and liquid phases are formulated. Besides the conservation equations for mass, momentum and energy of each phase, adequate constitutive equations and experimental correlations are required, describing mass, momentum and energy transfer between phases. The interaction terms can be determined using flow parameters and their derivatives coupled with empirical correlations for various flow regimes. The classic six-equation version of the two-fluid model is a single pressure model. Here, both phases are in mechanical equilibrium, but not in chemical and thermal equilibrium. The seven-equation flow model simulates each phase separately and does not assume, by contrast to the six-equation flow model, pressure equilibrium between phases. The five-equation or four-equation version of the two-fluid models supposes mechanical and thermal equilibrium, but not chemical equilibrium. The selection of the appropriate flow model depends mainly on the computational effort and the desired level of accuracy. Although the mixture flow model entails a less accurate description of the two-phase flow phenomena, in many cases it presents the most efficient approach. This is due to the fact that the response of the mixture behaviour and not of each phase is often sufficient. Accordingly, the mixture flow model can be found in all process simulation programmes such as APROS, ASPEN PLUS DYNAMICS, PPSD and DYMOLA. For a more detailed dynamic simulation of thermal power plant components (e.g. evaporator and condenser), the two-fluid models are used since they offer the possibility to include non-equilibrium thermodynamics state into the formulation.

In the previous sections, different process components such as thin-walled tube, thick-walled tube and turbomachines are presented. Base modelling of thermal power plants is carried out by selecting the process components and combining them to generate a model of an existing system and/or a new process for research. Although the building of a model is generally limited by the programme libraries, a specific component may be modelled by combining various existing process components. All necessary specifications can be inserted in the process components that are linked by flows of either material, energy or heat properties to finish construction. Here, two different examples are selected, showing detailed dynamic models of a combined-cycle power plant built in ASPEN ASPEN DYNAMICS and a pulverised coal-fired power plant built in APROS.

In Fig. 2.7, a combined-cycle power plant model that comprises a gas turbine connected to a vertical heat recovery steam generator and a steam turbine is

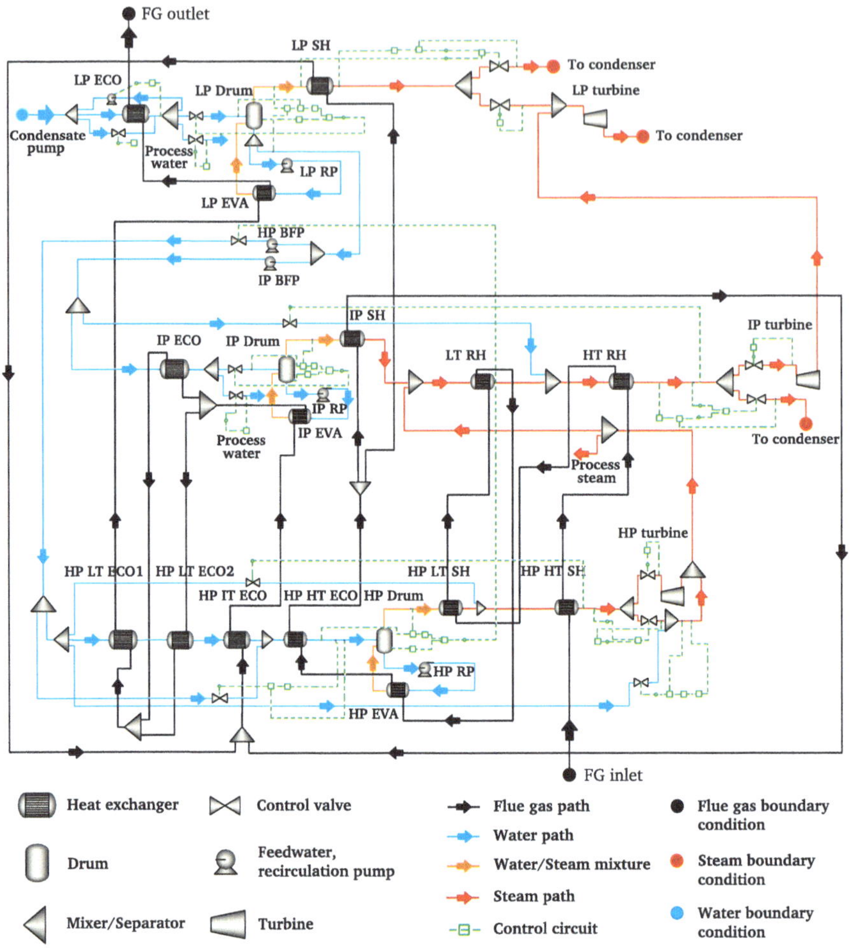

Fig. 2.7 Combined-cycle power plant modelled in ASPEN ASPEN DYNAMICS (Alobaid et al. 2014)

illustrated. The heat recovery steam generator that is arranged downstream of the gas turbine is unfired. The water/steam circuit is a three-pressure-stage with forced circulation in high, intermediate and low-pressure evaporator paths as well as a reheater section after the high-pressure turbine. Process components used to build the model include heat exchangers, drums, pumps, turbines, pipes, valves and attemperator. The LP feedwater directed to the LP economiser is supplied by the condensate pumps. A control valve for the LP drum level is located between the LP economiser outlet and the LP drum. In the LP drum, the water circulates through the LP evaporator tube bundle, where it is heated by the gas turbine exhaust gas and converted into saturated steam. The saturated water in the LP drum is separated and a significant portion of the flow is directed into the HP and IP circuits via high and intermediate boiler feedwater pumps (HP BFP and IP BFP). The dry steam exits the LP drum and flows through the LP superheater. After leaving the superheater, the superheated steam enters the LP turbine. Process water is extracted from the LP circuit for fuel gas preheating, before it is returned to the condenser. The feedwater flows into the HP drum via the HP economisers. The water in the HP drum circulates through the HP evaporator with the help of the forced circulation pump (HP RP). The circulated water is heated by the flue gas and converted into saturated steam in the HP drum. While the liquid stays in the drum and mixes with water coming from the HP economisers, the steam exits the drum and flows to the HP superheaters. The steam flows through the high-pressure superheaters, where it absorbs additional heat from the flue gas. The HP superheated steam exits the HRSG and enters the HP turbine section. A high-pressure attemperator is provided at the inlet of the last superheater to control the temperature at the inlet of the HP turbine. The water for the HP attemperator is taken from the high-pressure boiler feedwater pumps. The high-pressure feedwater mass flow is controlled by the HP drum level control valve that is located upstream of the HP economisers. The structure of the HP circuit applies to the IP circuit and thus there is no justification to explain it.

In Fig. 2.8, a single-pass, super-critical, once-through pulverised coal-fired power plant modelled in APROS is presented. In the steam generator, the chemically bond energy of the hard coal (bituminous coal) is converted into heat that is transferred from the flue gas side to the water/steam circuit and used to generate live steam for a Rankine cycle. In the water/steam system, the feedwater enters the economiser (ECO) at the top of the steam generator. The water is then directed to the membrane wall evaporator, where the evaporation takes place. Due the once-through design, the evaporation zone is not defined and the two-phase region is load dependent. For low part load operation, during start-up or shut-down, a forced circulation is in operation. A cyclone is installed to separate the water droplets within the steam flow and a separator storage tank is used for phase separation. The circulation is forced by a circulation pump (CP) and the level is controlled by a circulation control valve. The circulated mass flow re-enters the steam generator at the economiser inlet. The saturated or partly superheated steam enters the steam cooled supporting tube system (superheater 1) of the convective heat exchangers and flows downwards to the superheater 2 (platen superheater). After the platen superheater, the steam flows through superheaters 3. In between the superheaters, different attemperator stages are arranged for steam temperature control. The steam leaves the once-through steam generator and flows through the high-pressure turbine, before re-entering the steam generator via reheater section. The latter is divided into two heat exchangers and has one attemperator in between. Afterwards, the steam fully expands in one intermediate turbine and two low-pressure turbines to condenser pressure. The condensed water is fed through several low and high-pressure feedwater preheaters back to the steam generator.

In the air/flue gas system, the primary air is supplied by speed controlled radial fans and flows through the regenerative air preheater into the mill (pulveriser). The coal dust is entrained with air, fed into the furnace and burned. The secondary air is supplied by forced-draft fans, which are specified as axial fans with variable blade

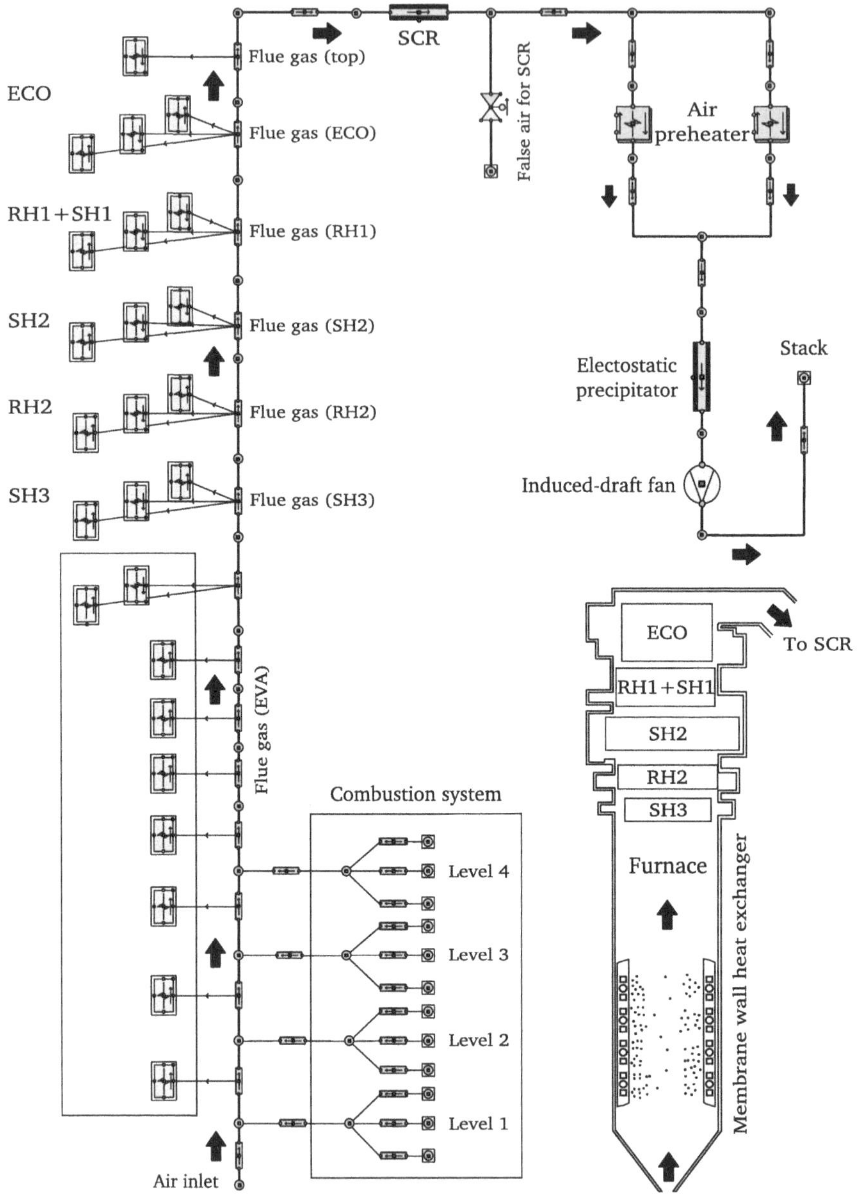

Fig. 2.8 Flue gas path of a single-pass, once-through pulverised coal-fired power plant modelled in APROS

angle. Afterwards, the secondary air passes a steam coil air preheater and the regenerative air preheater. In order to consider false air, which is impossible to avoid in a balanced-draft system, ambient air is introduced at the hopper of steam

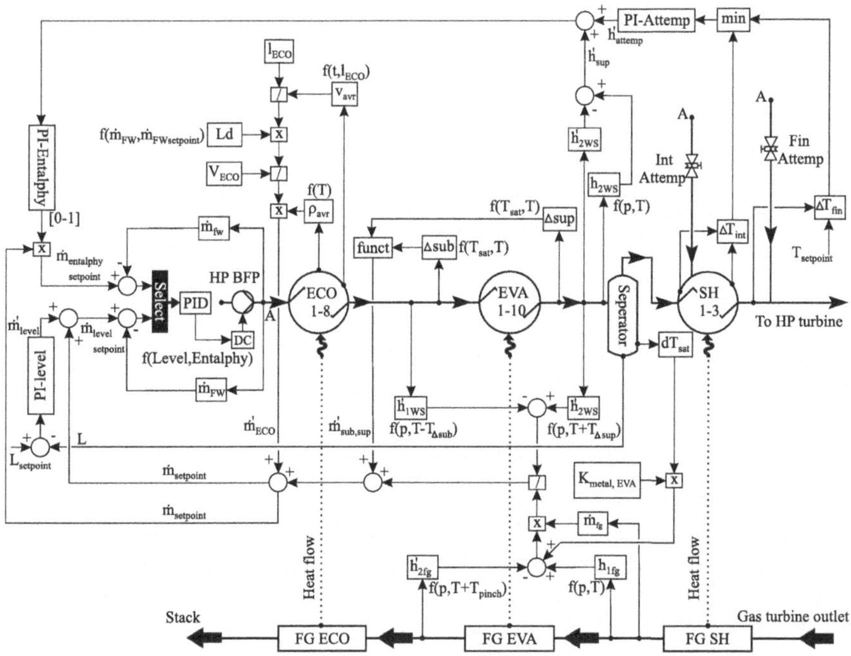

Fig. 2.9 Control circuit of the feedwater mass flow rate in a heat recover steam generator (Alobaid et al. 2012)

generator and adjusted for the nominal case. After exiting the steam generator, the now 350–400 °C hot flue gas flows through the selective catalytic reduction (SCR) unit, the regenerative air preheater, the electrostatic precipitator and the induced-draft fan that keeps furnace at a slight negative gauge pressure.

A thermal power plant contains an automation system, consisting of controllers, analogue and binary components. These automation components are combined together in order to satisfy certain control requirements. Here, a schematic representation of a feedwater control structure used in a heat recovery steam generator (HRSG) is illustrated in Fig. 2.9 as an example. This control circuit offers the possibility to operate the high-pressure system of the HRSG in level mode (nature circulation) or Benson mode (once-through). The idea behind the feedwater mass flow controller is to determine the mass flow rate that should enter the HRSG, depending on the heat flow of the flue gas that can be absorbed by the HP evaporators as well as the enthalpies of the working fluid at the inlet and the outlet of the HP evaporators. The obtained mass flow rate is then corrected by the mass flow rates of the attemperators, by the derivative element for considering additional heat output of the metal masses of the HP evaporators, by the derivative element for considering mass storage process of the working fluid in the economisers, by the degree of sub-cooling at the inlet of the HP evaporators, by the degree of superheating in case of Benson mode or the steam quality in case of level mode at the outlet of the HP

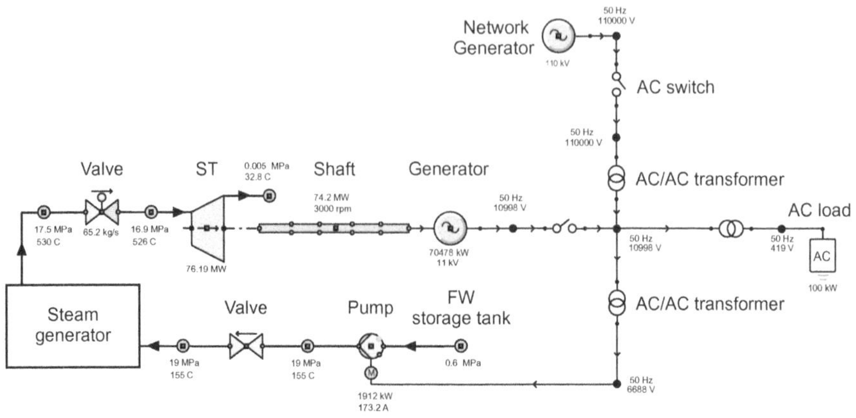

Fig. 2.10 Model of a thermal power plant with electrical circuits

evaporators. Further correction is carried out, depending on the operation mode (level or Benson). While in the level mode, the water level within the HP separator should be kept at a fixed level independent from the HRSG load, the enthalpy at the outlet of the HP separator in the Benson mode should meet the load dependent enthalpy setpoint.

The impact of possible failures in the electrical network on the energy system can be evaluated using the previously described electrical components such as node, line, switch, load and generator. Furthermore, the power consumption of process components can be assessed at steady-state and in transient operation cases. An example of an entire thermal power plant model is depicted in Fig. 2.10. The thermal power plant generates superheated steam that enters the steam turbine section. The mechanical power obtained by the steam turbine is transferred to the generator via a shaft. The electric power produced is fed to the power grid in addition to other generating units. Part of the electric power is used to cover the own electricity demand of the power plant and the rest is used to satisfy the load demand of the power grid.

References

Adánez J, Abad A, Mendiara T, Gayán P, de Diego LF, García-Labiano F (2018) Chemical looping combustion of solid fuels. Prog Energy Combust Sci 65:6–66

Alobaid F (2015a) An offset-method for Euler-Lagrange approach. Chem Eng Sci 138:173–193

Alobaid F (2015b) A particle–grid method for Euler-Lagrange approach. Powder Technol 286:342–360

Alobaid F, Karner K, Belz J, Epple B, Kim H-G (2014) Numerical and experimental study of a heat recovery steam generator during start-up procedure. Energy 64:1057–1070

Alobaid F, Mertens N, Starkloff R, Lanz T, Heinze C, Epple B (2017) Progress in dynamic simulation of thermal power plants. Prog Energy Combust Sci 59:79–162

Alobaid F, Pfeiffer S, Epple B, Seon C-Y, Kim H-G (2012) Fast start-up analyses for Benson heat recovery steam generator. Energy 46:295–309

Åström KJ, Hägglund T (1995) PID controllers: theory, design and tuning. Instrument Society of America, United States of America

Bequette BW (2003) Process control: modeling, design, and simulation. Prentice Hall Professional

Berenson P (1961) Film-boiling heat transfer from a horizontal surface. J Heat Transfer 83: 351–356

Bergman TL, Incropera FP, Lavine AS (2011) Fundamentals of heat and mass transfer. Wiley

Berndt G (1984) Mathematisches Modell eines Naturumlauf-Dampferzeugers zur St@: orfall-simulation und dessen experimentelle@: Uberpr@: ufung

Berry R, Peterson J, Zhang H, Martineau R, Zhao H, Zou L, Andrs D (2014) Relap-7 theory manual. Idaho National Laboratory, Tech. Rep. INL/EXT-14-31366

Berry R, Zou L, Zhao H, Andrs D (2013) RELAP-7: demonstrating seven-equation, two-phase. Idaho National Laboratory (INL)

Berry RA, Saurel R, LeMetayer O (2010) The discrete equation method (DEM) for fully compressible, two-phase flows in ducts of spatially varying cross-section. Nucl Eng Des 240:3797–3818

Bestion D (1990) The physical closure laws in the CATHARE code. Nucl Eng Des 124:229–245

Bhagwat SM, Ghajar AJ (2012) Similarities and differences in the flow patterns and void fraction in vertical upward and downward two phase flow. Exp Thermal Fluid Sci 39:213–227

Bhagwat SM, Ghajar AJ (2014) A flow pattern independent drift flux model based void fraction correlation for a wide range of gas–liquid two phase flow. Int J Multiph Flow 59:186–205

Blamey J, Anthony EJ, Wang J, Fennell PS (2010) The calcium looping cycle for large-scale CO2 capture. Prog Energy Combust Sci 36:260–279

Bouillon P-A, Hennes S, Mahieux C (2009) ECO2: post-combustion or oxyfuel—a comparison between coal power plants with integrated CO2 capture. Energy Procedia 1:4015–4022

Cooke DH (1984) On prediction of off-design multistage turbine pressures by Stodola's ellipse. In: 1984 joint power generation conference: GT papers. American Society of Mechanical Engineers, pp V001T004A004–V001T004A004

Diao Y-F, Zheng X-Y, He B-S, Chen C-H, Xu X-C (2004) Experimental study on capturing CO2 greenhouse gas by ammonia scrubbing. Energy Convers Manag 45:2283–2296

Epple B, Leithner R, Linzer W, Walter H (2012) Simulation von Kraftwerken und Feuerungen. Springer-Verlag

Fan GQ, Rees NW (1994) Modelling of vertical spindle mills in coal fired power plants, EEC 94: electrical engineering congress 1994; Preprints; Enabling technologies, developing industry; Congress and exhibition. Institution of Engineers, Australia, p 235

Geldart D (1973) Types of gas fluidization. Powder Technol 7:285–292

Groeneveld D, Snoek C (1986) A comprehensive examination of heat transfer correlations suitable for reactor safety analysis. Multiphase Science and Technology 2

Gryczka O, Heinrich S, Deen N, van Sint Annaland M, Kuipers J, Jacob M, Mörl L (2009) Characterization and CFD-modeling of the hydrodynamics of a prismatic spouted bed apparatus. Chem Eng Sci 64:3352–3375

Hänninen M (2009) Phenomenological extensions to APROS six-equation model. Non-condensable gas, supercritical

Hänninen M, Ahtinen E (2009) Simulation of non-condensable gas flow in two-fluid model of APROS—description of the model, validation and application. Ann Nucl Energy 36:1588–1596

Karimipour S, Pugsley T (2012) Application of the particle in cell approach for the simulation of bubbling fluidized beds of Geldart A particles. Powder Technol 220:63–69

Kuipers J, Prins W, Van Swaaij W (1991) Theoretical and experimental bubble formation at a single orifice in a two-dimensional gas-fluidized bed. Chem Eng Sci 46:2881–2894

Kunii D, Levenspiel O (1991) Fluidization engineering. Butterworth-Heinemann Boston

Lee K, Ryley D (1968) The evaporation of water droplets in superheated steam. J Heat Transfer 90:445–451

Li K, Leigh W, Feron P, Yu H, Tade M (2016) Systematic study of aqueous monoethanolamine (MEA)-based CO2 capture process: techno-economic assessment of the MEA process and its improvements. Appl Energy 165:648–659
Luo Y, Liu H, Jia L, Cai W (2011) Modeling and simulation of ball mill coal-pulverizing system. In: Industrial electronics and applications (ICIEA), 2011 6th IEEE conference on. IEEE, pp 1348–1353
Ohlemüller P, Alobaid F, Gunnarsson A, Ströhle J, Epple B (2015) Development of a process model for coal chemical looping combustion and validation against 100 kW th tests. Appl Energy 157:433–448
Palmer CA, Erbes MR (1994) Simulation methods used to analyze the performance of the GE PG6541B gas turbine utilizing low heating value fuels. American Society of Mechanical Engineers, New York, NY (United States)
Ransom VH, Hicks DL (1984) Hyperbolic two-pressure models for two-phase flow. J Comput Phys 53:124–151
Shah M (1979) A general correlation for heat transfer during film condensation inside pipes. Int J Heat Mass Transf 22:547–556
Song S, Xie L, Cai W-J (2002) Auto-tuning of cascade control systems. In: Proceedings of the 4th World Congress on intelligent control and automation, 2002. IEEE, pp 3339–3343
Stroh A, Alobaid F, Hasenzahl MT, Hilz J, Ströhle J, Epple B (2016) Comparison of three different CFD methods for dense fluidized beds and validation by a cold flow experiment. Particuology
Ströhle J, Junk M, Kremer J, Galloy A, Epple B (2014) Carbonate looping experiments in a 1 MW th pilot plant and model validation. Fuel 127:13–22
Ströhle J, Orth M, Epple B (2015) Chemical looping combustion of hard coal in a 1 MW th pilot plant using ilmenite as oxygen carrier. Appl Energy 157:288–294
Stuhmiller J (1977) The influence of interfacial pressure forces on the character of two-phase flow model equations. Int J Multiph Flow 3:551–560
Tan KK, Wang Q-G, Hang CC (2012) Advances in PID control. Springer Science & Business Media
Wallis GB (1969) One-dimensional two-phase flow. McGraw-Hill, New York
Walter H (2001) Modellbildung und numerische Simulation von Naturumlaufdampferzeugern. VDI-Verlag
Walter H, Epple B (2016) Numerical simulation of power plants and firing systems. Springer
Walter H, Hofmann R (2011) How can the heat transfer correlations for finned-tubes influence the numerical simulation of the dynamic behavior of a heat recovery steam generator? Appl Therm Eng 31:405–417
Wang J, Leithner R (1995) Konzepte und Wirkungsgrade kohlegefeuerter Kombianlagen. Brennstoff-Wärme-Kraft 47:11–17
Wu C, Berrouk A, Nandakumar K (2009) Three-dimensional discrete particle model for gas–solid fluidized beds on unstructured mesh. Chem Eng J 152:514–529
Zhou G, Si J, Taft CW (2000) Modeling and simulation of CE deep bowl pulverizer. IEEE Trans Energy Convers 15:312–322
Zuber N, Findlay J (1965) Average volumetric concentration in two-phase flow systems. J Heat Transfer 87:453–468

Chapter 3
Computational Fluid Dynamics

Computational fluid dynamics (CFD) offers a versatile and cost-efficient tool for the simulation of thermal power plants. Compared to experimental data, CFD results are capable of predicting qualitative and in many cases quantitative information. Accurate simulation results can be very useful in the design as well as in understanding the dynamic operation of reactors. In thermal power plants, the relevant applications of CFD models are mainly related to coal-fired power plants and boilers. Several numerical investigations of coal combustion have also been modified to make them available for the combustion and co-firing of biomass.

Some advantages of CFD simulations include:

- cost-effective insight into the hydrodynamic behaviour of the thermal power plant
- efficiency improvement and optimisation
- system behaviour under different operation conditions
- time and cost reduction by new design
- scale-up studies.

In the last decades, the numerical simulation of two-phase flows has been the focus of research and development. Here, the reactive gas-solid flow in terms of gasification and combustion of solid fuels in energy sector has attracted attentions in the scientific literature. By contrast to liquid or gaseous fuels, the combustion or gasification of solid fuels (coal, biomass, RDF or municipal solid waste) is a complex phenomenon, including different chemical and physical processes such as drying, pyrolysis and char combustion. The pyrolysis is a process, where the thermal degradation of organic substances takes place with absence of oxygen. During the pyrolysis of solid fuels, the products of char, tar, gases (e.g. carbon monoxide, carbon dioxide and nitrogen) and volatile organic compounds are basically formed, which are released in different proportions. The tar composes of various species, including heavy hydrocarbons, organic acids, aldehydes, alcohols and phenols. In subsequent reaction, the tar splits in char as well as in volatile

F. Alobaid, *Numerical Simulation for Next Generation Thermal Power Plants*,
Springer Tracts in Mechanical Engineering,
https://doi.org/10.1007/978-3-319-76234-0_3

components and gases. The char is the remaining solid that almost consists of pure carbon.

For the modelling of the gas-solid flow, there are basically two different approaches. In the Euler-Euler method, also known as two-fluid method, each phase is regarded as a continuum and is mathematically calculated by solving the balance equations. On the other hand, the Euler-Lagrange approach (single-particle method) combines the continuum descriptions of fluid phase with the Lagrange representation of dispersed phase on the basis of Newton's transport equations. This combined method for the modelling of granular flows provides detailed information regarding the behaviour of particle phase. The single particle-particle/wall collision in the Euler-Euler method is considered with the aid of additional model assumptions. In contrast, the collision detection between particle-particle and particle-wall in the Euler-Lagrange approach can be stochastically modelled or deterministically detected. The stochastic collision detection models present today the most efficient method to handle the practical gas-solid flows numerically by means of the single-particle method. However, these models are mainly appropriate for the simulation of dilute gas-solid flows. For the discrete particle simulations with deterministic collision detections, two models are widely used, namely the hard sphere model and the soft sphere model. In the hard sphere model, single binary collisions are modelled as instantaneous processes, which restrict the application of this model to dilute systems. The properties of particles after the collision are related to the properties of particles before the collision through momentum and energy balances. In the so-called soft sphere model, also known as discrete element method (DEM), force, velocity and displacement are determined during the collision. The particles in the DEM model can overlap each other or penetrate into the wall. Depending on the penetration depth, a resulting contact force can be modelled using a spring-damper-slider system.

In the following sections, the numerical methods for single-phase flow (particle and grid-based methods) and two-phase flows (quasi-single-phase, two-fluid and single-particle method) are explained in detail.

3.1 Numerical Methods for Single-Phase Flow

The physical values of the single-phase flow such as velocity, pressure, temperature and density can be determined using particle methods or grid-based methods (see Fig. 3.1).

In the particle methods such as smoothed particle hydrodynamic method or mesh-free Galerkin methods, the computational domain is divided into a finite number of nodes (mass particles), where the variations of physical values are described around each mass particle along its trajectory. In the grid-based method, the computational domain is discretised in finite objects, so-called numerical grid. The conservation equations are solved numerically based on different numerical methods, e.g. finite element method or finite volume method.

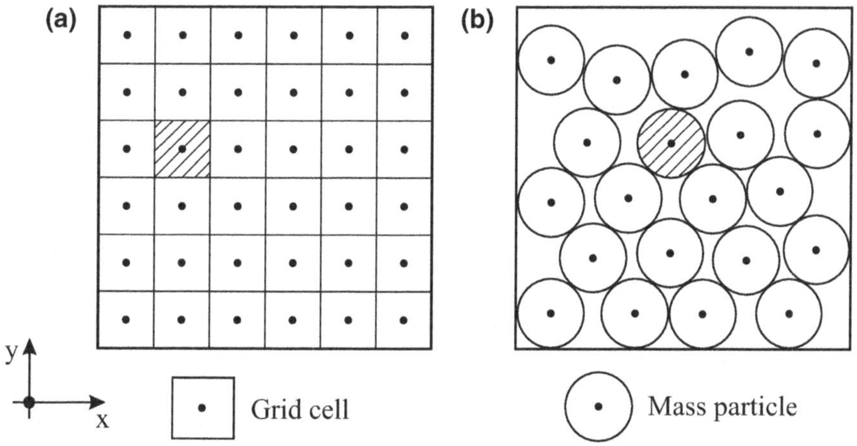

Fig. 3.1 Numerical approaches for single-phase flow modelling: **a** grid-based methods and **b** particle methods

3.1.1 Particle Methods

The particle methods, also known as mesh-free particle methods due to the absence of the numerical grid, are generally applied to fluids with high density. Some particle methods are associated with a numerical grid such as Lattice Boltzmann method and thus the particle methods cannot be completely considered as mesh-free methods. The majority of the particle methods are pure Lagrangian, but there are a few works on particle methods based on Eulerian formulation as well as on the arbitrary Lagrangian-Eulerian formulations. The following steps are necessary to obtain the solution:

- The continuum should be decomposed into a finite number of particles.
- At each particle, the functions, the derivatives and the integrals are approximated with the aid of all neighbour particles that exert on the investigated particle.
- The resulting ordinary differential equations are solved using numerical methods.

The particle methods are characterised by several advantages such as the possible dealing with very large domain deformations, the accurate representation of geometry due to the mesh-free, the possible using of higher-order continuous shape functions and the simple control of simulation accuracy since in the computational sub-domains, where higher accuracy is required, additional nodes (particles) can be added. The most important disadvantage is the computational time required for the simulation. Furthermore, some particle methods are related to numerical instabilities.

By replacing the numerical grid with particles, the conservation equations can significantly be simplified. For example, the particle equivalent of momentum equation for isothermal and incompressible single-phase flow can be expressed using the following equation:

$$\rho_i \frac{d\vec{u}_i}{dt} = \underbrace{\mu \nabla^2 u_i - \nabla p + \vec{f}}_{\vec{f}_{int} + \vec{f}_{ext}}, \tag{3.1}$$

where $\vec{u}_i$ is the velocity of the particle i and ρ_i denotes the density evaluated at the position of particle i. The symbols $\vec{f}_{int}$ and $\vec{f}_{ext}$ represent the internal and external force density acting on the particle, respectively. Typical force density is pressure, viscosity, gravitational, buoyancy and contact.

In the literature, there are many particle methods available, which are difficult to classify in separated groups. According to Li and Liu (2002), the particle methods can, however, be divided based on the computational formulation into three different categories: the particle methods with a strong form approximation of partial differential equations (PDEs), particle methods with a weak form approximation of the PDEs and particle methods with strong and weak form approximations of the PDEs. While the partial differential equations in the strong formulation are satisfied exactly at each point, the PDEs in the weak formulation are only satisfied in an averaged and integral sense over the computational domain. In the first approach (with a strong form approximation), the PDEs are discretised with the aid of specific collocation technique such as the smoothed particle hydrodynamic, the vortex methods and the generalized finite difference method. In the second group of particle methods that are generally summarised under mesh-free Galerkin methods, the Galerkin weak formulations are used. Examples of the mesh-free Galerkin methods are diffuse element method, element free Galerkin, h-p cloud method and many others. The last class of the particle methods is used strong and weak approximations of the PDEs, like particle-in-cell method and reproducing kernel particle method.

The particle methods can also be classified depending on the applied collision detection models, either stochastic or deterministic particle methods. The majority of the particle methods are based on stochastic collision detection model such as molecular dynamics methods, direct simulation Monte Carlo methods and Lattice Boltzmann method. The deterministic particle methods (e.g. smoothed particle hydrodynamic and vortex methods) are not applied as frequently as the stochastic particle methods by the reason of higher computational effort.

In the following sections, the particle methods is explained briefly and further information can be found for example in Belytschko et al. (1996), Liu (2009), Liu and Liu (2003, 2016).

3.1.1.1 Smoothed Particle Hydrodynamic Method

The smoothed particle hydrodynamic method (SPH) is an interpolation numerical method to approximate the values and derivatives of continuous fluid by means of discrete sample particles. The SPH method is based on a strong form approximation of partial differential equations. Originally, the SPH method dated back to Lucy (1977), who applied the SPH method to simulate astrophysical problems. In 1992, Monaghan modified this to deterministic mesh-free particle method and applied to solid and fluid mechanics simulations (Monaghan 1992).

The basic idea behind the SPH method is: the continuum has to be divided into discrete mass packets that are arbitrarily distributed in the computational domain and move according to the flow. The discrete objects, which are identified as smoothed SPH particles, represent the fluid entities such as mass, velocity, pressure and density. The spatial distance between the SPH objects is considered as the smoothing length. In the SPH model, the system of partial differential equations (PDE) is transformed into a system of time-dependent ordinary differential equations (ODE). The solution of the ODE system can be then preformed employing standard time integration methods. In order to achieve this transformation, the hydrodynamics equations (i.e. Navier-Stokes equations) should be approximated in two successive steps. In first phase, the continuum physical quantities and their derivatives are averaged by means of a kernel function that can serve as a smoothing interpolation field. In the next step, the obtained kernel integrals of each SPH object can be determined by a summation of the physical properties of all SPH particles that are existent within the range of the smoothing length. The physical meaning of the kernel function, often called window or weighting function, denotes the particle position probability, but is here just a smoothing technique. The main point of the SPH method is the selecting of a smooth kernel for the localisation of the strong form of the PDE using a convoluted integration.

Similar to the discrete element method (DEM), the SPH method uses a deterministic collision model to detect the contact between the SPH particles. Here, a contact is registered when a penetration between the collision partners occurs. The distance that the SPH particle has travelled inside its collision partner is defined as the penetration depth. The SPH method can be applied to different numerical problems such as compressible flow, incompressible flow, multiple phase flow, heat conduction, surface tension and many others. Recently, a lot of improvements have been achieved, for example the method efficiency, the correct selection of the interpolation kernel and the possibility of the parallel calculation, summarised in Liu and Liu (2016).

3.1.1.2 Mesh-Free Galerkin Methods

In the mesh-free Galerkin methods, the approximations (interpolants) are constructed among a set of particles that do not have any topological connections and are randomly distributed in the computational domain. A data-fitting algorithm formulates the most mesh-free approximations. This can be achieved using an inverse distance weighted principle that is deterministic multivariate interpolation for a set of scattered particles. Based on this principle, diverse mesh-free Galerkin methods can be found in the literature, for example diffuse element method, element free Galerkin method, reproducing kernel particle method, partition of unity methods, mesh-less local Petrov-Galerkin method and many others.

In 1992, the diffuse element method (DEM) is proposed. The basic idea behind the diffuse element method is to approximate the continuous function from a given set of points. The partial differential equations can then be solved employing this approximation method. In fact, the essence of this method is based on the moving least squares approximation that enables the reconstructing of continuous function from sample of unorganized points. The element free Galerkin method (EFG) that is devolved in 1994 is based on the moving least square approximation. However, it avoids the inconsistencies inherent of some smoothed particle hydrodynamic method formulations. Meanwhile, a similar approach, so-called the reproducing kernel particle method (RKPM), is developed in 1995. The only difference between EFG and RKPM is that the reproducing kernel particle method originates from the wavelets instead of the curve-fitting. The partition of unity methods (PUM) can be considered as a generalisation of classic finite element method. The basic difference among all partition of unity methods such as hp-cloud method, particle partition of unity method and extended finite element method is by the selection of partition of unity functions and/or the enrichment functions. One of well-known mesh-free Galerkin methods is the Mesh-less Local Petrov-Galerkin method (MLPG). This method differs from other methods since the local weak forms are generated by overlapping sub-domains instead using the global Galerkin weak formulations.

3.1.1.3 Other Particle Methods

Many particle methods have been developed recently and the research is still active on proposing new theories and computational formulations. However, each method owns its properties and thus it is difficult to found a method that is suitable for all application cases. In this context, there are important contributions that is mentioned here such as the molecular dynamics (MD), the vortex method (VM) and the Lattice Boltzmann method (LBM).

The MD methods are widely applied to simulate several scientific and engineering problems. In principle, the molecular dynamics methods can be classified in two categories, namely ab initio molecular dynamics and semi-empirical molecular dynamics. Another important particle method, especially for fluid mechanics applications is the vortex method. In the conventional grid-based methods, the numerical solution of the Navier-Stokes equations is achieved with a velocity-pressure coupling. The vortex method presents an alternative method based on the vorticity-velocity formulation. Similar to vortex method, the particle-in-cell method (PIC) is a dual Lagrangian and Eulerian descriptions. The basic idea behind the dual descriptions is to follow the motions of material particles that include information of physical values in a Lagrangian representation. The spatial discretisation (displacement interpolation) is carried out with respect to fixed spatial coordinate disregarded from the material particles in Eulerian representation. Other particle method that has been successful used for complex fluid calculation is the Lattice Boltzmann method (LBM). Here, the discrete Boltzmann equation is solved instead of the Navier-Stokes equations.

3.1.2 Grid-Based Methods

The fluid motion in the grid-based methods can be described in Eulerian or Lagrangian formulation. In Eulerian description (spatial description), the grid cells remain spatially fixed in space and do not change with time, while the material flows through these cells. In Lagrangian description (material description), the fluid materials are tracked as they move in the computational domain. Lagrangian grid-based methods are widely applied to solid mechanic problems, while Eulerian grid-based methods are almost employed in the field of fluid mechanics that because of the partial differential equations are mathematically simple to handle. A complete comparison between Lagrangian and Eulerian grid-based methods can be found for example in Epple et al. (2012). In both approaches, the basic mathematical formulations of the single-phase flow are the balance equations of mass, momentum, energy and species concentration as well as the constitutive relations and the thermodynamic equation of state. In Eulerian representation, the continuity equation and the momentum equations (Navier-Stokes equations) are expressed by:

$$\frac{\partial \rho}{\partial t} + \nabla \cdot (\rho \vec{u}) = S, \tag{3.2}$$

$$\frac{\partial (\rho \vec{u})}{\partial t} + \nabla \cdot (\rho \vec{u} \vec{u}) = \nabla \cdot (\mathbf{T}) - \nabla p + \rho \vec{g}, \tag{3.3}$$

with the stress tensor for Newtonian fluids (Bird 1976):

$$\mathbf{T} = \left(\lambda - \frac{2}{3}\mu\right)(\nabla \cdot \vec{u})\mathbf{I} + \mu\left((\nabla \vec{u}) + (\nabla \vec{u})^T\right). \tag{3.4}$$

Here, $\vec{u}$ represents the fluid velocity, μ is the fluid dynamic viscosity, $\mathbf{I}$ is the unit matrix, p and $\vec{g}$ denote the static pressure and the gravity acceleration, respectively. The bulk viscosity λ, also known as second or volume viscosity, may be neglected for incompressible fluid. The gas density ρ can be calculated with the aid of the ideal gas law as:

$$\rho = \frac{M}{RT} p, \tag{3.5}$$

where M is the molar mass and T represents the fluid temperature. The energy and species transport equations can be written as:

$$\frac{\partial(\rho h)}{\partial t} + \nabla \cdot (\rho \vec{u} h) = -\nabla \cdot \left(\frac{\mu}{\mathrm{Pr}} \nabla h\right) + S_h, \tag{3.6}$$

$$\frac{\partial(\rho c_{f,i})}{\partial t} + \nabla \cdot (\rho \vec{u} c_i) = -\nabla \cdot \left(\frac{\mu}{\mathrm{Sc}} \nabla c_i\right) + S_{c,i}. \tag{3.7}$$

The symbol h is the fluid enthalpy, c_i denotes the concentration of species i, Pr and Sc represent the Prandtl and Schmidt numbers, respectively. The heat source term S_h is composed of two parts: the heat source term due to the chemical reactions and the radiation source term. The concentration source term takes into account the formation or reduction of the species i by the reason of chemical reactions.

With the aid of previous coupled and non-linear partial differential equations, the spatial and temporal changes of single-phase flow variables such as velocity, pressure, temperature, concentration of species and density can be calculated. However, the analytical solution of these partial differential equations is only possible under extreme simplifications and assumptions. Generally, the exact solution is approximated by a finite number of values that can be determined through solving several algebraic equations instead of partial differential equations. In order to achieve this, three main steps (pre-processing, solving and post-processing) are necessary:

- In the first step, the problem is specified, including geometry and mesh generation. Here, the computational domain should be divided into sub-domains, also known as control volumes or grid cells.
- In the second step, the partial differential equations are converted into a system of algebraic equations employing one of discretisation approaches such as finite element or finite volume method. The obtained system of non-linear algebraic equations is numerically solved using Numerical solution, e.g. iterative techniques.
- In the third step, the simulation results are visualised and analysed.

In the following sections, the numerical approaches for the solution of balance equations of the single-phase flow in Eulerian formulation are explained according to Epple et al. (2012), Ferziger and Perić (2002), Schäfer (2006).

3.1.2.1 Numerical Grid

The numerical grid represents the discrete locations in the computational domain Ω, where the physical values of fluid phase are computed. In 2D cases, the grid cell is formed from grid lines, while in 3D cases it is formed from grid surfaces (grid curves) $\partial\Omega$ that consist of grid lines. The grid points identify the intersections of the grid lines (see Fig. 3.2).

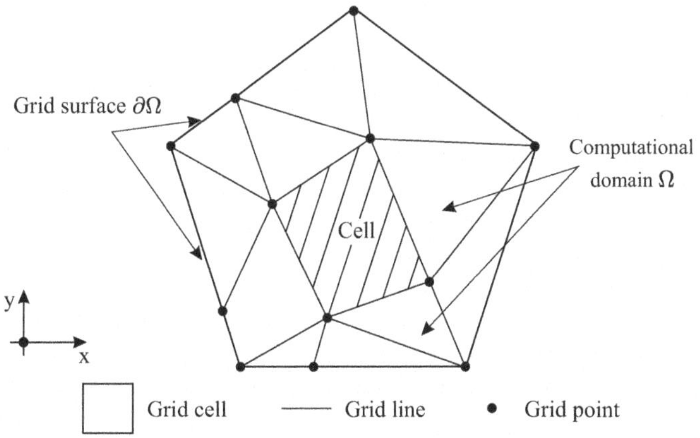

Fig. 3.2 Numerical grid

Depending on the logical arrangement of grid cells, the numerical grid can be classified as structured grids or unstructured grids. In the structured grids, the arrangement of the grid cells and thus the grid points are regular. Although the structured grids may be warped, basically they are rectangles in 2D or cuboids in 3D cases. Due to the regular arrangement of the grid cells, certain fixed relations can be derived between the neighbouring cells, which show important advantages for a simple discretisation and solution scheme. For the unstructured grids, by contrast, the arrangement of the grid points is irregular. The irregularity by distributing the grid cells over the computational domain offers higher flexibility for an accurate representation of the geometry. Furthermore, the unstructured grids can be easily fitted to computational domains with complex boundaries. However, the locations of the grid points as well as the relations between the neighbouring cells should be stored, which lead to additional data structure in comparison to the structured grids. The combination between structured and unstructured grids can result in obtaining the advantages of both grids (e.g. block-structured grids and hierarchically structured grids). The block-structured grids, although consist of different sub-blocks that all have structured grids, have irregular blocks arrangement. In the hierarchically structured grids, the grids in the sub-blocks are structured. In certain regions in the computational domain, the mesh is locally refined in a structured grid. In this context, these methods can be considered as a compromise between the structured grids with their inflexibility in the geometry representation and the unstructured grids with their high computational cost.

The numerical grid can also be classified regarding the grid type as Cartesian grid, boundary-fitted grid and overlapping grid. In Cartesian grids, the problem domain is discretised by a structured numerical grid. At the boundary, unstructured cells may be required, which need a special treatment. By contrast, all boundary parts are approximated with grid lines in the boundary-fitted grids, also known as body-fitted grids. In the overlapping grids, different regions of the computational domain can be meshed with independently grids. Here, the use of the structured grids can lead at the interfaces to overlapping areas, which require again a special treatment. Compared to Cartesian and overlapping grids that are applied only to special applications, the boundary-fitted grids are frequently used. A complete comparison between different grids types and their generation methods can be found for example in Ferziger and Perić (2002).

The linear equation's system resulting from the discretisation of partial differential equations can be solved with the aid of iterative solution methods, e.g. Jacobi or Gaus-Seidel methods. The following criterion is essential for the convergence: the finer is the grid the slower solution is converged. Almost all iterative solution methods converge on finer grids with higher accuracy, but require considerably more computing time. Therefore, it is advisable first to use a coarse grid to produce a starting solution for the fine grid in order to reduce the computational cost and to maintain the higher accuracy of the fine grid. This approach, which is known in the literature as multi-grid method, has received a lot of attention as it can solve the transport equations very efficiently. The idea behind the multi-grid methods is based on elimination of wavelength error components. The iterative algorithms eliminate efficiently the error components of an approximate solution if the wavelength error components correspond to the grid spacing. The errors with larger wavelengths can also be decrease but relatively slowly. The selection of the grid levels depends generally on the geometry, where 4–5 grid levels for 2D and 3–4 grid levels for 3D cases are sufficient. In the literature, there are many approaches for cycling through the different grid levels such as V-cycles and W-cycles. The computational effort per cycle in W-cycles is higher than V-cycles. However, the number of required cycles to reach a certain convergence criterion is lower. Detailed information can be found in Schäfer (2006).

3.1.2.2 Discretisation Methods

The second step in obtaining a numerical solution is to approximate the partial differential equations into a system of algebraic equations using discretisation methods. For the numerical solution of unsteady flows, the time must be also discretised besides the space discretisation.

Spatial Discretisation

The well-known spatial discretisation schemes are finite difference method, finite element method and finite volume method. Other discretisation schemes include hybrid method, spectral methods and boundary element method.

Finite difference method: The finite difference method (FD) represents the oldest discretisation approaches for the numerical solutions of partial differential equations. Generally, the numerical grid used in FD method is a structured grid, where the solution of differential equations is approximated at the grid points (see Fig. 3.3).

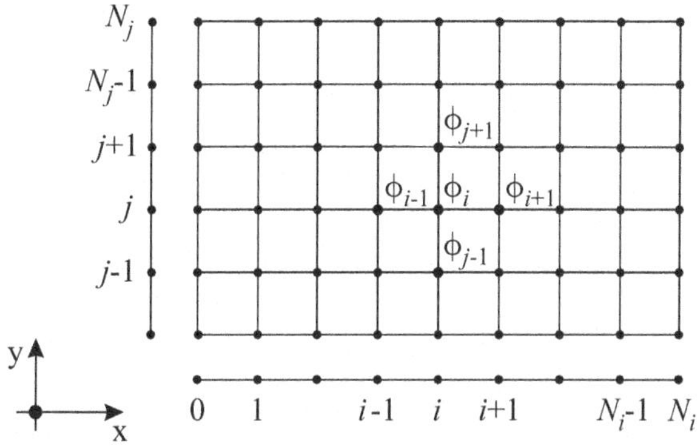

Fig. 3.3 Typical 2D mesh used in the finite difference method

The conservation equations are in differential forms and the derivatives are approximated at each grid point using finite difference equations. The differential form of the generic conservation equation in Cartesian coordinate and tensor notation is given as:

$$\underbrace{\frac{\partial(\rho\phi)}{\partial t}}_{\text{Instationary term}} + \underbrace{\frac{\partial(\rho u_j \phi_i)}{\partial x_j}}_{\text{Convective term}} = \underbrace{\frac{\partial}{\partial x_j}\left(\Gamma_\phi \frac{\partial \phi}{\partial x_j}\right)}_{\text{Diffusive term}} + \underbrace{S_\phi}_{\text{Source term}}, \tag{3.8}$$

where Γ_ϕ and S_ϕ denote the diffusion coefficient and the source term of the dependent variable ϕ, respectively. The variable ϕ refers to one of fluid physical values, namely velocity, enthalpy or species concentration. The first and second derivatives of the generic conservation equation can be approximated by means of different systematic approaches such as Taylor series expansion or polynomial fitting.

Any continuous differentiable function $\phi(x)$ can be expressed by Taylor series as an infinite sum of function derivatives at a single point i:

$$\phi(x) = \phi(x_i) + (x - x_i)\left(\frac{\partial \phi}{\partial x}\right)_i + \frac{(x - x_i)^2}{2!}\left(\frac{\partial^2 \phi}{\partial x^2}\right)_i + \cdots + \frac{(x - x_i)^n}{n!}\left(\frac{\partial^n \phi}{\partial x^n}\right)_i. \tag{3.9}$$

By neglecting the second and higher-order derivatives and replacing x with x_{i+1} or with x_{i-1}, the following forward and backward schemes for the first derivative in the convective term can be obtained:

$$\left(\frac{\partial \phi}{\partial x}\right)_i \approx \frac{\phi_{i+1} - \phi_i}{x_{i+1} - x_i}, \tag{3.10}$$

$$\left(\frac{\partial \phi}{\partial x}\right)_i \approx \frac{\phi_i - \phi_{i-1}}{x_i - x_{i-1}}. \tag{3.11}$$

Another approximation for the first derivative, so-called central difference scheme, can be obtained by replacing x with x_{i+1} and x_{i-1}:

$$\left(\frac{\partial \phi}{\partial x}\right)_i \approx \frac{\phi_{i+1} - \phi_{i-1}}{x_{i+1} - x_{i-1}}. \tag{3.12}$$

The second derivative in the diffusive term can be determined by using the first derivative approximation twice. The forward scheme for second derivative (outer derivative) is given as:

$$\left(\frac{\partial^2 \phi}{\partial x^2}\right)_i \approx \begin{cases} \frac{\left(\frac{\partial \phi}{\partial x}\right)_{i+1} - \left(\frac{\partial \phi}{\partial x}\right)_i}{x_{i+1} - x_i} & \text{Forward difference scheme (FDS)} \\ \frac{\left(\frac{\partial \phi}{\partial x}\right)_i - \left(\frac{\partial \phi}{\partial x}\right)_{i-1}}{x_i - x_{i-1}} & \text{Backward difference scheme (BDS)} \cdot \\ \frac{\left(\frac{\partial \phi}{\partial x}\right)_{i+1} - \left(\frac{\partial \phi}{\partial x}\right)_{i-1}}{x_{i+1} - x_{i-1}} & \text{Central difference scheme(CDS)} \end{cases} \tag{3.13}$$

Here, the inner derivative can also be estimated using diverse approximations, namely the forward, the backward or the central difference schemes. Furthermore, the Taylor series expansion may be applied to approximate the second derivative. By replacing x with x_{i+1} and x_{i-1} in Eq. (3.9), in addition to neglecting the third and higher-order derivatives, the following relation can be obtained:

$$\left(\frac{\partial^2 \phi}{\partial x^2}\right)_i \approx 2\frac{\phi_{i+1}(x_i - x_{i-1}) + \phi_{i-1}(x_{i+1} - x_i) - \phi_i(x_{i+1} - x_{i-1})}{(x_i - x_{i-1})(x_{i+1} - x_i)(x_{i+1} - x_{i-1})}. \tag{3.14}$$

The neglecting of higher-order derivatives by the calculation of the first and second derivatives leads to second and third-order truncation errors that measure the accuracy of approximations.

Besides the Taylor series expansion, the polynomial fitting represents an alternative method to approximate the derivatives. Generally, the polynomial fitting, also known as curve fitting, is a method of constructing a mathematical function (curve) that fits a set of data points. It can be achieved either with interpolation approaches, where new data points are generated within the range of a given discrete points or with smoothing methods, in which a smooth function that fits the data points is built. The derivatives can be estimated by fitting the function to an interpolation curve and differentiating the resulting curve. For example, the linear interpolation to approximate the function value at the point x can be expressed with the aid of x_{i+1} and x_i as:

$$\phi(x) = \frac{x - x_i}{x_{i+1} - x_i}\phi_{i+1} - \frac{x - x_{i+1}}{x_{i+1} - x_i}\phi_i. \tag{3.15}$$

The derivative of this function leads to the forward difference scheme (see Eq. 3.10). Likewise, the backward difference scheme can be obtained, if a linear interpolation is applied to interpolate the function between x_{i-1} and x_i. The use of quadratic interpolation to fit the function value at the points x_{i-1}, x_i and x_{i+1} results in the central difference scheme.

Taylor series expansion and polynomial fitting are not the only methods to approximate the derivatives. In the literature, there are many special schemes. For simple geometries, the FD method produces accurate results due to the possible use of higher-order schemes. However, the conservation cannot be guaranteed in the FD method unless the consideration of additional assumptions. Furthermore, the limitation of using structured grids and thus the limitation to simple geometries represents the main disadvantages of the FD method. Further information can be found for example in Ferziger and Perić (2002).

Finite element method: The finite element method (FE) is a numerical discretisation approach to approximate the solution of partial differential equations. Here, the computational domain is divided into a set of finite elements that are generally unstructured (see Fig. 3.4) and the solution is estimated at the grid points.

The basic idea behind the FE method is that the partial differential equations are multiplied by a weight function and then integrated over the entire computational domain. The solution of the differential equations is estimated by a linear combination of parameters, which are at the beginning unknown and approximation functions that is defined on the entire computational domain.

The solution can be approximated using Lagrange interpolation as:

$$\phi(x,y,z) \approx \phi^e(x,y,z) = \sum_{p=1}^{N_K^e} \phi_p^e \psi_p, \tag{3.16}$$

where ϕ_p^e is the unknown variable in the cell grid points and ψ_p is the local shape function. If the approximation function is inserted in the partial differential equations, the PDEs are no longer satisfied, resulting in residuals R. In the method of weighted residuals, the integrated residual over each element has to be disappeared:

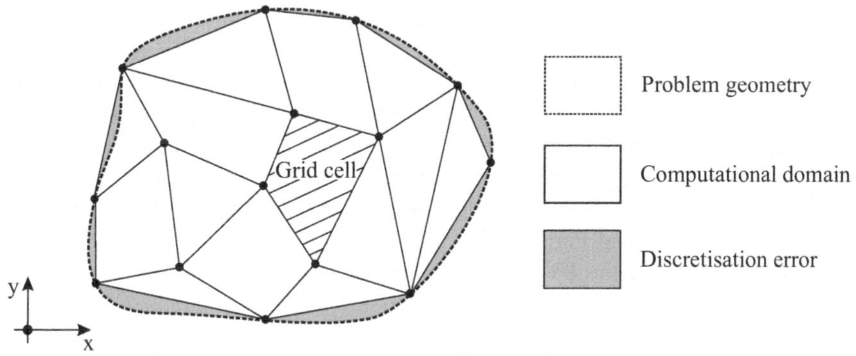

Fig. 3.4 Typical 2D mesh used in the finite element method

$$\int_{CV} R_{\phi}^{e} \psi_{w} dxdydz = 0. \tag{3.17}$$

Finally, a system of algebraic equations for the determination of variable ϕ can be obtained. The weighting function ψ_w that is used for the interpolation is identical to the local shape function ψ_p in Bubnow-Galerkin method. By Petrov-Galerkin method, the weighting function and the local shape function are different. The most important advantage of the finite element method is the possible dealing with complex geometries. The main disadvantage is that the matrices of the linearised equations system are unstructured compared to those of the structured grids, leading to difficulty by finding the efficient approach to obtain the solution. This drawback is also found in all discretisation methods that use unstructured grids. It should be mentioned here that the FE method is a special case of the discontinuous-Galerkin finite element method that uses polynomial approximation functions instead of the linear shape function. Detail information on the finite element method as well as the discontinuous-Galerkin finite element method and their applications can be found for example in Karakashian and Xing (2013), Schäfer (2006) Zienkiewicz et al. (2005).

Finite volume method: The finite volume method (FV) approximates the differential equations at discrete volumes. The differential equations of conservation equations are applied to each control volume, where they are approximated in the centre of grid cells. By using the divergence theorem, also known as Gauss's theorem, the volume integrals in a partial differential equation are converted to surface integrals. The variable at the control volume surface can be determined using interpolation techniques with the aid of nodal values of neighbour cells. The surface and volume integrals can then be estimated by main of mathematical formulae. For each control volume, an algebraic equation that contains the nodal values of the control volume and the neighbour cells is obtained. The number of algebraic equations is the same as the number of grid cells and thus the number of unknown values. These values can finally be calculated using numerical solution methods, e.g. iterative methods. Due to the fact that the flux entering a cretin control volume is equal to that leaving to the adjacent volume, the finite volume method is a conservative method. One of the important advantages of the FV method is the possible application to each grid type (structured or unstructured grid). The restriction of using and developing higher-order schemes, in particularly in 3D cases, represents the major disadvantage. The reason of this is that the FV method needs three level of approximation: interpolation, differentiation and integration. For unstructured grids, the use of higher-order schemes can result in numerical instabilities (Ferziger and Perić 2002).

As a start point for the FV method is the integral form of the generic conservation equation:

$$\underbrace{\frac{\partial}{\partial t}\iiint\limits_{CV} \rho\phi dV}_{\text{Instationary term}} + \underbrace{\iint\limits_{S} \rho\vec{u}\phi\cdot\vec{n}\, dS}_{\text{Convective term}} = \underbrace{\iint\limits_{S} \Gamma_\phi \nabla\phi\cdot\vec{n}\, dS}_{\text{Diffusive term}} + \underbrace{\iiint\limits_{CV} S_\phi dV}_{\text{Source term}}. \tag{3.18}$$

In 3D cases, each grid cell has six neighbours, unless it is located on the boundary of the computational region. The closed surface of a grid cell consists of sub-areas S_i. Each sub- surface has a normal vector n_i that is directed outwards of this area (see Fig. 3.5).

The surface and the volume integrals of the partial differential equations should be approximated in order to obtain a system of the algebraic equations. The approximation of the total flux surface integral can be formed by the sum of fluxes integrals over all sub-areas:

$$\iint\limits_{S} fdS = \sum_{i=1}^{k} \iint\limits_{S_k} f_k dS_k, \tag{3.19}$$

where the subscript k denotes the number of the control volume sub-surfaces; $k = 4$ for 2D and $k = 6$ for 3D case. The term f represents the component of convective or diffusive flux vector in the normal direction to the control volume surface k:

$$f_k = \begin{cases} (\rho\vec{u}\phi\cdot\vec{n})_k & \text{Component of convective flux} \\ (\Gamma_\phi\nabla\phi\cdot\vec{n})_k & \text{Component of diffusive flux} \end{cases}. \tag{3.20}$$

The exact determination of the surface integral requires the knowledge of the integrand f_k on the surface S_k. This cannot be achieved since only the values of ϕ in the centre of the control volumes are available. Therefore, the flux over a sub-area must be approximated using these nodal values, i.e. the values in the centre of the control volumes. The simplest approximation is the application of the centre rule method. In this context, the surface integral is estimated as a product of the mean value over the surface and the area of the control volume:

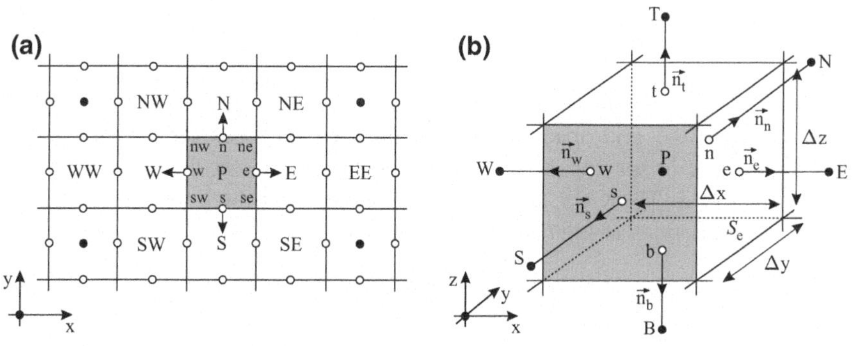

Fig. 3.5 Cartesian grid used in the finite volume method: **a** 2D mesh and **b** 3D mesh

$$\iint_{S_k} f_k dS_k = \bar{f}_k S_k \approx f_k S_k. \tag{3.21}$$

Now, the flux f_k at the centre of the surface S_k is required. This value is unknown and thus it has to be interpolated from the nodal value of the cell and the values of its neighbouring cells. The convective momentum flux can finally be obtained as:

$$\iint_{S_k} (\rho\vec{u}\phi \cdot \vec{n}\ dS)_k \approx (\rho\vec{u}\phi \cdot \vec{n})_k S_k = \dot{m}_k \phi_k. \tag{3.22}$$

The mass flow rate $\dot{m}_k$ that passes through the surface S_k is again determined by an interpolation. The non-linearity can be dissolved using a Picard-iteration. It is assumed that the velocity field, the fluid properties and the diffusion coefficient Γ_ϕ are known at all control volumes centres. The values of the last iteration are generally used. In order to calculate the convective and diffusive fluxes, the value ϕ_k and its gradient normal to the cell surface should be determined. They can be expressed employing different interpolation methods such as upwind interpolation, linear interpolation, quadratic upwind interpolation and high-order schemes.

In the upwind differencing scheme (UDS), the value ϕ_k at the control volume face is approximated depending on the flow direction using a backward or forward difference for the first derivative:

$$\phi_e = \begin{cases} \phi_P & \text{case}(\rho\vec{u} \cdot \vec{n})_e > 0 \\ \phi_E & \text{case}(\rho\vec{u} \cdot \vec{n})_e < 0 \end{cases}. \tag{3.23}$$

The UDS is a first-order interpolation method and distinguishes itself by its high stability. If the second-order linear interpolation (CDS) is used, the value ϕ_k is estimated between the two nearest cells as:

$$\phi_e = \phi_E \lambda_e + \phi_P (1 - \lambda_e), \tag{3.24}$$

with the linear interpolation factor:

$$\lambda_e = \frac{x_e - x_P}{x_E - x_P}. \tag{3.25}$$

The CDS approximation is of second-order accuracy and the error term is proportional to the grid spacing Δx for structured grid and proportional to the product of the grid spacing Δx and the grid expansion factor minus unity for the unstructured grid. The CDS, although characterises with high accuracy, leads to numerical oscillations at high local Peclet numbers:

$$\text{Pe} = \frac{\rho \Delta x |\vec{u}|}{\mu}. \tag{3.26}$$

A blending method that combines CDS and UDS can also be applied to approximate the value ϕ_k at the control volume face. Here, the advantages of both interpolation methods, i.e. the high numerical stability and the accuracy can be maintained.

$$\phi_e = (1 - \alpha_{UDS})\nabla\phi_P \cdot (\vec{r}_e - \vec{r}_P) + \alpha_{UDS}\begin{cases} \phi_P & \text{case } (\rho\vec{u}\cdot\vec{n})_e > 0 \\ \phi_E & \text{case } (\rho\vec{u}\cdot\vec{n})_e < 0 \end{cases}. \quad (3.27)$$

Here, α_{UDS} denotes the blending factor, $\vec{r}_e$ and $\vec{r}_P$ are the position vectors of the sub-surface $k = e$ and the centre of the grid cell, respectively. With the aid of the blending factor, CDS and UDS can be super-imposed.

The next logical improvement to estimate the value ϕ_k between the nodal cells (P and E) is to construct a parabola instead of straight line. For this purpose, more nodal values are required for the approximation:

$$\phi_e = \begin{cases} \phi_P + \lambda_1(\phi_E - \phi_P) + \lambda_2(\phi_P - \phi_W) & \text{case } (\rho\vec{u}\cdot\vec{n})_e > 0 \\ \phi + \lambda_1(\phi_P - \phi_E) + \lambda_2(\phi_E - \phi_{EE}) & \text{case } (\rho\vec{u}\cdot\vec{n})_e < 0 \end{cases}. \quad (3.28)$$

The coefficient λ_1 can be determined in terms of control volumes coordinates as:

$$\lambda_1 = \begin{cases} \frac{(x_e - x_P)(x_e - x_W)}{(x_E - x_P)(x_E - x_W)} & \text{case } (\rho\vec{u}\cdot\vec{n})_e > 0 \\ \frac{(x_e - x_E)(x_e - x_{EE})}{(x_P - x_E)(x_P - x_{EE})} & \text{case } (\rho\vec{u}\cdot\vec{n})_e < 0 \end{cases}, \quad (3.29)$$

and for the coefficient λ_2 as:

$$\lambda_2 = \begin{cases} \frac{(x_e - x_P)(x_E - x_e)}{(x_P - x_W)(x_E - x_W)} & \text{case } (\rho\vec{u}\cdot\vec{n})_e > 0 \\ \frac{(x_e - x_E)(x_P - x_e)}{(x_E - x_{EE})(x_P - x_{EE})} & \text{case } (\rho\vec{u}\cdot\vec{n})_e < 0 \end{cases}. \quad (3.30)$$

The determination of the weighting coefficients on the non-orthogonal and/or unstructured grids, by contrast to the structured grids, is not trivial. This scheme is of third-order accuracy and is known in literature as quadratic upwind interpolation for convective kinematics (QUICK). Interpolation of higher-order accuracy can be reached employing higher-order mathematical formulae e.g. the use of the fitting polynomial with at least degree three, which in turn leads to fourth-order truncation error. According to Schäfer (2006), the higher-order interpolations may not result in more accurate solution. High accuracy can only be reached if the grid is fine enough to get all details of the solution.

Besides the determination of the surface integral, some terms in the partial differential equations require integration over the volume. The exact calculation of the volume integral needs the knowledge of the integrand Q over the grid cell volume. This cannot be analytically achieved. Numerical approximation of the volume integral should therefore be performed. The simplest approximation is to estimate the volume integral by the product of the mean value of the integrand and the volume of the CV as follows:

$$\iiint_{CV} Q dV = \overline{Q}\Delta V \approx Q_P \Delta V. \tag{3.31}$$

Here, Q_P represents the value of the variable at the control volume centre. This approximation is of second-order and does not require any interpolation since the variable Q_P is available at each grid cell. Higher-order approximations need the nodal values of the variables at different locations.

Hybrid discretisation method: In form of a so-called hybrid discretisation approach, the finite volume method and the finite element method are combined. The hybrid discretisation approach, also known as control volume based finite element method (CVFEM), has the advantage of both FE and FV methods. Similar to the FV method, the conservation equations are in integral form and are applied to each control volume. The flux through the boundaries of control volumes are then determined in similar way like the FE method.

Spectral method: The spectral methods (SM) are used to solve numerically certain classes of partial differential equations. Compared to FE and FV schemes, the spectral methods are less suitable for general numerical purposes. However, they can be applied to important applications e.g. turbulence simulations. The basic idea behind SM is to approximate the spatial derivatives by means of Fourier series or one of their generalisations such as fast Fourier transform (FTT). The solution of partial differential equations is evaluated as a sum of certain basic functions as in FTT, where the signal is decomposed into sum of sinusoids (sine and cosine) of various frequencies. The functions coefficients are then calculated in order to satisfy the differential equations. The spectral methods and the finite element methods are based on the same idea. The major difference between both methods is that the SM uses non-zero basis functions over the entire computational domain, while the FE scheme uses these functions only in small sub-areas of the computational domain.

Boundary element method: The boundary element method (BEM), also known as boundary integral equation method, is one of the discretisation approaches for solving the partial differential equations. The boundary element method, although it can be used to solve different engineering problems, is rarely applied to fluid dynamics. The basic idea behind the BEM is to reformulate the partial differential equations into integral equations. Here, the given boundary conditions are used to fit values into the integral equations. Similar to the spectral methods, the main disadvantage of the boundary element method is that the integral equations reformulation can only be carried out for certain partial differential equations. For this reason, BEM is not employed as frequently as FD or FE methods. However, the BEM, where it is suited, is computationally more efficient compared to other competitive discretisation methods.

Temporal Discretisation

In many engineering applications, the investigated problems are unsteady, i.e. the physical values of the fluid phase depend on space and time. In order to obtain the time-dependent solution, initial conditions and boundary conditions that can also rely on time should be defined. Let the variable ϕ is depending on time, e.g. in a first-order differential equation:

$$\frac{d\phi(t)}{dt} = f(t, \phi(t)), \tag{3.32}$$

with the initial condition:

$$\phi(t^{t_0}) = \phi^{t_0}. \tag{3.33}$$

Now, the problem is to calculate the solution $\phi^{t^0+\Delta t}$ after a short time Δt from the start point t^0. By the same token, the solution $\phi^{t^0+\Delta t} = \phi^{t^1}$ can be considered as an initial condition for the new solution $\phi^{t^1+\Delta t} = \phi^{t^2}$ after a time step Δt. This problem can be solved by integrating Eq. (3.32) from t^n to $t^{n+\Delta t}$ as:

$$\int_{t^n}^{t^{n+1}} \frac{d\phi(t)}{dt} = \int_{t^n}^{t^{n+1}} f(t, \phi(t))dt. \tag{3.34}$$

Equation (3.34), although represents the exact solution, cannot be estimated without knowing the integral on the right-hand side. For this reason, an approximation should be introduced. The explicit method, also known as the forward Euler method, is obtained when the integral is evaluated using the value of the integrand at the initial point:

$$\phi^{n+1} = \phi^n + f(t^n, \phi^n)\Delta t. \tag{3.35}$$

Here, the shorthand notation for $\phi^n = \phi^{t^n}$ and $\phi^{n+1} = \phi^{t^n+1}$ is used. If, however, the final point is used to approximate the integral, the implicit or backward Euler method is extracted:

$$\phi^{n+1} = \phi^n + f(t^{n+1}, \phi^{n+1})\Delta t. \tag{3.36}$$

When the straight line interpolation between initial and final points is applied to estimate the integral, the trapezoid rule can be obtained:

$$\phi^{n+1} = \phi^n + \frac{1}{2}\left[f(t^n, \phi^n) + f(t^{n+1}, \phi^{n+1})\right]\Delta t. \tag{3.37}$$

This approximation is of second-order accuracy and gives the basic formula for the frequently used Crank-Nicolson method. Higher-order temporal discretisation schemes such as third or fourth-order Runge-kutta as well as Adams methods can be used when higher-order spatial discretisation schemes are also applied (Ferziger and Perić 2002).

Pressure Correction Methods

If the finite volume method is used, both mass and momentum balance equations must be satisfied at same time. The calculation of velocity field is not possible due to the unknown pressure, whose gradient contributes to all momentum equations. In compressible flow, the density is appeared as second variable, where the use of an additional equation (equation of state) is necessary. There is no explicit equation for the pressure that enables the pressure gradient to be linked with the momentum equations. The pressure can be indirectly defined through the continuity equation. This means, if the correct pressure field is inserted into the momentum equations, the obtained velocity field should also satisfy the continuity equation. One way to handle this indirect coupling is to eliminate the pressure using a penalty-function formulation. For incompressible flows, this procedure has many drawbacks. Other approaches eliminate the pressure from the entire formulation making the extension of these methods for compressible flows is not possible, in particularly when the boundary conditions are depending on the pressure. Other possibility is to combine the continuity and momentum equations. In this context, the continuity equation can be used to simplify the divergence of the momentum equation, resulting in Poisson equation for the pressure. In Cartesian coordinate and tensor notation, the Poisson equation is expressed by:

$$\frac{\partial}{\partial x_i}\left(\frac{\partial p}{\partial x_i}\right) = -\frac{\partial}{\partial x_i}\left[\frac{\partial}{\partial x_j}\left(\rho u_i u_j - \underbrace{\tau_{ij}}_{A}\right)\right] + \underbrace{\frac{\partial(\rho b_i)}{\partial x_i} + \frac{\partial^2 \rho}{\partial t^2}}_{B}, \tag{3.38}$$

where τ_{ij} is the viscous stress tensor and B the body force per unit mass. The terms A and B are zero in case of fluid with constant density and viscosity. The pressure equation can be solved using one of the discretisation methods such as FE or FV.

The Poisson equation is applied to determine the pressure in explicit and implicit solution methods. However, it is advisable to derive the equation of pressure from the discretised momentum and continuity equations rather than the discretisation of the Poisson equation in order to maintain consistency. In the literature, this procedure is known as implicit pressure-correction methods or projecting methods. For incompressible flows, the velocity field is constructed based on predefined pressure field. The obtained velocity field does not fulfil the continuity equation. It should then be corrected by subtracting the pressure gradient. The most popular projecting methods such as SIMPLE, SIMPLEC and PISO algorithms can be found for example in Epple et al. (2012), Ferziger and Perić (2002). Here, they are briefly explained.

The semi-implicit method for pressure-linked equations (SIMPLE) is a two-step algorithm. In the first step, a velocity field, which fulfils the Navier-Stokes equations, is determined. The pressure gradient is assumed to be known in this step; the values of the last iteration are used. Generally, the calculated values of the velocity field do not satisfy the continuity equation. In the second step, the mass defect is defined by using the continuity equation for each control volume. The pressure-correction equation is calculated in order to compensate this mass defect. From the pressure correction, the velocity correction can be estimated. Finally, the pressure as well as the velocity is corrected. Both steps are executed in an inner iteration loop until the L2- space of the mass defect m^* falls below a fixed limited value:

$$|m^*| = \sqrt{\sum_{i=1}^{N^{CV}} m_i^{*2}} < \text{Fixed value}. \quad (3.39)$$

The semi-implicit method for pressure-linked equations consistent (SIMPLEC) is a similar algorithm to SIMPLE method. The crucial difference between the two methods is in the formulation of the correction-equation for the velocity field. In SIMPLE algorithm, the term in the pressure-correction equation that considers velocity correction of neighbour cells is neglected, while in SIMPLEC algorithm is approximated. The pressure implicit with splitting of operators (PISO) calculates this term with aid of another correction step. In contrast to previous methods, the PISO algorithm uses besides the implicit also explicit correction steps in order to obtain the solution of mass and momentum equations.

In principle, there are two approaches for the calculation of consistent pressure and velocity fields, namely density-based and pressure-based methods. The density-based method assumes that the velocity field is known. The local density is then determined from the continuity equation by the given velocity field. The determination of the pressure can be carried out by the use of the thermal state equation for the ideal gas as:

$$p = \frac{\rho RT}{M}. \quad (3.40)$$

The symbol R denotes the ideal gas constant and M is the molar mass. Generally, the pressure and density fields do not fulfil the momentum equations, so the velocity field is corrected. The density-based method requires an explicit relationship between pressure and density. If, however, this method is applied to incompressible flows, a weak coupling between the pressure and the density should be formulated.

In the pressure-based method for determining the velocity and pressure fields, the local pressure is assumed to be known. For incompressible flows, the velocity field is then calculated from the momentum equations. The obtained velocity field does not satisfy the continuity equation, so that it should be corrected by subtracting the pressure gradient. Considering compressible flows, the density is calculated at a given pressure from the equation of the state followed then by solving the momentum equations to obtain the velocity field. The pressure-based method has the advantage that can basically be applied to compressible or incompressible flows without the need for the weak coupling between pressure and density.

The staggered grid arrangement was first proposed to avoid convergence problems and oscillations in the pressure and velocity fields. In the staggered agreement, the scalar variables such as pressure, density, total enthalpy, chemical species and turbulence characteristic values are stored in the cell centres, while the vector variable (velocity) are located at the cell faces of control volumes. For structured grids, the staggered grid can be applied to simulate compressible or incompressible flows. The extension of the staggered arrangement to unstructured grids results, however, in higher computational time by the reason of dealing with different control volumes for different variables. Therefore, the colocated grid arrangement has become more popular, where all variables share the same nodal point, i.e. all variables are stored in same position. The colocated arrangement, although has significant advantage in complex geometries, was rarely applied to incompressible flows due the difficulty in the pressure-velocity coupling. The increased use of the unstructured grids for the problems with complex geometries makes the colocated arrangement the standard method in the most modern codes, e.g. ANSYS-FLUENT and OpenFOAM.

Boundary and Initial Values

The continuity, the momentum, the species concentrations and the energy equations describe the behaviour of the fluid inside the computational domain. However, these partial differential equations can only be solved, if the boundary conditions and initial values are given. The boundary conditions specify the exchange of mass, momentum and heat between fluid and ambiance. Basically, there are three types of boundary conditions for steady-state and transient problems, namely Dirichlet, Neumann and Cauchy boundary conditions.

In Dirichlet boundary condition, the value of the computation quantity ϕ_f (velocity or enthalpy) is given on the surface of the computational domain $\partial\Omega$. Here, the computation quantity can be constant or variable depending on time and space. For example, the velocity value is constant in case of fixed walls $|\vec{u}| = 0$ or time depending in case of moving walls:

$$\phi = \phi^{bc}(\vec{x}, t) \quad \vec{x} \in \partial\Omega. \tag{3.41}$$

By Neumann boundary condition, the gradient value of the computation quantity in normal direction to the surface of the computational domain is given. In this type of boundary conditions, the derivative of the computation quantity is often defined at the outlet of the flow region, where the quantity does not change, for example $\nabla(\vec{u} \cdot \vec{n}) \cdot \vec{n} = 0$:

$$\frac{\partial \phi}{\partial \vec{n}} = \phi^{bc}(\vec{x}, t) \quad \vec{x} \in \partial\Omega. \tag{3.42}$$

The Neumann boundary condition can be used as a symmetry boundary condition, which in turn leads to reduce the measure of the computational domain.

In Cauchy boundary condition, also known as Robbins or mixed boundary condition, the value of the computation quantity ϕ and its derivative in the normal direction to the surface of the computational domain is given according to the following relation:

$$a \cdot \phi + \frac{\partial \phi}{\partial \vec{n}} = \phi^{bc}(\vec{x}, t) \quad \vec{x} \in \partial\Omega \; a > 0. \tag{3.43}$$

Turbulent Flow

Turbulent flow is a flow regime that its properties is characterised by high fluctuations in time and length. Compared to laminar flows, the turbulent flows are irregular and lead to increased rates of momentum, mass and heat transfer. The turbulent flow can be found in most of practical engineering applications. Reynolds number that is the ratio of inertial force to viscous force can be used as a measure of turbulence. If the Reynolds number increases above a certain value (critical Reynolds number), inertial forces and associated fluctuations are so great that instabilities are existing in the flow and thus turbulence occurs. In the literature, the numerical approaches for modelling the turbulent single-phase flow, including direct numerical simulation, large eddy simulation and Reynolds averaged Navier-Stokes simulation are adequately described, among others in Launchbury (2016), Oberlack et al. (2007), Pope (2000).

Direct numerical simulation (DNS), although represents the simplest methods regarding content and concept, is the most exact approach for the simulation of turbulent flow. The length and time scales (turbulence structures) are completely captured by the DNS. In order to achieve this, a very fine grid and adequate fluid time step should be applied. The required grid resolution and time step, which are necessary for resolving all turbulence structures, lead to an extreme computational cost. The computational time of the DNS behaves almost proportional to the number of grid cells. The real computational cost is equal to $O(N_f)^{1.1}$ by the reason of the overheads in the numerical algorithm. According to Lumley (1990), the required number of grid cells N_f for the numerical simulation of a turbulent channel flow depends on the fluid Reynolds number and can be expressed using the following equation:

$$N \approx \frac{\mathrm{Re}^{2.7}}{10000}. \tag{3.44}$$

Large eddy simulation (LES) is particularly suited to flows with high Reynolds numbers and/or for computational domain with complex geometries. The computational effort of LES is much lower than DNS. The largest scale motions (eddy elements) of the flow are solved by the computational grid, while the small scale motions are approximated. Like DNS also LES is a numerical approach for unsteady flow simulations, i.e. all physical values of the flow are spatially and temporally resolved. The earliest and the most known LES model dated back to Smagorinsky (1963). Since that time, LES have been further improved, e.g. the dynamic model of Germano (Germano et al. 1991) that does not require any model parameters.

Reynolds averaged Navier-Stokes simulation (RANS) is another possibility to determine the turbulent values of a flow. Here, the equations of motions are averaged over the time, i.e. each fluid variable is written as the sum of a time-average value and a fluctuation part. By inserting this in the continuity and the Navier-Stokes equations, the time averaged continuity and Navier-Stokes equations are obtained. In order to close the equation system, there are different levels of approximations available as algebraic, one-equation, two-equation (e.g. K-epsilon and K-omega models) and Reynolds stress models. Generally, the RANS models are characterised by their numerical stability and high efficiency. As a result, they are used in large-scale industrial applications since many years ago.

Radiation

In the combustion chamber of boilers, the heat transfer from flue gas to surrounding walls takes place mainly through radiation. The mathematical modelling of the heat transfer by radiation requires a radiative transfer equation that describes the intensity change along the radiation propagation direction. The balance equation for the total radiation intensity I can be expressed using the following equation:

$$\underbrace{\frac{1}{c}\frac{\partial I(\vec{r},\vec{s})}{\partial t}}_{\substack{\textit{Temporal charge of}\\ \textit{radiation intensity}}} + \underbrace{\frac{\partial I(\vec{r},\vec{s})}{\partial s}}_{\substack{\textit{Radiation intensity}\\ \textit{change along the path}}} + \underbrace{(a_{\mathrm{rad}}+\sigma_{sca})I(\vec{r},\vec{s})}_{\substack{\textit{Intensity reduction by}\\ \textit{absorption and out-scattering}}}$$
$$= \underbrace{a_{rad}n_{rad}^{2}\frac{\sigma T^{4}}{\pi}}_{\substack{\textit{Increase}\\ \textit{by emission}}} + \underbrace{\frac{\sigma_{sca}}{4\pi}\int_{0}^{4\pi} I(\vec{r},\vec{s}_{sca})\Phi(\vec{s},\vec{s}_{sca})d\Omega_{sca}}_{\substack{\textit{Intensity increase by}\\ \textit{in-scattering}}}. \tag{3.45}$$

Here, T is the medium temperature in the control volume, Φ denotes the phase function, Ω_{sca} is the solid angle of the scattered radiation, s is the path length, n_{ref} is the refractive index, $\vec{r}$ is the position vector of the control volume in space, the vectors $\vec{s}$ and $\vec{s}_{sca}$ represent the propagation direction of radiation and scattered radiation, respectively. The scattering coefficient σ_{sca} can often be treated as a constant, while the absorption coefficient a_{rad} is a function of representative layer thickness, total pressure and local concentrations of all polyatomic molecules, which are able to absorb radiation.

The temporal change of radiation intensity is neglected for engineering applications since this change arises with $1/c$, where c is the speed of light and thus $I(\vec{r}, \vec{s})$ responds almost instantaneously to temperature changes. The term $(a_{rad} + \sigma_{sca})$ refers to the medium optical thickness. The phase function $\Phi(\vec{s} \cdot \vec{s}_{sca})$ that is a probability density function describes how much radiant energy in the direction of radiation $\vec{s}$ is in-scattered. The integral of the phase function is normalised to unity:

$$\frac{1}{4\pi} \int_{0}^{4\pi} \Phi(\vec{s} \cdot \vec{s}_{sca}) d\Omega_{sca} = 1. \tag{3.46}$$

The analytical solution of the radiative transfer equation can only be achieved for simple geometries and with simplifications for the properties of the radiation-active medium such as the radiation behaviour of a medium with grey body between two parallel, infinitely large and flat plates. Here, the spectral intensity of radiation emission of the medium is proportional to Planck's law of black body radiation. For complex geometries, numerical methods for the solution of the radiative transport equation have to be introduced. In the literature, there are a lot of numerical methods available: discrete transfer radiation model (DTRM), P1 radiation model, discrete ordinates (DO) radiation model and many others that are already built-in in most CFD programmes like ANSYS-FLUENT or OpenFOAM. Further information can be found for example in Epple et al. (2012), Howell et al. (2015).

3.1.2.3 Solution Methods

The discretisation methods result in an algebraic equation for each fluid variable ϕ at each grid cell, containing the value at that cell as well as the nodal values of neighbouring cells. Depending on the nature of partial differential equation, the discretised equation can be linear or non-linear. The obtained algebraic equation of each control volume has the following form:

$$\underbrace{A^P \phi^P}_{\text{Considered cell}} + \underbrace{\sum_{k=1}^{l} A^k \phi^k}_{\text{Neighbouring cells}} = \underbrace{Q^P}_{\text{Source term}}, \tag{3.47}$$

where the index P represents the node, at which the differential equation is discretised and the subscript l denotes all neighbour cells that involved in the approximation scheme. The coefficients $A^{P,k}$ depend on the variable values, the geometrical data and the fluid properties. The variables $\phi^{P,k}$ are unknown, while the symbol Q^P includes all known terms.

The matrix system of the entire computational domain with a finite number of grid cells N for the fluid variable ϕ is then constructed as:

$$A\phi = Q. \tag{3.48}$$

In this system of algebraic equations, the number of equations and the unknown are equal. Most elements in the matrix system are zero, i.e. the matrix is spares, which simplifies the task to obtain the solution. In practical engineering applications, the number of the algebraic equations is very large, so that it cannot be calculated analytically and instead it is solved numerically. For solving a linear system of algebraic equations, the direct methods that deliver an exact solution are preferred. In this context, Gauss elimination represents the basic direct algorithm. The idea behind Gauss elimination is the systematic reduction of large systems of linear equations to smaller ones through a sequence of operations performed on the matrix. Based on Gauss elimination algorithm, different methods have been developed such as LU decomposition and cyclic reduction. The previous algorithms can be applied efficiently to solve any matrix of linear equations. However, the computational effort of the direct methods is very high, particularly when the triangular factors of sparse matrices are not sparse.

The iterative methods are generally applied for solving non-linear system of algebraic equations, but they can also be applied to linear systems. In the iterative methods, the solution is first suggested and then the equation is used to improve the solution. This includes guessing a solution, linearisation the equations, improving the solution and repeating the process till converged results are reached. Although the direct methods can deliver the exact solution of the problem in the absence of rounding errors, the computational effort of iterative methods is relatively lower. Some basic iterative algorithms are Jacobi method, Gauss-Seidel method, successive over-relaxation SOR, incomplete LU decomposition and conjugate gradient method (Epple et al. 2012; Ferziger and Perić 2002; Schäfer 2006).

3.2 Numerical Methods for Gas-Solid Flow

In previous sections, the numerical methods for a single-fluid flow are described. In contrast, the modelling of the fluid phase in a gas-solid flow is more complicated due to the interaction between both phases. The control volumes in the computational domain are not completely occupied with fluid, but rather both fluid and solid phases share the grid cells. In this context, the parameter void fraction is defined, which is a measure of the solid content in the gas-solid flow. Various definitions are used for specifying the void fraction, namely local void fraction, chordal void fraction, cross-section void fraction and volumetric void fraction. The letter is frequently used to characteristic the gas-solid flow. It is given as:

$$\varepsilon_f^V = \frac{V_f}{V^{CV}} = \frac{V_f}{V_f + V_P}, \tag{3.49}$$

where V_f and V_P are the volume of the grid cell occupied by fluid and solid, respectively. The volumetric void fraction can also be expressed to the solid volume, the so-called solid volumetric void fraction, as:

$$\varepsilon_P^V = \frac{V_P}{V^{CV}} = \frac{V_P}{V_P + V_f} = 1 - \varepsilon_f^V. \tag{3.50}$$

Similar to the volumetric void fraction, the cross-section void fraction is defined as:

$$\varepsilon_f^s = \frac{S_f}{S^{CV}} = \frac{S_f}{S_f + S_P} = 1 - \varepsilon_P^s. \tag{3.51}$$

Stiess (1994) proved that the summation of cross-section void fractions for n surfaces when $n \to \infty$ is equal to the volumetric void fraction:

$$\overline{\varepsilon_f^s} = \lim_{n\to\infty} \frac{1}{n} \sum_{i=1}^{n} \varepsilon_{f,i}^s s = \varepsilon_f^V = \varepsilon_f. \tag{3.52}$$

The volumetric void fraction consists of inner and outer void fractions, also known as inner and outer porosities (see Fig. 3.6):

$$\varepsilon_f^V = \varepsilon_f^{V,out} + \varepsilon_f^{V,in} - \varepsilon_f^{V,out} \varepsilon_f^{V,in}. \tag{3.53}$$

The inner porosity is formed by the cavities in the disperse phase itself. The cavities between the particles form the outer porosity, also known as bulk porosity. In most of practical engineering applications, it is assumed that the fluid can flow between particles, but cannot flow through the particles. In this case, the volumetric void fraction is equal to the outer porosity.

When the particles are packed randomly, an empirical parameter (random-close packing limit) that defines the maximum volumetric void fraction of solid is introduced. The random-close packing limit is independent of the grain size for regular spheres packing (monodisperse system) and semi non-cohesive particles (adhesive forces that act between the particles are smaller than volume forces, see Sect. 3.2.3.2.3). For monodisperse, fine-grained and cohesive particles, the random-close packing limit is significantly larger than the random-close packing limit of monodisperse, coarse-grained and non-cohesive particles of same material.

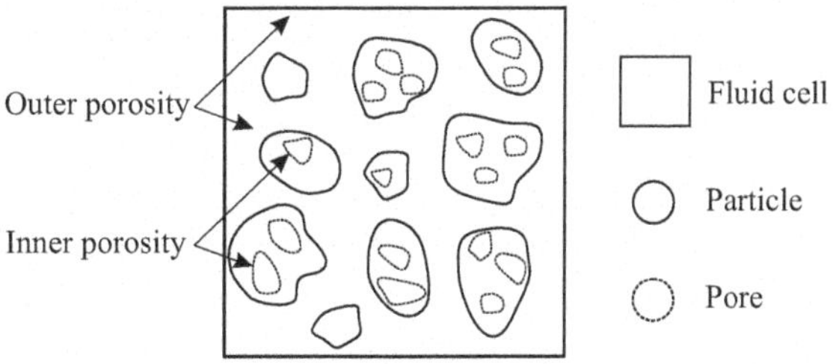

Fig. 3.6 Volumetric void fraction

In case of polydisperse systems with non-spherical particles, the random-close packing limit depends on the particle size distribution and the particle shape.

Different numerical methods at diverse levels of accuracy have been applied to simulate the hydrodynamic behaviour of gas-solid flows (see Fig. 3.7). While the solid phase can be generally described by two different levels of detail: a quasi-continuum or discrete phase, the fluid phase is presented in each case as a continuum. The simplest approach for modelling of the the gas-solid flows is possible under the assumption of homogeneity. In the quasi-single-phase method (homogeneous model), the physical values of fluid and solid phases are obtained as volume average values by solving the balance equations for a single-phase flow. A particulate flow can be considered as homogeneous, if the volume distribution of particles in the fluid phase is uniform and the velocity differences between two phases are negligibly small. In practical flows, these requirements are not fulfilled, so that the quasi-single-phase method cannot be used. An alternative method is the two-fluid method (Euler-Euler approach) that distinguishes itself from the quasi-single-phase model with higher accuracy and a wide range of applications. In principle, the gas phase in the fluid-solid flow is modelled with the balance equations of a single-phase. For the disperse phase, the conservation equations for

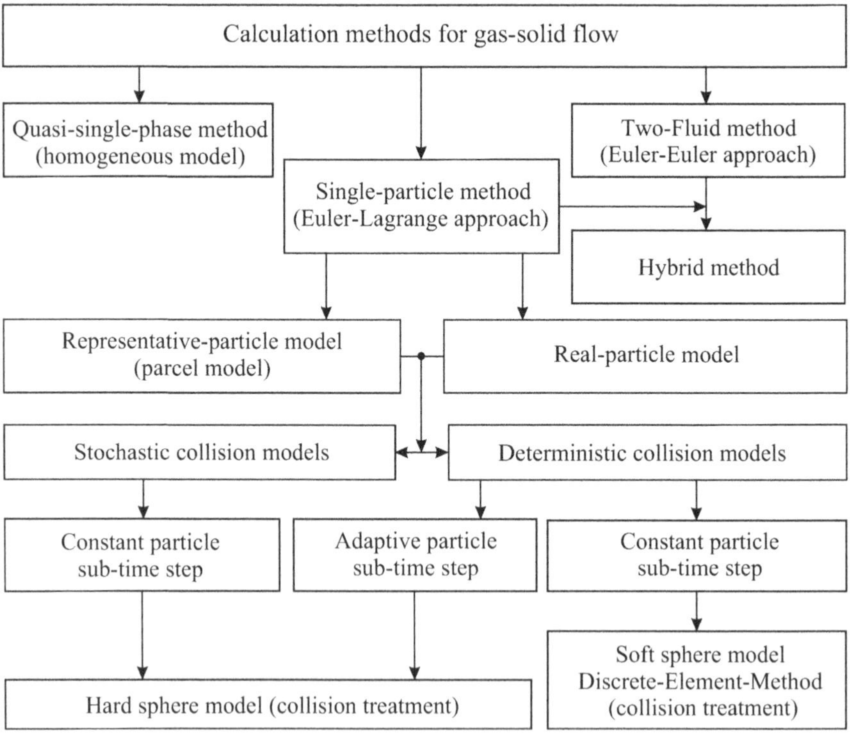

Fig. 3.7 Computation methods for gas-solid flow (Alobaid 2013)

mass, momentum and energy are solved, too. The impact of solid phase on fluid phase is considered by additional terms in the balance equations, which can be determined using the kinetic theory of dense gases. In the single-particle method (Euler-Lagrange approach), each particle in the solid phase is followed employing the Newtonian equations of motions, while the gas phase is described in Eulerian framework as a continuum. Here, the collisions between particle-particle and particle-wall can be detected using stochastic or deterministic collision models. The single-particle method is computationally expensive when tracking a few hundred thousand of particles. In order to reduce the computational cost and simulate gas-solid systems with millions of particles, the real particles are represented by so-called computational particles (also known as parcels). In the last group of models, the two-fluid method and the single-particle-method are combined, resulting in so-called hybrid method.

In the following sections, the computation methods for gas-solid flows, including quasi-single-phase, two-fluid, single-particle and hybrid methods are explained in detail. The modelling of turbulence and radiative heat transfer in gas-solid flows are not taken into account in this work and are only explained briefly below.

Compared to the single-phase flow, the modelling of the turbulent gas-solid flows is more complicated. Strong interactions take place between turbulence structures (eddy) and solid particles in addition to the interactions between the turbulence structures themselves. In this context, the turbulence is damped by small particles and strengthened by large particles. The simulation of turbulence is often neglected for flows with a solid content higher than 0.1% in each control volume (Elgobashi 2006). In practical applications such as circulating fluidized beds or pulverised coal-fired power plants, high velocities are present and/or there is simultaneous occurrence of zones with low volumetric void fractions of solid (in worst case, with a zero solid loading), which in turn means that the modelling of turbulence is of importance for the simulation of gas-solid flows. The influence of the solid phase on the turbulence and thus on the fluid velocity field can be summarised as:

- The particle movements in the flow lead to an additional disturbance (boundary layer separation and vortex formation around the particle), especially when the size of particles relatively big.
- Particles may mechanically influence the turbulence structures. The particles motions result in destroying the large-scale turbulence structures to smaller-scale ones (accelerating the dissipation of turbulent kinetic energy), especially when the size of particles and fluid turbulence elements are approximately the same.
- Disturbance in the flow field may arise due to particle-particle interactions.

The influence of the turbulent flow on the motion of the solid phase is also of relevance. Here, the temporal and spatial gradients of the fluid phase should be considered during the calculation of particle forces and moments of force through additional models. The turbulent dispersed flow including the turbulence-particle interactions are actual state of research and further experimental and theoretical investigations are required.

The radiative heat transfer of the solid phase has both receiving and emitting portions. If a radiant heat flux strikes a particle surface, the incident radiation, also called irradiation, is partly absorbed, while the remaining part is scattered out of the surface. The reflection, refraction, diffraction and transmission are high reliant on the optical properties of the particle and the particle size relative to the wavelength of the incident radiation (Fan and Zhu 2005). In the literature, the modelling of heat transfer through radiation in gas-solid flows is usually not considered because of the high complexity of the problem. Great efforts in terms of meaningful experimental and numerical investigations must therefore be carried out in order to gain a better understanding of radiative heat transfer between gas and solid phases.

3.2.1 Quasi-single-phase Method

The quasi-single-phase method assumes that a gas-solid flow is homogeneous and no difference in velocities between both phases is present. Under these conditions, the gas-solid flow can be treated as a quasi-single-phase flow. The conservation equations for mass, momentum and energy of a single-phase flow are applied to model the mixture physical values.

The continuity equation for the mixture is:

$$\frac{\partial \rho_m}{\partial t} + \nabla \cdot (\rho_m \vec{u}_m) = S_m. \tag{3.54}$$

The momentum equations for the mixture are:

$$\frac{\partial (\rho_m \vec{u}_m)}{\partial t} + \nabla \cdot (\rho_m \vec{u}_m \vec{u}_m) = \nabla \cdot (\mathbf{T}) - \nabla p + \rho_m \vec{g}. \tag{3.55}$$

The specific enthalpy equation for the mixture is expressed by:

$$\frac{\partial (\rho_m h_m)}{\partial t} + \nabla \cdot (\rho_m \vec{u}_m h_m) = -\nabla \cdot \left(\frac{\mu_m}{\mathrm{Pr}} \nabla h_m \right) + S_h. \tag{3.56}$$

The transport equation of species i is given as:

$$\frac{\partial (\rho_m c_{m,i})}{\partial t} + \nabla \cdot (\rho_m \vec{u}_m c_{m,i}) = -\nabla \cdot \left(\frac{\mu_m}{\mathrm{Sc}} \nabla c_{m,i} \right) + S_{c,i}. \tag{3.57}$$

In these conservation equations, $\vec{u}_m$, h_m and $c_{m,i}$ represent the mixture velocity, the mixture specific enthalpy and the concentration of the species i in the mixture, respectively. The source term S_m denotes injection or leakage of the mixture, while S_h and $S_{c,i}$ are heat and concentration source terms due to chemical reactions. The mixture density can be defined using the following relation:

$$\rho_m = \frac{\rho_s}{1 + \varepsilon_f\left(\frac{\rho_s}{\rho_f} - 1\right)}, \tag{3.58}$$

where ε_f is the volumetric void fraction of fluid phase. Although the quasi-single-phase method shows relevant advantages such as high computing stability and low computing cost, the obtained results are generally associated with a low level of accuracy. Furthermore, the configurations of practical gas-solid systems do not fulfil the method criterions in most of engineering applications.

3.2.2 Two-Fluid Method

In the two-fluid method, also known as Euler-Euler approach or multi-fluid method, the fluid and solid phases are modelled with the balance equations of the single-phase flow, where the solid phase is considered as a second heavy gas phase. Here, the continuity, the Navier-Stokes and the energy equations of both phases are extended. The modified equations contain transport values of dispersed phase such as granular pressure, granular viscosity and granular stress. The description of these values is based on the concept of kinetic theory of granular fluids (KTGF) that allows for a given solid to own all physical gas properties. The KTGF theory is described in detail by Gidaspow (1994), which is related to the kinetic theory of dense gases (KTDG) (Chapman and Cowling 1970). The basic idea behind KTGF is: the particles are in continuous and chaotic motion due to the interaction within the fluid phase for dilute system and with the particles/walls or both for dense flows. A temperature, known as granular temperature, is defined that is a measure of particle velocity fluctuations. The granular viscosity, the granular pressure and the granular stress are a function of granular temperature that varies in time and position within the entire computational domain. The modelling of the granular temperature can be achieved using an additional balance equation, similar to the energy equation of the fluid phase.

The governing equations of the two-fluid model available in ANSYS-FLUENT are here briefly explained. The gas and solid phases are modelled mathematically as fully interpenetrating continua by introducing the concepts of the solid viscosity and the solid pressure. The mass conservation equation for the phase q is written as:

$$\frac{\partial\left(\varepsilon_q\rho_q\right)}{\partial t} + \nabla\cdot\left(\varepsilon_q\rho_q\vec{u}_q\right) = \sum_{p=1}^{n}\left(\dot{m}_{pq} - \dot{m}_{qp}\right). \tag{3.59}$$

Here, ε_q is the phasic volume fraction, which represents the space occupied by each phase. The symbol n is the number of phases, ρ_q and $\vec{u}_q$ denote the density and the velocity of phase q. The term $\dot{m}_{pq}$ describes the inter-phase mass transfer from phase p to phase q, while the term $\dot{m}_{qp}$ characterises the inter-phase mass transfer from phase q to phase p.

The momentum conservation equation for the phase q is formulated as:

$$\frac{\partial\left(\varepsilon_q \rho_q \vec{u}_q\right)}{\partial t}+\nabla\cdot\left(\varepsilon_q \rho_q \vec{u}_q \vec{u}_q\right)=-\varepsilon_q \nabla p+\nabla\cdot\left(\mathbf{T}_q\right) \\ +\varepsilon_q\left(\vec{F}_{ext,q}+\vec{F}_{lift,q}+\vec{F}_{vm,q}+\vec{F}_{Bas,q}\right)+\sum_{p=1}^{n}\left(\vec{R}_{pq}+\dot{m}_{pq}\vec{u}_{pq}-\dot{m}_{qp}\vec{u}_{qp}\right), \tag{3.60}$$

where $\vec{F}_{ext,q}$, $\vec{F}_{lift,q}$, $\vec{F}_{vm,q}$ and $\vec{F}_{Bas,q}$ represent external body force per volume, lift force per volume (Saffman force), virtual mass and Basset forces per volume, respectively. The term $\vec{R}_{pq}$ describes the inter-phase force between phases (drag force per volume), while $\vec{u}_{pq}$ and $\vec{u}_{qp}$ stand for the inter-phase velocities, which are defined as $\vec{u}_p$ or $\vec{u}_q$ depending on the inter-phase mass transfer. The stress-strain tensor $\mathbf{T}_q$ is determined using Eq. (3.4). The momentum conservation equations must be closed by appropriate expressions for the inter-phase forces per volume $\vec{R}_{pq}$ and $\vec{R}_{qp}$. Generally, simple interaction terms can be applied:

$$\vec{R}_{pq}=K_{pq}\left(\vec{u}_p-\vec{u}_q\right), \tag{3.61}$$

$$\vec{R}_{qp}=K_{qp}\left(\vec{u}_q-\vec{u}_p\right), \tag{3.62}$$

with the inter-phase momentum exchange coefficient $K_{pq}=K_{qp}=\beta$. Different drag models for determining the momentum exchange coefficient of gas-solid flows are proposed in the literature. The general equation, describing the inter-phase momentum transfer coefficient, is expressed by:

$$\beta=\frac{\left(1-\varepsilon_f\right)\rho_s D_{dra}}{\tau_s}, \tag{3.63}$$

where ρ_s is the density of solid. The relaxation time of particle τ_s is given by:

$$\tau_s=\frac{\rho_s d_s^2}{18\mu_f}. \tag{3.64}$$

The symbol d_s denotes the diameter of particle and μ_f is the fluid dynamic viscosity. The drag function D_{dra} can be defined by applying different drag models such as Gidaspow et al. (1992), Syamlal and O'Brien (1989), Wen and Yu (1966). In the model proposed by Syamlal and O'Brien (1989), the drag function is written as:

$$D_{dra}=C_{dra}\frac{1}{24}\frac{\mathrm{Re}_s\varepsilon_f}{\left(v_s^{ter}\right)^2}. \tag{3.65}$$

The drag coefficient C_{dra} can be defined according to Dalla Valle equation (Dalla Valle 1948) as:

$$C_{dra} = \left(0.63 + \frac{4.8}{\sqrt{\mathrm{Re}_s / v_s^{ter}}}\right)^2. \tag{3.66}$$

The term v_s^{ter} stands for the terminal velocity correlation for the solid phase, given with the following expression:

$$v_s^{ter} = \begin{cases} 0.5\left(\varepsilon_f^{4.14} - 0.06\mathrm{Re}_s + \sqrt{(0.06\mathrm{Re}_s)^2 + 0.12\mathrm{Re}_s\left(1.6\varepsilon_f^{1.28} - \varepsilon_f^{4.14}\right) + \varepsilon_f^{8.28}}\right) & \text{case } \varepsilon_f \leq 0.85 \\ 0.5\left(\varepsilon_f^{4.14} - 0.06\mathrm{Re}_s + \sqrt{(0.06\mathrm{Re}_s)^2 + 0.12\mathrm{Re}_s\left(2\varepsilon_f^{2.65} - \varepsilon_f^{4.14}\right) + \varepsilon_f^{8.28}}\right) & \text{case } \varepsilon_f > 0.85 \end{cases}, \tag{3.67}$$

with the solids Reynolds number Re_s:

$$\mathrm{Re}_s = \frac{\rho_f d_s \left|\left(\vec{u}_s - \vec{u}_f\right)\right|}{\mu_f}. \tag{3.68}$$

The resulting inter-phase momentum transfer coefficient can be written as:

$$\beta_{Sya} = \frac{3}{4} C_{dra} \frac{\rho_f \varepsilon_f (1 - \varepsilon_f)}{d_s \left(v_s^{ter}\right)^2} \left(\frac{\mathrm{Re}_s}{v_s^{ter}}\right) \left|\left(\vec{u}_s - \vec{u}_f\right)\right|. \tag{3.69}$$

In the model of Wen and Yu (1966) that is appropriate for dilute systems, the inter-phase momentum transfer coefficient has the following form:

$$\beta_{Wen} = \frac{3}{4} C_{dra} \frac{\rho_f \varepsilon_f^{-1.65} (1 - \varepsilon_f)}{d_s} \left|\left(\vec{u}_s - \vec{u}_f\right)\right|. \tag{3.70}$$

In Eq. (3.70), the drag coefficient is computed as:

$$C_{dra} = \frac{24}{\varepsilon_f \mathrm{Re}_s} \left(1 + 0.15\left(\varepsilon_f \mathrm{Re}_s\right)^{0.687}\right). \tag{3.71}$$

Gidaspow model (Gidaspow et al. 1992) that combines the model of Wen and Yu and the model of Ergun is recommended for dense systems. For higher volumetric void fraction $\varepsilon_f \geq 0.8$, the inter-phase momentum transfer coefficient is computed from the model of Wen and Yu (see Eq. 3.70). For $\varepsilon_f < 0.8$, the inter-phase momentum transfer coefficient is calculated by Ergun's equation for monodisperse particles:

$$\beta_{Erg} = 150 C_{dra} \frac{\mu_f (1-\varepsilon_f)^2}{\varepsilon_f d_s^2} + 1.75 \frac{\rho_f (1-\varepsilon_f) \left| (\vec{u}_s - \vec{u}_f) \right|}{d_s}. \tag{3.72}$$

The conventional drag models such as Wen and Yu, Syamlal and O'Brien as well as Gidaspow assume a uniform particle distribution (homogeneous gas-solid flow) and thus calculate the inter-phase momentum transfer coefficient without considering the different flow structures within the control volume. These drag models were proved to overestimate the momentum exchange between gas and solid phases (Li and Kwauk 2003). Therefore, several modifications on the conventional drag models have been performed. For example, a three-zone or a four-zone drag model calculates the inter-phase momentum transfer coefficient by selecting universal drag laws for each zone to match the experiment data (Gao et al. 2012). In addition, advanced drag models that are based on flow structures are also developed such as the energy minimization multi-scale (EMMS) model (Li and Kwauk 2003), showing more accurate results regarding the modelling of circulating fluidized beds (Atsonios et al. 2011; Zeneli et al. 2015).

In order to balance the energy of the phase q, a transport equation for the specific enthalpy is solved:

$$\frac{\partial(\varepsilon_q \rho_q h_q)}{\partial t} + \nabla \cdot (\varepsilon_q \rho_q \vec{u}_q h_q) = \varepsilon_q \frac{\partial p}{\partial t} + \nabla \cdot \left(\frac{\varepsilon_q \mu_q}{\mathrm{Pr}} \nabla h_q \right) + S_h. \tag{3.73}$$

For each chemical component i, a substance transport equation is also solved:

$$\frac{\partial(\varepsilon_q \rho_q c_{q,i})}{\partial t} + \nabla \cdot (\varepsilon_q \rho_q \vec{u}_q c_{q,i}) = \nabla \cdot \left(\frac{\varepsilon_q \mu_q}{\mathrm{Sc}} \nabla c_{q,i} \right) + S_i. \tag{3.74}$$

In energy and species transport equations, the dimensionless quantities Pr and Sc are Prandtl and Schmidt numbers. The source term S_i takes into account the formation/reduction of component i by the reason of chemical reactions. The heat source term S_h considers the heat released by reactions, the radiation source term of the phase q, the heat transfer between two phases by convection, radiation and mass transfer due to the phase transformation from phase p to phase q. The temperature of the phase q is then calculated from the specific enthalpy and the mean specific heat capacity.

In the two-fluid method, the modelling of the particle motion is based on the kinetic theory of granular flow. In this theory, the gas-particle drag and the non-ideal particle-particle collisions are considered. Furthermore, particles are allowed to travel freely and to collide with neighbouring particles. Here, the granular temperature Θ_s is introduced, representing the solid fluctuating energy:

$$\frac{3}{2}\left(\frac{\partial}{\partial t}\left((1-\varepsilon_f)\rho_s\Theta_s\right)+\nabla\cdot\left((1-\varepsilon_f)\rho_s\Theta_s\vec{u}_s\right)\right)$$
$$=\underbrace{(-p_s\mathbf{I}+\mathbf{T}_s):\nabla\vec{u}_s}_{1}+\underbrace{\nabla\cdot(\Gamma_{\Theta_s}\nabla\Theta_s)}_{2}-\underbrace{\gamma_{\Theta_s}}_{3}+\underbrace{\phi_{ls}}_{4}. \tag{3.75}$$

In Eq. (3.75), the term 1 describes the energy generation by the solid stress tensor, the term 2 stands for the diffusion of energy, the term 3 represents the rate of energy dissipation within the solid phase due to inelastic collisions and the term 4 denotes the energy exchange between the lth fluid or solid phase and the sth solid phase. The diffusion coefficient for granular energy Γ_{Θ_s} is given by the expression derived by Gidaspow et al. (1992):

$$\Gamma_{\Theta_s}=\frac{150\rho_s d_s\sqrt{\Theta_s\pi}}{384 g_{0,ss}(1+e_{ss})}\left(1+\frac{6}{5}g_{ss}^{rad}(1-\varepsilon_f)(1+e_{ss})\right)^2$$
$$+2\rho_s d_s g_{ss}^{rad}\sqrt{\frac{\Theta_s}{\pi}}(1-\varepsilon_f)^2(1+e_{ss}). \tag{3.76}$$

The rate of energy dissipation within the solid phase γ_{Θ_s} and the energy exchange between phases ϕ_{ls} can be written according to Lun et al. (1984) as:

$$\gamma_{\Theta_s}=\frac{12\rho_s g_{ss}^{rad}(1-e_{ss}^2)(1-\varepsilon_f)^2\Theta_s^{3/2}}{d_s\sqrt{\pi}}, \tag{3.77}$$

$$\phi_{ls}=-3\beta\Theta_s, \tag{3.78}$$

with the particle-particle restitution coefficient e_{ss} and the radial distribution function g_{ss}^{rad} that is a correction factor modifying the probability of collisions between particles for dense systems.

The solid pressure is composed of a kinetic term and a second term due to particle collisions. It can be expressed using the following equation:

$$p_s=\underbrace{\rho_s(1-\varepsilon_f)\Theta_s}_{Kinetic}+\underbrace{2\rho_s(1+e_{ss})(1-\varepsilon_f)^2 g_{ss}^{rad}\Theta_s}_{Particle\ collisions}. \tag{3.79}$$

In order to calculate the stress-strain tensor of solid phase Eq. (3.4), the shear and bulk viscosities should be determined. The granular viscosity (solid shear viscosity) is a summation of three contributions (collisional, kinetic and frictional parts):

$$\mu_s=\mu_s^{col}+\mu_s^{kin}+\mu_s^{fri}. \tag{3.80}$$

The collisional and kinetic viscosities can be determined according to Gidaspow et al. (1992), while the frictional viscosity may be computed according to Schaeffer (1987):

$$\mu_s^{col} = \frac{4}{5}\rho_s d_s (1 + e_{ss})(1 - \varepsilon_f)^2 g_{ss}^{rad} \sqrt{\frac{\Theta_s}{\pi}}, \tag{3.81}$$

$$\mu_s^{kin} = \frac{10\rho_s d_s \sqrt{\Theta_s \pi}}{96(1 - \varepsilon_f)(1 + e_{ss}) g_{ss}^{rad}} \left(1 + \frac{4}{5} g_{ss}^{rad} (1 - \varepsilon_f)(1 + e_{ss})\right)^2, \tag{3.82}$$

$$\mu_s^{fri} = \frac{p_s \sin\varphi}{2\sqrt{\mathbf{I}_{2D}}}. \tag{3.83}$$

The angle φ represents the internal friction angle and the symbol $\mathbf{I}_{2D}$ is the second invariant of the deviatoric stress tensor.

The solid bulk viscosity can be expressed according to Lun et al. (1984) by:

$$\lambda_s = \frac{4}{3}\rho_s d_s (1 + e_{ss})(1 - \varepsilon_f) g_{ss}^{rad} \sqrt{\frac{\Theta_s}{\pi}}. \tag{3.84}$$

At walls, the tangential component of the solid velocity and the granular temperature have to be defined, considering the particle-wall restitution coefficient and the specularity coefficient. The particle-wall coefficient stands for the turbulent kinetic energy of the solid, while the specularity coefficient is used to quantify the nature of particle-wall collisions. The specularity coefficient that is an empirical parameter characterises the friction between the particles and the wall. The value of the specularity coefficient varies between 0 and 1. A value of zero implies a specular collision (free-slip conditions) and a value of unity corresponds to diffuse collision (no-slip conditions).

The valid application of the two-fluid method to gas-solid flows is associated with the satisfaction of continuum hypothesis, i.e. the number of particles and fluid molecules in each control volume is sufficiently large. If this criterion fulfils, the physical variables of each phase (gas or solid) can then be modelled in every control volume as statistical average values. With the aid of Knudsen number Kn, which is defined as the ratio of mean free path $\bar{l}$ to characteristic length l_{char}, the continuum hypothesis is checked for its validity:

$$Kn = \frac{\bar{l}}{l_{char}}. \tag{3.85}$$

The mean free path represents the average distance that can cover by a moving particle or gas molecular till the next collision with other object occurs, leading to modify the physical properties of particles such as position, velocity and temperature. The characteristic length is a function of the grid cell volume:

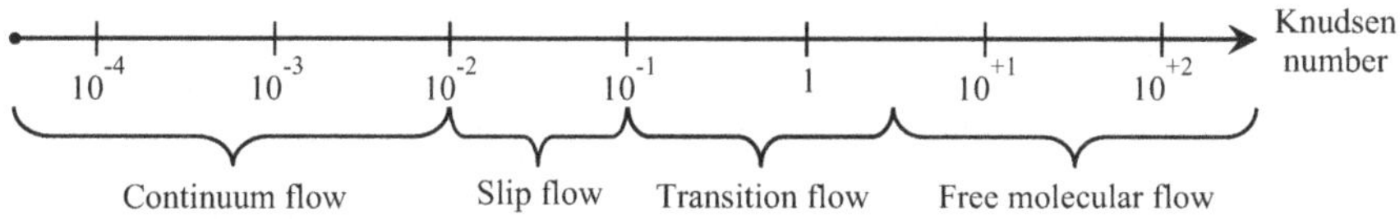

Fig. 3.8 Gas-solid flow regimes as a function of Knudsen number

$$l_{char} = \left(V^{CV}\right)^{1/3}. \tag{3.86}$$

According to the definition of Knudsen number, the Knudsen domain $0 < Kn < \infty$ is often divided into four regimes, namely continuum flow, slip flow, transient flow and free molecular flow (see Fig. 3.8).

Flows with smaller Knudsen numbers $kn \leq 10^{-2}$ is considered to be continuum, while flows with higher Knudsen numbers $kn \geq 3$ is seen as free molecular flows. For slip and transient regimes $0.01 < kn < 3$, the gas-solid flow can be characterised as continuum or as free molecular flows.

It is clear that flows with high solid content have smaller average free paths and thus result in Knudsen numbers less than 0.01. As a consequence of this, the two-fluid method is suitable for the simulation of the dense gas-solid flow. However, the solid distribution in the fluid phase shows in most of the gas-solid flow systems high measure of inhomogeneity, e.g. stationary and circulating fluidized beds. For strands and gas bubbles, in which the solid mass fraction is very low or the solid does not exist, the continuum assumption by inadequate spatial resolution is not fulfilled. Reducing the grid resolution in order to meet the continuum hypothesis leads on the one hand to inaccurate resolution of local inhomogeneities and on the other hand to imprecise numerical resolving of both phases. Other restriction of the two-fluid method application to dense gas-solid flow simulation is related to the kinetic theory of granular fluids. This theory can only consider a certain range of solid velocity fluctuations as a function of granular temperature. Götz (2006) and Kanther (2003) investigated the Maxwell-Boltzmann distribution of particle velocity in order to draw conclusions about the suitability of the two-fluid method for the dense fluid-solid flow simulation (e.g. fluidized bed). Employing a coupled CFD/DEM programme, Götz carried out a fluidized bed simulation, in which the distribution of particle velocity can be determined numerically. The resulted distributions show in large areas high deviations from Maxwell-Boltzmann distribution. The maximum deviations are noticed at the interface between gas bubbles and fluidized dispersed particles. In these boundary layers might be raised bimodal distribution functions that are not summarised under the KTGF theory. How the impact of these distribution functions on the quality of numerical simulations is still not clear and further examinations are required. Another difficulty is that the two-fluid model is generally applied to gas-solid flows, assuming that the solid phase is monodispersed, whereas in many applications it is characterised by a particle size distribution (PSD). For polydisperse flows, the formulation of the two-fluid method is complicated by the reason of solving separate conservation equations for each particle size or each particle shape. Here, the population balance equation (PBE) can be coupled to the continuity and momentum balance equations, but the consideration of different particle sizes leads to an increase in computing time. In contrast to the two-fluid method, the single-particle method offers an efficient approach for the modelling of granular flows with a wide range of particle size distribution, but the computational cost of collision detections between particles as well as particles and walls is very high.

3.2.3 Single-Particle Method

In the single-particle method, the solid phase is considered as dispersed phase, i.e. for each particle a set of transport equations is solved, while the gas phase is modelled using modified conservation equations of the single-phase flow (Alobaid 2013; Epple et al. 2012; Walter and Epple 2016).

The extended continuity equation of fluid phase can be expressed by:

$$\frac{\partial(\varepsilon_f \rho_f)}{\partial t_f} + \nabla \cdot (\varepsilon_f \rho_f \vec{u}_f) = S^{m,P}_{P\to f}. \tag{3.87}$$

The extended Navier-Stokes equations are given as follows:

$$\frac{\partial(\varepsilon_f \rho_f \vec{u}_f)}{\partial t_f} + \nabla \cdot (\varepsilon_f \rho_f \vec{u}_f \vec{u}_f) = \nabla \cdot (\varepsilon_f \mathbf{T}_f) - \varepsilon_f \nabla p + \varepsilon_f \rho_f \vec{g} + \vec{S}^{u,P}_{P\to f}, \tag{3.88}$$

where ρ_f is the gas density, $\vec{u}_f$ represents the fluid velocity, p and $\vec{g}$ are the static pressure and the standard gravity, respectively. The size of the fluid time step t_f can be arbitrarily selected and remains generally constant during the calculation (see Sect. 3.2.3.4). The source terms $S^{m,P}_{P\to f}$ and $S^{u,P}_{P\to f}$ on the right side of both continuity and momentum equations consider the mass and momentum transfers with the solid phase i.e. interaction between solid and fluid phases.

The extended transport equations for the specific enthalpy and for each chemical component i can be formulated as:

$$\frac{\partial(\varepsilon_f \rho_f h)}{\partial t_f} + \nabla \cdot (\varepsilon_f \rho_f \vec{u}_f h_f) = \varepsilon_f \frac{\partial p_f}{\partial t_f} + \nabla \cdot \left(\frac{\varepsilon_f \mu_f}{\mathrm{Pr}} \nabla h_f\right) + \underbrace{\left(S^{h,P}_{P\to f} + S^h_f\right)}_{S^h_{P\to f,f}}, \tag{3.89}$$

$$\frac{\partial(\varepsilon_f \rho_f c_{f,i})}{\partial t_f} + \nabla \cdot (\varepsilon_f \rho_f \vec{u}_f c_{f,i}) = \nabla \cdot \left(\frac{\varepsilon_f \mu_f}{\mathrm{Sc}} \nabla c_{f,i}\right) + S^{c,i}_f. \tag{3.90}$$

Here, the symbol μ_f is the dynamic viscosity, Pr and Sc are Prandtl and Schmidt numbers. The heat source term $S^h_{P\to f,f}$ consists of three parts: the heat released by the homogeneous gas phase reactions, the radiation source term of the fluid phase and the heat transfer between two phases by convection, radiation and mass transfer due to the phase transformation from solid to fluid phase. The fluid temperature is then calculated from the specific enthalpy and the mean specific heat capacity. The source term $S^{c,i}_f$ takes into account the formation/reduction of gaseous components by the reason of chemical reactions.

The simulation of the solid phase can be divided in two sequential steps. In the first step (the free flight phase), the simultaneous calculation of particle position and angular displacement changes for a particle-time step is taken place, which it is usually based on the explicit Euler method:

$$\vec{x}_P(t+\Delta t_P) = \vec{x}_P(t) + \int\limits_t^{t+\Delta t_P} \vec{u}_P(t)dt, \tag{3.91}$$

$$\vec{\varphi}_P(t+\Delta t_P) = \vec{\varphi}_P(t) + \int\limits_t^{t+\Delta t_P} \vec{w}_P(t)dt. \tag{3.92}$$

The changes of particle translational velocity $\vec{u}_P$ and angular velocity $\vec{w}_P$ for a particle time step Δt_P are calculated with the aid of forces $\sum_k \vec{F}_{P,k}(t)$ and moments of force $\sum_k \vec{M}_{P,k}(t)$:

$$\vec{u}_P(t+\Delta t_P) = \vec{u}_P(t) + \int\limits_t^{t+\Delta t_P} \vec{b}_P(t)dt, \tag{3.93}$$

$$\vec{w}_P(t+\Delta t_P) = \vec{w}_P(t) + \int\limits_t^{t+\Delta t_P} \vec{a}_P(t)dt, \tag{3.94}$$

with translation and angular accelerations:

$$\vec{b}_P(t) = \frac{1}{m_P}\sum_k \vec{F}_{P,k}(t), \tag{3.95}$$

$$a_P(t) = \frac{1}{I_P}\sum_k \vec{M}_{P,k}(t). \tag{3.96}$$

The symbol k represents the number of acting forces or moments of force on the investigated particle. The values m_P and I_P denote the particle mass and the moment of inertia, respectively.

In the second step, the unconsidered short-range forces between particles in the first step can now be calculated. The short-range forces (contact and adhesive forces) result from particle-particle and/or particle-wall collisions (see Sect. 3.2.3.2.2). The collision process is divided into collision detection and collision treatment, where the collision detection plays a special role for the efficiency of the entire algorithm. Generally, the collision detection between particle-particle/wall can be stochastically modelled or deterministically detected. For the collision treatment, two approaches are widely used: the hard sphere model and the soft sphere model. In the first model, single binary collisions are considered as instantaneous processes. The properties of particles after the collision are correlated to the properties of particles before the collision through momentum and energy balances. In the so-called soft sphere model, also known as discrete element method (DEM), the particles can overlap each other or penetrate into the wall. Depending on the penetration depth, a contact force is determined, resulting in changing the motions of particles.

The collision treatment in the stochastic method is only carried out with the hard sphere model, while the collision in the deterministic collision models can be treated with both hard and soft sphere models. Which model is more appropriate to describe the particle-particle/wall collisions, it depends on the average particle velocity and the solid loading in the gas-solid flow. At low particle velocities, as for example in bubbling or circulating fluidized beds, the hard sphere model causes significant difficulties in considering multiple simultaneous collisions of a particle with other collision partners. By contrast to the hard sphere model, the soft sphere model has the disadvantage that at high particle velocities relatively small particle time steps should be applied and thus the computational time further increases. Flows with high solid content require an accurate description of multiple simultaneous particle-particle/wall collisions and therefore the soft sphere model is preferred. The hard sphere model is more suitable for dilute gas-solid flows, where the simultaneous collisions occur less frequently.

In the literature, the distinction between dense and dilute gas-solid flows may follow different criteria. A distinguishing feature can be the mass flow ratio of solid to fluid, where flows with values greater than 15 (Marcus et al. 1990), 20 (Leva 1959) or 80 (Kunii and Levenspiel 1991) are regarded as dense. Another criterion is the volumetric void fraction, which describes the ratio of fluid volume to entire control volume. In this context, dense gas-solid flows have fluid volumetric void fraction values less than 0.9 (Epple et al. 2012). Crowe (1982) classified whether the gas-solid flow is dense or dilute by the influence of particle-particle/wall collisions on the particle trajectories. Similar to Knudsen number, Stokes number Stk is introduced, which is a measure of particle abilities to follow the surrounding flow streamlines. It is the ratio of particle relaxation time t_{relax} to characteristic time of fluid flow t_{char}:

$$Stk = \frac{t_{relax}}{t_{char}}. \tag{3.97}$$

The relaxation time is defined as:

$$t_{relax} = \frac{l_{stop}}{|\vec{u}_P|}. \tag{3.98}$$

The stopping distance l_{stop} is obtained when a particle with initial velocity $|\vec{u}_P|$ is released in a quiescent fluid and comes to rest after a certain time t_{relax}. Generally, the stopping distance is calculated under the assumption of laminar flow. The characteristic time is determined as:

$$t_{char} = \frac{l_{char}}{\left|\vec{\overline{u}}_f\right|}. \tag{3.99}$$

The parameters l_{char} and $\left|\vec{\overline{u}}_f\right|$ are the characteristic length of flow apparatus and the mean fluid velocity, respectively. Stokes numbers bigger than unity represent dense gas-solid flows, i.e. the particle trajectories deviate from the flow streamlines, while Stokes numbers smaller than unity denote to dilute flows, i.e. the particle trajectories are strongly oriented to flow lines.

3.2.3.1 Collision Detection Models

For gas-solid flows with Stokes numbers bigger than unity, the short-range forces that rise between particle-particle and particle-wall such as the contact force have a relevant influence on the obtained results. The considering of the short-range forces requires a specific description of collision process, which is divided in collision detection and collision treatment. Basically, the stochastic or deterministic collision model can be applied in the frame of the single-particle method to detect the particle-particle/wall collisions (see Fig. 3.7).

Stochastic Collision Detection Models

Because of the stochastic nature of collision detections, the stochastic collision models are often grouped under the so-called Monte Carlo methods. The earliest research in this field is carried out by O'Rourke (1981), who developed a collision detection model based on a stochastic method. In this model, the collision probability is determined for each pair of particles in the grid cell using a Poisson distribution. The computation effort of O'Rourke model is proportional to the square of particle number. A linear dependence between the computation effort and the particle number has been reached by Oesterlé et al. (Oesterle and Petitjean 1993). The idea behind Oesterlé model is: the trajectories of particles are sequentially determined. In this context, sequentially means that the motion of each particle remains uninfluenced by the presence of remaining particles. Nevertheless, information on other particles should be available, which enables the calculation of the collision between the observed particle and their partners. For this purpose, mean values as well as standard deviations of certain variables such as particle velocity and particle position are determined in each grid cell. The local statistical data serves as a basis for generating virtual collision partners whose properties are given by the local average values of neighbouring particles of the investigated particle. The probability of a collision between the real particle and the virtual particles can be met by means of random numbers. If the collision event occurs, the subsequent collision treatment is then performed by the hard sphere model. The physical properties of real particle after the collision are calculated using the properties of real and virtually particles before the collision. While the obtained velocity of the investigated particle is stored in the memory, the virtual particle is eliminated. If the collision event does not taken place, the velocity of the real particle remains then unaltered. The major disadvantage of Oesterlé collision model is that the kinetic energy conservation cannot be guaranteed even during completely elastic collisions. The extension of Oesterlé model by Sommerfeld (1996), in which the local velocity fluctuations of the neighbouring particles are considered using additional partial correlations, cannot eliminate the model weakness, too (Hußmann 2009). In contrast to Oesterlé model, the modified Babovsky-Nanbu model (Babovsky 1989; Nanbu 1980) ensures on the one hand the conservation of the kinetic energy and has on the other hand a linear computational effort to the number of particles. Although the stochastic collision models can efficiently simulate the granular flows with a wide range of particle size distribution, the collision frequency is extremely high for practical applications with high solid content in the fluid phase. Nevertheless, the computational cost for the collision detections is still less than the deterministic collision models.

Deterministic Collision Detection Models

The deterministic collision methods detect each single particle for possible collision with other particles and walls. The particle time step can be selected constant or adaptive in opposite to the stochastic collision models, where the particle time step is assumed to be constant (see Fig. 3.7).

In the adaptive deterministic collision model, the particle time step is varied during the simulation. No collision will occur in the period of time between t and $t + \Delta t_P$, but only at time $t + \Delta t_P$ one particle collides with other particle or wall. The size of the particle time step is determined by fixing the intersection points between the particle trajectories and the wall (see Sect. 3.2.3.4). This event-oriented method was presented for the first time by Hoomans et al. (1996). If the collision partner moves away from each other as it will not collide, the scalar product of relative velocity $\vec{u}_{ij} = \vec{u}_i - \vec{u}_j$ and position vector $\vec{r}_{ij} = \vec{r}_i - \vec{r}_j$ has then a positive value (see Fig. 3.9a). A collision takes place only when the collision partners move towards each other, i.e. the scalar product of the relative velocity and the position vector has a negative value (see Fig. 3.9b, c).

The computation effort of the deterministic collision detection model with a variable particle time step behaves quadratic to number of particles. The particle properties after the collision is subsequently determined with the hard sphere model (see Sect. 3.2.3.2.1). The use of variable particle time step is characterised by high simulation accuracy since the collision detections between partners are described in detail. The event-oriented method can be efficiently applied to dilute fluid-solid flows, in which the mean free path of particles is relatively large, resulting in coarser particle time steps. In dense fluid-solid flows, by contrast, the efficiency advantage cannot be achieved due to the smaller mean free paths and thus the smaller particle time steps are required. In case of parallel computing, the event-oriented method may result in efficiency loss due to the asynchronous parallel load between the computer nodes (Götz 2006).

In deterministic collision detection model with constant particle time step, the particles can overlap each other or penetrate into the wall (see Fig. 3.10a, b).

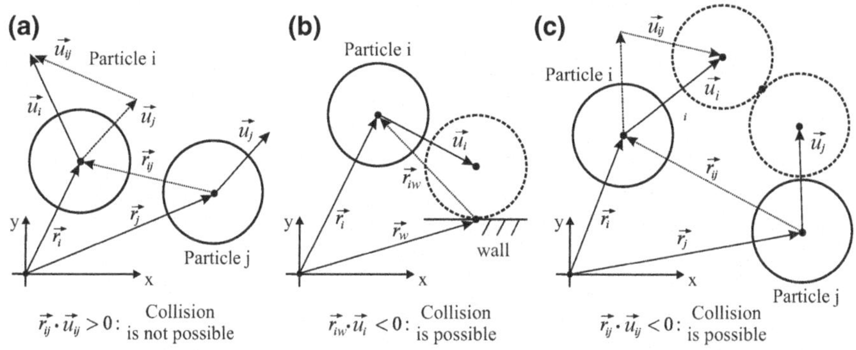

Fig. 3.9 Collision occurrence in event-oriented method: **a** the collision is not possible, **b** the particle-wall collision is possible and **c** the particle-particle collision is possible

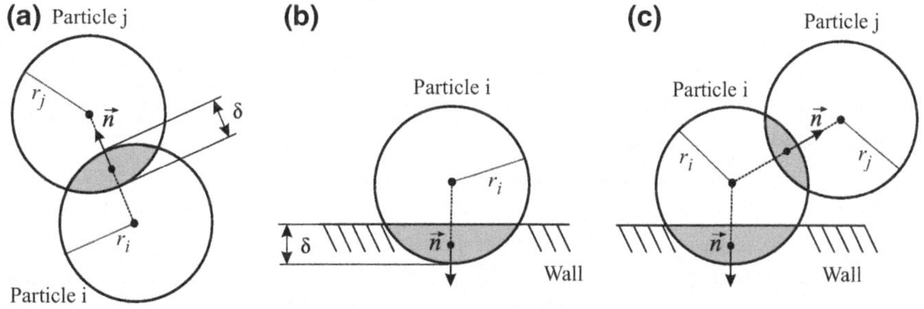

Fig. 3.10 Collision occurrence in the DEM model: **a** a particle-particle collision, **b** a particle-wall collision and **c** particle-particle/wall collisions

Depending on the penetration depth δ, a contact force is modelled using a spring-damper-slider system. This approach has been introduced first by Cundall and Strack (1979) and in the literature is often referred to as discrete element method (DEM). Due to the penetration behaviour both collision partners, this model is also known as soft sphere model. Although DEM dated back originally to Cundall and Strack, Tsuji et al. (1992, 1993) are often citied since they have been successfully employed the DEM model to simulate the hydrodynamic behaviour of the dense gas-solid flow in a fluidized bed. One of the most important advantages related to the DEM model is the possibility to calculate simultaneously the multiple collisions between particles-particles and particles-walls (see Fig. 3.10c). In addition to contact forces, other short-range forces such as adhesive forces can also be determined in frame of the DEM model. In order to achieve that, it is necessary first to detect the collision between particles using a specific criterion. For the contact criterion as an example, the distance between the centre points of two particles or the distance between the particle centre point and the wall surface should be smaller than the sum of both radii or smaller than the particle radius, respectively. Another example for the short-range forces is the criterion of liquid bridge formation, where the distance between the centre points of two particles has to be smaller than the critical distance of liquid bridge formation.

The collision detection in the DEM model is based on a simple principle: each particle is tested for possible collision events with other particles or walls. Similar to the deterministic detection model with variable particle time step, the computational effort of the DEM model is a quadratic function to the number of particles. Here, there is a substantial difference between one-dimensional, two-dimensional and three-dimensional cases. In 1D case, only the particles on left and right of the investigated particle are eligible as collision candidates. Furthermore, the neighbouring particles remain unchanged during the calculation. In 2D and 3D cases, by contrast, the number of possible collision partners is significantly higher and the continuous changing of neighbouring particles leads to extra computation time. In the literature, different algorithms to determine the collision between particle-particle/wall in granular flows can be found. The most applied deterministic collision algorithms, described below, are standard, neighbourhood and bounding-boxes methods.

Standard method: The standard method presents the simplest algorithms to examine the possible collisions between particles. The collision detection of spherical particles runs relatively trivial in this method. Here, the investigated particle is tested for possible collisions with all remained particles and walls in the computational domain according to a predefined criterion. The collision probabilities between particles are indirectly given by the corresponding particle spacing. In concrete term, this means that the algorithm computes the collision probabilities between the tested particle and all other particles regardless of their relative positions. Distant particles, even their contact probabilities are clearly impossible, are examined for possible contact with the investigated particle (see Fig. 3.11).

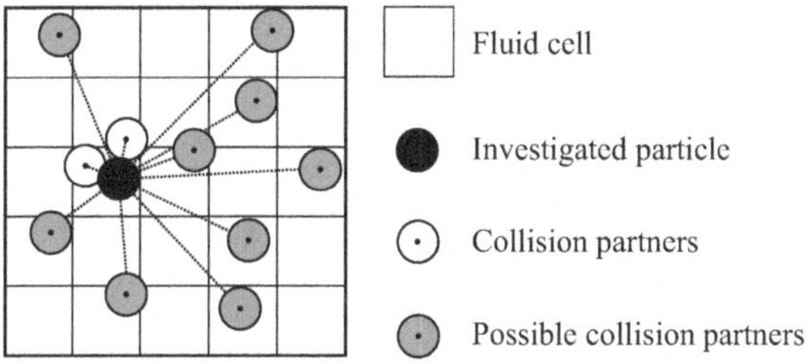

Fig. 3.11 Collision detection process in the standard method

Since the collision detections between particles are computed independent of the corresponding particle positions, the computational effort behaves approximately quadratic to the number of particles:

$$T_{tot} = T_{col}(N-1)N \approx T_{col}N^2. \tag{3.100}$$

The symbol T_{col} denotes the computing time that is required for a single collision detection. Although the programming and the implementation complexity of the standard method is comparatively low, the computational time for the collisions detections is enormous, even at low solid content in the fluid phase.

Neighbourhood method: The neighbourhood method, in contrast to the standard method, shows a significant improvement regarding the collision detection efficiency. Here, the collision detection consists of two consecutive steps. The searching for possible collision partners is achieved in the first step, while in the second step the accurate collisions between the investigated particle and their potential candidates are detected. The well-known and frequently applied neighbourhood methods are the Verlet algorithm and the search-grid method.

In Verlet algorithm (Allen and Tildesley 1989), a circular (2D) or spherical (3D) collision region that includes the possible collision partners is defined for each particle (see Fig. 3.12).

The radius of the collision region $R_{i,\max}$ depends on the selected fluid time step as well as on the properties of the observed particle and their adjacent particles. According to Deen et al. (2007), Hoomans et al. (1996), the collision region radius $R_{i,\max}$ can be determined using the following equation:

$$R_{i,\max} = r_i + r_{P,\max} + 2\left|\vec{u}_{P,\max}\right|\Delta t_f, \tag{3.101}$$

with:

$$r_{P,\max} = \max(r_i) \quad \forall i \in A, \tag{3.102}$$

$$\left|\vec{u}_{P,\max}\right| = \max(|\vec{u}_i|) \quad \forall i \in A. \tag{3.103}$$

The list $A = \{0, 1, \ldots, N\}$ contains all particles and walls of the computational domain. The distances between the investigated particle i and all collision partners are computed and compared with the radius $R_{i,\max}$. The particles including walls

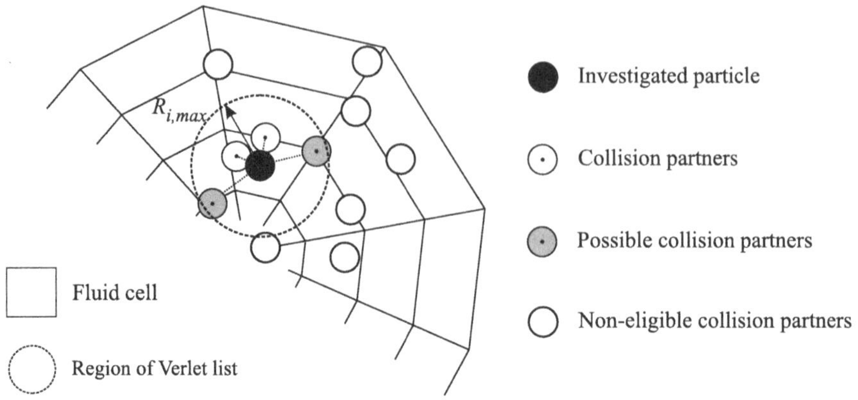

Fig. 3.12 Relationship between particles in Verlet algorithm

that are within the collision area of the particle i will be stored in its neighbour list (Verlet list). The exact collision detection can then be performed after building all neighbour lists, where each considered particle is only tested with its potential collision partners in the Verlet list. Similar to the standard method, the computation effort of the Verlet algorithm is a quadratic function to the number of particles by the reason of the building of the particle neighbour lists. In practice, the neighbour lists should not be updated for each time step. The actualisation of the particle i neighbour list is performed only when a particle leaves its collision area. This requires that the distance between the particle i and the particles in its neighbour list at every time step to be computed and compared with the radius $R_{i,\max}$. If one or more particles left the Verlet list of the particle i, then a new list should be rebuilt. In dilute gas-solid flows, in which the neighbour lists can be updated once in a while, the Verlet algorithm can be employed. This efficiency advantage is no longer valid for the dense gas-solid flows due to the fact that the neighbour lists may be actualised for each time step.

In the search-grid method after Hockney and Eastwood (1988), the computational domain is divided into finite cells, which contain indexed particles. A particle may come in contact only with particles of its own cell or with those of its direct neighbouring cells. By neglecting any further collisions with particles that are located in far cells, the computational time is reduced. In Fig. 3.13, a particle in cell 12 (red), its potential collision partners in cells 12, 6, 7, 8, 11, 13, 16, 17, 18 (white and grey) and its non-eligible collision partners in remaining cells (black) are illustrated.

In contrast to the Verlet algorithm, the search-grid method can be applied to dilute as well as dense gas-solid flows without efficiency loss. The application of the search-grid method necessitates a virtual orthogonal grid, which is placed over the entire computational domain and is independent of the computational grid. The computational complexity of the search-grid method depends on the total number of particles N in the computational domain as well as on the particle numbers in the

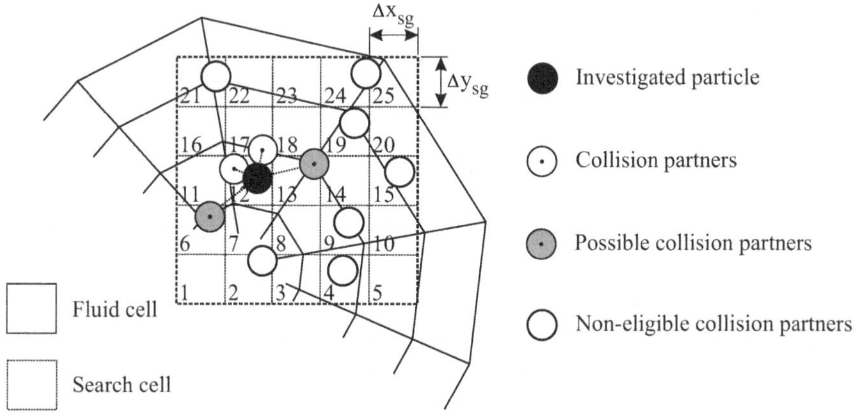

Fig. 3.13 Relationship between particles in the search-grid method

search-grid cell N_{sgc}. It consists of two parts: the computing time T_{loc} that is required for the allocation of each individual particle in the search cells and the computing time T_{col} that is needed for the collision detection between the investigated particle and its potential collision partners:

$$T_{tot} = N\left(T_{loc} + D \cdot N_{sgc} T_{col}\right). \tag{3.104}$$

The constant D relates to the computational domain dimension. For 1D case, it has the value 3, while it corresponds to values 9 and 27 for 2D and 3D cases, respectively. The computing time T_{loc} for allocating the particles in the orthogonal cells is much smaller than the required time for the collision detection T_{col} (Götz 2006). For a monodisperse system with uniform particle distribution, the computational effort in ideal case is almost proportional to the number of particles. This can occur if the size of the search-grid cell is selected equal to the particle diameter and in each search cell one particle is allocated. Equation (3.104) can then be reduced to:

$$T_{tot} = N\left(\underbrace{T_{loc}}_{very\ small \approx 0} + D \cdot \underbrace{N_{sgc}}_{=1} T_{col}\right) \approx N\left(\underbrace{D \cdot T_{col}}_{Constant}\right) \sim N. \tag{3.105}$$

Since the computing time is a function of the search cell size $(\Delta x_{sg}, \Delta y_{sg})$, the correct selection of the search-grid size is of crucial importance. If the size of the search-grid is selected equal to the particle diameter for monodisperse system, the computational effort behaves almost proportional to the number of particles. A further reduction of the search-grid size beyond the particle diameter results in a reasonable reduction in the computing time, but this can lead to undetected contacts between the particles. For example, in Fig. 3.14, the contact between particles i and j

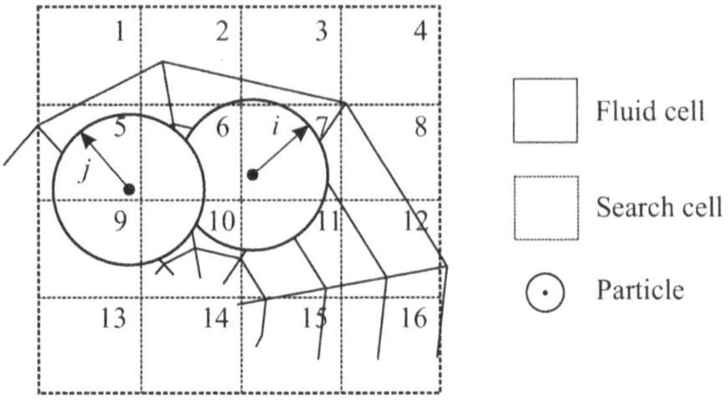

Fig. 3.14 Undetected contacts between particles by selecting a search-grid size smaller than the particle diameter

is not detected. The centre of mass of the particle *i* is located in the search-grid cell 7, so that the particle *i* can only collide with particles that their centre of masses are in the search-grid cells 2, 3, 4, 6, 8, 10, 11 and 12, but not in the search-grid cell 5.

The computing time dependence on the size of the search cell and accordingly on the particle diameter is the major weakness of this procedure. For a polydisperse system with a broad range of particle size distribution, the determination of the search-grid size is correlated to the largest particle (see Fig. 3.15). As a consequence, the average particle number per search cell increases and thus the computational effort rises.

Bounding-boxes method: In the bounding-boxes method (Baraff 1995; Cohen et al. 1995), each particle is enclosed with a cuboid whose edges are aligned parallel to the coordinate axes. The enveloping cuboid is identified by the positions of its projected four vertices in 2D or six vertices in 3D cases on the correspond axes, respectively. For each axis, a particle list is generated, in which the vertices of all cuboids and thus the geometrical properties of particles are stored. Similar to neighbourhood methods, the collision detection consists of two consecutive steps. A possible collision between two particles will occur, once their cuboids vertices overlap in the x-, y-and z-directions. If this happens, an addition condition should

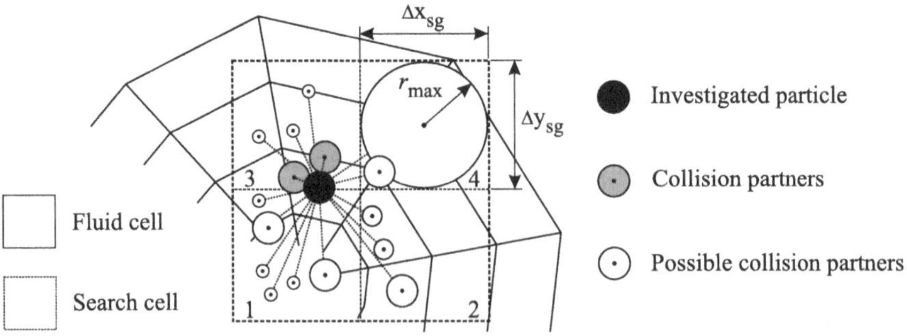

Fig. 3.15 Determination the size of the search-grid cell in polydisperse system

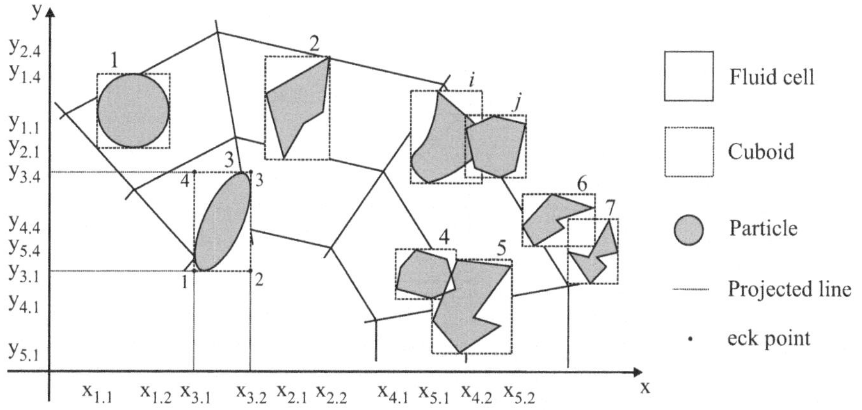

Fig. 3.16 Collision detection process in the bounding-boxes algorithm under the consideration of sorting method

be verified in order to specify, which surfaces of the colliding partners have come into contact (see Fig. 3.16: particles 4 and 5). Here, it should be mentioned that a collision can be detected between cuboids, but no collision takes place on particle level (see Fig. 3.16: particles 6 and 7).

The bounding-boxes method is characterised by two relevant advantages. The non-spherical particles can be treated in this method with a similar computational effort as spherical particles. Furthermore, the particle size distribution has no impact on the efficiency of procedure. This is due to the fact that each particle is enclosed with a suitable cuboid independent of its shape or geometry. The computational complexity of the bounding-boxes method behaves quadratic to particle number. However, the computational effort can significantly be reduced by a special sort and insertion methods, so-called incremental sort and update algorithm (Schinner 1999). Here, the cuboids vertices are sorted in the particle lists (x-, y-and z-directions) in ascending order for each time step. The sorting process permutes the elements within a list based on simple mathematical methods. According to Sedgewick (1992), the required time for the sorting process has the order of $O(N \log N)$ per coordinate list. The computational effort of the bounding-boxes algorithm under the consideration of the sorting method behaves approximately linear to the particle number, which is more or less as the neighbourhood methods.

In dilute gas-solid flows, where the cuboids vertices are far apart from each other, the bounding-boxes method can be applied satisfactory. In dense granular flows, by contrast, a significant increase in the computational effort arises. This efficiency loss is caused due to the small distances between the sorted vertices. In worst case, the order of the vertices between two time steps may completely reverse, which results in a computational effort for the sorting process of the second-order of the particle number (see Fig. 3.17). It should be acknowledged that even under the assumption of low particle velocities and smaller particle time steps, the sorted lists within two successive time steps can be quite different. As results of

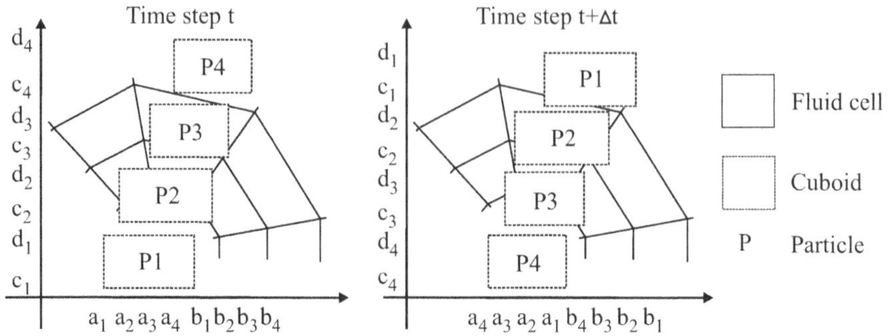

Fig. 3.17 The order of the sorted vertices between successive time steps

this, the computational complexity of the bounding-boxes algorithm can highly fluctuate from time step to other time step.

3.2.3.2 Collision Treatment Models

Depending on whether a stochastic or a deterministic collision detection model is applied, the subsequent collision treatment is performed either with the soft sphere model or the hard sphere model. In the following sections, the collision treatments using the hard and soft sphere models are explained in detail.

Hard Sphere Model

If the particle-particle and/or particle-wall collisions are detected stochastically with constant particle time step or deterministically with adaptive particle time step, the collision treatment is then carried out using the hard sphere model. Here, the properties of particles after the collision are correlated to the properties of particles before the collision through momentum and energy balances. In the hard sphere model, single binary collisions are modelled as instantaneous processes, where the contact between the collision partners occurs at a point. This model assumes that the particles are perfect spherical as well as all other finite forces that act on the collided particles during the collision are neglected. The first hard sphere model was developed by Campbell and Brennan (1985) and has been successfully employed to study a variety of complex granular systems since. The hard sphere model used by Sommerfeld (1996) and Hoomans et al. (1996) is briefly described here.

Figure 3.18 shows two colliding particles i and j with position vectors $(\vec{r}_i, \vec{r}_j)$, radii (r_i, r_j) and masses (m_i, m_j). The relative velocity at the contact point can be calculated using the translational as well as the angular velocities of collision partners as:

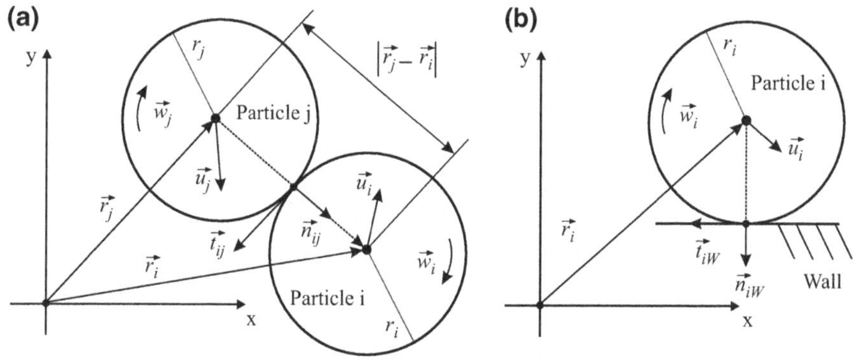

Fig. 3.18 Hard sphere model: **a** a particle-particle collision and **b** a particle-wall collision

$$\vec{u}_{ij}^{(0)} = \left(\vec{u}_i^{(0)} - \vec{u}_j^{(0)}\right) + \left(r_i \vec{w}_i^{(0)} + r_j \vec{w}_j^{(0)}\right) \times \vec{n}_{ij}, \tag{3.106}$$

where the index (0) identifies the situation before the collision. The normal unit vector $\vec{n}_{ij}$ that is directed from the centre of the particle j to the centre of the particle i is defined with the following expression:

$$\vec{n}_{ij} = \frac{\vec{r}_i - \vec{r}_j}{\left|\vec{r}_i - \vec{r}_j\right|}. \tag{3.107}$$

The tangential unit vector can be calculated with the aid of the relative velocity before the collision and the normal unit vector as follows:

$$\vec{t}_{ij} = \frac{\vec{u}_{ij}^{(0)} - \vec{n}_{ij}\left(\vec{u}_{ij}^{(0)} \cdot \vec{n}_{ij}\right)}{\left|\vec{u}_{ij}^{(0)} - \vec{n}_{ij}\left(\vec{u}_{ij}^{(0)} \cdot \vec{n}_{ij}\right)\right|}. \tag{3.108}$$

By applying Newton's second and third laws, the translational and angular velocities of particles after the collision, which result directly from the properties of collided particles before the collision, can be determined for the particle i:

$$\vec{u}_i = \vec{u}_i^{(0)} + \frac{\vec{J}_{ij}}{m_i}, \tag{3.109}$$

$$\vec{w}_i = \vec{w}_i^{(0)} + \frac{r_i \vec{n} \times \vec{J}_{ij}}{I_i}, \tag{3.110}$$

and for the particle j:

$$\vec{u}_j = \vec{u}_j^{(0)} - \frac{\vec{J}_{ji}}{m_j}, \tag{3.111}$$

$$\vec{w}_j = \vec{w}_j^{(0)} + \frac{r_j \vec{n} \times \vec{J}_{ji}}{I_j}. \tag{3.112}$$

The symbols $\vec{J}_{ij}$ represents the impulse vector. Rearranging Eqs. (3.109–3.112) under the consideration of Eq. (3.107) and the following mathematical relationship:

$$\left(\vec{n} \times \vec{J}\right) \times \vec{n} = \vec{J} - \vec{n}\left(\vec{J} \cdot \vec{n}\right), \tag{3.113}$$

one can obtain:

$$\vec{u}_{ij} - \vec{u}_{ij}^{(0)} = \frac{7\vec{J}_{ij} - 5\vec{n}_{ij}\left(\vec{J}_{ij} \cdot \vec{n}_{ij}\right)}{2m_{ij}}, \tag{3.114}$$

$$\vec{w}_{ij} - \vec{w}_{ij}^{(0)} = \frac{5\left(\vec{J}_{ij} \times \vec{n}_{ij}\right)}{2m_{ij}}, \tag{3.115}$$

with the moment of inertia for spherical particles:

$$I_P = \frac{2}{5} m_P r_P^2. \tag{3.116}$$

The reduced mass m_{ij}, also known as effective mass, allows the treating of two collided particles as one particle. The reduced mass is defined using the following formula:

$$m_{ij} = \frac{m_i m_j}{m_i + m_j} = \left(1/m_i + 1/m_j\right)^{-1}. \tag{3.117}$$

For the particle-wall collision, the mass of the particle j (i.e. the wall) is infinitely large and therefore the reduced mass is equal to the mass of particle j $\left(m_{ij} = m_i\right)$.

In order to determine the translational and angular velocities of collided particles after the collision, the impulse vector $\vec{J}_{ij}$ should be known. This requires an exact physical description of binary collisions. For the fluidized bed simulation employing the hard sphere model as an example, a large number of collisions should be considered, which is typically between 10^6 and 10^9 collisions per time step (Link 2006). In order to increase the calculation efficiency for processing the collisions together with maintaining the simulation accuracy, the physical description of collisions should be simplified to some extent. In this context, three constitutive relations (parameters) are introduced to the model in order to close the set of equations as follows:

The restitution coefficient in normal direction:

$$e^n = -\frac{\vec{u}_{ij}^{(0)} \cdot \vec{n}_{ij}}{\vec{u}_{ij} \cdot \vec{n}_{ij}} \quad 0 \le e^n \le 1, \tag{3.118}$$

the restitution coefficient in tangential direction:

$$\beta^t = -\frac{\vec{u}_{ij}^{(0)} \cdot \vec{t}_{ij}}{\vec{u}_{ij} \cdot \vec{t}_{ij}} \quad 0 \le \beta^t \le 1, \tag{3.119}$$

and the dynamic coefficient of friction:

$$\mu_{dyn} = -\frac{\vec{n}_{ij} \cdot \vec{J}_{ij}}{\left|\vec{n}_{ij} \times \vec{J}_{ij}\right|} \quad \mu_{dyn} \ge 0. \tag{3.120}$$

The impulse vector can be calculated employing the relative velocity between collision partners before the collision $\vec{u}_{ij}^{(0)}$ Eq. (3.106) as well as the known collision parameters e^n, β^t and μ_{dyn} Eqs. (3.118–3.120) that consider the dissipation of kinetic energy during the collision.

The normal component of the impulse vector is obtained by combining Eqs. (3.114) and (3.118):

$$J_{ij}^n = -(1+e^n)m_{ij}\left(\vec{u}_{ij}^{(0)} \cdot \vec{n}_{ij}\right). \tag{3.121}$$

For the tangential component of the impulse vector, two types of collisions have to be considered: sticking and sliding collisions. The collision is sliding, if gross sliding takes place throughout the duration of contact. Sticking collision occurs when the restitution coefficient in tangential direction has the value zero (Link 2006). The distinction between static and sliding collisions is verified with the aid of Coulomb's law of friction. By inserting Eqs. (3.119) and (3.120) in Eq. (3.114), the tangential component of the impulse vector is obtained:

$$J_{ij}^t = \begin{cases} \underbrace{-\frac{2}{7}(1+\beta^t)m_{ij}\left(\vec{u}_{ij}^{(0)} \cdot \vec{t}_{ij}\right)}_{\textit{Sticking collision}} & \text{case } \mu_{dyn}J_{ij}^n \ge \frac{2}{7}(1+\beta^t)m_{ij}\left(\vec{u}_{ij}^{(0)} \cdot \vec{t}_{ij}\right) \\ \underbrace{-\mu_{dyn}J^n}_{\textit{Sliding collision}} & \text{case } \mu_{dyn}J_{ij}^n < \frac{2}{7}(1+\beta^t)m_{ij}\left(\vec{u}_{ij}^{(0)} \cdot \vec{t}_{ij}\right) \end{cases}. \tag{3.122}$$

The resulting impulse vector is then expressed by:

$$\vec{J}_{ij} = J_{ij}^n \cdot \vec{n}_{ij} + J_{ij}^t \cdot \vec{t}_{ij}. \tag{3.123}$$

By knowing the impulse vector $\vec{J}_{ij}$, the translational velocities of colliding objects after the collision can be defined, applying Eqs. (3.109) and (3.111). Similar procedure may be carried out for the determination of angular velocities using Eqs. (3.115), (3.118), (3.119) and (3.120). In order to obtain the new position and the new angular displacement of collided particles, Eqs. (3.91) and (3.92) are used, where the particle time step must be defined in advance (see Sect. 3.2.3.4).

Soft Sphere Model

If the collisions between particle-particle and particle-wall are detected deterministically with a constant particle time step, the soft sphere model is applied to the collision treatment. Here, the determination of the single particle trajectory and the angular displacement is achieved by establishing balances of forces and moments of force on each particle.

Force Balance

The translational velocity of a single particle is calculated through the integrating of Newton's second law of motion:

$$\underbrace{m_P \frac{d\vec{u}_P}{dt_P} = \sum \vec{F}_i = \underbrace{\vec{F}_{gra} + \vec{F}_{buo} + \vec{F}_{ele} + \vec{F}_{mag}}_{Volume\ forces} + \underbrace{\vec{F}_{dra} + \vec{F}_{vir} + \vec{F}_{Bas} + \vec{F}_{pre} + \vec{F}_{Saf} + \vec{F}_{rad} + \vec{F}_{Mags}}_{Aerodynamic\ forces} + \underbrace{\underbrace{\vec{F}^n_{con} + \vec{F}^t_{con}}_{Contact\ forces} + \underbrace{\underbrace{\vec{F}_{van} + \vec{F}_{elst}}_{Without\ physical\ contact} + \underbrace{\vec{F}_{liq} + \vec{F}_{sol}}_{With\ physical\ contact}}_{Adhesive\ forces}}_{Short-range\ forces\ between\ the\ particles}}_{Surface\ forces} \tag{3.124}$$

Here, $\vec{u}_P$ is the particle translational velocity, t_P represents the particle time step, m_P is the particle mass and $\vec{F}_i$ denotes the forces acting on the particle. These forces are classified as volume and surface forces. The particle volume forces are gravitational, buoyancy, electric and magnetic forces. The surface forces can be divided into two categories: the aerodynamic surface forces that result from the interactions between fluid and particles (drag, virtual mass, Basset, pressure, Saffman, radiometric and Magnus forces) and the short-range surface forces that result from interactions between particles (contact and adhesive forces). The adhesive force is divided into adhesive forces associated either with or without physical contact. The adhesive forces without physical contact include van der Waals and electrostatic forces, while the adhesive forces with physical contact are liquid bridge and sintering forces.

In the following sections, the acting forces on particles including the volume forces, the aerodynamic surface forces and the short-range surface forces are explained according to Alobaid (2013), Epple et al. (2012), Fan and Zhu (2005), Götz (2006), Paschedag (2004), Schubert (1979).

I. **Volume forces**

The volume forces, also known as body or field forces, are imposed by external fields outside the flow such as gravity, magnetic or electric field. These forces act throughout the volume of object regardless of the flow configuration. Typical volume forces in the gas-solid flows are gravitational force, buoyancy force, electric and magnetic forces.

Gravitational force: The gravitational force is determined for spherical particles from the product of particle mass m_P and acceleration of gravity $\vec{g}$ as follows:

$$\vec{F}_{gra} = m_p\vec{g} = \frac{\pi}{6}\rho_p d_p^3 \vec{g}, \tag{3.125}$$

where d_P and ρ_P represent the diameter and the density of particle, respectively.

Buoyancy force: According to Archimedes buoyancy principle, the static buoyancy force acting on a floating particle is equal to the fluid weight m_{dis} that is displaced by this object. The static buoyancy force occurs in liquids and gases and counteracts the gravitational force. For spherical particles, it is given by:

$$\vec{F}_{bau} = -m_{dis}\vec{g} = -\frac{\pi}{6}\rho_k d_p^3 \vec{g}. \tag{3.126}$$

Here, the subscript k is either g(gas) or liq (liquid).

Electric force: If an electric field is applied to charged particles in a gas-solid flow, the particles are subjected to an electric force that is calculated by the product of the particle charge q and the electric field intensity $\vec{E}$:

$$\vec{F}_{ele} = q\vec{E}. \tag{3.127}$$

While particles with positive charges move to cathodic side, the particles with negative charges tend to move to anodic side. In case of static charged particles, the electric force is also known as Coulomb force.

Magnetic force: In a gas-solid flow, where the particles are ferromagnetic (attracted to magnets, e.g. iron and nickel), a magnetic force is exerted once the particles are exposed to a magnetic field. It can be calculated using the following relation:

$$\vec{F}_{mag} = n\varepsilon_{rel}\vec{B}_{vac}, \tag{3.128}$$

where n represents the number of north magnetic poles, ε_{rel} is the relative permeability of the material and $\vec{B}_{vac}$ denotes the magnetic flux density in vacuum. In an electromagnetic field, the forces acting on charged particles consist of three parts: the force due to the net charge, the force due to the electric dipole in the electric field and the force due to the magnetic dipole in the magnetic field.

The resulted Lorentz force on a charged particle that moves with velocity $\vec{u}_P$ through an electromagnetic field can be determined by neglecting the effect of the magnetic dipole as follows:

$$\vec{F}_{mag,ele} = q\left(\vec{E} + \vec{u}_P \times \vec{B}\right). \tag{3.129}$$

II. Surface forces

While the volume forces act on the entire body of the particle, the surface forces (aerodynamic surface forces and short-range surface forces) exert on the surface of the particle.

(1) Aerodynamic surface forces

The aerodynamic surface forces result from the interactions between fluid and solid phases. These forces can be identified through analytical methods, leading to a significant amount of time and effort by the reason of complex nature of gas-solid interactions. The most common way to describe this complex behaviour is to perceive the particle movement as a superposition of four simple motions with the assumption that they are non-interactive (see Fig. 3.19):

(1) Particle moving in an isotropic flow field (uniform) with constant velocity (i.e. drag force)
(2) Particle accelerating or decelerating in an isotropic flow field (i.e. virtual mass and basset forces)
(3) Particle moving in an anisotropic flow field (non-uniform) with constant velocity (i.e. pressure gradient, radiometric and concentration gradient forces)
(4) Particle rotating in an isotropic flow field with constant angular velocity (i.e. Magnus force).

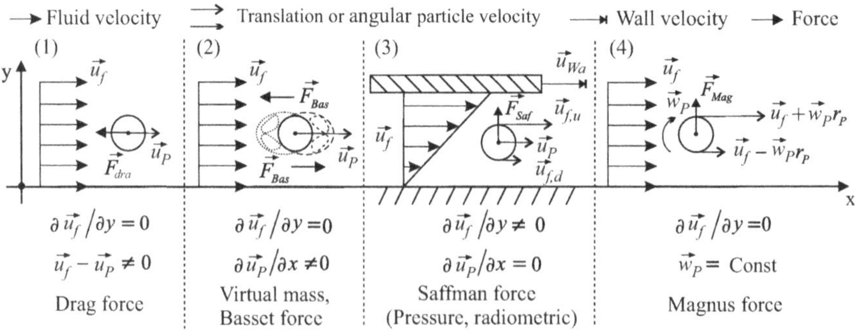

Fig. 3.19 Description of gas-solid interaction as superposition of four simple motions of sphere (Alobaid 2013)

The aerodynamic surface forces resulted from these four simple motions are explained in detailed below.

(1) When a particle moves in an uniform flow field, where the particle velocity is constant, a force that resists the particle motion is induced:

Drag force: In practical gas-solid flows, the particle velocity $\vec{u}_P$ differs generally from the gas velocity $\vec{u}_f$. The relative velocity (slip velocity) causes a viscous stress and an unbalanced pressure distribution on the particle surface, which yields a force, known as the drag force. In an isotropic flow field, the drag force of a moving particle with a constant velocity acts opposite to the direction of particle velocity. It can be determined as follows:

$$\vec{F}_{dra} = \frac{1}{2} C_{dra} A_P \rho_f \left(\vec{u}_f - \vec{u}_P\right)\left|\vec{u}_f - \vec{u}_P\right|. \tag{3.130}$$

Here, A_P represents the cross-section area of the particle (exposed area to incoming flow), ρ_f is the gas density and C_{dra} denotes the drag coefficient that is a function of the particle Reynolds number:

$$\mathrm{Re}_P = \frac{\rho_f d_P \left|\vec{u}_f - \vec{u}_p\right|}{\mu_f}. \tag{3.131}$$

The symbol μ_f is the fluid dynamic viscosity.

In the literature, there are a lot of experimental works and analytical studies on the drag coefficient for a single sphere at various particles Reynolds numbers. These data have been collected into a drag coefficient curve for a fixed sphere particle (see Fig. 3.20) by Schlichting and Gersten (2000).

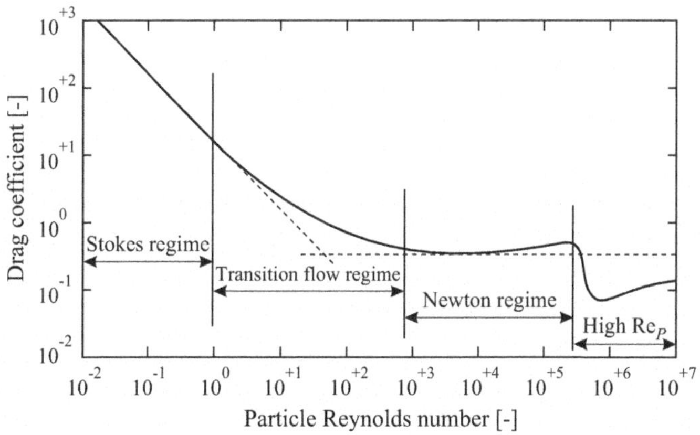

Fig. 3.20 Drag coefficient curve for fixed sphere particle (Schlichting and Gersten 2000)

For creeping flow regime, also known as Stokes flow regime $\mathrm{Re}_P < 1$, the viscous effect dominates the flow and the inertia is negligible. The drag coefficient is determined using the following expression:

$$C_{dra} = \frac{24}{\mathrm{Re}_P}. \tag{3.132}$$

In the transition flow regime $1 \leq \mathrm{Re}_P \leq 10^3$, there are a lot of approximations with sufficient accuracy available, for example, the relationship of Schiller and Naumann (1933):

$$C_{dra} = \frac{24}{\mathrm{Re}_P}\left(1 + 0.15\mathrm{Re}_P^{0.687}\right). \tag{3.133}$$

In the creeping flow and the transition regimes, the drag force decreases with increasing the particle Reynolds number. For the flow regime $10^3 \leq \mathrm{Re}_P \leq 3 \times 10^5$, where the inertia is a dominant factor, the drag coefficient is defined from Newton's experiments $C_{dra} = 0.44$ independent of the Re_P. For higher particle Reynolds numbers, the drag coefficient drops to about 0.07 for a spherical particle in incompressible flow and then rises again to about 0.3.

(2) When a particle moves in an uniform flow field, where a particle velocity changes, additional forces related to particle acceleration or decelerating can arise:

Virtual mass force: During the acceleration or the deceleration of a particle in a fluid, the surrounding fluid (boundary layer) due to the adhesion to the particle surface must be accelerated or decelerated, too. Here, a wake vortex is generated, which its vorticity depends highly on particle Reynolds number. Consequently, an additional form of resistance referred to as virtual mass, also called as carried mass or added mass, is induced. The virtual mass force that is directed opposite to the gradient of the particle velocity is proportional to the relative acceleration or deceleration and to the density ratio between fluid and solid phases. Generally, it can be expressed by:

$$\vec{F}_{ver} = C_{ver}\frac{\rho_f}{\rho_P} m_P \left(\frac{D\vec{u}_f}{dt} - \frac{d\vec{u}_P}{dt}\right), \tag{3.134}$$

where C_{ver} is a correction factor that considers the disproportionate increase in the virtual mass force by the reason of generated wake vortex. At lower particle Reynolds numbers, this factor has the value of unity, while it rises significantly at higher relative velocities (Sommerfeld 1996):

$$C_{ver} = 1.05 - \frac{0.066}{A^2 + 0.12}, \tag{3.135}$$

with:

$$A = \frac{\left|\vec{u}_f - \vec{u}_P\right|^2}{d_P\left|\frac{d}{dt}\left(\vec{u}_f - \vec{u}_P\right)\right|}. \tag{3.136}$$

Basset force: Basset force, also known as the history force, is caused due to relative acceleration or deceleration of particle to fluid. During acceleration or deceleration process, the shape and volume of the boundary layer is subjected to continuous changes. Basset force occurs during the unsteady processes as well as by fluctuations of particle movement in quasi-steady-state flows. It describes the temporal delay in boundary layer development when the relative velocity changes. The resulting force depends highly on the density ratio between fluid and particle as well as on the relative acceleration or deceleration. The Basset force is directed opposite to the gradient of the relative velocity and can be calculated as follows:

$$\vec{F}_{Bas} = 9\sqrt{\frac{\rho_f \mu_f}{\pi}}\frac{m_P}{\rho_P d_P} C_{Bas} \int_{t_0}^{t} \frac{\frac{D\vec{u}_f}{dt} - \frac{d\vec{u}_P}{dt}}{\sqrt{t - t_{relax}}} dt_{relax}. \tag{3.137}$$

The time t_{relax} denotes Stokes relaxation time (see Eq. 3.98) and the correction factor C_{Bas} describes the dependence of Basset force on the particle Reynolds number.

$$C_{Bas} = 2.88 + \frac{3.12}{(1 + A)^3}. \tag{3.138}$$

Basset force is of relevance, if the relative acceleration or deceleration rate is high, the time change is much greater than the Stokes relaxation time and/or the density difference between fluid and solid phases has a considerable value.

(3) When a particle moves at constant velocity in an anisotropic flow field, where velocity, pressure or temperature gradients are present, additional forces correlated to these gradients can arise:

Pressure gradient force: The pressure gradient force (dynamic buoyancy force) acts on the particle surface due to the existence of pressure gradient in the fluid. The resulting force is in the opposite direction of the pressure gradient (i.e. from higher pressure region to lower pressure region). The force due to the pressure gradient can be determined using the following equation:

$$\vec{F}_{pre} = -\frac{1}{6}\pi d_P^3 \nabla p. \tag{3.139}$$

Saffman force: When a particle moves at a constant velocity in a flow field, where a velocity gradient exists (e.g. near a wall or in a shear flow), the opposite sides of particle undergo different velocities. Subsequently, a pressure gradient on the particle surface is formed due to the velocity distribution. While a positive pressure arises at the side with lower velocity, a negative pressure forms at the faster side. This leads to a lift force (Saffman force) perpendicular to the relative velocity and in the opposite direction to the pressure gradient. Saffman force depends on the particle diameter, the velocity gradient and the relative velocity of both phases.

For three-dimensional flows, Saffman force can be expressed as follows:

$$\vec{F}_{Saf} = 1.615 d_P^2 \sqrt{\frac{\rho_f \mu_f}{|\vec{w}_f|}} (\vec{u}_f - \vec{u}_P) \times \vec{w}_f f(\mathrm{Re}_P, \mathrm{Re}_S), \tag{3.140}$$

with the angular velocity vector of fluid phase:

$$\vec{w}_f = \frac{1}{2} \nabla \times \vec{u}_f. \tag{3.141}$$

The empirical correction factor $f(\mathrm{Re}_P, \mathrm{Re}_S)$ that ensures the dependence of Saffman force on particle Reynolds number is computed with the aid of the following threshold function (Götz 2006; Sommerfeld 2003):

$$f(\mathrm{Re}_P, \mathrm{Re}_S) = \begin{cases} \left(1 - 0.3314\sqrt{\frac{\mathrm{Re}_S}{2\mathrm{Re}_P}}\right) \exp\left(-\frac{\mathrm{Re}_P}{10}\right) + 0.3314\sqrt{\frac{\mathrm{Re}_S}{2\mathrm{Re}_P}}, & \text{case } \mathrm{Re}_P \leq 40 \\ 0.0524\sqrt{\frac{\mathrm{Re}_S}{2}}, & \text{case } \mathrm{Re}_P > 40 \end{cases}, \tag{3.142}$$

where Re_S represents the Reynolds number of the shear flow, defined as:

$$\mathrm{Re}_S = \frac{\rho_g d_P^2 |\vec{w}_f - \vec{w}_P|}{\mu_f}. \tag{3.143}$$

Radiometric force: In a region, where a fluid temperature gradient or non-uniform radiation field exists, a particle is subjected to radiometric force. It is directed towards the region of lower temperature (i.e. opposite to the temperature gradient). If the mean free path of fluid (the average distance that can cover by a moving gas molecular till the next collision with other object occurs) is much smaller than the particle diameter, the radiometer force can be calculated using the following formula:

$$\vec{F}_{rad} = -\frac{3}{2} \pi \eta_f^2 d_P \frac{R}{p} \nabla T_P. \tag{3.144}$$

The constant R is the ideal gas constant, ∇T_P the temperature gradient on the particle surface, which is proportional to the temperature gradient of the fluid ∇T_f:

$$\nabla T_P = \frac{3\lambda_f}{2\lambda_f + \lambda_P} \nabla T_f. \tag{3.145}$$

The parameters λ_f and λ_P denote the thermal conductivities of fluid and solid phases, which are material properties. For flows with low temperature gradient, the radiometer force should only be considered for particles of submicron size. At higher temperature gradients, the radiometer force is relevant for larger particles.

Concentration gradient force: If the particle is found in a flow, where a concentration gradient is existing, then the particle will accelerate in the direction of negative concentration gradient.

(4) When a particle rotates with a constant angular velocity in an uniform flow field, an addition lift force related to this rotation can arise:

Magnus forces: A similar effect as in the shear flow can occur when the particle rotates by itself. The particle rotation leads also to an irregular pressure distribution on the particle surface as a result of relative velocity difference between the particle surface and the incoming flow. The resulting shear/lift force is called Magnus force and can be determined according to Crowe et al. (2011), Tsuji et al. (1985) as:

$$\vec{F}_{Mags} = \frac{\pi}{8} d_P^2 C_{Mags} \rho_g \frac{\left|\vec{u}_f - \vec{u}_P\right|}{\left|\vec{w}_f - \vec{w}_P\right|} \left[\left(\vec{u}_f - \vec{u}_P\right) \times \left(\vec{w}_f - \vec{w}_P\right)\right]. \tag{3.146}$$

For the lift coefficient C_{Mags}, the following relation is applied (Alobaid 2013; Götz 2006):

$$C_{Mags} = \begin{cases} 0.4G, & \text{if } G \leq 1 \\ 0.4, & \text{if } G > 1 \end{cases}, \tag{3.147}$$

with:

$$G = \frac{d_P}{2} \frac{\left|\vec{w}_f - \vec{w}_P\right|}{\left|\vec{u}_f - \vec{u}_P\right|}. \tag{3.148}$$

Magnus force depends on the particle diameter, the relative translational and angular velocities both phases. The main reason for particle rotation is the particle-wall collisions. Therefore, Magnus force is of relevance for granular flows, in particularly when the wall has a significant effect on the particle movements (e.g. particle flow in a pipe).

The influence of the particle rotation on the flow behaviour can be neglected at lower particle Reynolds numbers. At high particle Reynolds numbers, Magnus force becomes important and should be considered.

(2) **Short-range surface forces**

The short-range forces that act between particle-particle and particle-wall can be classified as adhesive and contact forces. The contact force is determined using Voigt-Kelvin model, which describes the viscoelastic and the time-dependent behaviour of collision. The Voigt-Kelvin model is composed of a Hooke's spring and a Newton's damper, which are connected in parallel. The spring is used for modelling the reversible deformation, while the damper for time-dependent and irreversible deformation. Since the Voigt-Kelvin model can only describe the normal interaction, a sliding element is introduced to determine the tangential interaction. The adhesive forces depend mainly on the particle diameter, the particle density and the mechanical properties of contacting objects. The adhesive forces can be divided into two categories: adhesive forces associated either with or without physical contact. The adhesive forces without physical contact include van der Waals and electrostatic forces, while the adhesive forces with physical contact are based on the formation of a physical bond between particles and/or particles and wall. This connection can occur as solid bridges (sintering, melting or bonding) or by wetting the particle surface (adsorption or movable liquid bridges). It should be mentioned here that the short-range forces cannot be calculated, before the collision partners are identified. This can be achieved employing a detection collision model with a specific criterion. One example is the contact criterion, where the distance between the centres points of two particles should be smaller than the sum of both radii (see Section Contact force). Another example is the criterion of liquid bridge formation. A liquid bridge will build when the distance between the centres points of two particles falls below the bridge critical length (see Section Adhesive force due to liquid bridge).

- **Contact force**

In the discrete element method, it is assumed that the particles can overlap each other or penetrate into the wall. The contact point is seen as a line in 2D or as a surface in 3D cases. Depending on the penetration depth, the resulting contact force can be determined and is calculated here for particle-particle collision, as an example. Similar procedures are used to the contact force due to the particle-wall collision with the assumptions that the wall has an infinite radius and its values of translational and angular velocities are zero. The contact between two spherical particles i and j with the position vectors $\vec{r}_i$ and $\vec{r}_j$, the radii r_i and r_j, the masses m_i and m_j, the translational and angular velocities $\vec{u}_i$, $\vec{u}_j$, $\vec{w}_i$, $\vec{w}_j$, is illustrated in Fig. 3.21.

The relative velocity at the contact point is obtained under the consideration of translational and angular velocities of collided particles, employing the following relation:

$$\vec{u}_{ij} = \vec{u}_{ij}^{n} + \vec{u}_{ij}^{t}, \tag{3.149}$$

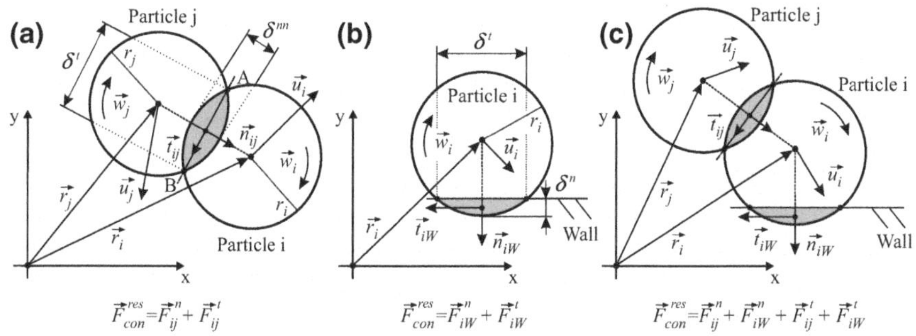

Fig. 3.21 Contact forces: **a** a particle-particle collision, **b** a particle-wall collision and **c** particle-particle/wall collisions

with:

$$\vec{u}_{ij}^{n} = \left(\left(\vec{u}_i - \vec{u}_j\right) \cdot \vec{n}_{ij}\right) \cdot \vec{n}_{ij} = \left|\vec{u}_{ij}^{n}\right| \vec{n}_{ij},$$
$$\vec{u}_{ij}^{t} = \left(\vec{u}_i - \vec{u}_j\right) + \left(r_i \vec{w}_i + r_j \vec{w}_j\right) \times \vec{n}_{ij} - \vec{u}_{ij}^{n} = \left|\vec{u}_{ij}^{t}\right| \vec{t}_{ij}. \quad (3.150)$$

While the normal unit vector is calculated using the radii and the position vectors of collision partners, the tangential unit vector is determined using the tangential relative velocity as:

$$\vec{n}_{ij} = \frac{\vec{r}_i - \vec{r}_j}{\left|\vec{r}_i - \vec{r}_j\right|},$$
$$\vec{t}_{ij} = \frac{\vec{u}_{ij}^{t}}{\left|\vec{u}_{ij}^{t}\right|}. \quad (3.151)$$

The overlap area between two particles or particle-wall is characterised by normal and tangential penetration depths. The penetration depth in normal direction δ^n is defined as follows:

$$\delta^n = r_i + r_j - \left|\vec{r}_i - \vec{r}_j\right|. \quad (3.152)$$

The tangential penetration depth cannot be explicitly determined from the particle positions, but can be calculated through time integration of the relative tangential velocity at the contact point:

$$\delta^t = \int_{t_0}^{t} \left|\vec{u}_{ij}^{t}\right| dt_P \quad (3.153)$$

By knowing the distribution of the force on the contact point, it is possible to calculate the resultant contact force. The determination of the contact force distribution needs, however, an exact physical description of the collision, which is a complex process. This is due to the dependence on various geometrical and physical factors, in particular the mechanical properties and the motion state of particles in contact. For this reason, the description of the collision process should be simplified to some extent. In this context, the distribution of the force over the contact area can be modelled employing a mechanical analogue system. Based on the well-known viscoelastic Voigt-Kelvin model that describes the time-dependent behaviour of the collision, the physical effects of particle interaction during the collision can be described. The Voigt-Kelvin model consists of a Hookean spring (with stiffness k) and a Newtonian dashpot (with viscosity η), connected in parallel. While the spring describes the purely elastic collisions (i.e. the reversible deformation), the damper represents the time-dependent irreversible deformation. According to the Voigt-Kelvin model, a collision can occur elastically or inelastically. If the resulted deformation during the collision is stored as potential energy and is completely transformed back into kinetic energy, the collision considers then as elastic. In elastic collisions, the sum of kinetic energies of colliding particles before and after the collision is the same (i.e. neither damping nor friction forces are regarded). Consequently, the damping element has to be omitted in these cases. In inelastical collisions, by contrast, a part of the stored potential energy is dissipated in other energy forms, like for example internal energy.

The contact force is calculated as the sum of both tangential and normal components. The tangential contact force is assumed to be either a static force or a sliding force. The static friction is modelled using the damper element, while the sliding friction that occurs due to the relative velocity between the collision partners is modelled employing the sliding element. The Voigt-Kelvin model can describe the normal contact force (see Fig. 3.22a), but the model should be extended by integration a sliding component in order to determine the tangential contact force (see Fig. 3.22b).

The resulting contact force on the particle i is obtained by summing up all normal and tangential contact forces that act simultaneously between the particle i and other particles and walls:

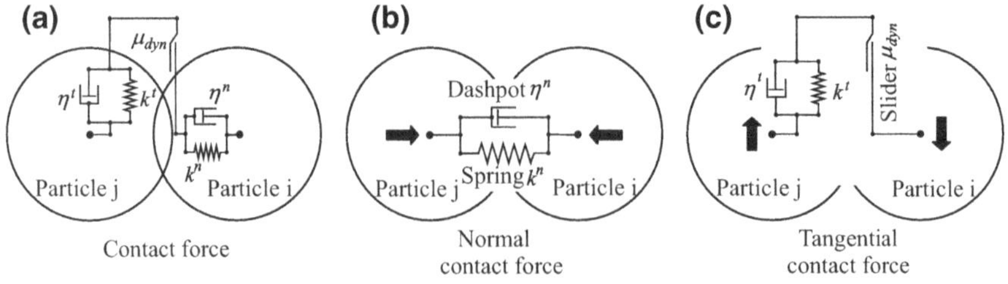

Fig. 3.22 Mechanical analogue according to Voigt-Kelvin model: **a** the resultant contact force, **b** the normal contact force and **c** the tangential contact force

$$\vec{F}^{res}_{con} = \sum_{\substack{k = 1 \\ k \neq i}}^{k=N} \left(\vec{F}^{n}_{con,ik} + \vec{F}^{t}_{con,ik} \right). \tag{3.154}$$

Here, the index N represents the number of contacts for the particle i.

The normal contact force that consists of an elastic force $\vec{F}^n_{ela}$ and a damping force $\vec{F}^n_{dam}$ is described by the differential equation of the spring-damper system as:

$$\vec{F}^n_{con,ij} = m_{ij} \frac{d\vec{u}^n_{ij}}{dt_P} = \underbrace{-k^n (\delta^n)^a \vec{n}_{ij}}_{\vec{F}^n_{ela}} - \underbrace{\eta^n \vec{u}^n_{ij}}_{\vec{F}^n_{dam}}, \tag{3.155}$$

where k^n is the normal stiffness constant, η^n is the normal coefficient of damping, m_{ij} is the reduced mass [effective mass; see Eq. (3.117)], δ^n and $\vec{u}^n_{ij}$ represent the displacement and the relative velocity in the normal direction. Specifying the exponent a in Eq. (3.155) allows the selection between several spring contact models. The linear approach $(a = 1)$ proposed by Hooke and also used by Cundall and Strack (1979) shows a satisfactory accuracy in most applications. Non-linear approaches proposed by Hertz (1882) and Maw et al. (1976) $(a \neq 1)$ improve the accuracy of the force calculation, but lead to increase in the computational costs. This is due to the consideration of the changes in the geometrical and physical properties of the colliding particles. A detailed comparison of different contact force models was carried out by Di Renzo et al. and Kruggel-Emden et al. (Di Maio 2004; Kruggel-Emden et al. 2007).

In order to calculate the normal contact force, the stiffness coefficient and the damping parameter in the normal direction must be determined. For the approach of Hooke with a linear dependence between the spring force and the penetration depth, the stiffness coefficient is calculated from the ratio of maximum normal contact force to maximum penetration depth:

$$k^n = \frac{\left| \vec{F}^n_{con,\max} \right|}{\delta^n_{\max}}. \tag{3.156}$$

Depending on the maximum relative velocity between the colliding objects, the maximum normal contact force and the maximum penetration depth can analytically be determined using a complex non-linear model, developed for example by Di Renzo et al. (Di Renzo and Di Maio 2004). The normal stiffness coefficient can also be defined depending on the properties of the collision partners as:

$$k^n = \left(\frac{1}{\gamma_{ij}} \right)^{\frac{3}{2}}, \tag{3.157}$$

with:

$$\gamma_{ij} = \frac{3}{4} \sqrt[3]{\left(\frac{r_i + r_j}{r_i r_j}\right)^2 \left(\frac{1 - v_i}{G_i} + \frac{1 - v_j}{G_j}\right)^2}. \tag{3.158}$$

The shear modulus G, which describes the linear elastic deformation of a body due to a shear stress, and Poisson's ratio v that enables the calculation of the transverse contraction of the body, are material properties and can be defined experimentally. The resulting stiffness coefficient is in the range of (10^4–10^9 N/m) and depends to a large extent on the material properties of the contacting partners.

The dissipation of the energy during an inelastic collision can be defined through the coefficient of restitution in the normal direction e^n, which is also known as the collision coefficient. The coefficient of restitution is defined as the ratio of the normal relative velocity at the contact point after the contact to that before the contact [index: (0)]:

$$e^n = \frac{\left|\vec{u}_{ij}^n\right|}{\left|\vec{u}_{ij}^{n(0)}\right|} = \sqrt{\frac{h}{h^{(0)}}}. \tag{3.159}$$

The coefficient of restitution can be experimentally measured. Here, the particle is dropped from the height $h^{(0)}$ to a large plate that has the same material properties as the contacting objects, and then the height of the rebound is measured. The coefficient of restitution depends on several parameters, e.g. material paring, collision velocity and/or particle shape. At lower relative velocities, the coefficient of restitution is considered to be constant, while additional part of kinetic energy is irreversibly dissipated in form of deformation energy at higher relative velocities, which results in decreasing the restitution coefficient.

The determination of damping parameter in normal direction, which describes the energy dissipation during the collision, is expressed by the relation:

$$\eta^n = \begin{cases} \alpha_{dam}(e^n)\sqrt{m_{ij}k^n}(\delta^n)^b & \text{for } e^n \neq 0 \\ 2\sqrt{m_{ij}k^n}(\delta^n)^b & \text{for } e^n = 0 \end{cases}. \tag{3.160}$$

The empirical constant $\alpha_{dam}(e^n)$ can be calculated using the following expression:

$$\alpha_{dam}(e^n) = -\frac{2 \ln e^n}{\sqrt{\pi^2 + \ln^2 e^n}} \quad \text{for } e^n \neq 0. \tag{3.161}$$

The empirical constant $\alpha_{dam}(e^n)$ is a function of the restitution coefficient in the normal direction e^n. This correlation can be applied to particle-particle and particle-wall collisions. The influence of dissipation on the normal contact force can be adjusted with the exponent b. By combining the linear spring model ($a = 1$) with a linear visco-elastic damping element $b = 0$, a complete linear model can be obtained. A partial non-linear visco-elastic contact model that combines a non-linear spring approach by Hertz $a \neq 1$ with a linear visco-elastic damping element $b = 0$ shows at lower relative velocities a reduction in the coefficient of restitution. In contrast to the partial non-linear visco-elastic contact model, a fully non-linear visco-elastic model tends to increase the coefficient of restitution at lower relative velocities. A detailed comparison of different contact force models can be found for example in Di Renzo and Di Maio (2004) and Zhu et al. (2007). If the stiffness coefficient value in the normal direction is underestimated or the relative collision velocities are very high, great penetration depths raise between the collision partners. As a result of this, the elastic force becomes much bigger than the damping force, which can lead to unrealistic contact forces and consequently to energy imbalance. In contrast, when the elastic force becomes smaller than the damping force by the reason of high damping coefficient used, tensile forces between the collision partners may be established. However, the influence of the tensile forces on the determination of contact force can be disregarded due to the fact that rigid objects can only transmit the compressive forces. For the determination of the contact forces in gas-solid flows, Lungfiel (2002) introduced a maximum function that neglects the effect of tensile forces as:

$$\left|\vec{F}^{n}_{con,ij}\right| = \max\left[0\,;\ -k^{n}(\delta^{n})^{a} - \eta^{n}\left|\vec{u}^{n}_{ij}\right|\right]. \tag{3.162}$$

The modelling of the tangential contact force is based on the spring-damper-slider system, where the tangential contact force is assumed to be either a static force or a sliding force. The tangential contact force is given by the following differential equation:

$$\vec{F}^{t}_{con,ij} = m_{ij}\frac{d\vec{u}^{t}_{ij}}{dt_P} = \begin{cases} \underbrace{\underbrace{-k^{t}\left(\vec{\delta}^{t}\right)^{a}}_{\vec{F}^{t}_{ela}} - \underbrace{\eta^{t}\vec{u}^{t}_{ij}}_{\vec{F}^{t}_{dam}}}_{\vec{F}^{t}_{static}}, & \text{for } \left|\vec{F}^{t}_{static}\right| \leq \left|\vec{F}^{t}_{slide}\right| \\ \underbrace{-\mu_{dyn}\left|\vec{F}^{n}_{con,ij}\right|\vec{t}_{ij}}_{\vec{F}^{t}_{slide}}, & \text{for } \left|\vec{F}^{t}_{static}\right| > \left|\vec{F}^{t}_{slide}\right| \end{cases}. \tag{3.163}$$

In order to determine the tangential contact force, the tangential stiffness coefficient k^t, the tangential damping parameter η^t and the dynamic friction coefficient μ_{dyn} have to be known. The tangential stiffness coefficient is calculated using the normal stiffness coefficient and the mechanical properties of colliding objects:

$$k^{t} = k^{n}\frac{(1 - \nu_i)/G_i + (1 - \nu_j)/G_j}{(1 - 0.5\nu_i)/G_i + (1 - 0.5\nu_j)/G_j}. \tag{3.164}$$

The shear modulus G and Poisson's ratio ν are material properties, while the damping parameter in the tangential direction is given by:

$$\eta^t = \begin{cases} \alpha_{Dam}(\beta^t)\sqrt{\frac{2}{7}m_{ij}k^t(\delta^t)^b}, & \text{for } \beta^t \neq 0 \\ 2\sqrt{\frac{2}{7}m_{ij}k^t(\delta^t)^b}, & \text{for } \beta^t = 0 \end{cases}. \tag{3.165}$$

The empirical constant $\alpha_{Dam}(\beta^t)$ is obtained from:

$$\alpha_{dam}(\beta^t) = -\frac{2\ln\beta^t}{\sqrt{\pi^2 + \ln^2\beta^t}}, \tag{3.166}$$

with the coefficient of restitution in the tangential direction β^t:

$$\beta^t = \frac{\left|\vec{u}_{ij}^t\right|}{\left|\vec{u}_{ij}^{t(0)}\right|}. \tag{3.167}$$

The symbols $\vec{u}_{ij}^{t(0)}$ and $\vec{u}_{ij}^t$ represent the relative velocity in the tangential direction before and after the contact, respectively. The dynamic friction coefficient μ_{dyn} can be obtained from experiments. The tangential contact force can be computed using linear or non-linear models, depending on the required accuracy of the contact force curve. Non-linear approaches of varying complexity are published and tested by several authors. In general, non-linear models tend to higher computational complexities as well as to higher costs regarding the implementation phase. In the literature, detailed comparison studies of different contact force models can be found (e.g. Zhu et al. 2007).

The tangential penetration depth is determined with the aid of the following threshold function:

$$\vec{\delta}^t = \delta^t\vec{t}_{ij} = \begin{cases} \vec{\delta}^{t(0)}\mathrm{R} + \int\limits_{\tau_0}^{\tau} \vec{u}_{ij}^t dt_P, & \text{if } \left|\vec{F}_{static}^t\right| \leq \left|\vec{F}_{slide}^t\right| \\ \frac{\mu_{dyn}}{k^t}\left|\vec{F}_{con,ij}^n\right|\vec{t}_{ij}, & \text{if } \left|\vec{F}_{static}^t\right| > \left|\vec{F}_{slide}^t\right| \end{cases}. \tag{3.168}$$

The rotation matrix R that is applied to perform a rotation in three-dimensional space of Euclidean geometry can be defined as:

$$\mathrm{R} = \begin{bmatrix} (1-\cos\varphi)h_x^2 + \cos\varphi & (1-\cos\varphi)h_xh_y - h_z\sin\varphi & (1-\cos\varphi)h_xh_z + h_y\sin\varphi \\ (1-\cos\varphi)h_xh_y + h_z\sin\varphi & (1-\cos\varphi)h_y^2 + \cos\varphi & (1-\cos\varphi)h_yh_z - h_x\sin\varphi \\ (1-\cos\varphi)h_xh_z - h_y\sin\varphi & (1-\cos\varphi)h_yh_z + h_x\sin\varphi & (1-\cos\varphi)h_z^2 + \cos\varphi \end{bmatrix}, \tag{3.169}$$

with the rotation unit vector:

$$\vec{h} = \frac{\vec{n}_{ij} \times \vec{n}_{ij}^{(0)}}{\left|\vec{n}_{ij} \times \vec{n}_{ij}^{(0)}\right|}, \tag{3.170}$$

and the rotation angle:

$$\varphi = \arcsin\left(\left|\vec{n}_{ij} \times \vec{n}_{ij}^{(0)}\right|\right). \tag{3.171}$$

As can be seen from Eq. (3.168), the tangential penetration depth cannot be explicitly determined, but through time integration of the relative tangential velocity at the contact point. This requires the application of complex numerical algorithm, leading to additional computational effort. Therefore, explicit approaches for calculating the static force should be introduced, e.g. the proposed equation by Lungfiel (2002):

$$\vec{F}^t_{static} = \mu_{dyn} \frac{\left|\vec{u}^t_{ij}\right|}{\vec{u}_f} \left|\vec{F}^n_{con,ij}\right|. \tag{3.172}$$

Götz (2006) investigated the influence of Eq. (3.172) on the tangential contact force with respect to the absolute values after the collision. Here, the calculated tangential restitution coefficient and the angular velocities after the collision are compared with the experimental data at different collision angles α. For the simulations, the tangential restitution coefficient is defined using the ratio of tangential velocities before and after the collision. The contact force is determined employing a linear model after Hooke and a non-linear model based on Hertz theory, where the damping force is disregarded. The applied particle diameter and density are d_P = 5 mm and ρ_P = 4000 kg/m^3, respectively. Furthermore, the relative collision velocity is 3.9 m/s that equates to particle velocity before the collision due to the zero values of wall translational and angular velocities. The comparison shows a

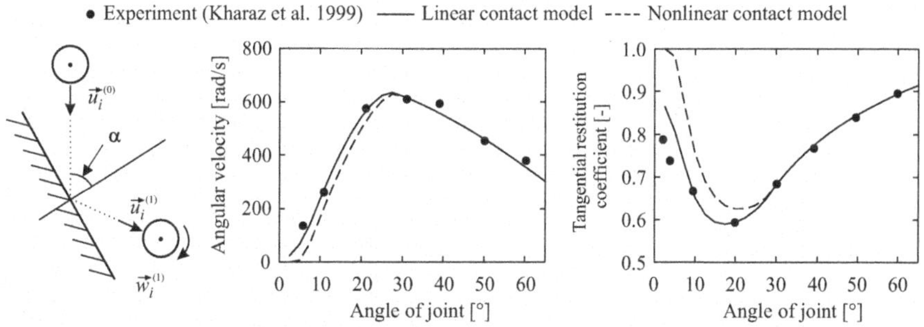

Fig. 3.23 Angular velocity after the collision and tangential restitution coefficient (Götz 2006); comparison between simulation [considered Eq. (3.172)] and experiment

very good agreement with the experimental data over a large range of collision angles (see Fig. 3.23). The coefficient of restitution in the tangential direction is overestimated at lower collision angles, where the particle collides the wall almost vertically. As previously mentioned, the impact of the contact model selection is insignificant on the macroscopic physical values after the collision. Compared to complex, non-linear models, the efficient linear approach provides better accuracy, especially at lower collision angles. Similar results are also obtained by Di Renzo et al. (Di Renzo and Di Maio 2004). It should be noted finally that these results are only valid to completely elastic collision models. In case of partial or fully non-linear visco-elastic models, this matter is still an open issue, which needs to be clarified through further investigations.

All in all, the resulting contact force can be determined with the sum of normal and tangential components by means of the restitution coefficient, the stiffness coefficient and the dynamic coefficient of friction. These constants that have a decisive influence on the accuracy and the efficiency of simulations are functions of mechanical properties of the collision partners.

- **Adhesive forces without physical contact**

If the particle size decreases, the adhesive forces without physical contact between particles or particle and wall will then increase up to the level that should be taken into account (see Fig. 3.24). The van der Waals' interaction and the electrostatic effect are the main sources of short-range adhesive forces without physical contact (see Fig. 3.25).

Van der Waals force: Van der Waals referred in 1873s to an electromagnetic force between real gas molecules or atoms, resulting from the dipole interaction. The van der Waals force consists of three types of attractive intermolecular forces: the force caused by the interactions between two permanent dipoles, the force

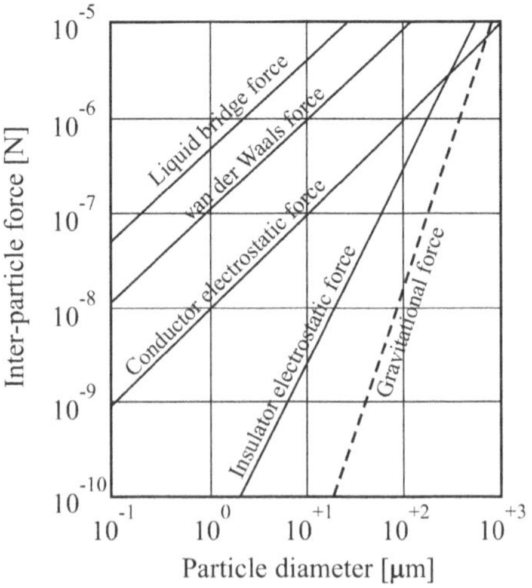

Fig. 3.24 Magnitude comparison of adhesive forces either with or without physical contact (Schubert 1979)

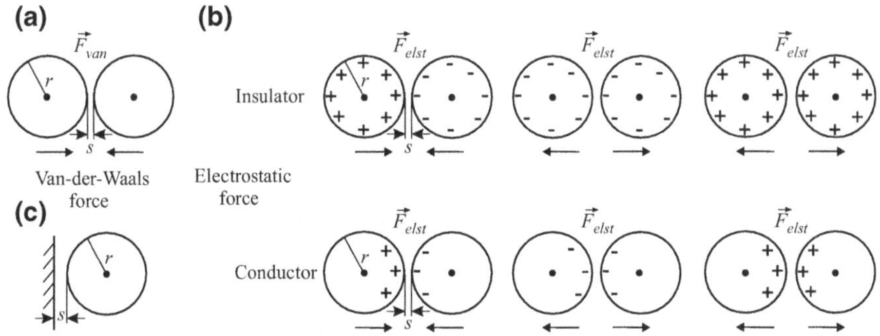

Fig. 3.25 Short-range adhesive forces without physical contact: **a** the van der Waals force (particle-particle), **b** the electrostatic force (particle-particle: insulator or conductor) and **c** the van der Waals or electrostatic force (particle-wall)

caused by two instantaneously induced dipoles and the force caused by dipole-induced dipole. The interactions of permanent dipoles of molecules lead to electrostatic force such as a dipole-dipole interaction in hydrogen chloride HCl, where the polar molecule end with positive charge will attract the negative end of other molecules. The instantaneous dipole-induced dipole force, also known as London force, results from the interacting molecules, where the movements of electrons of both molecules influence each other. The induced dipole force is caused by the fast changing in atom dipole. These dipole fluctuations result in an electric field that affects the neighbouring atoms polarizability. The polarisability is the measure of changing electron cloud distribution of molecule by an external electric field. The induced dipole of neighbouring atom tends to move in phase with original state, which leads to an attractive atomic force (Fan and Zhu 2005).

In general, van der Waals force describes the dipole-dipole interactions between atoms, molecules and surface and can occur between two microscopic molecules or between two macroscopic bodies (e.g. particle-particle or particle-wall). Here, different approaches can be utilised to determine the van der Waals force. While the theory of London and van der Waals is used for microscopic bodies, the theory of Lifshitz and van der Waals is applied to macroscopic bodies.

The van der Waals force between two atoms or molecules (i and j) separated by a distance s can be defined using the following relation:

$$\vec{F}_{van} = \frac{\partial E}{\partial s} \vec{s}, \tag{3.173}$$

where $\vec{s}$ represents the unit vector and E denotes the potential energy of interaction that can be calculated employing the theory of London and van der Waals as:

$$E = -\frac{\lambda_{ij}}{s_{ij}^6}. \tag{3.174}$$

The constant λ_{ij} that is generally referred to as London constant depends on the characteristic of interacting molecules or atoms. For microscopic bodies, the energy of interaction between two objects i and j can be obtained according to Hamaker by integration all interaction energies of molecules that exist in the objects (concept of additively):

$$E = -\int_{V_i}\int_{V_j} \frac{\lambda_{ij}}{s_{ij}^6} n_i n_j dV_i dV_j. \tag{3.175}$$

Here, V_i and V_j are the volume of the objects, n_i and n_j represent the atom densities, also known as the number density (i.e. number of atoms per unit volume in body). For two spherical bodies of radii r_i and r_j, van der Waals force can be calculated by inserting Eq. (3.175) in Eq. (3.173) as:

$$\vec{F}_{van,ij} = \frac{A_{ij} r_{ij}}{6 s_{ij}^2} \vec{r}_{ij}, \tag{3.176}$$

with the position vector $\vec{r}_{ij}$, the relative radius (reduced radius) that allows the treating of two bodies as a one object:

$$r_{ij} = \frac{r_i r_j}{r_i + r_j} = \left(1/r_i + 1/r_j\right)^{-1}, \tag{3.177}$$

and the Hamaker constant:

$$A_{ij} = \pi^2 n_i n_j \lambda_{ij}. \tag{3.178}$$

Similar procedures are followed to calculate the van der Waals force between a spherical body with a radius r_j and a wall. With the assumption that the wall has an infinite radius, Eq. (3.177) is reduced to:

$$r_{iwal} = (1/r_i + 1/\infty)^{-1} = (1/r_i)^{-1} = r_i. \tag{3.179}$$

Van der Waals force can then be determined as:

$$\vec{F}_{van,iwal} = \frac{A_{ij} r_i}{12 s_{iwal}^2} \vec{r}_{iwal} = \frac{\pi^2 n_i n_j \lambda_{ij} r_i}{6 s_{iwal}^2} \vec{r}_{iwal}. \tag{3.180}$$

The Hamaker constant between two different materials in air is calculated with the aid of Hamaker constant for each material as:

$$A_{ij} = \left(A_{ii}A_{jj}\right)^{\frac{1}{2}}. \tag{3.181}$$

If the materials i and j interact through a third material k, the Hamaker constant is then expressed using the following relation:

$$A_{ijk} = (A_{ii} - A_{kk})\left(A_{jj} - A_{kk}\right). \tag{3.182}$$

In case of macroscopic rigid bodies, where atoms are closely packed, the concept of additivity is insufficient to calculate the van der Waals force. The additively problem can be eliminated by using the macroscopic theory of Lifshitz and van der Waals. This can be realised by regarding the solid as a continuum, where the van der Waals force can be determined in terms of bulk properties, e.g. the dielectric constants and the refractive indices of the substances. The modified Hamaker constant is then defined as:

$$A = \frac{3}{4\pi} h\varpi. \tag{3.183}$$

The Lifshitz-van der Waals constant $h\varpi$ depends on the substance properties; but is independent of the material geometry. For two bodies of materials i and j separated by a medium k, the Lifshitz-van der Waals constant is given by:

$$h\varpi = h \int\limits_0^\infty \left(\frac{\varepsilon_{rel,i} - \varepsilon_{rel,k}}{\varepsilon_{rel,i} + \varepsilon_{rel,k}}\right)\left(\frac{\varepsilon_{rel,j} - \varepsilon_{rel,k}}{\varepsilon_{rel,j} + \varepsilon_{rel,k}}\right) d\zeta, \tag{3.184}$$

with Planck constant h. The dielectric constant $\varepsilon_{rel,i}$ describes the relative permittivity of the dielectric substance i along the imaginary frequency axis ζ. Due to the fact that the solving of Eq. (3.184) is extremely difficult, many approximations have been suggested in order to enable the calculation of Lifshitz-van der Waals constant (Fan and Zhu 2005). By knowing the Lifshitz-van der Waals constant, the modified Hamaker constant and thus the resulted van der Waals force can be calculated.

Electrostatic force: The electrostatic effect may arise due to different reasons, summarised as follows:

- Particle-particle and Particle-wall collisions: during the contact, the interaction between the collided objects can lead to charge the particle surfaces.
- Corona charging and scattering (electrostatic contact potential difference): if the current flows from an electrode with a high potential into a neutral fluid, the fluid will be ionized and generates an areas of plasma around the electrode. This plasma field enables accordingly charge to be passed from high potential electrode to lower potential region.
- Thermionic emission: the thermionic emission in high temperature environment results in charged surfaces.

- Other charging mechanisms: e.g. colloidal propulsion for dielectric substances in intense electric field (Soo and Deyan 1990), also known as image-charge effect.

In 1785s, Coulomb reports "the force of attraction or repulsion between two point charges is directly proportional to the product of magnitude of each charge and indirectly proportional to the square of distance between them". While particles with opposite charges attract each other, particles with identical polarity are repulsive.

In gas-solid flows, the electrostatic force resulted from charged particles can highly influence the particles motions. The electrostatic force between two charged bodies, although the basic mechanism of electrostatic effect has not fully understood, is given according to Coulomb's law by:

$$\vec{F}_{elst,ij} = \frac{1}{4\pi\varepsilon_0} \frac{q_i q_j}{s_{ij}^2} \vec{r}_{ij}, \tag{3.185}$$

where q_i represents the charge carried by the object i, s_{ij} is the distance between centres of both bodies, ε_0 denotes to the electric constant (vacuum permittivity) of the surrounding medium. The electrostatic force can be determined for two charged particles with radii r_i and r_j made of electrical insulating materials as:

$$\vec{F}_{elst,ij} = \frac{\pi}{\varepsilon_0 \varepsilon_{rel}} \frac{\varphi_i \varphi_j r_{ij}^2}{\left(1 + \frac{s_{ij}}{r_i + r_j}\right)^2} \vec{r}_{ij}, \tag{3.186}$$

and for charged particles made of electrical conduction materials as:

$$\vec{F}_{elst,ij} = \frac{\pi}{2} \varepsilon_0 \varepsilon_{rel} U_{ij}^2 \frac{r_{ij}}{s_{ij}} \vec{r}_{ij}. \tag{3.187}$$

Here, ε_{rel} denotes the relative permittivity that is equal to unity in vacuum, φ_i is the surface charge of object i and U_{ij} is the contact potential of two particles.

The electrostatic force between a charged particle with radius r_i and a wall, both of them made of electrical insulation materials, is expressed using the following expression:

$$\vec{F}_{elst,iwal} = \frac{2\pi}{\varepsilon_0 \varepsilon_{rel}} \varphi_i \varphi_j r_{iwal}^2 \vec{r}_{iwal}. \tag{3.188}$$

For a charged particle and a wall, both of them made of electrical conducting materials, the electrostatic force is written as:

$$\vec{F}_{elst,iwal} = \pi \varepsilon_0 \varepsilon_{rel} U_{iwal}^2 \frac{r_{iwal}}{s_{iwal}} \vec{r}_{iwal}. \tag{3.189}$$

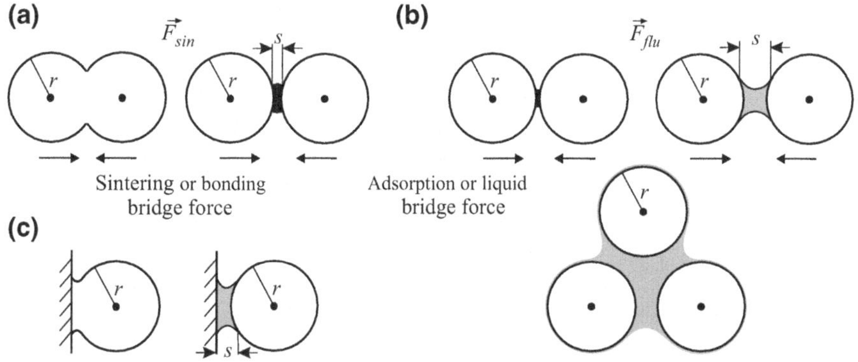

Fig. 3.26 Short-range adhesive forces with physical contact: **a** a solid bridge (particle-particle), **b** a liquid bridge (particle-particle) and **c** a solid bridge or liquid bridge (particle-wall)

- **Adhesive forces with physical contact**

Compared to the adhesive forces without physical contact, the adhesive forces with physical contact are characterised by a material connection between particle-particle/wall, resulting in a strong adhesion (see Fig. 3.26). Depending on the material properties of the physical bond, these forces can be classified into two categories: solid or liquid bridge. While the adhesive forces with solid connection occur due to sintering, melt adhesion, thermochemical reaction or crystallisation, the adhesive forces with liquid connection form high viscosity binding mechanism and/or capillary binding force by the reason of water content in the granular flow.

Adhesive force due to liquid bridge: In gas-solid flows with a certain amount of water content in gas phase, adsorption process that acts at the interface surfaces can be occurred. If the distance between two particles, whose surfaces are wetted with liquid films due to adsorption process, falls below a critical distance, a liquid bridge is then formed, resulting in adhesive and cohesive forces. The adhesive force denotes attractive force between two objects of different materials, while the cohesive force represents the attractive forces between molecules of same substance. In Fig. 3.27, the liquid bridges between particle-particle and particle-wall are shown at low moisture content.

For particles with diameters less than 1 mm, the gravity and buoyancy forces acting on the liquid bridge can be neglected (Schubert 1979). Based on this assumption, the resultant adhesive force between two spherical particles with radius r_P can be determined by summing up the contributions of contact and capillary actions. The contact force is caused by the surface tension effect of liquid at the contact line of three phases. It occurs by the reason of inter-molecular force between the gas, liquid and solid surfaces. The capillary effect arises due to the pressure difference between surrounding fluid p_{sur} and the phase boundary between the liquid bridge and the particle p_{liq}. The capillary force results then by multiply the pressure difference with the cross-section area A_{liq}:

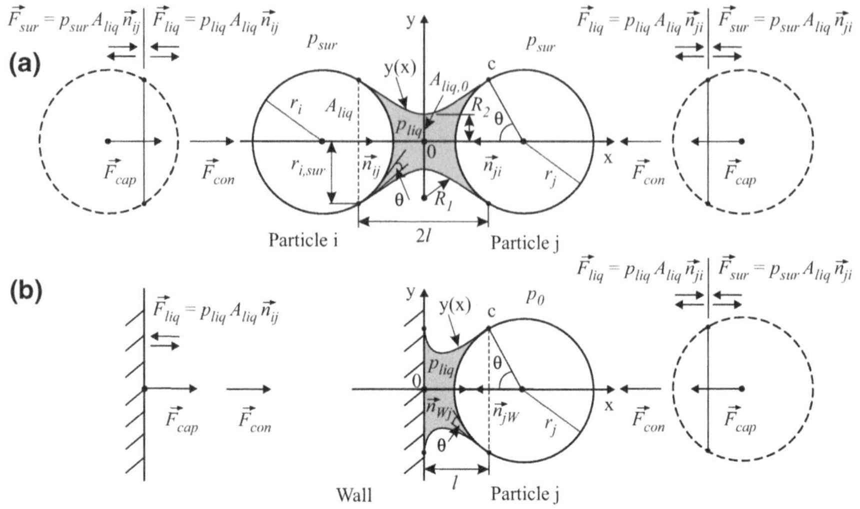

Fig. 3.27 Adhesive force due to liquid bridge (same contact and fill angles): **a** a particle-particle collision and **b** a particle-wall collision

$$\vec{F}_{cap,ij} = A_{liq}(p_{sur} - p_{liq})\vec{n}_{ij}. \tag{3.190}$$

If the liquid bridge has biconcave shape as is often the case in the water bridges, a capillary negative pressure p_{liq} is present inside the liquid of the bridge relative to gas phase p_{sur}. Here, the resultant adhesive force between two particles, which acts in the direction of the collision normal vector $\vec{n}$, can be calculated according to the relation proposed by Lian et al. (1993) as:

$$\vec{F}_{liq} = \left(\underbrace{2\pi r_P \sigma \sin(\xi)\sin(\theta + \xi)}_{\vec{F}_{con}} + \underbrace{\pi r_P^2 \Delta p \sin^2(\xi)}_{\vec{F}_{cap}} \right)\vec{n}. \tag{3.191}$$

Here, σ denotes the surface tension of liquid, r_P is the particle radius, ξ and θ are the fill and contact angles, l, R_1 and R_2 represent the length and the radii of bridge, respectively. For the particle i, the adhesive force due to the liquid bridge is in the direction of the collision normal vector $\vec{n}_{ij}$ and for the particle j is in the direction of $\vec{n}_{ji}$.

A liquid bridge can only be formed when the distance between two particles falls below the critical bridge length. If this occurs, the calculation of the adhesive force requires then the knowledge of the liquid bridge geometry. One possibility to determine the geometry of the bridge and thus the adhesive force due to the liquid bridge is based on the assumption of so-called toroidal shape. Here, the meridian profile of phase boundary is considered as a circular between the liquid and the surrounding fluid and as a circular segment among the liquid, the surrounding fluid and the solid surface of particle. Under this assumption, the resulted adhesive force can be calculated using three different approaches, namely boundary pressure method, boundary method and Gorge method, also known as neck method. In the boundary pressure and boundary methods, the adhesive force is estimated at the solid-liquid phase interface, while in Gorge method is determined at the smallest cross-section area of liquid bridge $A_{liq,\min} = \pi R_2^2$.

Lian et al. (1993) compared the obtained adhesive force due to the liquid bridge with the aid of toroidal shape assumption with experimental data, showing a good agreement. Furthermore, he found a relative error between the analytical solution taking into account the exact geometry of the liquid bridge (e.g. hyperboloid or nodoid) and the toroidal shape assumption less than 10%. However, the approximation that a liquid bridge is toroid results in a liquid bridge surface with a non-constant mean curvature. This leads to inconsistent with the Laplace-Young equation that is applied to determine the contribution of the capillary force. Other disadvantages by using the toroidal shape assumption are on the one hand that the critical length of the bridge cannot be estimated. On the other hand, the adhesive force cannot be expressed as an explicit function of the liquid bridge volume and the separated distance between two collided particles. An alternative approach to the toroidal shape assumption is to calculate the capillary force by reducing the Laplace-Young equation to a non-dimensional form. The Laplace-Young equation is a non-linear partial differential equation that describes the relationship between the mean curvature of phase boundary (i.e. liquid bridge) and the capillary pressure difference.

Mikami et al. (1998) developed a DEM simulation programme that considers the adhesive force due to liquid bridges between particles as well as between particles and walls. Employing this code, the agglomeration processes of wet particles in a fluidized bed were investigated. The results of numerical simulations are presented in Fig. 3.28 for particle-particle and particle-wall collisions. An explicit correlation between the absolute value of the non-dimensional adhesive force and the non-dimensional distance are proposed by Mikami with the aid of a regression analysis:

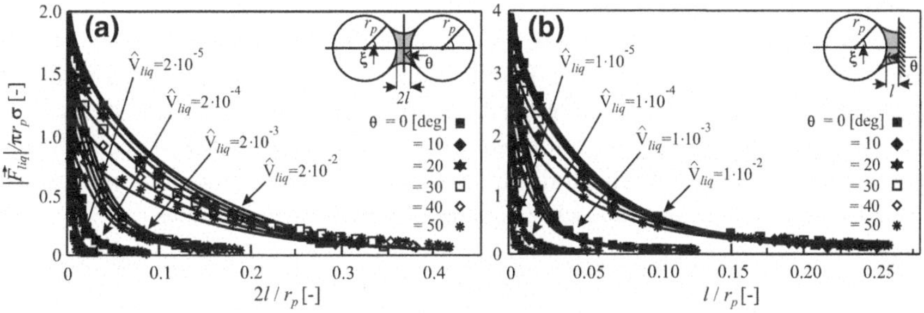

Fig. 3.28 Dimensionless liquid bridge force against dimensionless distance as a function of contact angle: **a** a particle-particle with same radius and **b** a particle-wall (Mikami et al. 1998)

$$\frac{\left|\vec{F}_{liq}\right|}{\pi r_P \sigma} = \exp\left(A\frac{l}{r_P} + B\right) + C, \tag{3.192}$$

where the parameters A, B and C are highly reliant on the water content in the gas phase.

The following correlations are used to determine these parameters for particle-particle collisions:

$$A_{PP} = -1.1\hat{V}_{liq}^{-0,53}, \tag{3.193}$$

$$B_{PP} = \left(-0.34\ln\hat{V}_{liq} - 0.96\right)\theta^2 - 0.019\ln\hat{V}_{liq} + 0.48, \tag{3.194}$$

$$C_{PP} = 0.0042\ln\hat{V}_{liq} + 0.078, \tag{3.195}$$

and for particle-wall collisions:

$$A_{Pwal} = -1.9\hat{V}_{liq}^{-0.51}, \tag{3.196}$$

$$B_{Pwal} = \left(-0.016\ln\hat{V}_{liq} - 0.76\right)\theta^2 - 0.12\ln\hat{V}_{liq} + 1.2, \tag{3.197}$$

$$C_{Pwal} = 0.0013\ln\hat{V}_{liq} + 0.18, \tag{3.198}$$

with the non-dimensional volume of the liquid bridge:

$$\hat{V}_{liq} = \frac{V_{liq}}{r_P^3}. \tag{3.199}$$

Another important parameter for determination the adhesive force is the non-dimensional critical bridge length $\hat{l}_{cri}$:

$$\hat{l}_{cri} = \frac{l_{cri}}{r_P}. \tag{3.200}$$

The non-dimensional value l_{cri} specifies the maximum distance between two particles or particle and wall, at which a stable liquid bridge still exists. If the critical distance is exceeded, the liquid bridge will be destroyed as a result of instabilities and accordingly the related adhesive force will be vanished. By contrast, a liquid bridge will be reformed when the distance between the collision partners falls short of the critical distance (see Fig. 3.29). For the non-dimensional critical bridge length, the following equations between two particles:

$$\hat{l}_{cri,PP} = (0.62\theta + 0.99)\hat{V}_{liq}^{0.34}, \tag{3.201}$$

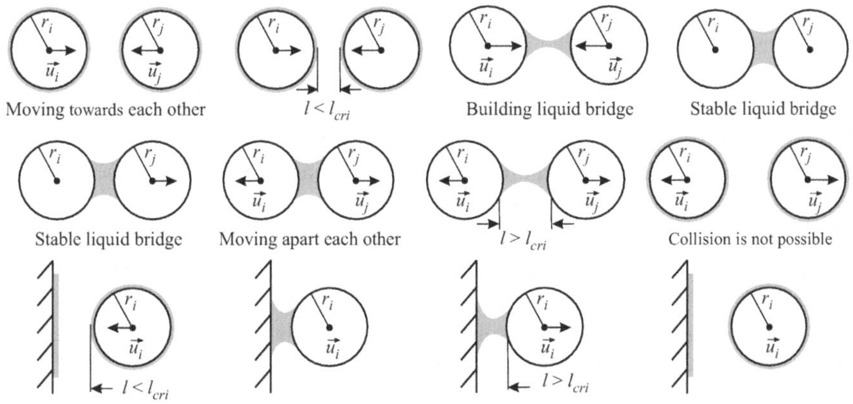

Fig. 3.29 Formation and destroying mechanisms of liquid bridges

as well as between a particle and a wall:

$$\hat{l}_{cri,Pwal} = (0.22\theta + 0.95)\hat{V}_{liq}^{0.32}, \tag{3.202}$$

are valid.

For modelling the adhesive force due to the liquid bridge using previous equations, the following assumptions should be considered according to Mikami et al. (1998):

- The liquid is evenly dispersed in the packed bed among all particles.
- The particles are spherical in shape, uniform in diameter and have the same material properties (i.e. monodisperse system and homogeneous material).
- The dynamic force due to viscosity in comparison to the static adhesive force by the reason of surface tension is negligibly small.
- The maximum number of liquid bridges that can be formed by each particle is related to the number of coordinate axes. This means, two liquid bridges in 2D cases and three liquid bridges in 3D simulations.
- All particles are wetted with a liquid film with constant thickness. However, the liquid volume should be sufficient for building three liquid bridges in case of 3D problems.
- A stable bridge will be formed, once the separation distance between particles falls below the critical length. The required liquid of the bridge is equally taken from both collision partners.
- When the critical length of the bridge is exceeded, it comes to the separation of the liquid bridge between the collision partners. Here, the liquid is distributed in equal parts to each collision partner.
- For particle-wall liquid bridge, the assumption only the points 3, 6 and 7 are taken into account.

At low volume fraction of water in the gas phase, Mikami model shows a compromise between meaningful results and acceptable computational effort. By increasing the water content in the gas flow, transition regions are built, which are filled with liquid. Isolated liquid bridges are not possible in this case due to the fact that several particles are agglomerated to each other. Further increasing in the volume fraction of the water can finally lead to unifying some of transition regions. In these regions, the particles may be completely surrounded with water and accordingly no movable liquid bridges are existent (see Fig. 3.30).

The resulted adhesive force on particles in the transition regions cannot be determined by Mikami model. In this context, a saturation level S_{liq}, which is defined as the ratio of the water volume V_{tra} in the transition regions to the cavity volume V_{cav}, is defined to check Mikami model for its validity:

$$S_{liq} = \frac{V_{tra}}{V_{cav}}. \tag{3.203}$$

According to Schubert (1979), the transition region starts forming from $S_{liq} \geq 0.2$ and thus the Mikami model can be applied for $S_{liq} < 0.2$.

During the formation of a stable liquid bridge, each partner should contribute to a cretin amount of the bridge liquid. If the bridge is broken by the reason of exceeding the critical length, the liquid should redistribute again between the collision partners. The redistribution of the liquid in the bridge is currently based on three different principles. Mikami et al. (1998) presupposes that the particles are wetted with constant liquid film and the formed bridge is axially symmetrical in direction of x coordinate and rotationally symmetrical with respect to $x - y$ plane. In case of separation, the liquid of the bridge will be evenly distributed over the collided objects. Muguruma et al. (2000) assumes that the particles can be wetted with different amounts of liquid. Here, the collision partners contribute to different amounts of liquid for building the bridge. During the destroying of the bridge, each partner receives the same amount of liquid, which is already participated during the building of the bridge. In the above two approaches, the meridian profile of phase boundary between the liquid of the bridge and the surrounding fluid is considered to

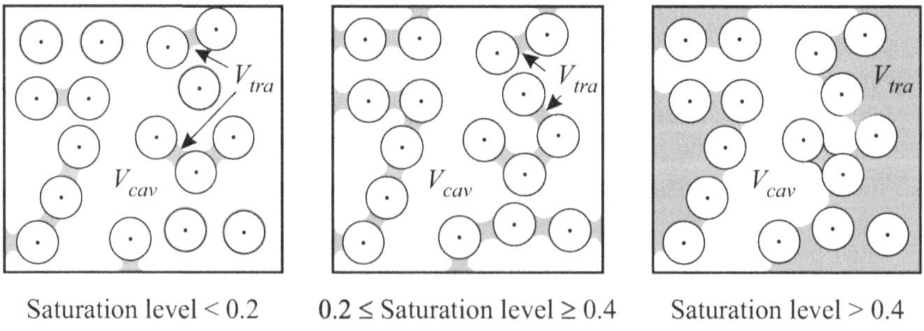

Fig. 3.30 Saturation level

have a symmetrical geometry. This is the case in granular flows at low liquid content, where the particles have spherical geometry, constant diameter and from same material properties (i.e. each collision partner has the same contact and fill angels). The separation of the liquid bridge due to its symmetrical form occurs exactly in the middle. The redistribution of the bridge liquid is taken place equally to each collision partner (no liquid transport is considered).

Diverse contact and fill angles can arise during the formation of liquid bridges between monodisperse particles with different material properties or between two polydisperse particles of identical or different material properties. Here, the bridge neck is not anymore in the centre between the collision partners, but closer to the particle with larger contact angle. In case of separation, the particle with smaller contact angle holds more proportion of liquid back, while the particle with larger contact angle receives the reaming liquid (see Fig. 3.31). Shi and McCarthy (2008) calculated the adhesive force due liquid bridge under the consideration of polydisperse particles and unequal liquid distribution. In their study, the single particle surfaces can be wetted by different amounts of liquid (i.e. liquid film with diverse thickness). For each collision, the volume of the liquid bridge should be determined separately, which is relatively time consuming. The entire liquid volume of the bridge is analytically calculated by summing up the liquid contributions of the collision partners as:

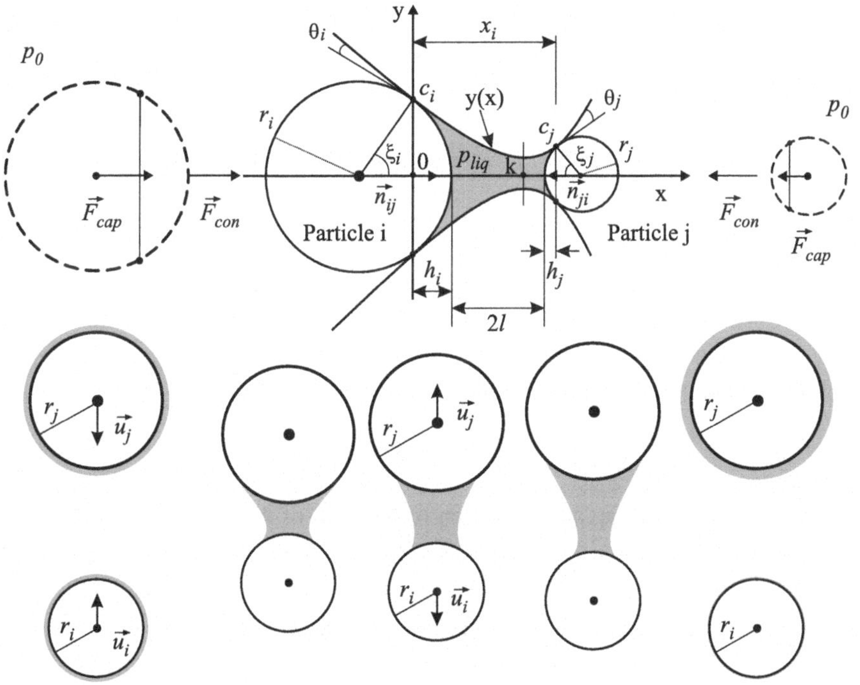

Fig. 3.31 Adhesive force due to liquid bridge (diverse contact and fill angles)

$$V_{liq} = V_{liq,i} + V_{liq,j} \tag{3.204}$$

The liquid volumes are determined for the particle i as:

$$V_{liq,i} = \pi \int_{x_k}^{x_{c,i}} y^2(x)dx - \frac{1}{3}\pi(1 - \cos\xi_i)^2(2 + \cos\xi_i), \tag{3.205}$$

and for the particle j as:

$$V_{liq,j} = \pi \int_{x_k}^{x_{c,j}} y^2(x)dx - \frac{1}{3}\pi(1 - \cos\xi_j)^2(2 + \cos\xi_j). \tag{3.206}$$

Simpler expression to compute the entire liquid volume of the bridge can be achieved using the spherical cap that is a portion of a particle cut off by a plane:

$$V_{liq} = \pi \int_{x_{c,i}}^{x_{c,j}} y^2(x)dx - \frac{\pi}{6}\left(3y_{c,i}^2 h_i + h_i^3 + 3y_{c,j}^2 h_j + h_j^3\right). \tag{3.207}$$

By exceeding the critical length of the bridge, the bridge is broken and the liquid is not distributed over the collided objects evenly, but according to the following equations:

$$V'_{liq,i} = \pi \int_{x_{\min}}^{x_{c,i}} y^2(x)dx - \frac{\pi}{6}\left(3y_{c,i}^2 h_i + h_i^3\right), \tag{3.208}$$

$$V'_{liq,j} = \pi \int_{x_{\min}}^{x_{c,j}} y^2(x)dx - \frac{\pi}{6}\left(3y_{c,j}^2 h_j + h_j^3\right), \tag{3.209}$$

with the mass conservation of the liquid bridge:

$$V_{liq} = V'_{liq,i} + V'_{liq,j}. \tag{3.210}$$

The heights of the cap are a function of the radii of the colliding particles:

$$h_i = r_i - \sqrt{r_i^2 - \left(\frac{r_i r_j}{r_i + r_j}\right)^2}, \tag{3.211}$$

$$h_j = r_j - \sqrt{r_j^2 - \left(\frac{r_i r_j}{r_i + r_j}\right)^2}. \tag{3.212}$$

The bridge length is expressed by:

$$2l = x_{c,i} + x_{c,j} - h_i - h_j. \tag{3.213}$$

In order to determine the new distribution of the liquid between the particles, the thinnest point of the liquid bridge shortly before the separation should be known:

$$x_{\min} = x_{k,sep}. \tag{3.214}$$

This can be numerically solved with the aid of following boundary conditions:

$$y(x_{c,i}) = \sqrt{r_i^2 - (r_i - h_i)^2}, \tag{3.215}$$

$$y(x_{c,j}) = \sqrt{r_j^2 - (r_i - h_j)^2}, \tag{3.216}$$

$$y(x_{\min}) = 0, \tag{3.217}$$

$$\frac{dy(x_{\min})}{dx} = 0, \tag{3.218}$$

where the meridian profile of the phase boundary between the liquid bridge and the surrounding fluid is described as a polynomial of degree two (quadratic polynomial). If the volume ratio V_i'/V_i is greater than unity, the liquid transport occurs during the bridge separation in direction of the particle i, while the liquid transport takes place towards the particle j when the volume ratio is less than unity. Is the volume ratio equal to unity, each collision particle holds the same amount of the liquid back, which has contributed with its.

Adhesive force due to solid bridge: The adhesive force due to solid bridge results in a strong and a permanent bond between collision partners. On the one hand, this effect can be the desired procedure in industrial applications such as briquette or powder coating, but on the other hand also can have negative consequences for most combustion systems by the reason of the ash deposition on boiler walls and the resulted worse heat transfer between flue gas path and water/steam side. At a sufficiently high temperature and a long contact time, sintered bridges can arise between particles. A stable solid bridge can be formed when the collided particles temperature exceeds 60% of the absolute melting temperature. The more the temperature falls below this limit the slower the sintering neck will growth, till finally the resulting adhesive force can be neglected. The building of the sinter neck depends on the surrounding pressure and the interfacial energies. While in case of low-pressure and small particles, the interfacial tension is the main mechanism in building the solid

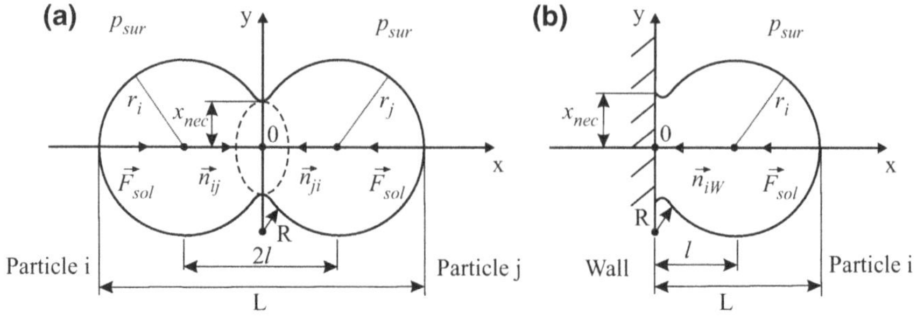

Fig. 3.32 Adhesive force due to solid bridge: **a** a particle-particle collision and **b** a particle-wall collision

bridge, the pressure sintering is crucial at higher pressure level $p > 1$ bar and relatively big particles $r_P \geq 10^{-3}$ mm (Schubert 1979). As an example, a solid bridge of two colliding objects i.e. particle-particle or particle-wall is illustrated in Fig. 3.32.

As a reference study for metallic bridging, Kuwagi et al. (2000) investigated the sintering process of iron particles at atmospheric pressure. Here, the pressure sintering effect can be neglected, while the sintering due to the interfacial tension is the main mechanism. The neck growth of solid bridge between collision partners is based on Kuczynski sintering model (Kuczynski 1949). This model assumes that the surface diffusion including the effect of surface roughness and the volume diffusion mechanisms have a relevant influence on the metal powders sintering. The neck growth, i.e. the neck radius x_{nec} is a function of neck growth time t_{nec} and can be calculated for surface diffusion cases as:

$$x_{nec} = \left(\frac{56\sigma a^4}{k_B T_P} D_s R_{nec}^3 t_{nec} \right)^{\frac{1}{7}}, \tag{3.219}$$

and for volume diffusion cases as:

$$x_{nec} = \left(\frac{10\sigma a^3}{k_B T_P} D_V R_{nec}^2 t_{nec} \right)^{\frac{1}{5}}, \tag{3.220}$$

where R_{nec} denotes the neck curvature radius, k_B is the Boltzmann constant, T_P refers to the particle temperature, σ is the surface tension, a is the lattice constant, D_s and D_V represent the surface and volume diffusion coefficients. The neck growth time t_{nec} that corresponds to each neck is reset to zero when the neck is broken during collisions. The diffusion coefficients can be calculated with the aid of Arrhenius equation as:

$$D_k = A_{0,k} \exp(-E_k / RT_P), \tag{3.221}$$

with the pre-exponential factor or frequency factor $A_{0,k}$, the universal gas constant R and the activation energy E_k. The subscript k is either surface s or volume V. The pre-exponential factor as well as the activation energy can be defined by experiments.

According to Kuczynski sintering model, the selection between the diffusion mechanisms depends on the temperature of particles. While the surface diffusion is the main factor that plays an important role in sintering of iron particles at relative lower temperature $T_P < 1050$ °C, the volume diffusion is the main mechanism at higher temperature $T_P \geq 1050$ °C. The resulted adhesive force due to solid bridge is related to the neck radius and the surface or volume energy of the neck $\sigma_{nec,k}$. It can be expressed between two spherical particles by:

$$\vec{F}_{sol} = \begin{cases} \pi x_{nec}^2 \sigma_{nec,s} \vec{n} = \pi \sigma_{nec,s} \left(\frac{56\sigma a^4}{k_B T_P} D_s R_{nec}^3 t_{nec} \right)^{\frac{2}{7}} \vec{n} & \text{case } T_P < 1050\,°\text{C} \\ \pi x_{nec}^2 \sigma_{nec,V} \vec{n} = \pi \sigma_{nec,V} \left(\frac{10\sigma a^3}{k_B T_P} D_V R_{nec}^2 t_{nec} \right)^{\frac{2}{5}} \vec{n} & \text{case } T_P \geq 1050\,°\text{C} \end{cases}. \tag{3.222}$$

Here, the adhesive force for the particle i is in the direction of the collision normal vector $\vec{n}_{ij}$ and for the particle j is in the direction of $\vec{n}_{ji}$.

The surface roughness is an important parameter for estimating the neck radius and thus the sintering behaviour. Kuwagi et al. (2000) investigated the roughness effect by introducing spherical sub-grains into the particle surface. For smooth surface, the contact area has one grain. A grain can be broken down to multiple micro-contact points. As a result, the number of sub-grains between particles with a surface roughness can be changed then to three or nine micro-contact points (see Fig. 3.33).

The resulted adhesive force due to solid bridge between two spherical particles at lower temperature under the consideration of surface roughness is determined for a smooth surface (one grain) as:

$$\vec{F}_{sol,1} = \pi \sigma_{nec} \left(\frac{56\sigma a^4}{k_B T} D_s R_{nec}^3 t_{nec} \right)^{\frac{2}{7}} \vec{n}, \tag{3.223}$$

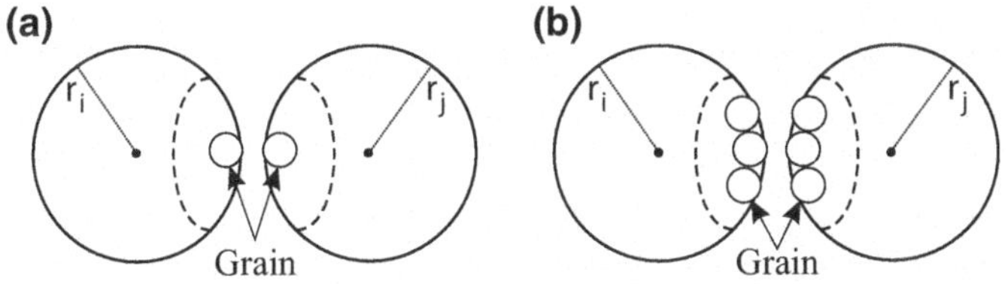

Fig. 3.33 Surface roughness: **a** a smooth surface and **b** a rough surface (Kuwagi et al. 2000)

and for three sub-grains as:

$$\vec{F}_{sol,3} = 3\pi\sigma_{nec}\left(0.056\frac{\sigma a^4}{k_B T}D_s R_{nec}^2 t_{nec}\right)^{\frac{2}{7}}\vec{n}, \tag{3.224}$$

and for nine sub-grains as:

$$\vec{F}_{sol,9} = 9\pi\sigma_{nec}\left(0.056\frac{\sigma a^4}{k_B T}D_s R_{nec}^3 t_{nec}\right)^{\frac{2}{7}}\vec{n}. \tag{3.225}$$

According to Kuwagi, the surface roughness leads to an increase in the solid bond strength between the collided particles.

The neck is assumed to continue growing, i.e. the neck radius and neck growth time will be further accumulated. This will occur when the absolute value of the contact force in normal or tangential direction is smaller than the sum of neck cross-sectional areas of all micro-contact points $A_{nec,k}$. However, the neck is broken if the condition:

$$\left|\vec{F}_{con}^{n}\right| \geq \sigma_{nec}\sum_{k=1}^{k=n} A_{nec,k}, \tag{3.226}$$

or:

$$\left|\vec{F}_{con}^{t}\right| \geq t_{nec}^{acc}\sum_{k=1}^{k=n} A_{nec,k}, \tag{3.227}$$

is met, where n represents the number of micro-contact points.

For modelling the adhesive force due to solid bridge with the aid of Kuwagi model (Kuwagi et al. 2000), the following assumptions should be considered:

- The stiffness coefficient of the solid bridge has the same value as collision partners.
- The neck can be formed between any particles in contact.
- The neck radius will only be increased during the period of contact.
- For each contact, the neck between the collision partners can be developed or separated.
- The neck will be broken when the absolute value of normal or tangential contact force exceeds the sum of neck cross-sectional areas of all micro-contact points.
- In case of neck breakage, the neck growth time and thus the neck radius is reset to zero.
- The dynamic friction coefficient μ_{dyn} is set equal to infinity so that the resulting tangential contact force is supposed to be a static force, i.e. no sliding is allowed between the collision partners (see Eq. 3.163).

The stiffness coefficient k is generally set at smaller values in order to reduce the computation effort. The use of smaller stiffness coefficients increases the value of the particle time step compared to the real particle time step (see Sect. 3.2.3.4). Sequentially, the period of collision (contact time) using the modified stiffness coefficients will be smaller than the collision period of non-modified stiffness constants. The neck radius at the end of collision should be calculated based on the real time of the collision, i.e. without modifying the stiffness coefficient. Other important point is the spring contact models that are applied to calculate the contact force. The linear approach proposed by Hooke that shows a satisfactory accuracy is frequently used. However, the neck radius at the end of a linear Hookean collision should have the same value as that at the end of a non-linear Hertzian collision. This is very important to have the correct judgement whether the neck is broken after collision or not. The judgement should not be influenced with the modification of stiffness coefficients by the reason of computation time as well as the using of a linear collision model. For this reason, a correction factor is considered and accordingly the neck radius can be determined using the following equation:

$$x_{nec} = x_{nec}^0 \left(\frac{t_{nec}}{t_{nec}^0} \right)^{\frac{1}{7}}, \tag{3.228}$$

with:

$$t_{nec} = t_{nec}^0 + t_{nec}^+. \tag{3.229}$$

Here, t_{nec}^0 represents the neck growth time based on the modified collision period, x_{nec}^0 is the neck radius resulted from t_{nec}^0 and t_{nec}^+ is the time correction factor. This correction factor is an increment of the modified neck growth time, which considers the difference between the real collision time t_{col}^{Hertz} and the adopted collision time t_{col}^{Hooke}. It can be expressed with the aid of the applied collision time t_{col} as follows:

$$t_{nec}^+ = t_{col} \frac{t_{col}^{Hertz}}{t_{col}^{Hooke}}. \tag{3.230}$$

The duration of the collision is calculated using Hertz and Hooke theories as:

$$t_{col}^{Hertz} = 2.44 \left(\frac{m_{ij}^2}{\xi^2 u_{ij}} \right)^{\frac{1}{5}}, \tag{3.231}$$

$$t_{col}^{Hooke} = \pi \sqrt{\frac{m_{ij}}{k}}, \tag{3.232}$$

where m_{ij} and u_{ij} represent the reduced mass and the relative velocity of the collision partners (i and j), respectively. The constant ξ is a function of the particle diameter and the material properties (Young's modulus E and Poisson's ratio ν):

$$\xi = E\frac{\sqrt{d_P}}{3(1-\nu^2)}. \tag{3.233}$$

The Young's modulus, also known as elastic modulus, describes the stiffness of an elastic material.

Based on the above model, the adhesive force due to solid bridge for a monodisperse system with particles of same materials (homogenous) can be numerically determined. However, the simulation of adhesive force due to solid bridge requires significantly more computational time than other adhesive forces such as liquid bridge or van der Waals. In real granular systems, the numerical calculation of adhesive force due to solid bridge is more complicated by the reason of particle size distribution as well as the inhomogeneity in the material composition. Generally, the particles consist of different materials, which in turn lead to diverse surface and volume diffusion coefficients. At high temperature, these materials may react with each other, resulting in particles with different species compositions. For a polydisperse system with particles of different materials (inhomogeneous materials), the calculation of adhesive force due to solid bridge is still an open issue, which needs to be clarified via additional theoretical and numerical works.

Momentum Balance

The particle rotation can result from a velocity gradient in an anisotropic flow field (e.g. near a wall or in a shear flow) or from particle-particle/wall collisions. The resulted moment of force consists of two contributions: moment of force due to the fluid-particle interaction and moment of force due to the short-range forces between particles (see Fig. 3.34).

The moments of force $\vec{T}_i$ acting on a particle i is given according to the following formula:

$$I_P\frac{d\vec{w}_P}{dt_P} = \sum \vec{T}_i = \underbrace{\vec{M}_{f\to P}}_{Fluid-Particle interaction} + \underbrace{\underbrace{\vec{M}^t_{con}}_{Moment\ of\ tangential\ contact\ force} + \underbrace{\vec{M}^r_{con}}_{Moment\ of\ friction\ force}}_{Moment\ of\ short-range\ forces}. \tag{3.234}$$

Here, $\vec{w}_P$ is the angular velocity, I_P and t_P denote the moment of inertia and the particle time step, respectively.

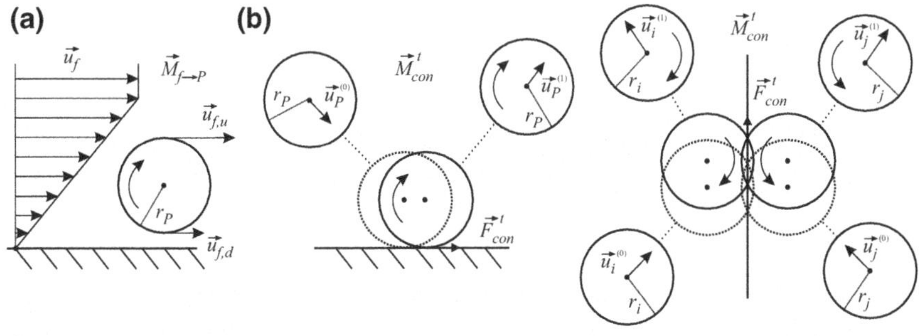

Fig. 3.34 Moment of force acting on a particle: **a** a moment of force due to fluid-solid interaction and **b** a moment of force due to short-range forces

I. Moment of force due to fluid-particle interaction

The resulting moment of force due to the fluid-particle interaction can be defined for lower particle Reynolds numbers using the relation proposed by Rubinow and Keller (1961) and for higher particle Reynolds numbers using the relation proposed by Sommerfeld (1996) based on the theoretical and numerical results of Dennis et al. (1980):

$$\vec{M}_{f\to P} = \begin{cases} \pi\mu_f d_P^3 \vec{w}_{rel} & \text{case } \mathrm{Re}_P \le 1 \\ \frac{1}{64}\rho C_R d_P^5 |\vec{w}_{rel}|\vec{w}_{rel} & \text{case } \mathrm{Re}_P > 1 \end{cases}, \tag{3.235}$$

where C_R is a rotation coefficient and $\vec{w}_{rel}$ represents the angular velocity between solid and fluid phases. The rotation coefficient depends on the rotating Reynolds number Re_R and can be defined using a step function as:

$$C_R = \begin{cases} \frac{16\pi}{\mathrm{Re}_R} & \text{case } 0 < \mathrm{Re}_R < 10^1 \\ \frac{6.45}{\sqrt{\mathrm{Re}_R}} + \frac{32.1}{\mathrm{Re}_R} & \text{case } 10^1 \le \mathrm{Re}_R < 10^3 \\ \frac{6.8}{\sqrt{\mathrm{Re}_R}} & \text{case } 10^3 \le \mathrm{Re}_R < 4\times 10^4 \\ \frac{0.058}{\sqrt[20]{\mathrm{Re}_R}} & \text{case } 4\times 10^4 \le \mathrm{Re}_R < 4\times 10^5 \\ \frac{0.397}{\sqrt[5]{\mathrm{Re}_R}} & \text{case } 4\times 10^5 \le \mathrm{Re}_R < 10^7 \end{cases}. \tag{3.236}$$

The particle Reynolds number and the rotating Reynolds number are identified using the following relations:

$$\mathrm{Re}_P = \frac{\rho_f d_P |\vec{u}_{rel}|}{\mu_f}, \tag{3.237}$$

$$\mathrm{Re}_R = \frac{0.25\rho_f d_P^2 |\vec{w}_{rel}|}{\mu_f}, \tag{3.238}$$

with the relative translation velocity between solid and fluid phases $\vec{u}_{rel}$.

It is clear that the rotation coefficient increases with decreasing the rotating Reynolds number. Consequently, the moment of force due to the fluid-particle interaction should not be neglected even at lower relative angular velocities.

II. **Moment of force due to short-range forces between particles**

The moment of force due to the short-range forces between particles results from the forces acting at the contact point. It includes two parts as follows:

Tangential moment of force: The resulting tangential moment of force on the particle i is explicitly calculated by multiplying the particle radius with the tangential contact forces that act simultaneously between the particle i and other collision partners:

$$\vec{M}^t_{con,i} = \sum_{\substack{k=1 \\ k \neq i}}^{k=N} r_i \vec{F}^t_{con,ik} \times \vec{n}_{ik}, \tag{3.239}$$

where the index N represents the number of collision partners, i.e. particles and walls, which are in contact with the particle i.

Moment of force due to asymmetrical normal traction distribution: The moment of force due to the asymmetrical normal traction distribution, also known as rolling friction torque, results from the rolling resistance between two particles or a particle and a wall in contact due to the relative rotation. The normal traction distribution is symmetrical and thus this additional torque is neglected when the particles are rigid and the contact occurs only at a point or a relative small penetration depth between collision partners. For inelastic particles, the normal traction distribution is asymmetrical. This produces a higher traction on the front half of the contact area than ,the rear (see Fig. 3.35).

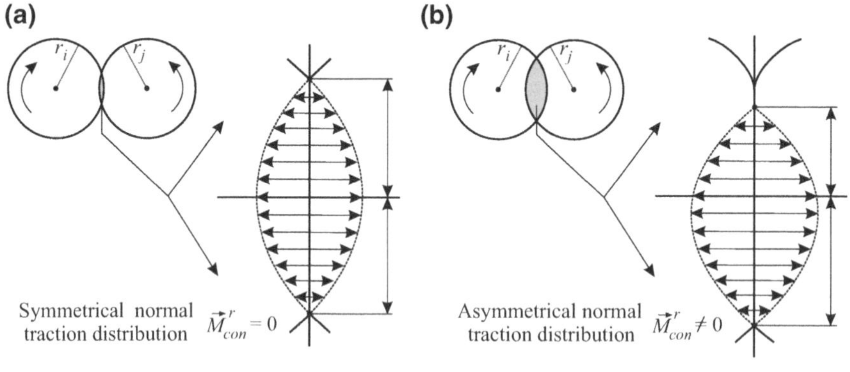

Fig. 3.35 Moment of force due to short-range forces: **a** a symmetrical normal traction distribution and **b** an asymmetrical normal traction distribution

Compared to the tangential moment of force, the determination of rolling friction torque is very difficult and still remains an open research issue. Generally, the torque due to the asymmetrical normal traction distributions can be neglected for rigid as well as viscoelastic particles. However, it plays a significant role in some cases, where the transition between static and dynamic states is present such as the formations of shear band, the heaping and the movement of a single particle on a plane (Zhu et al. 2007).

Different approaches are proposed to calculate the moment of force due to asymmetrical normal traction distribution acting on the particle i, which are listed below:

$$\vec{M}^{r}_{con,i} = \begin{cases} \sum\limits_{\substack{k=1\\k\neq i}}^{k=N} \mu_r \left|\vec{F}^{n}_{con,ik}\right| \vec{\mathrm{w}}^{n}_{ik} & \text{(Beer 1967)} \\ \sum\limits_{\substack{k=1\\k\neq i}}^{k=N} \left(-k_r \vec{\varphi}_{ik,rel} - \eta_r \frac{d\vec{\varphi}_{ik,rel}}{dt_P}\right) & \text{(Iwashita etal.1998)/(Iwashita etal. 2000)} \\ \sum\limits_{\substack{k=1\\k\neq i}}^{k=N} k_R \left|\vec{F}^{n}_{Con,ik}\right| \left|\vec{w}^{n}_{ik}\right| \vec{\mathrm{w}}^{n}_{ik} & \text{(Brilliantov etal. 1998)} \\ \sum\limits_{\substack{k=1\\k\neq i}}^{k=N} -\min\begin{bmatrix} \mu_r \left|\vec{F}^{n}_{Con,ik}\right|; \\ k_R \left|\vec{F}^{n}_{Con,ik}\right| \left|\vec{w}^{n}_{ik}\right| \end{bmatrix} \vec{\mathrm{w}}^{n}_{ik} & \text{(Zhou etal. 1999)/(Zhu etal. 2002)} \end{cases}, \tag{3.240}$$

where $\vec{\varphi}_{ik,rel}$ represents the relative angular displacement and $\vec{\mathrm{w}}^{n}_{ij}$ is the unit vector of the component of relative angular velocity in contact plane:

$$\vec{\mathrm{w}}^{n}_{ik} = \frac{\vec{w}^{n}_{ik}}{\left|\vec{w}^{n}_{ik}\right|}, \tag{3.241}$$

and $\left|\vec{w}^{n}_{ik}\right|$ is the magnitude of the component of the relative angular velocity between the particles (i and k):

$$\left|\vec{w}^{n}_{ik}\right| = \left|(r_i \vec{w}_i + r_k \vec{w}_k) \times \vec{n}_{ik}\right|. \tag{3.242}$$

The rolling friction coefficient μ_r, the rotational stiffness coefficient k_R, the rolling stiffness constant k_r and the rolling damping constant η_r are material properties and can be determined by experiments.

The main difference among these models is whether the relative angular velocity for calculation the rolling friction torque is considered or not. While the model of Beer et al. (Beer and Johnston 1967) is independent of the relative angular velocity, the approaches by Iwashita et al. (Iwashita and Oda 1998, 2000) and Brilliantov et al. (Brilliantov and Pöschel 1998) are a function of relative angular velocity.

Zhou et al. and Zhu et al. (Zhou et al. 1999; Zhu and Yu 2002) combined the approaches of Beer and Brilliantov (Beer and Johnston 1967; Brilliantov and Pöschel 1998) into a dynamic model by means of a minimum function. Due the complexity of describing the asymmetrical normal traction distribution, there is no universal model that can be applied to define the rolling friction torque. Which approach is suitable for a specific problem, further numerical and experimental studies must be carried out.

3.2.3.3 Inter-phase Coupling

In gas-solid flows, the modelling of fluid phase requires the extension of the conservation equations of the single-phase flow. In order to solve the extended conservation equations (see Eqs. 3.87–3.90), the impact of solid phase on fluid phase in each grid cell, so-called interaction values should first be determined. These interaction values can be obtained using an averaging process, applied to the physical properties of particles that are located in the investigated control volume:

$$\psi_{P \to f} = f\left(\sum_{\forall i \in CV} f_{P,i}^{CV} \Phi_{P,i} \right). \tag{3.243}$$

The symbol $\psi_{P \to f}$ represents one of the interaction values, i.e. volumetric void fraction, momentum and heat transfers (ε_f, $S_{P \to f}^{u,P}$, $S_{P \to f}^{h,P}$). The parameter $\Phi_{P,i}$ denotes the particle volume, the particle velocity or the particle temperature. The spherical section $f_{P,i}^{CV}$ indicates the fraction of the particle volume, which exists in the control volume (see Fig. 3.36a). The most accurate but also the most complex method to determine the interaction values in a grid cell is to consider all particles that are completely located inside the control volume and other particles that are partly located in this CV. This is a complex task, so that numerical approximations should be introduced. Here, the particle centre method represents the simplest approach since it assumes that the entire volume of the particle is allocated to the fluid cell, where the centre of this particle is found. If the particle centre method is applied to approximate the interaction values, then the complex spherical section $f_{P,i}^{CV}$ to be determined in

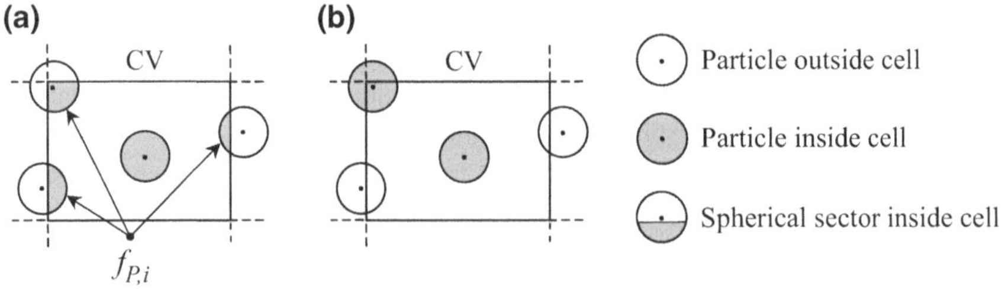

Fig. 3.36 Determination of interaction values: **a** the exact method and **b** the particle centre method

Eq. (3.243) is omitted (see Fig. 3.36b). The efficiency gained in this method is associated with high uncertainty for the determination of the interaction values.

Volumetric Void Fraction

The volumetric void fraction is a measure of the solid content in each control volume of the computational domain. The volumetric void fraction in a grid cell ε_f^{CV} can be defined as:

$$\varepsilon_f^{CV} = 1 - \varepsilon_P^{CV} = 1 - \frac{1}{V^{CV}} \sum_{\forall i \in CV} f_{P,i}^{CV} V_{P,i}, \tag{3.244}$$

where V^{CV} is the volume of the grid cell, $V_{P,i}$ denotes the volume of the particle i and $f_{P,i}^{CV}$ is the spherical section.

Momentum Transfer

The influence of particles motions on the fluid phase is calculated for each control volume by determining the change in the particle velocity along their trajectories based on all particles crossing this CV during the time interval between two consecutive fluid time steps:

$$\vec{S}_{P \to f}^{u,P} = \frac{\vec{F}_{P \to f}}{V^{CV}}. \tag{3.245}$$

According to the third Newtonian axiom, the exerted force from dispersed phase on fluid phase $\vec{F}_{P \to f}$ must be equal to the force acting from fluid phase on particles $\vec{F}_{f \to P}$, but in the opposite direction. The momentum interaction between fluid and solid phases can be performed in three different approaches (Feng and Yu 2004). In the first scheme, the transferred force from fluid to solid phase is separately calculated for each particle depending on the particle velocity as:

$$\vec{F}_{f \to P,i} = \left(\vec{F}_{dra} + \vec{F}_{vir} + \vec{F}_{bas} + \vec{F}_{pre} + \vec{F}_{saf} + \vec{F}_{rad} + \vec{F}_{mag}\right)_i, \tag{3.246}$$

while the transferred force from particles to fluid phase is determined using the local arithmetic averaging as follows:

$$\vec{F}_{P \to f} = \beta\left(\overline{\vec{u}}_P - \vec{u}_f\right)V^{CV} = \beta\left[\left(\frac{1}{N^{CV}} \sum_{i=1}^{k=N^{CV}} \vec{u}_{P,k}\right) - \vec{u}_f\right]V^{CV}. \tag{3.247}$$

Here, the variables N^{CV} and $\overline{\vec{u}}_P$ represent the number of particles and the mean particle velocity in the investigated grid cell, respectively. The symbol β denotes the inter-phase momentum transfer coefficient (resistance coefficient). In this scheme, the compliance of third Newtonian axiom is not guaranteed.

In the second scheme, the transferred force from particles to fluid phase $\vec{F}_{P\to f}$ is performed by a local averaging method such as in the first scheme. Then, the resultant force is distributed on the particles located in the control volume (uniformly in case of monodisperse systems and non-uniformly in case of polydisperse systems):

$$\vec{F}_{f\to P,i} = \begin{cases} \frac{\vec{F}_{P\to f} V^{CV}}{N^{CV}} & \text{For monodisperse system} \\ w(r)\frac{\vec{F}_{P\to f} V^{CV}}{N^{CV}} & \text{For polydisperse system} \end{cases}, \tag{3.248}$$

where $\vec{F}_{f\to P,i}$ corresponds to the fluid force acting on each particle. The weight function $w(r)$ allows big particles to have more influence than smaller ones. The scheme 2, although it guarantees the third Newtonian axiom, is inaccurate from a physical view.

In the third scheme, the fluid force acting on individual particle is calculated according to Eq. (3.246), while the particle force acting on the fluid phase is determined through the sum of all particles forces:

$$\vec{F}_{P\to f} = \sum_{i=1}^{N^{CV}} \vec{F}_{f\to P,i} = \beta\left[\sum_{i=1}^{k=N^{CV}} \left(\vec{u}_{P,k} - \vec{u}_f\right)\right] V^{CV}. \tag{3.249}$$

Here, $\vec{u}_f$ represents the fluid velocity in the considered control volume and $\vec{u}_{P,k}$ is the velocity of the particle k in the CV. The above-described stability and accuracy problems, which occur in schemes 1 and 2, no longer take place in scheme 3 and this has made it very popular (Alobaid and Epple 2013; Deen et al. 2007; Link 2006; Stroh et al. 2016).

For determination the momentum transfer, the use of completely explicit method may result in convergence problems, while solving the conservation equations of fluid phase. The reason for this is that the Navier-Stokes equations can be stiffly deflected by large velocity differences between phases. Therefore, a semi-implicit method is generally applied to define the momentum transfer. The semi-implicit treatment of momentum transfer has the advantage that the inter-phase momentum transfer coefficient β can be treated as constant during each fluid time step (Götz 2006).

In the literature, there are a lot of models for the calculation of the inter-phase momentum transfer coefficient in the frame of single-particle method. These drag models differ from each other in their application validity. In this context, Ergun and Wen model is the most frequently model used. It consists of Ergun equation for the high loading sub-areas and the correlation of Wen for low loading sub-areas (Ergun 1952; Wen and Yu 1966):

$$\beta_{Erg} = \frac{\mu_f}{d_P^2}\frac{(1-\varepsilon_f)}{\varepsilon_f}\left[150(1-\varepsilon_f) + 1.75\,\mathrm{Re}_P\right] \quad \text{case}\ \varepsilon_f < 0.8, \tag{3.250}$$

$$\beta_{Wen\&Yu} = \frac{3}{4} C_{dra} (1 - \varepsilon_f) \varepsilon_f^{-2.65} \mathrm{Re}_P \quad \text{case } \varepsilon_f > 0.8. \tag{3.251}$$

The drag coefficient C_{dra} can be calculated for an isolated spherical particle as follows:

$$C_{dra} = \begin{cases} \frac{24}{\mathrm{Re}_P} \left(1 + 0.15 \mathrm{Re}_P^{0.687}\right) & \text{case } \mathrm{Re}_P < 10^3 \\ 0.44 & \text{case } \mathrm{Re}_P \geq 10^3 \end{cases}, \tag{3.252}$$

with the particle Reynolds number Re_P. The combined model as per Ergun and Wen shows a discontinuity at $\varepsilon_f = 0.8$.

The approach of Foscolo et al. (1983) predicts the inter-phase momentum transfer coefficient for $\varepsilon_f > 0.4$ using the following expressing:

$$\beta_{Fos} = \frac{\mu_f}{d_P^2} \frac{(1 - \varepsilon_f)}{\varepsilon_f^{2.8}} \left(\frac{17.3}{\mathrm{Re}_P} + 0.336 \right) \mathrm{Re}_P \quad \text{case } \varepsilon_f > 0.4, \tag{3.253}$$

whereas the approach of Garside et al. (Garside and Al-Dibouni 1977) calculates the inter-phase momentum transfer coefficient for all volumetric void fraction values, except the zone $0.75 < \varepsilon_f < 0.8$:

$$\beta_{Gar} = \frac{3}{4} \frac{\mu_f}{d_P^2} \frac{(1 - \varepsilon_f)}{\mathrm{Re}_t^2} \left(0.63 + 4.8 \sqrt{\frac{\varepsilon_f \mathrm{R}_t}{\mathrm{Re}_f}} \right)^2 \mathrm{Re}_P, \tag{3.254}$$

with

$$\mathrm{R}_t = 0.5 \left[A - 0.06 \frac{\mathrm{Re}_P}{\varepsilon_f} + \sqrt{\left(0.06 \frac{\mathrm{Re}_P}{\varepsilon_f}\right)^2 + 0.12 \frac{\mathrm{Re}_P}{\varepsilon_f} (2B - A) + A^2} \right]. \tag{3.255}$$

The coefficients A and B are given by:

$$A = \varepsilon_f^{4.14}, \tag{3.256}$$

$$B = \begin{cases} 0.8 \varepsilon_f^{1.28}, & \text{case } \varepsilon_f \leq 0.85 \\ \varepsilon_f^{2.65}, & \text{case } \varepsilon_f > 0.85 \end{cases}. \tag{3.257}$$

The influence of Ergun, Wen and Yu, Foscolo and Garside models on the characteristic of a gas-solid fluidized bed was assessed by Goldschmidt et al. (2004). The results show that the minimum fluidization velocity and the average bed height can be highly reliant on the drag model used. According to Goldschmidt, none of these models can be universally applied.

Based on the lattice Boltzmann method, Koch and Hill (Hill et al. 2001; Koch and Hill 2001) delivered from the simulation results the following relationship for the inter-phase momentum transfer coefficient:

$$\beta_{Koch/Hill} = \frac{\mu_f}{d_P^2}\left(A\frac{(1-\varepsilon_f)^2}{\varepsilon_f} + B(1-\varepsilon_f)\mathrm{Re}_P\right) \quad \text{case}\ \mathrm{Re}_P > 40, \tag{3.258}$$

with the coefficients A and B:

$$A = \begin{cases} 180, & \text{case}\ \varepsilon_f < 0.6 \\ \frac{18\varepsilon_f^3}{(1-\varepsilon_f)} \frac{1+\frac{3}{\sqrt{2}}(1-\varepsilon_f)^{\frac{1}{2}} + \frac{135}{64}(1-\varepsilon_f)\ln(1-\varepsilon_f) + 16.14(1-\varepsilon_f)}{1+0.681(1-\varepsilon_f) - 8.48(1-\varepsilon_f)^2 + 8.16(1-\varepsilon_f)^3} & \text{case}\ \varepsilon_f \geq 0.6 \end{cases}, \tag{3.259}$$

$$B = 0.6057(1-\varepsilon_f)^2 + 1.908(1-\varepsilon_f)\varepsilon_f^2 + 0.209\varepsilon_f^{-3}. \tag{3.260}$$

Based on the work of Koch and Hill (2001), Beetstra et al. (2007) developed a similar approach for the determination of the inter-phase momentum transfer coefficient, showing a good agreement with measurements:

$$\beta_{Beet} = \frac{\mu_f}{d_P^2}\left[A\frac{(1-\varepsilon_f)^2}{\varepsilon_f} + B(1-\varepsilon_f)\mathrm{Re}_P\right], \quad \text{case}\ \mathrm{Re}_P < 1000, \tag{3.261}$$

with the coefficients A:

$$A = 180 + \frac{18\varepsilon_f^4}{(1-\varepsilon_f)}\left(1 + 1.5\sqrt{(1-\varepsilon_f)}\right), \tag{3.262}$$

and B:

$$B = \frac{0.31\left(\varepsilon_f^{-1} + 3\varepsilon_f(1-\varepsilon_f) + 8.4\mathrm{Re}_P^{-0.343}\right)}{1 + 10^{3(1-\varepsilon_f)}\mathrm{Re}_P^{2\varepsilon_f - 2.5}}. \tag{3.263}$$

Additional information regarding the above mentioned drag models can be found for example in Deen et al. (2007), Epple et al. (2012). Recently, new drag models (e.g. three-zone and four-zone models) that determines the inter-phase momentum transfer coefficient by selecting universal drag laws for each zone to match the measurements as well as advanced drag models that are based on flow structures [e.g. energy minimization multi-scale (EMMS)] are developed, ensuring better accuracy.

Heat Transfer

The heat transfer from solid to fluid due to the temperature difference between both phases consists of three parts: the heat transfer through convection, the heat transfer through radiation and the heat transfer due to phase transformation from solid to fluid phase:

$$S_{P\to f}^{h,P} = \underbrace{S_{P\to f}^{conv}}_{\substack{\textit{Heat transfer}\\ \textit{through convection}}} + \underbrace{S_{P\to f}^{rad}}_{\substack{\textit{Heat transfer}\\ \textit{through radation}}} + \underbrace{S_{P\to f}^{mas}}_{\substack{\textit{Heat transfer}\\ \textit{through mass transfer}}} . \tag{3.264}$$

The convection heat transfer from particles to fluid phase between two consecutive time steps can be determined by considering the temperature change of all particles that are located in the control volume as:

$$S_{P\to f}^{conv} = \sum_{i=1}^{k=N^{CV}} \alpha_P A_{P,k} \left(T_{P,k} - T_f\right), \tag{3.265}$$

where α_P denotes the heat transfer coefficient, $A_{P,k}$ is the particle surface, $T_{P,k}$ is the temperature of the particle k, N^{CV} and T_f represent the number of particles in the control volume and the fluid temperature in the grid cell, respectively. The heat transfer through radiation is considered in the balance equation for the total radiation intensity (see Eq. 3.45).

The heat transfer due to phase transformation from solid to fluid phase occurs by the reason of heterogeneous reactions such as char-oxidation, Boudouard reaction and hydrogenating gasification.

The heat transfer from fluid to solid phase can be written as:

$$S_{f\to P,i} = \underbrace{S_{f\to P,i}^{conv}}_{\substack{\textit{Heat transfer}\\ \textit{through convection}}} + \underbrace{S_{f\to P,i}^{rad}}_{\substack{\textit{Heat transfer}\\ \textit{through radation}}} . \tag{3.266}$$

The convection and radiation parts are calculated for each particle that is located in control volume, depending on its temperature as:

$$S_{f\to P,i}^{conv} = \alpha_P A_{P,i} \left(T_f - T_{P,i}\right), \tag{3.267}$$

$$S_{f\to P,i}^{rad} = A_{P,i} q_{rad}, \tag{3.268}$$

where q_{rad} is the radiation heat flux. In Eq. (3.268), only the investigated control volume is considered, i.e. the particles that are located in the investigated grid cell are only taken into account. The determination of other grid cells' contributions is extremely difficult and thus additional theoretical and numerical works are necessary.

3.2.3.4 Particle Time Step

In the single-particle method, different time steps, i.e. the fluid time step and the particle time step, are used in order to calculate the time-dependent motion of particle and fluid phases. The size of the fluid time step can be arbitrarily set and remains constant during the calculation. The particle time step can be selected as variable or constant depending on the selected particle-particle/wall interaction model. When the single-particle method is combined with stochastic collision detection models, the particle time step is assumed to be constant. If the single-particle method is combined with deterministic collision detection models, the calculation of the particle phase takes place with adaptive particle time step size in the case of hard sphere model and with constant particle time step size in case of the soft sphere model.

Adaptive particle time step size: In the hard sphere model with adaptive particle time step size, a particle collide with an another particle or with a wall only at the time $t + \Delta t_P$. Starting from the last particle time step, the time that a particle i requires to come into contact with a possible collision partner j is calculated as follows:

$$\Delta t_{P,i}(j) = \frac{-\vec{r}_{ij} \cdot \vec{u}_{ij} - \sqrt{\left(\vec{r}_{ij} \cdot \vec{u}_{ij}\right)^2 - \vec{u}_{ij}^2 \left[-\vec{r}_{ij}^2 - \left(r_i - r_j\right)^2\right]}}{\vec{u}_{ij}^2}, \quad \forall j \in A. \tag{3.269}$$

Here, the list $A = \{1, 2..., N\}$ contains all particles and walls of the computational domain. If the scalar product of the relative velocity $\vec{u}_{ij} = \vec{u}_i - \vec{u}_j$ and the position vector $\vec{r}_{ij} = \vec{r}_i - \vec{r}_j$ has a positive value ($\vec{r}_{ij} \cdot \vec{u}_{ij} < 0$), the particles move towards one another. The particles move away from each other in the case of ($\vec{r}_{ij} \cdot \vec{u}_{ij} > 0$, see Fig. 3.9). The minimum of all $\Delta t_{P,i}(j)$ of the particle i with the possible collision partners j from the list A is called the collision time $\Delta t_{P,col}(i)$ of the particle i:

$$\Delta t_{P,col}(i) = \min\left(\Delta t_{P,i}(j)\right), \quad \forall j \in A. \tag{3.270}$$

The minimum collision time for all particles in the entire computational domain is then defined using the following formula:

$$\Delta t_{P,col}(i,j) = \min\left(\Delta t_{P,col}(i)\right), \quad \forall (i,j) \in A. \tag{3.271}$$

The minimum collision time for all particles in the entire computational domain is added to the last minimum collision time (within the current fluid time step size):

$$t_{P,acc} = t^0_{P,acc} + \Delta t_{P,col}(i,j). \tag{3.272}$$

In order to reduce the level of computational complexity, only the particle i and its collision partner j are moved to their collision positions at the point in time $t + t_{P,acc}$. Here, the distance travelled in each case is the result of the product of the old velocity and the current minimum collision time $\Delta t_{P,col}(i,j)$. After the collision has been determined, the translational and rotational velocities of the colliding particles can then be computed using Eqs. (3.109–3.112).

In the next step, a new minimum collision time is detected and the described sequence is repeated. This cycle runs until the accumulation time $t_{P,acc}$ is greater than or equal to the fluid time step size Δt_f. If this occurs, all the particles of the computational domain are moved to their new positions and the distance travelled by a particle k is defined as follows:

$$\Delta\vec{x}_k = \vec{u}_k\left[\Delta t_f - \left(t_{P,acc} - \Delta t_{P,col}(i,j)\right)\right], \quad \forall k \in A. \tag{3.273}$$

Constant particle time step size: In the soft sphere model, the fluid time step size Δt_f is subdivided into several equal particle time step sizes Δt_P. Using the particle time step size and the explicit Euler method for example, the translational and rotational velocities of the particles are determined by solving the conservation equations of momentum and angular momentum as:

$$\vec{u}_P(t+\Delta t_P) = \vec{u}_P(t) + \frac{1}{m_P}\int\limits_t^{t+\Delta t_P}\sum_k \vec{F}_k(t)dt \Rightarrow \vec{u}_P^{(j)} = \vec{u}_P^{(j-1)} + \frac{\Delta t_P}{m_P}\vec{F}_P, \tag{3.274}$$

$$\vec{w}_P(t+\Delta t_P) = \vec{w}_P(t) + \frac{1}{I_P}\int\limits_t^{t+\Delta t_P}\sum_k \vec{T}_k(t)dt \Rightarrow \vec{w}_P^{(j)} = \vec{w}_P^{(j-1)} + \frac{\Delta t_P}{I_P}\vec{T}_P. \tag{3.275}$$

The indices (j) and $(j-1)$ refer to the current and previous particle time steps. The particle's current position and angular displacement can be computed from the time integration of Eqs. (3.274) and (3.275) as follows:

$$\vec{x}_P(t+\Delta t_P) = \vec{x}_P(t) + \int\limits_t^{t+\Delta t_P}\vec{u}_P(t)dt \Rightarrow \vec{x}_P^{(j)} = \vec{x}_P^{(j-1)} + \Delta t_P\vec{u}_P^{(j)}, \tag{3.276}$$

$$\vec{\varphi}_P(t+\Delta t_P) = \vec{\varphi}_P(t) + \int\limits_t^{t+\Delta t_P}\vec{w}_P(t)dt \Rightarrow \vec{\varphi}_P^{(j)} = \vec{\varphi}_P^{(j-1)} + \Delta t_P\vec{w}_P^{(j)}. \tag{3.277}$$

The explicit Euler method is of the first-order regarding to the computational effort and is slightly time-consuming in terms of implementation. In the case of large particle time step sizes, unstable behaviour and inaccuracies in the conservation of energy may occur. Therefore, a higher order method such as the Runge-Kutta method is recommended.

The particle time step size can be determined using the following minimum function:

$$\Delta t_p = \min\left(\Delta t_P^{tur}; \Delta t_P^{con}; \Delta t_P^{col}\right), \tag{3.278}$$

where Δt_P^{tur}, Δt_P^{con} and Δt_P^{col} are the particle time steps due to turbulent flow, due to contact force curve and due to collision, respectively.

I. Particle time step due to turbulent flow

The particle time step size must be selected, so that the particle trajectory is determined during the free flight phase with a reasonable degree of accuracy. This is essential especially in the case of turbulent flows, where the aerodynamic surface forces acting on particles can be subjected to strong temporal and spatial fluctuations. In granular flows with high solid content, turbulence is often neglected and thus the influence of the fluid behaviour on the determination of the particle time step can be disregarded.

II. Particle time step due to contact force curve

The particle time step size must be selected, so that the approximation of normal and tangential contact force curves during a particle-particle or a particle-wall collision can be achieved with good accuracy. If the total contact time is resolved by N_T sub-time steps, the particle time step sizes in the normal direction and in the tangential direction are obtained from:

$$\Delta t_P^{n,con} = \frac{T_P^n}{N_T} = \frac{\pi}{N_T \lambda_P^n}, \tag{3.279}$$

$$\Delta t_P^{t,con} = \frac{T_P^t}{N_T} = \frac{\pi}{N_T \lambda_P^t}, \tag{3.280}$$

where T^n and T^t are the contact times that describe the collision duration. The time constants λ_P^n and λ_P^t can be computed using the following expressions:

$$\lambda_P^n = \sqrt{\frac{k^n}{m_{ij}} - (D_P^n)^2}, \tag{3.281}$$

$$\lambda_P^t = \sqrt{\frac{2}{7}\frac{k^n}{m_{ij}} - (D_P^t)^2}, \tag{3.282}$$

with:

$$D_P^n = \frac{\eta^n}{2m_{ij}}, \tag{3.283}$$

$$D_P^t = \frac{\eta^t}{2m_{ij}}. \tag{3.284}$$

By inserting Eqs. (3.281) and (3.282) in Eqs. (3.279) and (3.280) under the consideration of damping parameters in normal and in tangential direction Eqs. (3.160) and (3.165) for a linear spring model ($a = 1$) with a linear visco-elastic damping element ($b = 0$), the following relations can be obtained:

$$\Delta t_P^{n,con} = \frac{2\pi}{N_T}\sqrt{\frac{m_{ij,\min}}{k^n\left(4-\alpha_{dam}^2(e^n)\right)}}, \tag{3.285}$$

$$\Delta t_P^{t,con} = \frac{2\pi}{N_T}\sqrt{\frac{7}{2}\frac{m_{ij,\min}}{k^t\left(4-\alpha_{dam}^2(\beta^t)\right)}}. \tag{3.286}$$

The empirical damping constants $\alpha_{dam}(e^n)$ and $\alpha_{dam}(\beta^t)$ can be determined from Eqs. (3.161) and (3.166). The minimum reduced mass $m_{ij,\min}$ amounts to $0.5m_P$ in a monodisperse system and $0.5m_{P,\min}$ in a polydisperse system. If the absolute values after the collision are only of interest, the sub-time steps can be set in the range between 10 and 20. However, precise calculations of the contact force curves require a significantly higher number of sub-time steps with dependence on the stiffness coefficient. Götz (2006) compared the course of the normal penetration depth δ^n as a function of time for a linear spring model with a linear visco-elastic damping element with different damping constants $\alpha_{dam}(e^n)$. When the damping constant increases, the contact time T_P^n increases and consequently the particle time step size $\Delta t_P^{n,con}$ also increases.

In addition, the maximum penetration depth δ^n of the locus shifts to smaller values and is achieved at an earlier point in time (see Fig. 3.37a). If the particle time step size is too large, as shown in Fig. 3.37b, the collision partners can already exceed the physically meaningful, maximum penetration depth $\delta^n_{\max}$ by $\Delta\delta^n$ within the first time step. This leads on the one hand to an overestimated spring force (elastic force) due to the excessive depth of penetration $\delta^n_{\max} + \Delta\delta^n$ and on the other hand to an overestimated damping force caused by the excessive relative velocity $\vec{u}_{ij}^n + \Delta\vec{u}_{ij}^n$ (see Eq. 3.155). In this case, larger contact forces in the normal direction are calculated, which in turn result in a faulty energy balance since a violation of the third Newtonian axiom occurs.

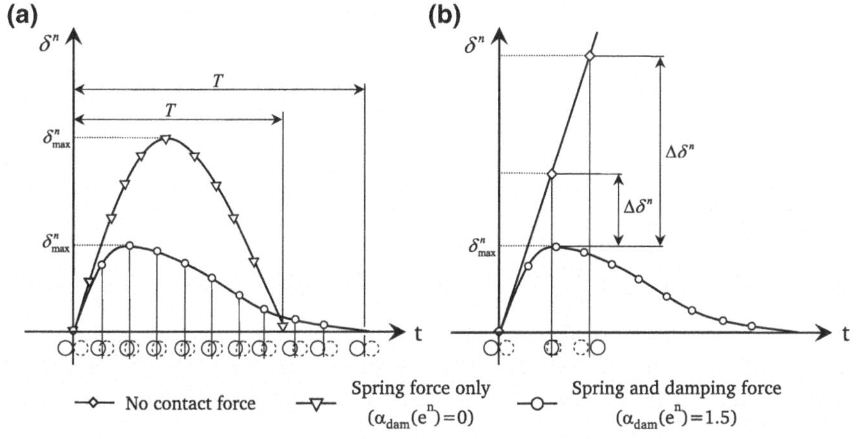

Fig. 3.37 Penetration depth course in the normal direction as a function of time

III. Particle time step due to collision

The particle time step size must be selected, so that each collision is registered in time. Regardless of the relative velocity of particles, the collision partners should not penetrate far away into each other before the collision is detected and the corresponding contact forces counteract the particles' movements. Large particle time steps cause great penetration depths that lead to unrealistic contact forces and consequently an erroneous in the energy balance. In order to detect the collision at the right time, a path length is defined, which both collision partners should not exceed it within the particle time step. Here, a maximum normal penetration depth is determined using the following formula after Götz (2006):

$$\delta^n_{\max} = \frac{\left|\vec{u}^{n(0)}_{ij,\max}\right|}{\sqrt{(\lambda^n)^2 + (D^n)^2}} \exp\left[-\left(\frac{D^n}{\lambda^n}\right)\arctan\left(\frac{\lambda^n}{D^n}\right)\right]. \tag{3.287}$$

The normal particle time step $\Delta t_P^{n,col}$, in which the maximum normal penetration depth is travelled, is given as follows:

$$\Delta t_P^{n,col} = \frac{\delta^n_{\max}}{\left|\vec{u}^{n(0)}_{ij,\max}\right|} = \frac{1}{\sqrt{(\lambda^n)^2 + (D^n)^2}} \exp\left[-\left(\frac{D^n}{\lambda^n}\right)\arctan\left(\frac{\lambda^n}{D^n}\right)\right]. \tag{3.288}$$

The maximum tangential penetration $\delta^t_{\max}$ and the tangential particle time step size $\Delta t_P^{t,col}$ are determined in a similar way.

In order to ensure the conservation of energy, the following conditions must be fulfilled (Link 2006):

$$\Delta t_P^{n,con} = \Delta t_P^{t,con}, \tag{3.289}$$

$$\Delta t_P^{n,col} = \Delta t_P^{t,col}, \tag{3.290}$$

which result in the following correlation between k^n and k^t:

$$\frac{k^t}{k^n} = \frac{7\left(4 - \alpha^2_{Dam}(e^n)\right)}{2\left(4 - \alpha^2_{Dam}(\beta^t)\right)}. \tag{3.291}$$

While the normal stiffness coefficient k^n is defined as per Eq. (3.157), the calculation of the tangential stiffness coefficient is carried out either with Eq. (3.164) or with Eq. (3.291). The previously described approach for calculating the normal and tangential stiffness coefficients is based on the material properties and provides values of high magnitude (10^4–10^9 N/m). Stiffness coefficients with high-order of magnitude require very small particle time steps, which can hardly be performed by the currently available computers. By selecting smaller stiffness coefficients, the particle time step can be increased and consequently the computing effort is reduced. However, there is a risk at sufficiently high relative velocity to raise unrealistic penetrations between the collision partners. In order to obtain the minimal stiffness coefficient that prevents the unphysical penetration between the particle-particle and particle-wall, an extra condition must be formulated. By definition, a penetration is unrealistic if the maximum penetration depth is greater than the smallest particle radius. Using this condition $\delta^n_{\max} \leq r_{P,\min}$, a minimal normal stiffness coefficient is derived:

$$k^n_{\min} = \frac{4}{3}\frac{\pi r^3_{P,\max}}{\left(\delta^n_{\max}\right)^2}\rho_P\left|\vec{u}^n_{ij,\max}\right|^2 \exp\left[\frac{-2\alpha_{Dam}}{\sqrt{4-\alpha^2_{Dam}}}\arctan\left(\frac{\sqrt{4-\alpha^2_{Dam}}}{\alpha_{Dam}}\right)\right], \tag{3.292}$$

where the maximum relative velocity $\vec{u}^n_{ij,\max}$ and the maximum penetration depth $\delta^n_{\max} \leq r_{P,\min}$ must be predefined before the simulation. If the normal stiffness coefficient is determined using Eq. (3.292), the calculated particle time step size ensures that both the courses of the locus (penetration depth) and the force curves can be approximated with a satisfactory degree of accuracy. Accordingly, unphysical penetrations between the particles and between the particles and the walls are avoided.

3.2.4 Hybrid Method

In order to reduce the computing time of collision detections, O'Rourke et al. (Andrews and O'Rourke 1996; O'Rourke et al. 1993) suggested a hybrid method that treats simultaneously the particles as a continuum and as a discrete phase. In their work, they extended the particle-in-cell (PIC) method to a multiphase particle-in-cell (MP-PIC) method. In the latter, the modelling of particle-particle stresses (particle-particle interactions) that is difficult to compute for each particle in dense flows is performed as a gradient on the Eulerian grid similar to the two-fluid method. In order to achieve that, the particle properties are interpolated to the Eulerian grid, where the particle-particle interactions are determined and the interaction values are then interpolated back to discrete particles. The calculation of particle-particle interactions on Eulerian grid increases the computational efficiency with no significant increase in the numerical error (Snider 2001).

The governing equations of MP-PIC numerical scheme, available in CPFD-BARRACUDA, are briefly explained here. The continuity, Navier-Stokes, specific enthalpy and species transport equations of the fluid phase in Eulerian representation can be expressed as in Eqs. (3.87–3.90). In these equations, the volumetric void fraction is defined as:

$$\varepsilon_f = 1 - \iiint \phi V_P dV_P d\rho_P d\vec{u}_P, \tag{3.293}$$

with the particle probability distribution function:

$$\phi(\vec{x}_P, \vec{u}_P, \rho_P, V_P, t). \tag{3.294}$$

Here, $\vec{x}_P$ refers to the particle position, $\vec{u}_P$ is the particle velocity, ρ_P and V_P represent the particle density and the particle volume, respectively. Integrating the particle probability distribution function over velocity and mass yields the probable number of particles per unit volume at location $\vec{x}$ and time t in the intervals $(\vec{u}_P, \vec{u}_P + d\vec{u}_P)$, $(\rho_P, \rho_P + d\rho_P)$ and $(V_P, V_P + dV_P)$. The inter-phase momentum transfer per volume can be expressed using the following formula:

$$\vec{S}^{u,P}_{P\to f} = -\iiint \phi V_P \rho_P \left[\frac{\beta}{\rho_P}\left(\vec{u}_f - \vec{u}_P\right) - \frac{1}{\rho_P}\nabla p\right] dV_P d\rho_P d\vec{u}_P, \tag{3.295}$$

where the symbol β denotes the inter-phase momentum transfer coefficient (resistance coefficient, see Sect. 3.2.3.3.2).

The movement of particles is calculated by solving the following equation:

$$\frac{\partial \phi}{\partial t} + \nabla \cdot (\phi \vec{u}_P) + \nabla_{u_P} \cdot \left(\phi \vec{G}\right) = 0. \tag{3.296}$$

The term ∇_{u_P} represents the divergence operator with respect to velocity and $\vec{G}$ is the discrete particle acceleration that can be given as:

$$\begin{aligned}\vec{G} = \; & \underbrace{\frac{\beta}{\rho_P}\left(\vec{u}_f - \vec{u}_P\right)}_{Acceleration\,due\,to\,aerodynamic\,drag\,force} - \underbrace{\frac{1}{\rho_P}\nabla p}_{Acceleration\,due\,to\,pressure\,gradient} \\ & + \underbrace{\vec{g}}_{Acceleration\,due\,to gravity} - \underbrace{\frac{1}{\left(1-\varepsilon_f\right)\rho_P}\nabla \tau_P}_{Acceleration\,due\,to gradient\,of\,the\,inter-particle\,stress}\end{aligned} \tag{3.297}$$

On the right side of Eq. (3.297), the first term denotes the acceleration due to drag force, while the second term represents the acceleration due to pressure gradient. The third and the fourth terms are the acceleration due to gravity and the gradient of inter-particle stresses (particle-particle stresses), respectively.

The Eulerian governing equations for particle phase can be obtained by multiplying Eq. (3.296) by $\rho_P V_P$ as well as $\rho_P V_P \vec{u}_P$ and integrating over particle density, volume and velocity coordinates. Accordingly, the particle continuity equation and the particle momentum equation are expressed by:

$$\frac{\partial\left(\bar{\rho}_P \overline{\left(1-\varepsilon_f\right)}\right)}{\partial t} + \nabla \cdot \left(\bar{\rho}_P \overline{\left(1-\varepsilon_f\right)} \bar{\vec{u}}_P\right) = 0, \tag{3.298}$$

$$\frac{\partial\left(\bar{\rho}_P\overline{(1-\varepsilon_f)}\vec{\bar{u}}_P\right)}{\partial t}+\nabla\cdot\left(\bar{\rho}_P\overline{(1-\varepsilon_f)\vec{\bar{u}}_P\vec{\bar{u}}_P}\right)=-\nabla\tau_P-(1-\varepsilon_f)\nabla p+\bar{\rho}_P\overline{(1-\varepsilon_f)}\vec{g}$$
$$+\iiint\phi V_P\beta(\vec{u}_f-\vec{u}_P)dV_Pd\rho_Pd\vec{u}_P$$
$$-\nabla\cdot\left[\iiint\phi V_P\rho_P(\vec{u}_P-\vec{\bar{u}}_P)(\vec{u}_P-\vec{\bar{u}}_P)dV_Pd\rho_Pd\vec{u}_P\right], \tag{3.299}$$

with the average particle velocity $\vec{\bar{u}}_P$ and the average particle density $\bar{\rho}_P$, defined as:

$$\vec{\bar{u}}_P=\frac{1}{\bar{\rho}_P(1-\varepsilon_f)}\iiint\phi V_P\rho_P\vec{u}_PdV_Pd\rho_Pd\vec{u}_P, \tag{3.300}$$

$$\bar{\rho}_P=\iiint\phi V_P\rho_PdV_Pd\rho_Pd\vec{u}_P. \tag{3.301}$$

The collisions between particles, where particles are modelled as a continuum, can be determined by an isotropic inter-particle stress tensor that its off-diagonal elements are neglected (Harris and Crighton 1994):

$$\tau_P=\frac{P_P(1-\varepsilon_f)}{\varepsilon_{P,cpl}-(1-\varepsilon_f)}, \tag{3.302}$$

where P_P is a positive constant that has the unit of pressure. The term $\varepsilon_{P,cpl}$ represents the volumetric fraction of solid at close packing limit. In order to remove the singularity at the close packing limit, Snider (2001) modified Eq. (3.302) by including the constant γ that has a value between 2 and 5 as well as the constant θ that is on the order of 10^{-7} as:

$$\tau_P=\frac{P_P(1-\varepsilon_f)^{\gamma}}{\max\left[\left(\varepsilon_{P,cpl}-(1-\varepsilon_f)\right);\theta\varepsilon_f\right]}. \tag{3.303}$$

This inter-particle stress tensor depends on the concentration of particles, but neglects the size and velocity of particles. If necessary, a more complex continuum inter-particle stress model based on the kinetic theory of dense gases can be applied:

$$\tau_P=\left[(1-\varepsilon_f)\overline{\rho_P}+(1-\varepsilon_f)^2(1+e_{ss})\overline{\rho_P}g_{ss}^{rad}\right]\Theta. \tag{3.304}$$

Here, Θ is the granular temperature, e_{ss} and g_{ss}^{rad} represent the restitution coefficient and the radial distribution function.

It should be mentioned here that in addition to the MP-PIC method there is another way to combine the two-fluid method and the single-particle-method. In so-called sub-domain method, the computational domain is divided into continuum and free molecular domains. While the two-fluid method is used for the sub-domains, in which the continuum hypothesis is satisfied, the single-particle method in combination with the stochastic or deterministic collision detection models is applied to the free molecular sub-domains including walls. Although the idea behind the sub-domain method is simple, it shows the advantages of both two-fluid and single-particle methods. The sub-domain method was successfully tested for the simulation of dilute gas-solid flows by Grüner (2004). Here, the two-fluid method and the single-particle in combination with a stochastic collision model were coupled. Grüner developed a criterion with the help of the equilibrium index proposed by Kanther (2003) that divided the whole computational domain adaptive into continuum and free molecular sub-regions. For dense granular flows, this hybrid method may face difficulties as a result of high inhomogeneities of solid distribution during the operation. This is why the sub-domain method has attracted les attention in the scientific literature compared to the MP-PIC method.

3.3 Conclusion

There are two approaches to describe the fluid motion, namely the particle methods and the grid-based methods. In both approaches, the determination of the fluid physical values involves three main steps: pre-processing, solving and post-processing. In the pre-processing step, the problem is specified including the geometry and the initial boundary conditions. In the second step, the partial differential equations are approximated in a set of algebraic equations that is numerically solved using numerical techniques. Finally, the simulation results are analysed in the post- processing phase.

In the particle methods, the continuum (generally fluids with high density, e.g. water) is decomposed into a finite number of particles. Here, the physical values of the fluid are determined without the need of numerical grids. The most of the particle methods are based on the Lagrangian representation, but there are some works based on the Eulerian formulation or the arbitrary Lagrangian/Eulerian representation. In order to obtain the solution, the following steps are required:

- The fluid is discretised into a set of particles, also known as mass packets.
- For each particle, the derivatives and the integrals are estimated by means of all adjacent particles.
- The time-depending ordinary differential equations are solved using numerical methods.

The absence of the numerical grids offers several advantages such as the exact representation of the geometry, the flexible dealing with large domain deformations, the easy parallelisation and the possible use of higher-order continuous shape functions. However, the major drawback of the particle methods is the high computational effort.

By contrast to particle methods, where the fluid phase is decomposed into discrete mass packets, the computational domain in grid-based methods is divided into a finite number of grid cells. The exact solution is estimated by solving a finite number of algebraic equations instead of partial differential equations. Here, the fluid motion can be described in the Eulerian or Lagrangian representation. In order to obtain the solution, the following steps are necessary:

- The computational domain is discretised into a finite number of control volumes, also known as the numerical grid.
- The partial differential equations are approximated into a system of algebraic equations using one of the discretisation approaches such as the finite difference method, the finite element method and the finite volume method. In the finite difference method, the conservation equations are in differential forms, where the first and second derivatives of the generic conservation equations are estimated at each grid point using different systematic approaches such as Taylor series expansion or polynomial fitting. The FD method can be applied to any grid type, but it can produce on the structured grids an accurate result due to the possible using of higher-order approximations. In the finite element method, the computational domain is divided into finite elements. The solution of the partial differential equations is approximated by a linear combination of parameters that are at the beginning unknown and shape functions that can be derived from its values at the grid points. The possibility to apply to complex geometries is an important advantage of the FE method. The finite volume method starts from the integral form of the conservation equations that are applied to each control volume. The surface and volume integrals are approximated in the centre of each grid cell. Here, various interpolations with different orders are available such as the upwind differencing scheme, the central differencing scheme and the quadratic upwind interpolation for convective kinematics. The result is algebraic equations for each control volume. Considering the entire computational domain, a system of algebraic equations is obtained, where the number of the unknowns is equal the number of these equations. The FV method is applied to the structured and unstructured grids and thus can be used for complex geometries. Compared to the FD, the using of higher-order interpolations is more difficult especially for 3D cases.
- If the investigated problems are unsteady, the time should also be discretised. The most well-known approximations are the forward Euler method, the backward Euler method and the trapezoid rule.
- The system of algebraic equations for steady-state or unsteady time-dependent problems can only be solved when the boundary conditions are defined. The boundary conditions consider the flow interaction with the environment. There are three different boundary conditions: Dirichlet, Neumann and Cauchy boundary conditions.
- The non-linear system of the algebraic equations is finally solved employing numerical techniques such as the iterative methods.

The accuracy of grid-based methods is highly related to the numerical grid including size, shape etc. Compared to the particle methods, the grid-based methods are widely applied due to their calculation efficiency and their better description of the fluid properties such as mass density and pressure field. The major disadvantage is the numerical grid, i.e. an additional effort for the grid generation is required. Furthermore, the numerical grid cannot represent the geometry exactly.

Most engineering problems involve gas-solid flows, where the particles of a wide range of shapes and sizes undergo time dependent turbulent motion, e.g. fluidized beds, ash deposition on boiler walls, gasification or combustion of biomass and coal as well as CO_2 capture methods. For the simulation of gas-solid flows, several approaches with different levels of detail are available, namely quasi-single-phase, two-fluid, single-particle and hybrid methods.

If the continuum hypothesis is fulfilled, the particle phase can be described as a fluid. The quasi-single-phase model presents in this context the simplest approach to simulate the behaviour of a gas-solid flow. Hence, the physical values of dispersed and gas phases can be represented as volume average values by solving the balance equations of a single-phase flow. The quasi-single-phase method assumes that the solid is evenly distributed in the fluid phase and the velocity differences between both phases are negligibly small. In many cases, the configurations of practical gas-solid flows do not meet the method requirements. Therefore, the more sophisticated continuum approach i.e. two-fluid method, which is related to a better accuracy and a wide application range, has been frequently employed. The basic idea of two-fluid method is that the fluid and solid phases are regarded as coexisting continua. Both phases are modelled with the balance equations of a single-phase flow, where the solid phase is considered as second dense fluid phase. In the single-particle model, the solid phase is treated as a disperse phase, i.e. the trajectory of a single particle is followed, showing much better agreement towards the measurements. The numerical simulation performed with the single-particle method is computationally expensive, especially when the number of particles exceeds one million. If certain numbers of computational particles (parcels) are tracked instead of modelling each single particle in the computational domain, the computational cost can be reduced significantly. In so-called hybrid methods such as the multiphase particle-in-cell (MP-PIC) method and the sub-domain method, the two-fluid method and the single-particle-method are coupled. The MP-PIC method describes the particle-particle interactions as a gradient on the Eulerian grid similar to the two-fluid method, resulting in considerable reduction in the computational time. In the sub-domain method, the two-fluid method is employed to the domains, where the continuum hypothesis is satisfied and the single-particle method is applied to the free molecular domains and walls.

Which approach is more appropriate to describe the behaviour of gas-solid flows, it can completely not be stated. Generally, it depends on the flow configurations, the mechanical properties of particles, the solid content in the flow and the level of accuracy that has to be accomplished. The continuum approach will likely remain the first choice for dense granular flows with particle size less than 1 mm. However, the use of continuum approach is limited to monodisperse gas-solid flows, but the calculation of polydisperse systems is possible. In this case, each particle size class should be seen as a separated phase. The consideration of different particle sizes leads to an increase in computing time up to the level, at which the single-particle method may offer a competitive approach.

In the single-particle method, the detection of particle-particle and particle-wall collisions is crucially important with respect to both computational effort and simulation accuracy. Basically, the collisions between particle-particle/wall can be stochastically or deterministically detected. The idea behind the stochastic collision models is: the motion of each particle is calculated independent of remaining particles. However, information on other particles should be available in order to generate virtual collision partners, whose properties are derived by the local average values of these particles. The occurrence probability of a collision between the investigated particle and the virtual particle can then be met by means of random numbers, where the particle time step is assumed to be constant. In the deterministic collision models, each particle is tested for a possible collision with other partners (particle, wall or both). The particle time step in deterministic collision models is either constant or adaptive. For the adaptive deterministic collision model, also known as event-oriented method, no collision during successive time steps will occur, but only at the time $t + \Delta t_P$ one particle collides with other particle or wall. In the deterministic collision model with a constant particle time step, a particle-particle collision takes place when the distance between the centres points of two particles is smaller than the sum of both radii. Likewise, a particle-wall collision occurs if the distance between the particle centre point and the wall surface is smaller than the particle radius.

In deterministic collision models, different algorithms are available for the collision determination between particle-particle and particle-wall such as standard, neighbourhood and bounding-boxes methods. In the standard method, all probabilities for possible collision between the investigated particle and other partners are considered. The computational cost of the standard method is a quadratic function to number of particles. In neighbourhood methods, a collision list is generated for each particle, in which only potential collision partners are saved. This means, a particle can only come in contact with particles in its direct neighbourhood. By neglecting any further collisions probabilities with particles that are far away from the investigated particle, the computational effort is reduced. For particles with a wide range of size distributions and/or complex geometries, the efficiency of the neighbourhood method decreases. In this context, the bounding-boxes algorithm offers a competitive procedure. Here, each particle is enclosed with a cuboid its edges are projected on the coordinate axes. A particle list, in which the vertices of all cuboids are saved, is generated for each axis. A possible collision between two objects will occur if their cuboids vertices overlap in the x-, y-and z-directions. The computational effort of the bounding-boxes method is more or less equated to the neighbourhood methods, by skilful use, approximately linear to the particle number. The bounding-boxes method allows the collision detection between particles with complex polygonal shapes. Furthermore, the particle size distributions have no influence on the method efficiency.

If collisions between particle-particle/wall are detected with stochastic collision models, the collision treatment is only performed with the hard sphere model. In this model, single binary collision is modelled, where the particles' properties after the collision are a function of the particles' properties before the collision. In case of deterministic collision models, the subsequent collision treatment can be achieved either with the hard sphere model or with the soft sphere model. In the latter, also known as discrete element method, the force, the velocity and the displacement are determined during the collision. In the DEM model, the change of the translational and angular velocities is determined by forces and momentums balances on each particle. The single particle trajectory and the angular displacement can then be calculated through integration Newtonian equations of motion.

The forces acting on particles are divided into two categories: volume and surface forces. The surface forces can be classified as aerodynamic surface forces and short-range surface forces that can also be decomposed into contact force, adhesive forces associated either with or without physical contact. Similar to the DEM model, the forces acting on particles can also be considered in the hard sphere model.

The volume forces result from external fields and act throughout the volume of object regardless of flow configuration. Typical volume forces are gravitational force, buoyancy force, electric and magnetic forces.

The aerodynamic surface forces that exert on the surface of particles result from the interactions between fluid and solid phases. They depend highly on the densities as well as the translational and angular velocities of both phases. In gas-solid flows, the aerodynamic surface forces include drag force, virtual mass force, Basset force, force due to velocity gradient (Saffman), force due to pressure gradient, force due to temperature gradient, force due to concentration gradient and Magnus force. The drag force that hinders the movement of a particle represents, in most cases, the dominant force on the particles in the granular flow. It is composed of two forces: the drag form force and the frictional resistance force that depends on the surface structure of the particle. With the acceleration or deceleration of a particle in a fluid, the surrounding fluid must be accelerated or decelerated, too. The unsteady effect due to the acceleration or the deceleration of particle with respect to fluid can be divided into two forces: the virtual mass force as an additional form of resistance and the Basset force due to the viscous effects when the relative velocity of both phases changes. The Basset force describes the temporal delay in boundary layer development during the acceleration or deceleration of particles. In anisotropic flows (i.e. flow with non-uniform characters), additional forces can arise by the reason of existing flow gradients such as velocity, pressure, temperature or concentration gradient. For a shear flow, the particle surface undergoes an unsteady pressure distribution that results in a force perpendicular to the relative velocity (Saffman force). The pressure gradient in the flow induces an additional force that acts in the opposite direction of the pressure gradient. If the particle locates in a flow with temperature or concentration gradients, an irregular pressure distribution will arise on the particle surface, leading to a force that acts in the opposite direction of this gradient. In isotropic flows (i.e. flow with uniform characters), the particle rotation leads also to an irregular pressure distribution on the particle surface. The resulting force on the particle is called Magnus force.

The short-range surface forces or inter-particle forces such as contact and adhesive forces result from the interactions between particle-particle and particle-wall. The determination of the short-range surface forces requires the identifying of the collision partners. This can be achieved employing a detection collision model with a specific criterion. For example, a contact between two particles will occur, if the distance between particle centres is smaller than the sum of their radii. Other example is the criterion of the liquid bridge formation. A collision is registered when the distance between particle centres falls below the bridge critical length.

In the DEM model, the collision partners are allowed to overlap slightly. As a sequence, the contact between two rigid particles is not represented as a single point like in the hard sphere model but as a finite area in 3D and as a line in 2D cases. Here, it is assumed that the particles undergo deformation during their contact. This deformation is equivalently considered to be the contact between the collision partners. Depending on the penetration depth, the resulting contact force can be modelled using a mechanical analogous model. The latter consists of a Hookean spring and a Newtonian damper, which are connected in parallel. Using this model, the collision can be modelled as either elastic or inelastic. In an elastic collision, the sum of kinetic energies of contact particles before and after the collision are equal (neither damping nor frictional forces is allowed). During an inelastic collision, by contrast, a part of kinetic energies of collided particles is converted into internal energy. The energy dissipation at inelastic collisions can be expressed by the collision coefficients. The well-known Kelvin-Voigt model (spring-damper-slider system) is generally applied to determine the contact force. The linear contact model by Hooke shows an adequate accuracy in most applications. The non-linear contact models based on Hertz theory improve the calculation accuracy, but lead to an additional computational cost. This is due to the fact that these models should consider the changes of geometric and physical properties of partners during the collision. The resulted contact force is determined as the sum of two components: the normal and the tangential contact forces. By the calculation of the tangential contact force, it has to be differed between static and sliding frictions, which can be achieved with the aid of Coulomb's friction law. The distinction between the static and sliding forces improves the realistic representation of the collision. The contact force is obtained finally by summing up all normal and tangential contact forces that act simultaneously between the particle and other particles and walls.

The hydrodynamic behaviour of the solid phase is also influenced by the formation of adhesive forces between particles or rather particles and walls. The adhesive forces depend highly on the material pairings such as diameter, density and surface roughness. Based on the adhesive forces formation, it is distinguished between two mechanisms. The first adhesive mechanism is formed without material connection (i.e. without physical connect) between the collision partners such as Van der Waals and electrostatic forces. The van der Waals force describes the dipole-dipole interactions between particle-particle and particle-wall. The electrostatic force that arises between two charged particles is directly proportional to the product of magnitude of each charge and indirectly proportional to the square of distance. The second mechanism is characterised by a material connection (i.e. with physical connect) between the collision partners. This connection can be formed as solid and liquid bridges. The liquid content in the gas-solid flow leads to adhesive force between the collision partners. The resulting adhesive force due to liquid bridge can be obtained as the sum of capillary force and force caused by the surface tension. While the capillary force results from the pressure difference between the bridge liquid and the surrounding fluid, the surface tension force results from the axial component of the surface tension of the bridge liquid. The definition of the liquid bridge geometry is crucial to enable the calculation of this adhesive force. Generally, it is assumed that the bridge geometry has toroidal shape, leading to explicit correlations between the absolute value of non-dimensional adhesive force due to liquid bridge and the non-dimensional distance between both collision partners. In case of bridge separation, the liquid of bridge is evenly distributed over the collided particles due to identical contact angles. Unequal contact angles appear, however, during the building of liquid bridges between monodisperse particles two different materials or between two polydisperse particles of identical or different materials. In this case, the neck of liquid bridges shifts to the particle with a large contact angle. During the bridge separation, the particle with a large contact angle holds less proportion of the liquid back compared to the particle with a small contact angle. At high temperatures, adhesive force due to solid bridge becomes more important, which leads to a strong and a permanent bond between collision partners. The building of the sinter neck is a function of the surrounding pressure and the interfacial energies of contact particles. Generally, the neck growth due to the pressure sintering can be neglected at atmospheric pressure. The neck growth due to the interfacial energies can be determined based on Kuczynski sintering model. It is assumed in this model that the surface diffusion including the effect of the surface roughness and the volume diffusion mechanisms have a relevant influence on the metal powders sintering.

The moments of force acting on the particles are classified as the moment of force due to the fluid-particle interaction and the moment of force due to the short-range forces that can also be divided into two parts: the tangential contact force and the force due to the asymmetrical normal traction distribution. In an anisotropic flow, e.g. with a velocity gradient, the moment of force due to the fluid-particle interaction leads to particle rotations. The moment of force due to the tangential contact forces is calculated by multiplying the particle radius with the tangential component of the contact force. The moment of force due to the asymmetrical normal traction distribution is caused by the rolling resistance of particles. Generally, the normal traction distribution is symmetrical for rigid particles with small penetration depth. Asymmetrical normal traction distribution can occur when the penetration depth of rigid particles is relative large or for inelastic particles regardless of their penetration depths. By contrast to the DEM model, the moment of force due to the tangential contact force and the asymmetrical normal traction distribution cannot be considered in the hard sphere model. This is due to the fact that the inter-particle forces (e.g. contact force) are calculated as impulsive actions.

However, the moment of force due to the fluid-particle interaction can be taken into the hard sphere model account.

Based on models, equations and correlations explained in this section, the volume, the aerodynamic surface forces and the short-range surface forces including the contact forces and the adhesive forces associated either with or without physical contact as well as the resulted moments of force acting on particles can be determined. However, the applied models are restricted to circular particles in 2D or spherical particles in 3D cases. The consideration of particles with complex geometries is limited not only by the modelling of mechanical behaviour of these particles, but also by the related additional computation time. Two approximation methods for the non-spherical particles are of relevance. In the first approach, the particle with complex geometry can be built as cluster of circles or spheres. The advantage of this approximation is that the mechanical behaviour of spherical particles has been actively investigated and thus the previous developed models can be applied with minor modifications. In the second approach, the predefined particle geometry is described by polynomial functions. The models that are used to calculate the volume and surface forces cannot be applied and new models should be developed based on the particle shapes. Since the computational complexity increases sharply with raising the order of polynomial function, the second approach is generally limited to simple geometries such as cylindrical or ellipsoidal particles.

The particle-fluid interaction describes the impact of solid phase on fluid phase and vice versa (volumetric void fraction, momentum and heat transfers). The measure of the solid loading in each control volume is described through the volumetric void fraction. It is calculated as the ratio of the fluid volume to the entire volume of the grid cell. The momentum transfer that describes the influence of particles motions on the fluid phase can be calculated by determining the change in the particle velocities along their trajectories. The momentum transfer between the solid and fluid phases can be carried out in three different approaches. In the first scheme, the transferred force from fluid to disperse phase is determined for each particle individually depending on the particle velocity. The transferred force from particles to fluid phase is then calculated with the aid of an arithmetic averaging. In the second scheme, the transferred force from particles to fluid phase is performed by a local averaging method like the first scheme, while the resultant force is distributed to the particles that are located in the control volume. In the third scheme, the fluid force acting on each particle is calculated according to the first scheme, while the particle force acting on the fluid phase is determined through the sum of all forces of particles that are existent in the control volume. Besides the particle velocity, the inter-phase momentum transfer coefficient should be defined to enable the calculation of the momentum transfer. The inter-phase momentum transfer coefficient can be modelled using different approaches such as Ergun/Wen, Foscolo, Beetstra, and Koch/Hill models. The use of a model needs, however, the knowing of the volumetric void fraction and the particle Reynolds number in the grid cell. The heat transfer from solid to fluid phase can be calculated based on the heat transfer through convection, the heat transfer through radiation and the heat transfer due to the phase transformation from solid to fluid phase.

The exact calculation of interaction values (volumetric void fraction, momentum and heat transfers) requires the knowledge of all particles that are completely or partly located in a control volume, their velocities and their temperatures. This can only be performed for simple cases, where the number of grid cells and/or particles is relatively low. By increasing the number of particles and/or improving the grid resolution, the number of particles that are allocated to different control volumes is sharply increased. The computational effort increases accordingly and numerical approximations should therefore be introduced such as the particle centre method. In this method, it is assumed that the entire volume of the particle is allocated to the fluid cell, where the centre of this particle is found.

The numerical simulation of the gas-solid flow using the single-particle method requires the definition of particle and fluid time steps. Generally, the size of the fluid time step is set constant and can be arbitrarily selected. The size of the particle time step can be variable or constant depending on the selected particle-particle/wall interaction model. The hard sphere model with deterministic collision detection model uses an adaptive particle time step size. The particle can collide with another particle or with the wall only at the time $t + \Delta t_P$, where the contact occurs at a point. In the DEM model, the particle time step size is assumed to be constant. However, the particle time step size should be set sufficiently small in order to prevent the unphysical penetrations between the collision partners. Larger particle time step sizes lead to unrealistic contact forces, while smaller particle time step sizes, although they improve the calculation accuracy, lead to higher computational efforts.

References

Allen MP, Tildesley DJ (1989) Computer simulation of liquids. Oxford University Press

Alobaid F (2013) 3D Modelling and simulation of reactive fluidized beds for conversion of biomass with discrete element method. tuprints, Germany

Alobaid F, Epple B (2013) Improvement, validation and application of CFD/DEM model to dense gas–solid flow in a fluidized bed. Particuology 11:514–526

Andrews M, O'rourke P (1996) The multiphase particle-in-cell (MP-PIC) method for dense particulate flows. Int J Multiph Flow 22:379–402

Atsonios K, Nikolopoulos A, Karellas S, Nikolopoulos N, Grammelis P, Kakaras E (2011) Numerical investigation of the grid spatial resolution and the anisotropic character of EMMS in CFB multiphase flow. Chem Eng Sci 66:3979–3990

Babovsky H (1989) A convergence proof for Nanbu's Boltzmann simulation scheme. Eur J Mech B Fluids 8:41–55

Baraff D (1995) Interactive simulation of solid rigid bodies. IEEE Comput Graph Appl 15:63–75

Beer FP, Johnston ER (1967) Mechanics for engineers: statics and dynamics. McGraw-Hill

Beetstra R, Van der Hoef M, Kuipers J (2007) Drag force of intermediate Reynolds number flow past mono-and bidisperse arrays of spheres. AIChE J 53:489–501

Belytschko T, Krongauz Y, Organ D, Fleming M, Krysl P (1996) Meshless methods: an overview and recent developments. Comput Methods Appl Mech Eng 139:3–47

Bird GA (1976) Molecular gas dynamics. Clarendon, Oxford

Brilliantov NV, Pöschel T (1998) Rolling friction of a viscous sphere on a hard plane. EPL (Europhys Lett) 42:511–516

Campbell CS, Brennan C (1985) Computer simulation of granular shear flows. J Fluid Mech 151:167–188

Chapman S, Cowling TG (1970) The mathematical theory of non-uniform gases. Cambridge University Press

Cohen JD, Lin MC, Manocha D, Ponamgi M (1995) I-COLLIDE: an interactive and exact collision detection system for large-scale environments. In: Proceedings of the 1995 symposium on interactive 3D graphics. ACM, pp 189–196

Crowe C (1982) Review-numerical models for dilute gas-particle flows. J Fluids Eng 104:297–303

Crowe CT, Schwarzkopf JD, Sommerfeld M, Tsuji Y (2011) Multiphase flows with droplets and particles. CRC Press

Cundall PA, Strack OD (1979) A discrete numerical model for granular assemblies. Geotechnique 29:47–65

Kunii D, Levenspiel O (1991) Fluidization engineering, 2nd edn. Butterworth-Heinemann, Stoneham, Boston

Dalla Valle J (1948) Micromeritics. Pitman, London, UK

Deen N, Van Sint Annaland M, Van der Hoef M, Kuipers J (2007) Review of discrete particle modeling of fluidized beds. Chem Eng Sci 62:28–44

Dennis S, Singh S, Ingham D (1980) The steady flow due to a rotating sphere at low and moderate Reynolds numbers. J Fluid Mech 101:257–279

Di Renzo A, Di Maio FP (2004) Comparison of contact-force models for the simulation of collisions in DEM-based granular flow codes. Chem Eng Sci 59:525–541

Elgobashi S (2006) An updated classification map of particle-laden turbulent flows. In: IUTAM symposium on computational approaches to multiphase flow. Springer, pp 3–10

Epple, B., Leithner, R., Linzer, W., Walter, H., 2012. Simulation von Kraftwerken und Feuerungen. Springer-Verlag

Ergun S (1952) Fluid flow through packed columns. Chem Eng Process 48:89–94

Fan L-S, Zhu C (2005) Principles of gas-solid flows. Cambridge University Press, Cambridge

Feng Y, Yu A (2004) Assessment of model formulations in the discrete particle simulation of gas-solid flow. Ind Eng Chem Res 43:8378–8390

Ferziger JH, Perić M (2002) Computational methods for fluid dynamics, 3rd edn. Springer, Berlin

Foscolo P, Gibilaro L, Waldram S (1983) A unified model for particulate expansion of fluidised beds and flow in fixed porous media. Chem Eng Sci 38:1251–1260

Gao X, Wu C, Cheng Y-W, Wang L-J, Li X (2012) Experimental and numerical investigation of solid behavior in a gas–solid turbulent fluidized bed. Powder Technol 228:1–13

Garside J, Al-Dibouni MR (1977) Velocity-voidage relationships for fluidization and sedimentation in solid-liquid systems. Ind Eng Chem Process Des Dev 16:206–214

Germano M, Piomelli U, Moin P, Cabot WH (1991) A dynamic subgrid-scale eddy viscosity model. Phys Fluids A 3:1760–1765

Gidaspow D (1994) Multiphase flow and fluidization: continuum and kinetic theory descriptions. Academic Press

Gidaspow D, Bezburuah R, Ding J (1992) Hydrodynamics of circulating fluidized beds, kinetic theory approach in fluidization. In: Proceedings of the 7th engineering foundation conference on fluidization, pp 75–82

Goldschmidt M, Beetstra R, Kuipers J (2004) Hydrodynamic modelling of dense gas-fluidised beds: comparison and validation of 3D discrete particle and continuum models. Powder Technol 142:23–47

Götz S (2006) Gekoppelte CFD-DEM-Simulation blasenbildender Wirbelschichten. Shaker

Grüner C (2004) Kopplung des Einzelpartikel- und des Zwei-Kontinua-Verfahrens für die Simulation von Gas-Feststoff-Strömungen. Shaker

Harris S, Crighton D (1994) Solitons, solitary waves, and voidage disturbances in gas-fluidized beds. J Fluid Mech 266:243–276

Hertz H (1882) Über die Berührung fester elastischer Körper. Journal für die reine und angewandte Mathematik 92:156–171

Hill RJ, Koch DL, Ladd AJ (2001) Moderate-Reynolds-number flows in ordered and random arrays of spheres. J Fluid Mech 448:243–278

Hockney RW, Eastwood JJW (1988) Computer simulation using particles. Taylor & Francis Group

Hoomans B, Kuipers J, Briels W, Van Swaaij W (1996) Discrete particle simulation of bubble and slug formation in a two-dimensional gas-fluidised bed: a hard-sphere approach. Chem Eng Sci 51:99–118

Howell JR, Menguc MP, Siegel R (2015) Thermal radiation heat transfer, 6th edn. CRC Press

Hußmann B (2009) Modellierung und numerische Simulation der zweiphasigen Strömungs- und Verbrennungsvorgänge in einem Staustrahltriebwerk mit Bor als Festtreibstoff. VDI-Verlag

Iwashita K, Oda M (1998) Rolling resistance at contacts in simulation of shear band development by DEM. J Eng Mech 124:285–292

Iwashita K, Oda M (2000) Micro-deformation mechanism of shear banding process based on modified distinct element method. Powder Technol 109:192–205

Kanther W (2003) Gas-Feststoff-Strömungen in komplexen Geometrien. Shaker

Karakashian O, Xing Y (2013) Recent developments in discontinuous Galerkin finite element methods for partial differential equations. Oak Ridge National Laboratory (ORNL)

Koch DL, Hill RJ (2001) Inertial effects in suspension and porous-media flows. Annu Rev Fluid Mech 33:619–647

Kruggel-Emden H, Simsek E, Rickelt S, Wirtz S, Scherer V (2007) Review and extension of normal force models for the discrete element method. Powder Technol 171:157–173

Kuczynski GC (1949) Self-diffusion in sintering of metallic particles. AIME Trans 185:169–178

Kuwagi K, Mikami T, Horio M (2000) Numerical simulation of metallic solid bridging particles in a fluidized bed at high temperature. Powder Technol 109:27–40

Launchbury DR (2016) Unsteady turbulent flow modelling and applications. Springer, Berlin

Leva M (1959) Fluidization. McGraw-Hill, New York

Li J, Kwauk M (2003) Exploring complex systems in chemical engineering—the multi-scale methodology. Chem Eng Sci 58:521–535

Li S, Liu WK (2002) Meshfree and particle methods and their applications. Appl Mech Rev 55:1–34

Lian G, Thornton C, Adams MJ (1993) A theoretical study of the liquid bridge forces between two rigid spherical bodies. J Colloid Interface Sci 161:138–147

Link JM (2006) Development and validation of a discrete particle model of a spout-fluid bed granulator. PrintPartners Ipskamp

Liu G-R (2009) Meshfree methods: moving beyond the finite element method. Taylor & Francis

Liu G-R, Liu MB (2003) Smoothed particle hydrodynamics: a meshfree particle method. World Scientific

Liu M-B, Liu G-R (2016) Particle methods for multi-scale and multi-physics. World Scientific

Lucy LB (1977) A numerical approach to the testing of the fission hypothesis. Astron J 82:1013–1024

Lumley JL (1990) Whither turbulence? Turbulence at the crossroads. Springer, Berlin

Lun C, Savage SB, Jeffrey D, Chepurniy N (1984) Kinetic theories for granular flow: inelastic particles in Couette flow and slightly inelastic particles in a general flowfield. J Fluid Mech 140:223–256

Lungfiel A (2002) Ermittlung von Belastungsgrößen mittels der Diskrete-Elemente-Methode für die Auslegung von Sturzmühlen. Eigenverlag der Technischen Universität Bergakademie Freiberg

Marcus RD, Leung LS, Klinzing GE, Rizk F (1990) Pneumatic conveying of solids. Chapman and Hall, New York

Maw N, Barber J, Fawcett J (1976) The oblique impact of elastic spheres. Wear 38:101–114

Mikami T, Kamiya H, Horio M (1998) Numerical simulation of cohesive powder behavior in a fluidized bed. Chem Eng Sci 53:1927–1940

Monaghan JJ (1992) Smoothed particle hydrodynamics. Ann Rev Astron Astrophys 30:543–574
Muguruma Y, Tanaka T, Tsuji Y (2000) Numerical simulation of particulate flow with liquid bridge between particles (simulation of centrifugal tumbling granulator). Powder Technol 109:49–57
Nanbu K (1980) Direct simulation scheme derived from the Boltzmann equation. I. Monocomponent gases. J Phys Soc Jpn 49:2042–2049
Oberlack M, Khujadze G, Barth S, Günther S, Frewer M, Weller T, Peinke J (2007) Progress in turbulence II: proceedings of the ITi conference in turbulence 2005. Physica-Verlag
Oesterle B, Petitjean A (1993) Simulation of particle-to-particle interactions in gas solid flows. Int J Multiph Flow 19:199–211
O'Rourke PJ (1981) Collective drop effects on vaporizing liquid sprays. Princeton University, Princeton
O'Rourke PJ, Brackbill J, Larrouturou B (1993) On particle-grid interpolation and calculating chemistry in particle-in-cell methods. J Comput Phys 109:37–52
Paschedag AR (2004) CFD in der Verfahrenstechnik: allgemeine Grundlagen und mehrphasige Anwendungen. Wiley VCH Verlag GmbH
Pope SB (2000) Turbulent flows. Cambridge University Press, Cambridge
Rubinow S, Keller JB (1961) The transverse force on a spinning sphere moving in a viscous fluid. J Fluid Mech 11:447–459
Schaeffer DG (1987) Instability in the evolution equations describing incompressible granular flow. J Differ Equ 66:19–50
Schäfer M (2006) Computational engineering—introduction to numerical methods. Springer, Berlin
Schiller L, Naumann AZ (1933) Über die grundlegenden Berechnungen bei der Schwerkraftaufbereitung. Verein Deutscher Ingenieure 77:318–320
Schinner A (1999) Fast algorithms for the simulation of polygonal particles. Granular Matter 2:35–43
Schlichting H, Gersten K (2000) Boundary-layer theory. Springer, Berlin
Schubert H (1979) Grundlagen des agglomerierens. Chemie Ingenieur Technik 51:266–277
Sedgewick R (1992) Algorithmen. Addison-Wesley Longman Verlag
Shi D, McCarthy JJ (2008) Numerical simulation of liquid transfer between particles. Powder Technol 184:64–75
Smagorinsky J (1963) General circulation experiments with the primitive equations, part I: the basic experiment. Mon Weather Rev 91:99–164
Snider D (2001) An incompressible three-dimensional multiphase particle-in-cell model for dense particle flows. J Comput Phys 170:523–549
Sommerfeld M (1996) Modellierung und numerische Berechnung von partikelbeladenen turbulenten Strömungen mit Hilfe des Euler,Lagrange-Verfahrens. Shaker
Sommerfeld M (2003) Analysis of collision effects for turbulent gas–particle flow in a horizontal channel: Part I. Particle transport. Int J Multiph Flow 29:675–699
Soo S-l, Deyan T (1990) Multiphase fluid dynamics. Science Press Brookfield
Stiess M (1994) Mechanische Verfahrenstechnik. Springer, Berlin
Stroh A, Alobaid F, Hasenzahl MT, Hilz J, Ströhle J, Epple B (2016) Comparison of three different CFD methods for dense fluidized beds and validation by a cold flow experiment. Particuology
Syamlal M, O'Brien TJ (1989) Computer simulation of bubbles in a fluidized bed. In: AIChE symposium series, pp 22–31
Tsuji Y, Kawaguchi T, Tanaka T (1993) Discrete particle simulation of two-dimensional fluidized bed. Powder Technol 77:79–87
Tsuji Y, Morikawa Y, Mizuno O (1985) Experimental measurement of the Magnus force on a rotating sphere at low Reynolds numbers. J Fluids Eng 107:484–488
Tsuji Y, Tanaka T, Ishida T (1992) Lagrangian numerical simulation of plug flow of cohesionless particles in a horizontal pipe. Powder Technol 71:239–250
Walter H, Epple B (2016) Numerical simulation of power plants and firing systems. Springer, Berlin

Wen C, Yu Y (1966) Mechanics of fluidization. AIChE J 62:100–111

Zeneli M, Nikolopoulos A, Nikolopoulos N, Grammelis P, Kakaras E (2015) Application of an advanced coupled EMMS-TFM model to a pilot scale CFB carbonator. Chem Eng Sci 138:482–498

Zhou Y, Wright B, Yang R, Xu B, Yu A (1999) Rolling friction in the dynamic simulation of sandpile formation. Physica A 269:536–553

Zhu H, Yu A (2002) Averaging method of granular materials. Phys Rev E 66:21–32

Zhu H, Zhou Z, Yang R, Yu A (2007) Discrete particle simulation of particulate systems: theoretical developments. Chem Eng Sci 62:3378–3396

Zienkiewicz OC, Taylor RL, Nithiarasu P (2005) The finite element method for fluid dynamics. Elsevier Science

Chapter 4
Results

Studies on dynamic process simulation and computational fluid dynamics applied to different thermal power plant technologies are presented in this chapter.

4.1 Dynamic Process Simulation

While the conventional design of thermal power plants is mainly focused on high process efficiency, market requirements increasingly target operating flexibility due to the continuing shift towards renewables. Dynamic process simulation is a cost-efficient tool for improving the power plant flexibility during transients such as load changes and start-up procedures. Specific applications include optimisation of control structures, stress assessment for critical components and plant safety analysis in malfunction cases. This section contains an overview on the dynamic process simulation applied to combined-cycle power plant, pulverised coal-fired power plant, concentrated solar power plant and municipal waste incineration.

4.1.1 Combined-Cycle Power Plant

According to World Energy Outlook, the gas-fired power generation in 2015 accounted for approximately 23% total share worldwide electricity generation, dominated by combined-cycle power plants (CCPP). The modern concept of the combined-cycle power plant is the result of an evolutionary process in the second half of last century, mainly driven by increasing performance of the gas turbine. As early attempts to combine a gas turbine and a steam cycle, the GT was installed to enhance the efficiency of existing large-scale steam power plants by using the hot exhaust gas for feedwater preheating instead of steam extractions or as a supply of

F. Alobaid, *Numerical Simulation for Next Generation Thermal Power Plants*, Springer Tracts in Mechanical Engineering,
https://doi.org/10.1007/978-3-319-76234-0_4

hot combustion air to the fully-fired steam generator. Korneuburg (A) power station, commissioned 1960 in Austria, is considered to be the first CCPP according to the modern definition of the combined-cycle power plant. The general idea is that the waste heat of a gas turbine is absorbed by a heat recovery steam generator (HRSG) installed downstream in the flue gas path. The generated steam is used in a Rankine bottoming cycle, which generates additional power in the steam turbine. The process efficiency of Korneuburg CCPP did not exceed 32.5%. The GT operating temperatures at the time were as low as 620 °C at the turbine inlet and 310 °C at the turbine outlet, so that the supplementary firing was still required to support the steam cycle. Major technological developments including high temperature resistant materials and thermal barrier coatings, low-NO_x combustion and innovative cooling methods significantly improved the gas turbine performance since. Whereas early CCPP configurations only used simple single-pressure HRSGs, additional pressure stages were introduced over time in order to increase the steam parameters and to reduce the temperature mismatch between flue gas and water/steam side. Today, a 1 + 1 arrangement of GT and ST units in combination with a triple-pressure reheat HRSG is state of the art (see Fig. 4.1). The supplementary firing is widely omitted due

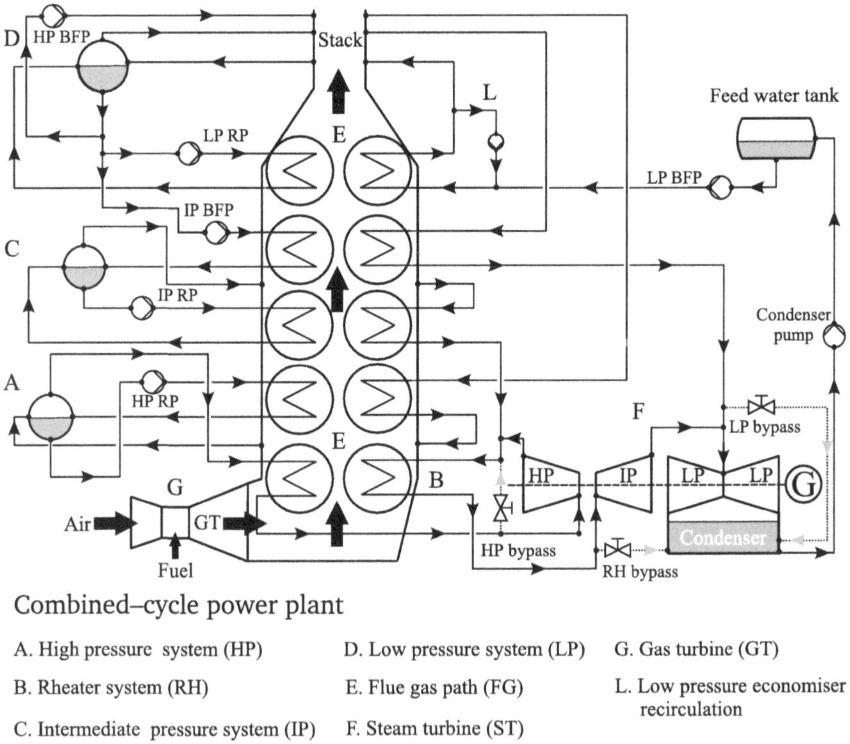

Fig. 4.1 Schematic diagram of 1 + 1 arrangement of gas turbine and steam turbine units in combination with a triple-pressure, reheat, vertical heat recovery steam generator (simplified)

to the high temperature at the turbine outlet (in the order of 600 °C). Accordingly, the nominal process efficiency can reach up to 60%. Also, combined-cycle power plants with efficiency levels greater than 60% are now running, for example Irsching 4 power station that is located in Irsching, Germany.

In addition to higher efficiency, combined-cycle power plants are also characterised by flexible unit dispatch. Fast response capability is a prerequisite for increasing shares of renewable feed-in and thus represents a competitive advantage for the operator in a changing market environment. Three criteria are typically considered to assess the practical flexibility of a power plant: start-up time, maximum load gradient (positive and negative) and minimum load. Only 20 min are required for starting a simple-cycle gas turbine, irrespective of its initial temperature. CCPP load transients are, however, limited by thermal stresses in the thick-walled components of the bottoming steam cycle, namely ST rotor, ST casing, high-pressure drum and outlet manifolds of HP superheater and final reheater. A modern CCPP can complete the start-up procedure in less than 50 min after an overnight shutdown (hot start-up) and sustain challenging load gradients up to ±60%/min, as e.g. stipulated by the Great Britain Grid Code for primary frequency response. For warm and cold start-ups, the HRSG can reach its nominal load in less than 90 and 150 min after the GT start. The minimum load of a combined-cycle is mainly determined by the gas turbine, where stable combustion as well as CO and NO_x levels in compliance with emission regulations must be preserved. Consequently, the corresponding operating point is dependent on the specific type of the gas turbine and possibly country-specific regulation. The operating load can be decreased to 40–50% of nominal load for typical gas turbines. This level may be further reduced to 20%, if a sequential-combustion design is used, so that one GT combustor can be shut-down entirely. The minimum load limit is relevant to flexible operation since it defines the lower boundary for negative load changes. If frequent cycling is anticipated, a CCPP capable of operation at low minimum load may also be an economically viable option to reduce the number of start-ups and shutdowns.

4.1.1.1 Load Changes and Off-Design Operation

Investigations into the capability of different process simulation codes to predict the behaviour of a real combined-cycle power plant during part loads and off-design operation are carried out using the process simulation software tools APROS and ASPEN PLUS DYNAMICS (Alobaid et al. 2008, 2014b, 2015b).

The thermal power plant (Prai Power Station, located in Malaysia) consists of three 50 Hz single shaft gas turbines manufactured by General Electric (GE) connected to three heat recovery steam generators built by Doosan Heavy Industries and Construction. The heat recovery steam generator for steam turbine manufactured by GE is unfired due to the high exhaust flue gas temperature of the gas turbine, approximately 628 °C. The exhaust gas will leave the HRSG heating surfaces through the main exhaust stack. The water/steam side of each sub-critical

HRSG has three pressure systems up to 100 bar in the high-pressure system and a reheater section after the high-pressure turbine, in detail as follows:

- Forced circulation evaporator path in high, intermediate and low-pressure circuits
- Reheater stage combining the expanded flow from the high-pressure turbine exit with the intermediate-pressure flow
- Three steam turbine stages (high, intermediate and low-pressure turbines)

The schematic process diagram, illustrated in Fig. 4.2, shows GT, ST and HRSG including high, intermediate, low-pressure systems and reheater.

The high-pressure system heats the feedwater and generates superheated steam for admission to the high-pressure steam turbine. The intermediate-pressure system heats the feedwater and generates superheated steam to be mixed with the cold reheat steam, admitted to the reheater section. Then, the reheated steam enters into the IP section of the steam turbine. The intermediate-pressure circuit provides also hot water from the IP economiser to preheat the natural gas fuel. The low-pressure system heats the feedwater and generates superheated steam for admission to the LP steam turbine. The HP and RH attemperators limit the temperature of the steam supplied to the HP and IP sections of the steam turbine, respectively. When the gas turbine is operated with a low sulphur natural gas fuel, the LP economiser and the LP economiser recirculation is in service. The recirculation circuit regulates the temperature at the inlet of the low-pressure economiser about 50 °C independent of the plant load. When operating with high sulphur natural gas or with distillate fuel, the LP economiser is completely bypassed. The metal temperature of the LP economiser in this case will not fall below the sulphur acid dew point, avoiding the occurrence of the corrosion. In this configuration, the LP main steam control valve is fully closed, allowing the LP drum equilibrium pressure to be below the minimum admission pressure.

Table 4.1 lists the characteristic technical data per single sub-critical HRSG provided at 100% nominal load.

HRSG Model

In APROS and ASPEN PLUS DYNAMICS, the flue gas path and the water/steam side are modelled from the gas turbine exit to the heat recovery steam generator exit. Here, the gas turbine and condenser sections are only considered as dynamic boundary conditions. On the water/steam side, all bundle heat exchangers have been implemented with real geometry data. The sub-critical HRSG model includes the followings major components:

- HPHT superheater, HPLT superheater, HP drum, HP evaporator, HPHT economiser, HPIT economiser, HPLT economiser1 and HPLT economiser2
- HT reheater, LT reheater, IP superheater, IP drum, IP evaporator and IP economiser

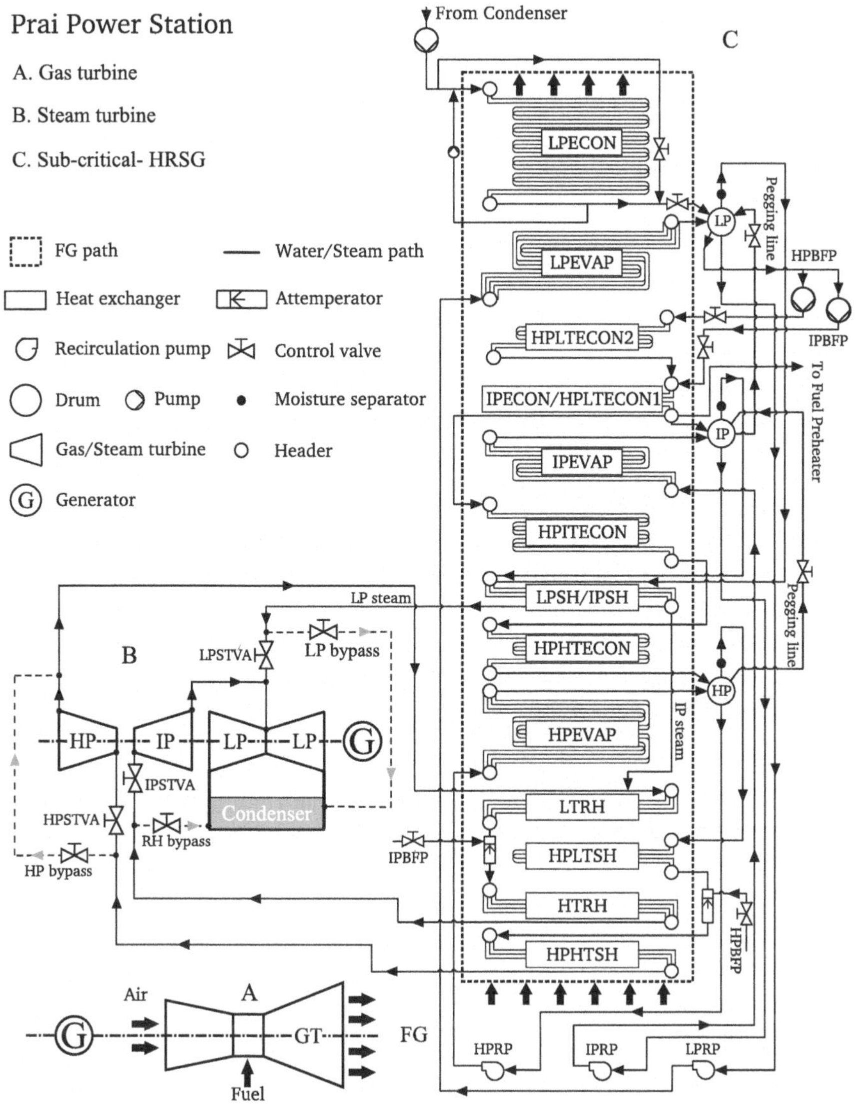

Fig. 4.2 Schematic flow diagram of Prai Power Station with a sub-critical heat recovery steam generator including high, intermediate and low-pressure circuits (simplified)

- LP superheater, LP drum, LP evaporator and LP economiser
- Connecting piping, headers, control valves, pumps, turbines, stream splitters, stream mixers and different controllers

In APROS, the sub-critical HRSG model is built with different nets (10 nets in total) in order to have a high level of order and flexibility. As examples for the

Table 4.1 Technical data of Prai Power Station located in Malaysia

Sub-critical heat recovery steam generator				Flue gas	ST power GT power
High-pressure system	Reheater section	Low-pressure system	Condenser inlet	$\dot{m}$ = 587 (kg/s) T_{inl} = 628 (°C) T_{out} = 81 (°C)	130 (MW_{el}) 220 (MW_{el})
p = 98 (bar) T = 566 (°C) $\dot{m}$ = 78 (kg/s)	p = 21 (bar) T = 567 (°C) $\dot{m}$ = 83 (kg/s)	p = 4.1 (bar) T = 293 (°C) $\dot{m}$ = 9.8 (kg/s)	p = 56 (mbar) T = 35 (°C) $\dot{m}$ = 92.8 (kg/s)		

model nets, the flue gas path, the high-pressure and low-pressure circuits are shown in Figs. 4.3, 4.4 and 4.5.

The exhaust gas leaves the gas turbine horizontally and enters the inlet duct of the HRSG. The exhaust gas will pass by the pressure sections (heat exchangers) in the following order: HPHTSH, HTRH, HPLTSH, LTRH, HPEVAP, HPHTECON, IPSH, LPSH, HPITECON, IPEVAP, HPLTECON1, IPECON, HPLTECON2, LPEVAP and LPECON (see Fig. 4.3). Hereby, each heat exchanger section causes a pressure drop to the flue gas stream. At the parallel arranged heat exchanger units (IPSH-LPSH and HPLTECON1-IPECON), the flue gas stream is split and combined again downstream of them.

The high-pressure system, illustrated in Fig. 4.4, is modelled from the outlet of the high-pressure boiler feedwater pump (HP BFP) to the superheater outlet (HPHTSH). The water flows into the HP drum via the HP economisers (HPLTECON1, HPLTECON2, HPITECON and HPHTECON). The high-pressure feedwater mass flow rate is controlled by the HP drum level control valve. It is located between the HPHT economiser outlet and the HP drum. The water of the HP drum circulates through the high-pressure evaporator tube bundle, and is heated

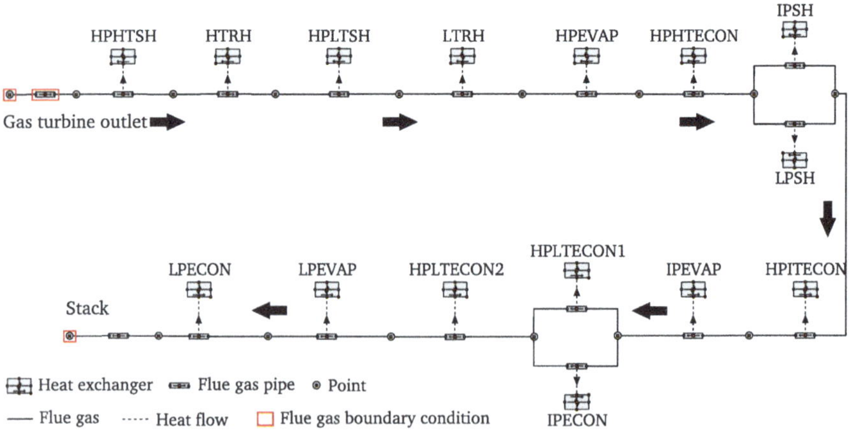

Fig. 4.3 Flue gas net of the HRSG model

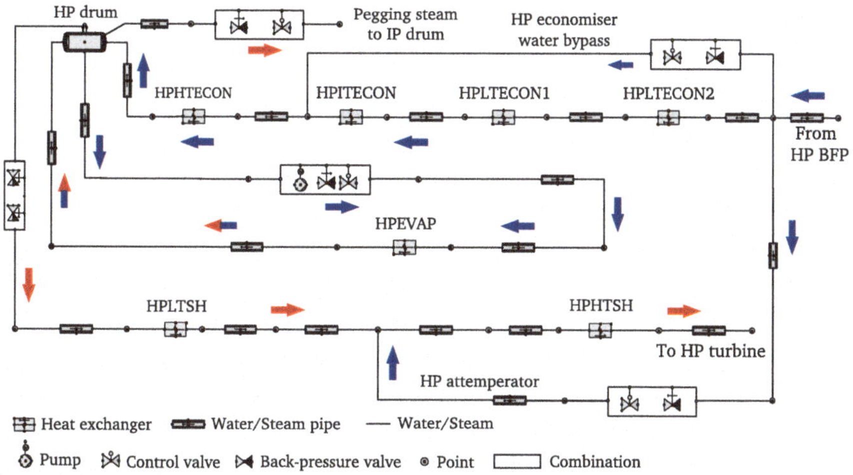

Fig. 4.4 High-pressure net of the HRSG model

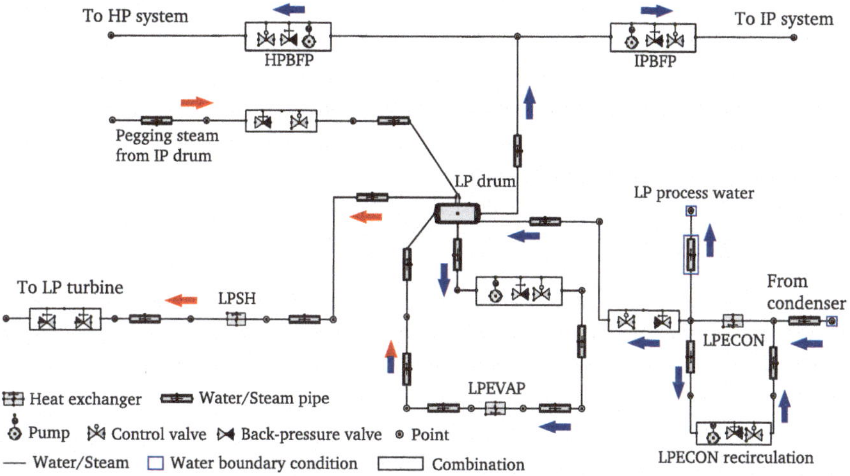

Fig. 4.5 Low-pressure net of the HRSG model

by the gas turbine exhaust gas and converted into saturated steam before it enters the HP drum again. The steam exits the HP drum passing through the moisture separator. Then, the steam is superheated by the high-pressure superheaters (HPLTSH and HPHTSH), which are of finned tube design. The superheated steam enters the HP turbine section. The sub-critical HRSG is equipped with a high-pressure attemperator provided at the inlet of the last superheater (HPHTSH). The aim of high-pressure attemperator is to adjust the temperature at the inlet of the HP turbine of about 566 °C. The attemperator uses the water from the high-pressure boiler feedwater pumps. During dynamic transients, the high-pressure economisers (HPLTECON2, HPLTECON1 and HPITECON) may partly be bypassed. The bypassed mass flow rate is varied to keep the outlet temperature of the HPHT economiser below boiling (sub-cooling of about 5 °C is recommended). This procedure will improve the flow stability in the high-pressure evaporator.

The intermediate-pressure circuit is modelled from the outlet of the IP feedwater pump to the IP superheater outlet. It consists of the following heat exchangers (IPECON, IPEVAP and LPSH). The IP feedwater mass flow rate to the IP drum is adjusted with the aid of a level control valve that is located upstream the IP economiser. The heated water in the IP drum circulates through the IP evaporator by forced circulation pump (IP CP), where it is heated by the gas turbine exhaust gas and converted into saturated steam in the IP drum. Here, the drum works as a separator. While the liquid stays in the drum and mixes with water coming from the IP economiser, the dry steam exits the intermediate-pressure drum and flows through the IP superheater. Some of the heated economiser water is taken out of the IP system to preheat the natural gas fuel, before it is returned to the condensate system at the condenser. The IP superheated steam joins with the partly expended steam coming from the high-pressure turbine outlet. The cold steam and the IP superheated steam provides the total steam to be admitted into the reheater section of the HRSG. The steam is reheated in the reheater tube bundle (LTRH and HTRH) by the exhaust gas of the gas turbine. The temperature at the reheater outlet is controlled by the reheater attemperator of about 566 °C using the water from the intermediate-pressure boiler feedwater pumps. Finally, the reheated steam enters the inlet of the intermediate-pressure turbine.

In Fig. 4.5, the low-pressure net of the HRSG model in APROS is presented. The source of feedwater flow into the LP circuit is from the condensate pump discharge. A level control valve is located between the low-pressure economiser outlet and the low-pressure drum. The water in the LP drum circulates through the evaporator tube bundle by a circulation pump, and is heated by the GT exhaust gas and converted into saturated steam in the low-pressure drum. The steam exits the low-pressure drum after passing through the moisture separator. The dry steam from the LP drum flows then through the LP superheater tube bundle and absorbs additional heat from the exhaust gas. The sub-critical HRSG is equipped with an LP economiser recirculation system. The LP economiser circulation pump recirculates the heated water from the discharge of the LP economiser to the inlet of the LP economiser. The control valve regulates the recirculation mass flow rate to maintain the LP economiser inlet temperature at the temperature setpoint (50 °C). During the operation with high sulphur fuel, the low-pressure economiser (LPECON) is totally bypassed. In this case, the LP drum equilibrium pressure is below the minimum admission pressure of the LP turbine and therefore the LP main steam control valve (LPMSCV) is fully closed. Furthermore, a pegging steam is supplied to the LP drum from the IP drum to maintain a minimum pressure of about 2 bar. The LP drum pegging steam system can also be applied to limit the LP drum pressure decrease during transients. The HP and IP boiler feedwater pumps take suction from the low-pressure drum through one downcomer.

In ASPEN PLUS DYNAMICS, the sub-critical HRSG model is built with a single net including the flue gas path, the water/steam side, the steam turbine stages and the associated control circuits (see Fig. 2.7).

Control Structure

In dynamic simulations, considerable work has to be achieved for building, improving and verifying the control circuits that must adjust all basic functions of the thermal power plant. Applying the correct control concepts and selecting the realistic parameters of the controllers are fundamental for getting the optimal response of those control circuits during transients. There are two basic approaches how to implement the control loops in dynamic simulation models. The first possibility is the adaption of the control structures from the real thermal power plant, if available. The other option is to use of published standard control concepts and modify them to the specific needs. Keeping the relevance of the control logics regarding the dynamic plant behaviour in mind, the first approach is preferred. However, real control schemas can be rather complex due to the fact that they take all possible eventualities during the power plant operation into account. Since the development of a power plant simulator is not the aim in most dynamic simulation models, the control schemas can be adapted from the real power plant to specific needs for the simulation scenarios.

For each pressure circuit (HP, IP and LP), the following control circuits are implemented: drum level control circuit, attemperator control circuit, economiser water bypass and turbine steam bypass control circuits. The control circuits of the HP system will be explained since the IP and LP control structures are almost similar to those of the HP system.

The control circuit of the HP pressure drum is based on three-element drum level control with a new feature that makes the controller more robust and more stable (see Fig. 4.6). The HP drum level controller regulates the HP feedwater mass flow rate by controlling the HP feedwater valve, which is located downstream of the high-pressure economisers. The operation algorithm is described as follows:

- The difference between the HP drum level setpoint and the actual value δL is measured and used as an input signal for the PI-controller. The output range of this controller is set between (0 and k), where k is a constant and can be defined by the user (k is equal to 100 for this model). The output signal of the PI-controller is then summed up to $(-k/2)$.
- The difference between the HP feedwater mass flow rate and the HP steam mass flow rate $\delta \dot{m}$ is measured.
- The summation of the signals $(\delta \dot{m}, \delta L)$ is used as an input signal for the PI-controller.
- The PI-controller commands the continuous device control (DC) that operates the HP drum level valve.
- If the pressure in the HP drum increases over a certain value (maximum pressure), the summation of the signals $(\delta \dot{m}, \delta L)$ will be replaced by the signal

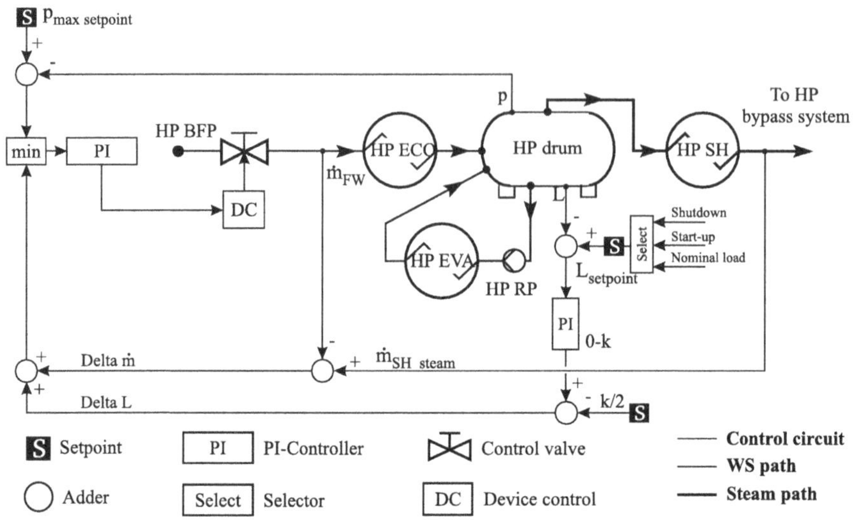

Fig. 4.6 Control circuit of the HP drum water level

resulted from the pressure difference between the HP drum and the maximum pressure setpoint in order to prevent the further raise in the HP drum pressure.

The high-pressure attemperator injects the feedwater mass flow at the inlet of the last high-pressure superheater to maintain a constant steam temperature at the inlet of the HP turbine. High vibrations or peaks in the superheated steam temperature should be avoided in order to protect the material of the HP turbine for a longer life span. The attemperator is configured to evaporate all water sprayed into the steam. The operation algorithm of the HP attemperator control is described as follows:

- The superheated steam temperature at the outlet of the HPSH is measured and then compared with the setpoint (here is 566 °C).
- The difference between these two signals is used as an input signal for the PI-controller.
- The PI-controller directs a DC continuous device that operates the HP attemperator valve.

The economiser water bypass controller of the HP system has two tasks as follows (see Fig. 4.4):

- When the gas turbine is operated with sulphur free fuel, a part of the water at the economiser inlet is partly bypassed to the drum inlet in order to prevent the last HP economiser from flashing. The bypassed mass flow rate is varied to keep the outlet temperature of the last economiser below the boiling point with a certain value (recommended sub-cooling is 5 °C).
- During the GT operation with heavy oil or high sulphur natural gas, the bypassed mass flow to the HP drum leads to a temperature decrease in the drum, which in turn

results in a reduction in the generated steam. In view of the available heat, the inlet temperature of the LP economiser can be increased from 55 to 120 °C, which is above the dew point of the sulphuric acid. This procedure will prevent the corrosive sulphuric acid in the flue gas from condensing at the LP economiser surfaces.

The high-pressure bypass circuit routes the high-pressure steam, which is not accepted by the high-pressure turbine into the cold reheater. In addition to the control circuits, the high-pressure bypass system includes two control valves and desuperheater. The high-pressure turbine valve (HPMSCV) controls the main steam mass flow rate to the HP turbine. It is operated by a time gradient function. The HP bypass valve (HPBPCV) adjusts the steam mass flow rate to the reheater section. A desuperheater has been implemented, which cools the steam behind the HP bypass control valve to 50 °C above the saturated steam temperature before it enters the cold reheater system. The injection water is delivered from the high-pressure boiler feedwater pumps. The high-pressure bypass circuit works during the HRSG start-up, as long as the HP steam quality has not matched the high-pressure turbine requirements. Furthermore, the high-pressure bypass system is used during the steam turbine trip and the gas turbine load rejection at any load. The working principle is described as follows:

- At the first stage of the start-up, both HP bypass valve (HPBPCV) and HP main steam valve (HPMSCV) are initially closed. This procedure enables the pressure in the high-pressure circuit to increase rapidly. When the minimum pressure setpoint is met, the HP bypass valve starts opening to counteract further rising of the pressure. Due to the continuous pressure increasing, the HPBPCV will be farther opened (see Fig. 4.7-(A)).
- Once the HPMSCV begins opening, the HPBPCV closes with the same ratio (function (B) in Fig. 4.7). During this process, the HP pressure should not exceed a maximum value. If it occurs, the function (C) will decelerate the closing of the bypass valve, resulting in releasing the pressure.
- In case of a steam turbine trip (function (D) in Fig. 4.7), the HP bypass control circuit is put in operation. While the HPMSCV is immediately closed, the HPBPCV controls the high-pressure to reach the pressure level, which was existent before the steam turbine trip. Holding the pressure at high level has the advantage that the HRSG is already prepared to the hot restart.
- In case of a gas turbine load rejection (function (E) in Fig. 4.7), the gas turbine exhaust temperature drops very rapidly to ambient temperature. The superheated steam temperature will decrease accordingly, but the HP pressure remains at its nominal value. In order to prevent the condensing in the HP superheater, the HP bypass valve will open discharging the pressure. The HP pressure is reduced to its saturation temperature below the present gas turbine exhaust temperature. This allows the medium in the HP superheater to stay in steam state and thus the undesired condensation is prevented.
- During the shutdown procedure (function (F) in Fig. 4.7), the pressure should be kept at its standing idle setpoint. In order to hold the pressure, there is a PI-controller that closes the main steam control valve (HPMSCV). If the

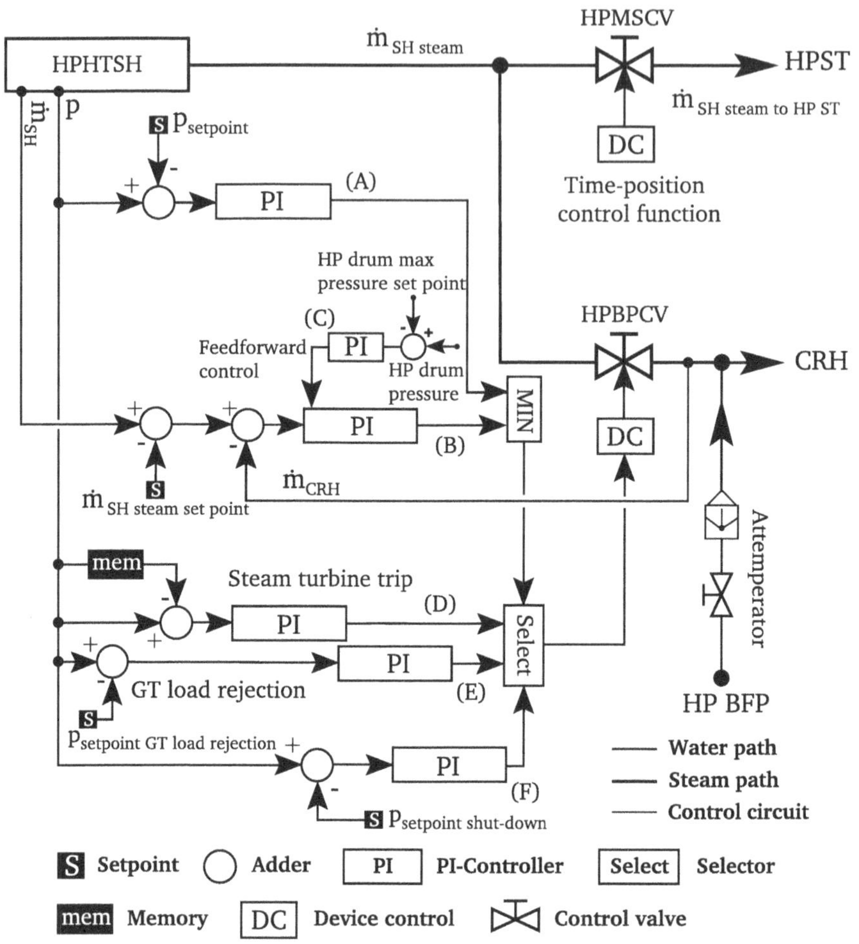

Fig. 4.7 High-pressure bypass control circuit

pressure exceeds the setpoint, the HPBPCV will be opened to control the pressure at its setpoint. Holding the pressure during the shutdown procedure has the advantage that during the start-up process the HRSG can reach its nominal operation values faster.

It should be noted here that the switching between different operation modes, e.g. shutdown procedure, steam turbine trip or gas turbine load rejection is performed with the help of the selector operator.

Figure 4.8 illustrates the changes in the valve positions of HPMSCV and HPBPCV during a start-up, which in turn leads to a smooth build-up of the pressure in the high-pressure system.

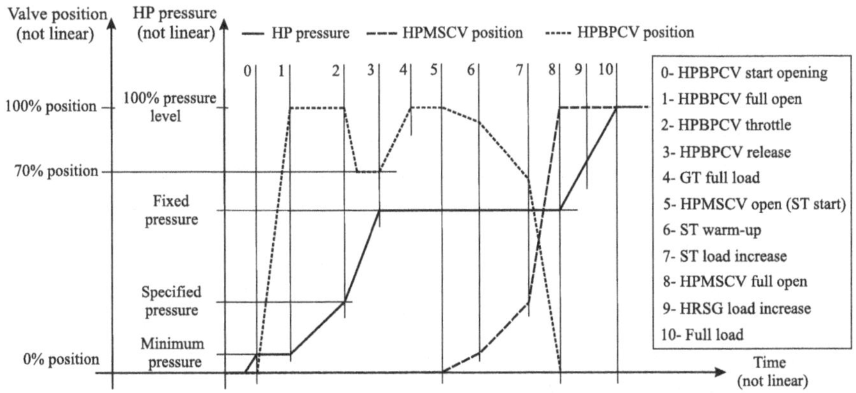

Fig. 4.8 Dynamic behaviour of HP pressure, HPMSCV and HPBPCV during a start-up procedure

Dynamic Simulation

The part load simulations are performed through the variation of the exhaust gas mass flow rate at the inlet of the HRSG models, while the flue gas temperature remains unaltered equal to 628 °C. Furthermore, the flue gas composition has been assumed to be constant. The sub-critical HRSG models will run a short time under full load conditions before ramping down to 60% load. After operating at this part load, the models will ramp up to an 80% load. The flue gas mass flow rate changes during the load conditions are described as follows:

- 100% steady-state for 30 min (flue gas mass flow rate is 587 kg/s)
- Load decrease of the GT (8%/min) from 100% load down to 60% load (flue gas mass flow rate is 419 kg/s)
- 60% steady-state for 120 min
- Load increase of the GT (8%/min) from 60% load to 80% load (flue gas mass flow rate is 490 kg/s)
- 80% load steady-state for 120 min
- Load increase of the GT (8%/min) from 80% load to 100% load (flue gas mass flow rate is 587 kg/s)

For the validation, the numerical results obtained from APROS and ASPEN PLUS DYNAMICS models at 80% and 60% load changes are compared with the measurements. Here, the steam mass flow rate, the temperature and the pressure for HP, RH and LP circuits are presented.

The comparison between the model predictions and the measurements at the outlet of the last HP superheater at different part loads is displayed in Table 4.2. Compared to operation data, most of the parameters have a relative error less than 1%. The HP steam mass flow rate obtained numerically agrees very well with the real power plant. The relative error is about 0.5% at the nominal load and 1.8% at 60% load. The HP superheated temperature is kept constant equal to 566 °C

Table 4.2 HPSH outlet at different part loads

HPSH outlet (%)	HPSH steam mass flow rate (kg/s)			HPSH temperature (°C)			HPSH pressure (bar)		
Load	Exp.	APROS	ASPEN	Exp.	APROS	ASPEN	Exp.	APROS	ASPEN
100	78	77.9	77.8	566	566	566	98	97.5	97.2
80	66.9	67.4	67.6	565.9	566	566	88	85.9	86.4
60	58.3	57.5	57.2	565.9	566	566	76.2	74.8	76.4

independent of the load because of the HPSH attemperator. The simulated superheated high-pressure matches well with the measured data. At 80% load, the pressure deviates slightly from the measurement with a relative error of about 1.8% and 2.3% in APROS and ASPEN PLUS DYNAMICS models, respectively.

As seen in Table 4.3, the numerical parameters of the reheater section show a good agreement with operation data. Only the pressure has an error in the order of magnitude of up to 6%. The simulated steam mass flow rate in APROS and ASPEN PLUS DYNAMICS model displays a maximum relative error of 2.4% at 60% load. Like the HP circuit, the reheated temperature is kept constant equal to 567 °C due to the RH attemperator. In the ASPEN PLUS DYNAMICS model, the pressure at 100% and 80% loads agrees very well with the real power plant. However, the model accuracy decreases at 60% load, showing a relative error of about 6%. The pressure in the APROS model displays at the nominal load a relative error of about 2%. At 80% and 60% loads, the pressure deviates from the measurement with a relative error of about 5%.

Table 4.4 presents the steam temperature, pressure, and steam mass flow rate at the outlet of the LP superheater at different loads. It is clearly that APROS model shows higher accuracy compared to the ASPEN PLUS DYNAMICS model regarding the numerical prediction of the LP circuit. The simulated LP steam mass flow rate in the APROS model has a maximum relative error of about 1%, while the error is up to 6% in the ASPEN PLUS DYNAMICS model. The LP superheated temperature shows in both programmes a very good agreement towards the measurement with a relative error less than 1%. Furthermore, the comparison between the simulated pressure and the measured pressure indicates a good correspondence. Here, the APROS model predicts an exact value, while the ASPEN PLUS DYNAMICS model has a relative error of about 3%.

Table 4.3 RH outlet at different part loads

RH outlet (%)	RH steam mass flow rate (kg/s)			RH temperature (°C)			RH pressure (bar)		
Load	Exp.	APROS	ASPEN	Exp.	APROS	ASPEN	Exp.	APROS	ASPEN
100	83	83.1	82.9	567	567	567	21	20.9	21.6
80	74.7	74.1	74.8	566.9	567	567	20.6	19.4	20.7
60	64.9	63.3	63.5	566.9	567	567	17.7	16.8	18.9

Table 4.4 LPSH outlet at different part loads

LPSH outlet (%)	LPSH steam mass flow rate (kg/s)			LPSH temperature (°C)			LPSH pressure (bar)		
Load	Exp.	APROS	ASPEN	Exp.	APROS	ASPEN	Exp.	APROS	ASPEN
100	9.8	9.8	9.9	293	292.7	292.5	4.1	4.1	4.1
80	7.7	7.8	8.3	289.9	288	289	3.9	3.9	3.8
60	6.4	6.4	6.8	281	279	283.5	3.3	3.3	3.2

According to the comparison between APROS and ASPEN PLUS DYNAMICS models, it can be concluded that both programmes can represent the real power plant at different loads with good accuracy. The maximum relative error amounts to 5%, although several parameters have a relative error of less than 1%.

In a next step, the variations of the gas turbine exhaust gas mass flow rate and the temperature for 400 min are investigated (see Fig. 4.9). The off-design operation is described as follows:

- The flue gas mass flow rate and temperature are at their nominal values for a period of time equal to 40 min.
- A ramp occurs, where the flue gas mass flow rate decreases to about 75% its nominal load together with a considerable increasing in the flue gas temperature.
- Minor oscillations take place in the remaining time in view of the load demand.

The response of feedwater and steam mass flow rates, temperature and pressure for HP, IP/RH and LP circuits during the off-design phase is presented in the following Figures (dotted line red in colour for the APROS model and dashed line blue in colour for the ASPEN PLUS DYNAMICS model). The base data for the comparison is the dynamic measurements from the real combined-cycle power plant (solid line black).

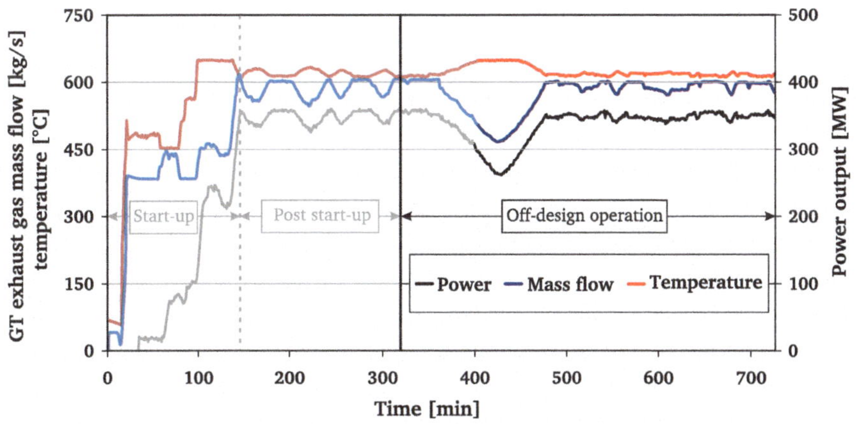

Fig. 4.9 Gas turbine exhaust gas mass flow rate and temperature during off-design operation

The feedwater and steam mass flow rates of the high-pressure system are displayed in Fig. 4.10. The computed feedwater mass flow rate in both programmes shows a good agreement with the operation data with a maximum relative error of about 6% (observed at t = 108 min). However, the numerical models underestimate the measured steam mass flow rate with higher relative error of about 11% during the complete off-design phase. This deviation is of minor relevance due to the difference in the measured HP feedwater and steam mass flow rates. This difference (3–6 kg/s) can only be caused by the HP attemperator. However, the injected water mass flow rate appears to be very high in the real power plant, which raises the question of whether this deviation is related to the HRSG models. As further evidence is that the measured steam mass flow rate at 100% nominal load (see Table 4.2) amounts to 78 kg/s, while it is approximately 81 kg/s during the off-design operation.

In Fig. 4.11, the dynamic behaviour of the HP steam temperature and the pressure is shown. The measured temperature oscillates in the range between 568 °C and 545 °C according to exhaust flue gas variations during the off-design phase. In the period of time between t = 75 min and t = 135 min, the superheated temperature obtained numerically has an almost opposite behaviour compared to the operation data. APROS as well as ASPEN PLUS DYNAMICS models overestimate the measured superheated temperature with about 25 °C. From t = 135 min, a quantitative developing of the temperature with a maximum difference of 2 °C between the numerical models and the measurement can be observed. The simulated pressure shows a very good correspondence with the real power plant. In the period

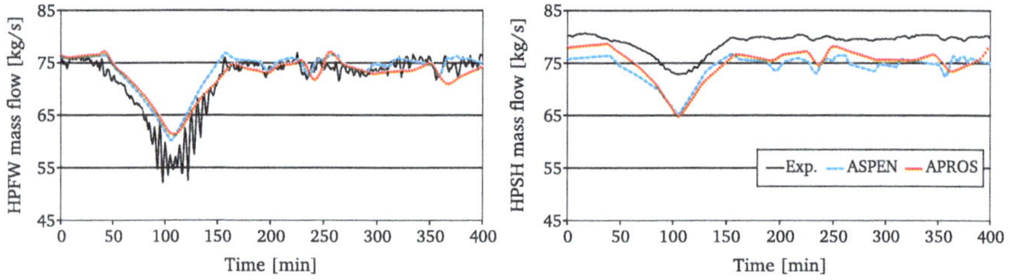

Fig. 4.10 Measured and numerically obtained high-pressure feedwater mass flow rate and high-pressure steam mass flow rate during the off-design operation

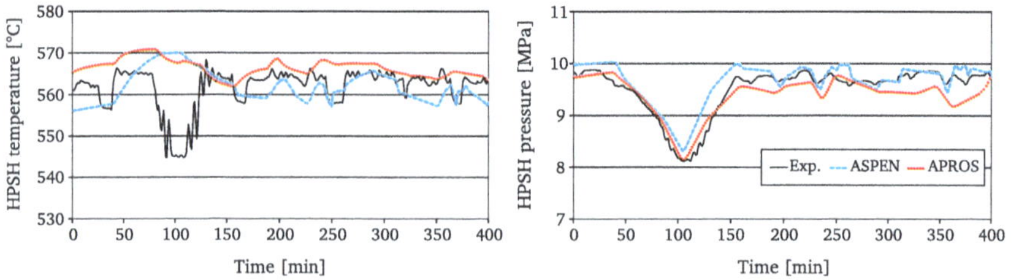

Fig. 4.11 Measured and numerically obtained high-pressure superheated temperature and high-pressure superheated pressure during the off-design operation

of time between t = 0 and t = 180 min, the APROS model follows the measured development of the pressure with higher accuracy compared to the ASPEN PLUS DYNAMICS model. From t = 300 min, the APROS model underestimates the real pressure with a considerable relative error of about 6%, while the ASPEN PLUS DYNAMICS model reproduces the measured HP pressure with a maximum relative error of about 3%.

In Figs. 4.12 and 4.13, the dynamic simulation results of the IP circuit including the feedwater and steam mass flow rates, the pressure as well as the superheated temperature are presented. The IP feedwater mass flow rate obtained numerically agrees well with the operation data. Here, APROS and ASPEN PLUS DYNAMICS models reproduce the real IP feedwater mass flow rate with a high accuracy during the entire time of the off-design operation phase. Similar to the IP feedwater mass flow rate, the simulated superheated steam mass flow rate follows accurately the dynamic behaviour of the measured data. However, a slight deviation that lasted for 30 min can be detected between the numerical models and the real power plant in the period of time between t = 90 min and t = 120 min. Here, APROS and ASPEN PLUS DYNAMICS models underestimate the experiment with about 0.7 kg/s and 1 kg/s, respectively. From t = 120 min, both models show high accuracy with a maximum relative error of approximately 2%.

The superheated temperature and pressure obtained numerically, depicted in Fig. 4.13, display qualitative agreements with the measurements during the

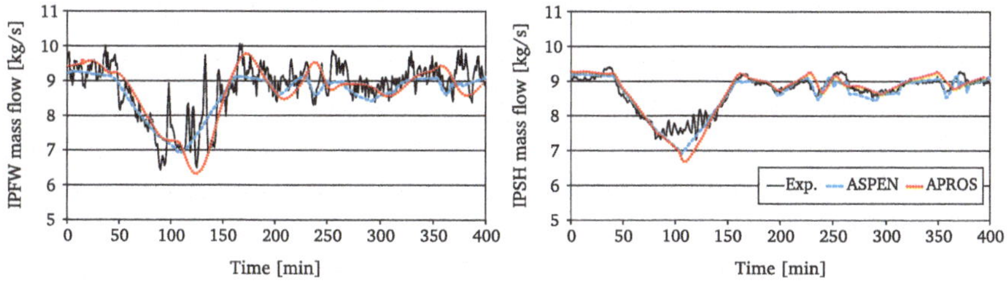

Fig. 4.12 Measured and numerically obtained intermediate-pressure feedwater mass flow rate and intermediate-pressure steam mass flow rate during the off-design operation

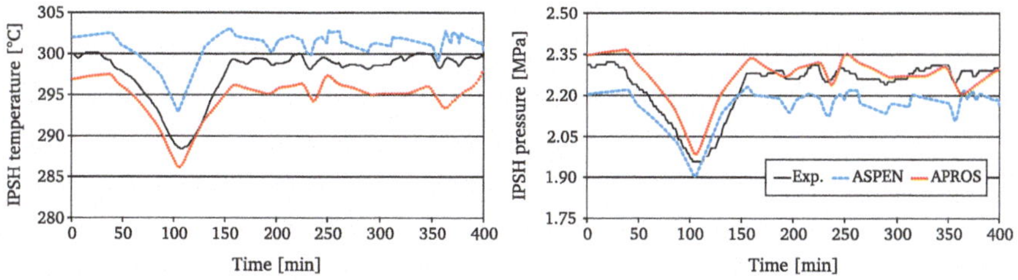

Fig. 4.13 Measured and numerically obtained intermediate-pressure superheated temperature and intermediate-pressure superheated pressure during the off-design operation

off-design operation. The superheated temperature calculated in the ASPEN PLUS DYNAMICS model overestimates the measured one, while the simulated temperature in the APROS model underestimates the measurement. In the range of time between t = 0 and t = 180 min, a minor discrepancy between the ASPEN PLUS DYNAMICS model and the measured temperature with a relative error of about 2% can be detected. From t = 200 min, this relative error decreases to approximately 1%. In the APROS, the relative error within the first 200 min is, by contrast, about 1% and then increases to 2% in the remaining time of the off-design operation.

The nominal value of the measured pressure is about 23 bar. From t = 30 min, the pressure starts decreasing and reaches 80% its nominal value within 75 min due to the reduction in the flue gas mass flow rate. Starting from t = 115 min, the measured pressure increases gradually and then experiences several oscillations around its nominal value. The numerical models reproduce qualitatively the dynamic behaviour of the IP pressure. Compared to the ASPEN PLUS DYNAMICS model, the APROS model simulates the dynamic change of the pressure with higher accuracy, especially in the period of time between t = 150 min and t = 400 min. Here, APROS and ASPEN PLUS DYNAMICS models show maximum relative errors of about 3 and 7%, respectively.

The IP superheated steam and the partly expended steam coming from the HP turbine are combined together and then sent to the reheater section. The reheated temperature and pressure calculated by the numerical models are compared with the operation data (see Fig. 4.14). The reheated temperature obtained numerically agrees well with measured data. Similar to the HPSH temperature, the reheated temperature simulated in APROS and ASPEN PLUS DYNAMICS models show confusing characteristic in the period of time between t = 60 min and t = 130 min. While the reheated temperature decreases from 565 to 548 °C in the measurement, it increases to approximately 575 °C in both programmes with a maximum relative error of 5%. This disagreement is of relevance, although the relative error between the measurement/simulation is relatively small. An explicit explanation regarding this deviation cannot be provided. This is due to the fact that the simulated reheated pressure shows on the one hand an excellent compliance with the measured data and on the other hand the flue gas temperature increases in the period of time between t = 75 min and t = 150 min about 6% above its nominal value.

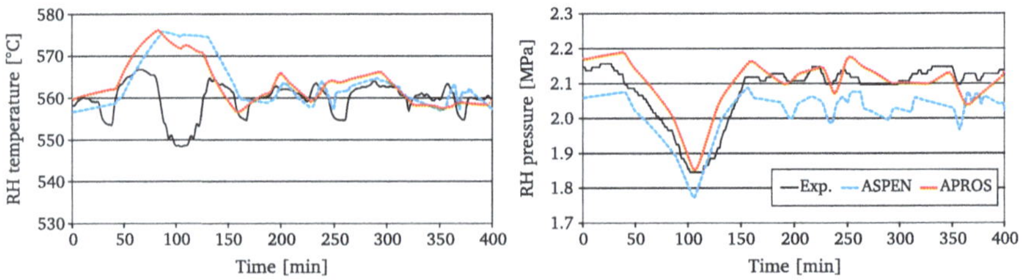

Fig. 4.14 Measured and numerically obtained reheated temperature and reheated pressure during the off-design operation

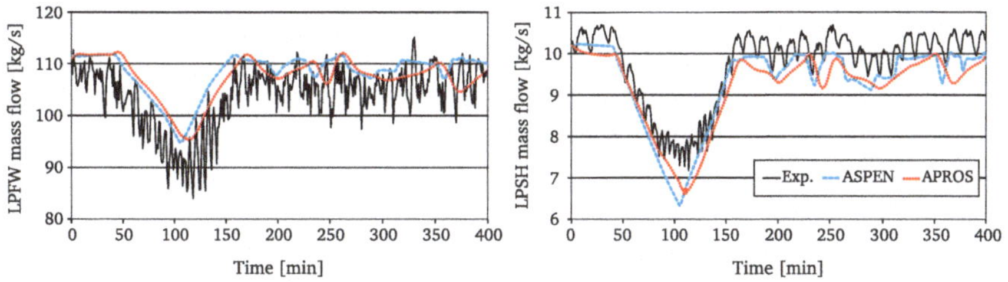

Fig. 4.15 Measured and numerically obtained low-pressure feedwater mass flow rate and low-pressure steam mass flow rate during the off-design operation

The dynamic behaviour of the low-pressure circuit during the off-design operation is depicted in Fig. 4.15. The low-pressure feedwater mass flow rate represents the entire mass flow rate entering into the heat recovery steam generator. The simulated LP feedwater mass flow rate shows a good agreement with the operation data. APROS and ASPEN PLUS DYNAMICS models predict the dynamic change of the LP feedwater mass flow rate with almost same accuracy. Compared to numerical models, the measured feedwater mass flow rate oscillates highly with an altitude of approximately 15 kg/s. This behaviour could not be simulated by APROS and ASPEN PLUS DYNAMICS models in view of the following reasons. One indication is the less mass inertia of the models compared to the real power plant. Some sub-systems in the real HRSG, which are not included in the numerical models, are for example the vent and drain circuits. Other reason is related to the HP, IP and LP feedwater controllers, which respond smoothly due to optimised values of three-term controller (proportional, integral and derivative).

The LP superheated steam mass flow rate obtained numerically by APROS and ASPEN PLUS DYNAMICS models follows qualitatively the measured mass flow rate. A considerable deviation between the numerical models and the real power plant is observed for 30 min in the range of time between t = 90 min and t = 120 min. Here, APROS and ASPEN PLUS DYNAMICS models underestimate the real measurement with maximum relative errors of about 10% and 12%, respectively. In the remaining time, the computed superheated steam mass flow rate is slightly below the measured data with approximately 0.8 kg/s.

4.1.1.2 Start-up and Shutdown Procedures

The validation of dynamic simulation models during shutdown and start-up procedures are of relevance since the GT exhaust gas mass flow rate and temperature vary with great values. Here, the ability of the numerical models to represent the real power plant behaviour can be closely evaluated. Depending on the standing idle time after the GT shutdown, it is generally distinguished among three start-up procedures: hot, warm and cold start-ups. In hot start-up, the standing idle time is up to approximately 8 h, while it is more than 48 h in the cold start-up. In this section, dynamic simulations for a sub-critical three-pressure-stage heat recovery steam generator are carried out during shutdown as well as during hot and warm start-ups. A wide set of operation data from an existing power plant (Prai Power Station) is applied to validate the generated dynamic models.

Warm Start-up

Figure 4.16 shows the warm start-up curves for exhaust gas mass flow rate and temperature after 26 h standing idle. The variation of the flue gas mass flow rate and the temperature at the inlet of the HRSG can be split into two phases: start-up phase (0–150 min) and post start-up (150–320 min). The start-up phase can also be divided into three main parts (the venting process, the start-up/hold point and the load increase). The procedure begins with the start-up venting process, where the GT purges the HRSG with ambient air to remove the residual combustible gases. During the shutdown, the combustible substances accumulate in the HRSG and can cause an explosion during the GT start-up, which results in great damage in the facility. The venting process lasts about 15 min, using an air mass flow rate of 42 kg/s (7% of the nominal gas turbine mass flow rate). The gas turbine is then put into operation and accordingly jumps in the flue gas mass flow rate and the temperature can be detected. From t = 25 min, the flue gas mass flow rate and the temperature are almost kept more or less at the same level (70% of the gas turbine full load). This process lasts 30 min and aims to warm-up the critical components of the HRSG such the high-pressure superheaters and the high-pressure drum. In the period of time between t = 55 min and t = 150 min, the gas turbine load increases to the nominal load. At t = 150 min, the post start-up phase starts. Here, the load of the GT does not change as steeply as in the start-up phase. Each reduction in the flue gas mass flow rate is related to a slight increase in the flue gas temperature. It

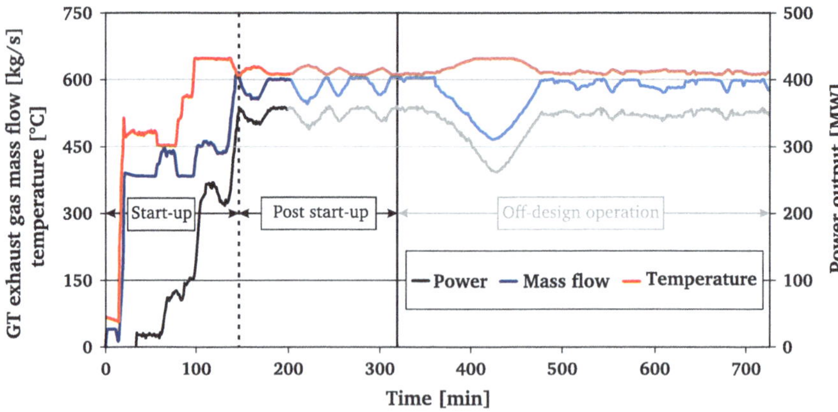

Fig. 4.16 Gas turbine exhaust gas mass flow rate and temperature during the warm start-up procedure

should be mentioned here that the gas turbine load in the post start-up phase is varied for purpose of measurements.

As dynamic boundary conditions for the start-up procedure, the mass flow rate and the temperature of the exhaust gas at the inlet of the sub-critical HRSG models (see Sect. HRSG Model) are set according to Fig. 4.16. Here, the flue gas composition has been assumed to be constant over the load range of the gas turbine. The response of the feedwater and steam mass flow rates, the temperature and the pressure as well as the drum level for each system is presented in the following figures (dotted line red in colour for the APROS results and dashed line blue in colour for the ASPEN PLUS DYNAMICS results) (Alobaid et al. 2008, 2014b, 2015c). The base data for the comparison is the dynamic measurement from the real combined-cycle power plant (solid line black).

In Fig. 4.17, the computed feedwater and steam mass flow rates of the high-pressure system are compared with the operation data. The simulated feedwater mass flow rates in both programmes show a good agreement for start-up and post start-up phases. However, the measured feedwater mass flow rate oscillates highly with the amount of approximately 50 kg/s, especially in the start-up phase. This behaviour is correlated with the HP drum level reduction prior to start-up process. Generally, the level of drums in the HRSG is lowered to about 70% of the nominal drum level (see Fig. 4.19). After 25 min from the start-up, the start-up drum level setpoint is switched over to the nominal drum level setpoint, which causes the measured feedwater mass flow rate to oscillate accordingly. Such behaviour is underestimated in APROS and ASPEN PLUS DYNAMICS models, although the drum level is decreased in both models to about 1.1 m. This can be attributed to several reasons. The most important one is the less mass inertia of the models compared to the real power plant (some sub-systems in the real HRSG are not included in the numerical models such as vent and drain circuits). Other indication is related to the HP feedwater controller that responds roughly due to the inaccurate input signals and/or the wrong setting of three-term controller (proportional, integral and derivative values). In the time interval between t = 150 min and t = 320 min, the simulated feedwater mass flow rate, especially the ASPEN PLUS DYNAMICS model, agrees very well with the measured data.

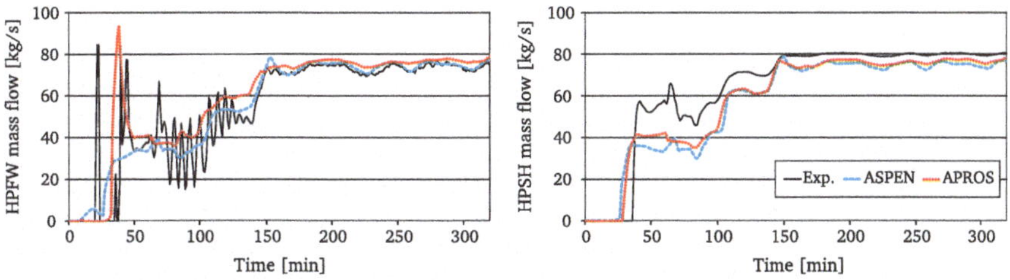

Fig. 4.17 Measured and numerically obtained high-pressure feedwater mass flow rate and high-pressure steam mass flow rate during the warm start-up procedure

The high-pressure steam mass flow rate obtained numerically shows a qualitative agreement during the start-up phase and a quantitative agreement in the remaining time. Prior to the start-up, the HP bypass valve and the HP steam valve are initially closed. As the gas turbine rumps up, the pressure in the HP drum increases rapidly to reach its fixed pressure (here is 44 bar). The high-pressure bypass valve will open rapidly, avoiding any further increase in the HP pressure. As a result, the HP steam mass flow rate jumps steeply in the power plant to about 70% of the nominal load. The numerical models failed to reproduce this behaviour. The steam mass flow rate starts earlier 10 min in the ASPEN PLUS DYNAMICS model and jumps to about 35 kg/s with a relative error of about 38%, while the steam is generated 8 min earlier in the APROS model with a relative error of about 30%. In the time period between t = 40 min and t = 80 min, the simulated steam mass flow rate in Apros and Aspen Plus Dynamics models oscillates around average values of 35 kg/s and 40 kg/s, respectively because of the flue gas hold point. Beyond 80 min, the gas turbine load increases to nominal load within 70 min, leading to higher steam production. Although the numerical models underestimate the measurement during the start-up phase, but qualitative behaviour can be clearly observed. In the time interval between t = 150 min and t = 320 min, the gas turbine operates in post start-up phase. The numerical models reproduce reasonably the measured steam mass flow rate with a discrepancy of about 8 kg/s. When looking at the measured HP feedwater and steam mass flow rates from t = 150 min, a mass flow rate difference of 5–10 kg/s is observed that can only be resulted from the HP attemperator, but seems to be very high. Furthermore, the measured steam mass flow rate at 100% nominal load (see Table 4.2) is 78 kg/s, while it is about 81 kg/s during the post start-up phase.

The dynamic behaviour of the high-pressure steam temperature and pressure during the gas turbine start-up shows a good agreement with the operation data (see Fig. 4.18). During the standing idle process (26 h), the temperature of the high-pressure circuit decreases gradually due to the natural cooling to about 150 °C. The superheated temperature is almost unchanged during the venting process. In the period of time between t = 20 and t = 50 min, the superheated temperature increases sharply with a thermal gradient of about 11 °C/min. Beyond 70 min, the temperature remains constant equal to 455 °C by the reason of the flue gas hold

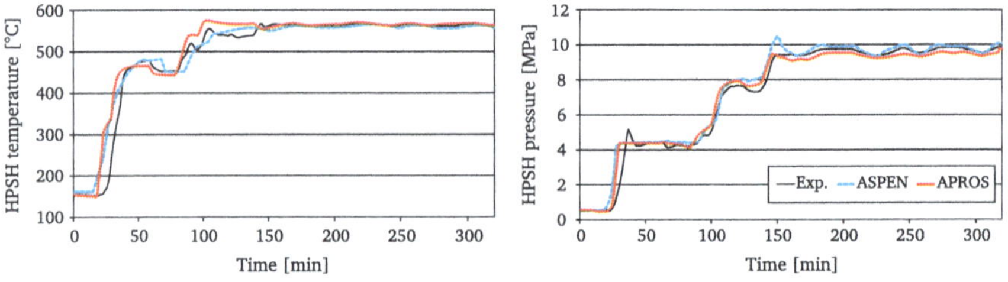

Fig. 4.18 Measured and numerically obtained high-pressure superheated temperature and high-pressure superheated pressure during the warm start-up procedure

point (see Fig. 4.16), followed with a second jump but with lower thermal gradient (5 ° C/min). From t = 150 min, the high-pressure attemperator controls the injected mass flow rate at the inlet of the last superheater (HPHTSH) to maintain a constant steam temperature of approximately 567 °C at the inlet of the high-pressure steam turbine.

In order to enable faster start-up procedures, the HP pressure is kept equal to 5.5 bar, while the HRSG shutdowns. Like the superheated temperature, the HP pressure displays similar behaviour during the venting process. Then, the pressure reaches fast to its fixed value (44 bar) with a pressure gradient of 14 bar/min. At this point, the HP bypass valve starts opening gradually in order to prevent further increase in the pressure. Holding the pressure and temperature at their fixed values are of relevance to warm-up the critical components of the HRSG. In the time interval between t = 85 and t = 150 min, the pressure increases steadily to it nominal value. Beyond t = 150 min, the computed pressure in the HP circuit follows the flue gas boundary condition and shows a good agreement with the measurement.

From Fig. 4.18, APROS and ASPEN PLUS DYNAMICS models respond to the flue gas changing similarly. One notable difference can be detected in the time period between t = 145 and t = 165 min. In APROS model, the simulated temperature exceeds the measured data, although the computed pressure agrees very well. In the ASPEN PLUS DYNAMICS model, an opposite behaviour is registered. The reason of this different behaviour is related to the HP attemperator controller. In case of the ASPEN PLUS DYNAMICS model, the controller responds on the one hand rigidly. As a result, the SH temperature is not allowed to exceed the setpoint (566 °C). On the other hand, the HP pressure goes beyond the measurement by the reason of the additional injected mass flow rate.

The characteristics of the high-pressure and intermediate drum levels during the warm start-up are displayed in Fig. 4.19. While the gas turbine ramps up, the temperature and the pressure of the fluid in the drum will increase accordingly. This can lead to a volume expansion of the saturated water up to 30% in the drum. In order to prevent the formation of bubbles in the drum and the risk to transfer water drops in superheater section, the drum level is decreased to about 70% its nominal load prior to the HRSG start-up. In the time interval between t = 0 and t = 25 min, the simulated HP drum level by APROS has almost the same trend with

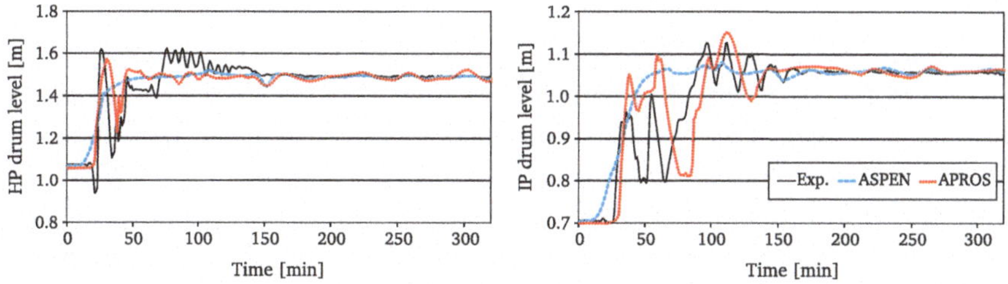

Fig. 4.19 Measured and numerically obtained high-pressure drum level and intermediate-pressure drum level during the warm start-up procedure

measurement. By contrast, the drum level obtained by ASPEN PLUS DYNAMICS model starts 5 min earlier than measurement in case of the HP drum and 15 min earlier in case of the IP drum.

From t = 25 min, the drum level setpoint is switched over to the nominal start-up drum level setpoint. Due to the large difference between these setpoints ($\Delta L = 0.4$ m for the HP drum and $\Delta L = 0.35$ m for the IP drum), the feedwater mass flow rate control circuit will respond fast, resulting in drum level oscillations in experiment and in the APROS model. By contrast, the ASPEN PLUS DYNAMICS model displays smooth developments of the HP and IP drums' levels by the reason of the rigid setting of the feedwater mass flow rate controller values. Generally, both simulation models show good qualitative agreements with real power plant. During the post start-up phase, the numerically obtained and measured drum levels fluctuate around their setpoints with relatively smaller values (0.05 m for the HP drum and 0.01 m for the IP drum).

The dynamic behaviour of the intermediate-pressure feedwater mass flow rate during the warm start-up procedure is illustrated in Fig. 4.20. After 8 min from the GT start-up, the measured feedwater mass flow rate jumps and oscillates around an average value of about 2.5 kg/s for 50 min. The mass flow rate raises then rapidly to 12 kg/s within a few minutes. In the period of time between t = 60 and t = 160 min, considerable oscillations in the measured mass flow rate can be observed. In the remaining time, the value of the IP mass flow rate is around 9 kg/s. The simulated feedwater mass flow rate shows a noticeable difference during the start-up phase. Similar to the measurement, the IP mass flow rate obtained numerically by ASPEN PLUS DYNAMICS model starts increasing at t = 8 min. However, the simulated mass flow rate exceeds the measured data and reaches a maximum value of about 9 kg/s at t = 30 min, before it decreases again to 6.5 kg/s at t = 65 min. The IP mass flow rate computed by Apros starts increasing 30 min later than the measurement and ASPEN PLUS DYNAMICS model. Here, a peak can be observed, in which the feedwater mass flow rate reaches a maximum value of 5.5 kg/s, before it decreases again to zero. It can be clearly detected that both numerical models failed to reproduce the dynamic behaviour of the IP feedwater mass flow rate during the first phase of the start-up. The reason of this discrepancy with the operation data is that the numerical models underestimate the thermal inertia of the real HRSG, since auxiliary systems such as safety vents, draining systems and steam sampling lines as well as supporting structures in thermal contact with the primary water/steam circuits and ambient losses are not considered. The difference in the developments of the mass flow rate in APROS and ASPEN PLUS DYNAMICS models is caused by the different thermal hydraulic flow models used. The two-fluid flow model (six-equation) is applied to the APROS model, while the mixture flow model (three-equation) is used for the ASPEN PLUS DYNAMICS model. In the period of time between t = 65 and t = 160 min, the feedwater mass flow rate obtained numerically is in good agreement with the measurement. The mass flow rate oscillates between 5.2 kg/s and 8 kg/s in case of the APROS model and between 5.2 kg/s and 9.5 kg/s in case of the ASPEN PLUS DYNAMICS model. The both numerical models, although failed to reproduce quantitatively the high fluctuations measured in the real power plant, show a qualitative behaviour. In the post start-up phase, APROS and ASPEN PLUS DYNAMICS models follow the dynamic changes of the flue gas mass flow rate and temperature with high accuracy.

The simulated IP steam mass flow rate, displayed in Fig. 4.20, is compared with the measurement during the start-up and post start-up phases. After 50 min for the gas turbine start-up, the superheated steam mass flow rate increases sharply in the real power plant to about 10 kg/s within 4 min. By contrast, the IP steam mass flow rate obtained numerically is generated about 10 min and 30 min earlier in APROS and ASPEN PLUS DYNAMICS models, respectively. Sub-systems existing in the real power plant such as vent and drain systems are not included in the HRSG models. The associated less thermal inertia leads to an earlier steam generation in the IP evaporator and thus to a faster increase in the system pressure compared to the real HRSG. Accordingly, the fixed pressure (here is 12 bar) is rapidly reached by the numerical models (see Fig. 4.21), leading to start opening of the IP bypass valve in order to counteract the further pressure increase. The APROS model that used the six-equation flow model follows the behaviour of the measured steam mass flow rate with higher accuracy compared to the ASPEN PLUS DYNAMICS model. In the period of time between t = 60 and t = 160 min, an acceptable qualitative behaviour between APROS and ASPEN PLUS DYNAMICS models are observed, where the simulated steam mass flow rates are slightly below the measurement. During the post start-up phase, the superheated steam mass flow rate simulated with ASPEN PLUS DYNAMICS model shows a quantitative agreement towards the real data, while the APROS model reproduces the superheated steam mass flow rate with lower accuracy. Here, a minor deviation between the measurement and simulation in the amount of approximately 0.2 kg/s can be detected in the time interval between t = 160 and t = 320 min.

Figure 4.21 shows the response of the IP steam pressure and temperature during the start-up and post start-up phases in APROS and ASPEN PLUS DYNAMICS models. The measured steam pressure and temperature from the real power plant is also plotted. Generally, the warm start-up is taken place when the power plant is in standing idle for more than 20 h after the shutdown process. Due to the natural cooling, the initial temperature in the IP circuit is equal to 150 °C in simulations and experiment. At t = 15 min, the IP superheated temperature calculated in the

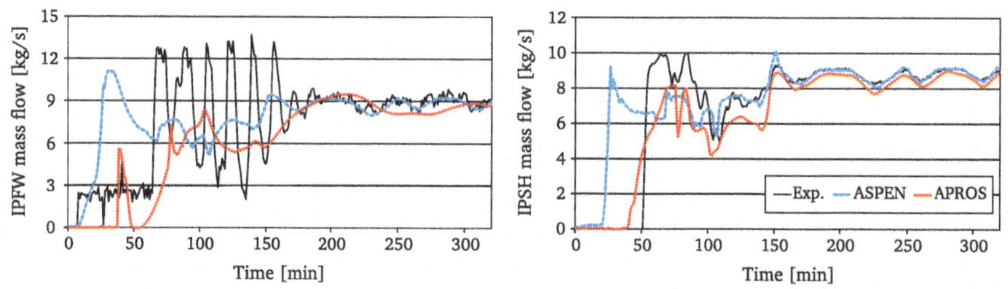

Fig. 4.20 Measured and numerically obtained intermediate-pressure feedwater mass flow rate and intermediate-pressure superheated steam mass flow rate during the warm start-up procedure

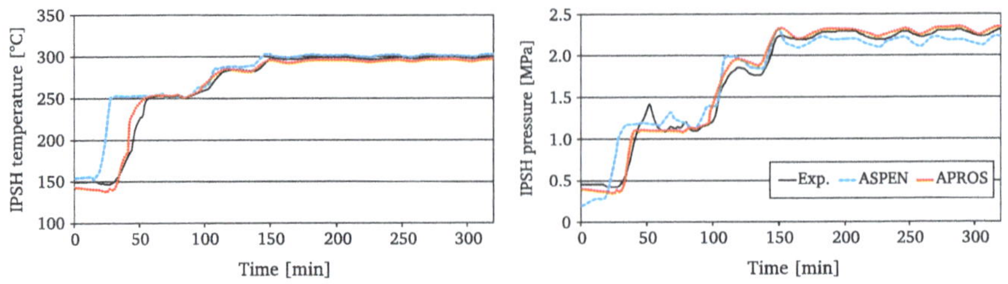

Fig. 4.21 Measured and numerically obtained intermediate-pressure superheated temperature and intermediate-pressure superheated pressure during the warm start-up procedure

ASPEN PLUS DYNAMICS model increases 15 min earlier than the measured temperature and the simulated one in the APORS model. In the period of time between t = 15 and t = 60 min, the APROS model reproduces the real power plant with high accuracy, while the ASPEN PLUS DYNAMICS model shows a great disparity. Although this discrepancy is relatively high, the ASPEN PLUS DYNAMICS model predicts the thermal gradient curve similar to those in experiment and the APROS model with a value of about 4 °C/min. In the time range between t = 65 and t = 160 min, the superheated temperature obtained numerically agrees very well with measured data. Here, it can be detected that APROS and ASPEN PLUS DYNAMICS models overestimate slightly the measurement with relative errors of about 1% and 2%, respectively. In the post start-up phase, the measured steam temperature remains almost constant (300 °C). The predicted temperatures in both numerical models demonstrate a quantitative behaviour compared with the measured data.

The development of the IP steam pressure shows a similar pattern as the HP steam pressure. Depending on the start-up types (hot, warm or cold), the pressure is hold at a certain value. This enables the HRSG to reach faster its nominal pressure. Prior to the HRSG start-up, the measured pressure in the IP system is equal to 4.5 bar, while it is 4.2 and 2 bar in APROS and ASPEN PLUS DYNAMICS models, respectively. At t = 30 min, the measured pressure increases to reach its fixed pressure within 20 min. Then, the pressure is kept almost unchanged around 12 bar by the reason of the IP bypass control circuit. In the period of time between t = 50 and t = 100 min, the IP bypass valve (IPBPCV) will gradually open, preventing any further jump in the IP pressure. From t = 100 min, the IPBPCV is fully opened and thus the IP pressure rises to 18 bar within 30 min. The pressure remains then almost constant for 10 min before it continues increasing to its nominal value by the reason of the GT load increase.

During the start-up phase, the trend of the IP pressure simulated by APROS model displays a quantitative agreement with the operation data. The APROS model predicts the start point of the pressure jump (t = 30 min) correctly, but underestimates slightly the altitude of the maximum pressure measured with about 10 bar. The reason of this discrepancy is associated to the intermediate-pressure bypass control circuit that responds rigidly in the APROS model. As explained before, the intermediate-pressure bypass valve will open when the fixed pressure setpoint (12 bar) is reached. In the power plant, the IP bypass valve opens slowly, resulting in the pressure peak. In the time interval between t = 60 min and t = 100 min, the IPBPCV maintains the pressure at its fixed value. Due to the complete opening of the intermediate-pressure bypass valve, the IP pressure obtained numerically by APROS rises then gradually. From t = 130 min, the gas turbine load increases. At the same time, the intermediate-pressure main steam control valve starts opening and the IPBPCV starts closing. The IP pressure obtained numerically by ASPEN PLUS DYNAMICS model shows a good qualitative agreement with the measurement. Approximately 20 min after the GT start-up, the pressure increases and reaches its fixed pressure value. Indeed, this is about 10 min earlier than the measurement, but the gradient curve of the pressure is similar to the measured data. In the time range between t = 50 min and t = 150 min, ASPEN PLUS DYNAMICS model shows a similar behaviour like the APROS model. During the post start-up phase, the simulated pressure follows qualitatively the measured pressure with a maximum relative error of about 5%.

The simulation results of the reheater system allow similar conclusions as obtained from the high-pressure system. Here, the plots of the reheated steam temperature and pressure, illustrated in Fig. 4.22, are only presented. The reheated steam mass flow rate combines the steam exiting from the HP turbine with the superheated steam of the IP circuit. The reheated steam mass flow rate shows a similar behaviour like the HP superheated steam mass flow rate (see Fig. 4.17) due to the fact that the large portion of the reheated steam mass flow comes from the HP circuit (up to 94%). The reheated temperature obtained numerically by ASPEN PLUS DYNAMICS model displays a considerable deviation during the start-up phase, but a good agreement in the post start-up phase. After 26 h standing idle, the simulated reheated temperature prior to the HRSG start-up is with about 160 °C clearly below the measurement. During the standing idle phase, no fluid mass flow exists in the real power plant and the temperature of the plant decreases, while the pressure is kept constant. This is not possible in the numerical models due to the fact that the temperature is calculated as a function of the pressure. As explained before, the pressure during the standing idle phase can be hold at a certain value. As a result, the saturation temperature is calculated according to this pressure. At t = 15 min, the reheated temperature increases sharply to about 470 °C

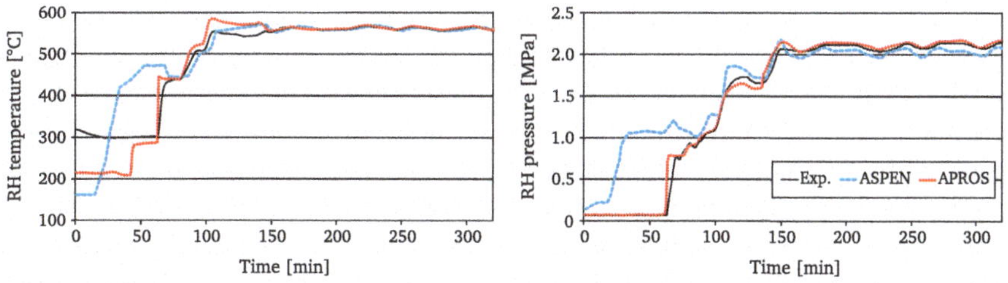

Fig. 4.22 Measured and numerically obtained reheated temperature and reheated pressure during the warm start-up procedure

within 35 min. This occurs almost 40 min earlier than the temperature jump in the real power plant. This noticeable deviation is caused by early steam production in the numerical models. In the period of time between t = 70 min and t = 150 min, the reheated temperature matches very well with the measured data with a maximum relative error of about 4%. Starting from t = 150 min, an increase of the reheated temperature above 567 °C is prevented with the aid of the reheater attemperator.

Compared to the ASPEN PLUS DYNAMICS model, the APROS model reproduces the reheated temperature with a better accuracy, especially during the first phase of the start-up procedure. At the beginning, the estimated temperature is 100 ° C below the measured data. A small jump in the reheated temperature occurs at t = 40 min, making the simulated reheated temperature in a good agreement with the measurement. At t = 65 min, a temperature jump can be observed in the experiment and simulation. The starting point as well as the value of this temperature jump is predicted by Apros model with high accuracy. In the time range between t = 80 and t = 150 min, the simulated reheated temperature deviates slightly from the measured data. The maximum difference can be seen at t = 105 min, which is correlated to the RH attemperator control circuit. This control circuit responds slowly, allowing to the reheated temperature to increase above the 565 °C. During the post start-up phase, the reheated temperature agrees quantitatively with the real power plant.

The reheated pressure prior to the HRSG start-up shows an almost identical value in the APROS model and in the measurement. At t = 60 min, the pressure jump can be numerically captured with high accuracy. In the remaining time, the simulated reheated pressure has a similar trend like in the real power plant. By contrast, the reheated pressure obtained by ASPEN PLUS DYNAMICS model shows the greatest discrepancy in comparison to other computed thermodynamic parameters. In the numerical model, the pressure build-up occurs very fast because of early steam generation. The reheated pressure exceeds the fixed pressure value (12 bar) and the reheat bypass valve opens accordingly. A hold point can be therefore observed in the period of time between t = 35 min and t = 85 min. During the post start-up phase, the computed pressure by ASPEN PLUS DYNAMICS model displays a qualitative response with the measured data with a maximum relative error of about 6%.

The dynamic behaviour of the LP system during the start-up and post start-up phases is presented in Figs. 4.23 and 4.24. The entire feedwater mass flow rate entering the power plant, illustrated in Fig. 4.23, matches very well with the measurement. In the period of time between t = 20 min and t = 150 min, different dynamic behaviours of the low-pressure feedwater mass flow rate in the simulation and the real power plant are observed. The considerable oscillations in the measured feedwater mass flow rate with an amount of an average value of 75 kg/s are not predicted by the numerical models. The vibrations in the feedwater mass flow rate result from the oscillations in the HP and IP feedwater mass flow rates, which are caused by the level reduction in the HP and IP drums prior to the start-up as previously explained (see Fig. 4.19). The HRSG models display an almost smooth development of the feedwater mass flow rate since the feedwater mass flow controller is optimised, so that a limited deviation from level setpoint is tolerated.

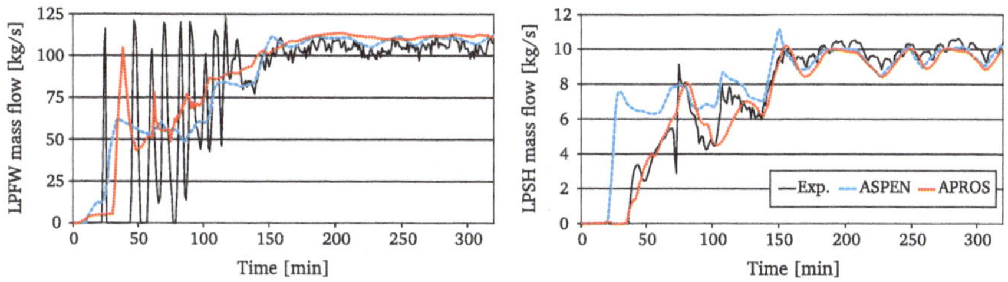

Fig. 4.23 Measured and numerically obtained low-pressure feedwater mass flow rate and low-pressure steam mass flow rate during the warm start-up procedure

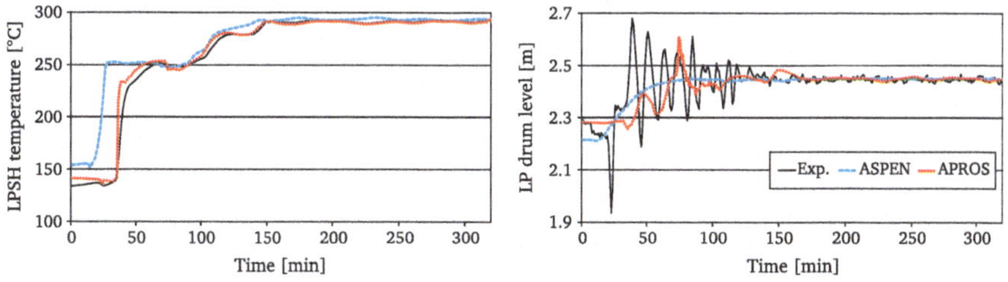

Fig. 4.24 Measured and numerically obtained low-pressure superheated temperature and low-pressure drum level during the warm start-up procedure

By contrast to the ASPEN PLUS DYNAMICS model, the feedwater mass flow rate obtained by APROS model shows, however, minor variations. Starting form t = 150 min, the feedwater mass flow rate lies marginally above the measured data with a relative error of 2%.

The computed steam mass flow rate exiting from the LP superheater has a good agreement with the real power plant. At t = 36 min, the starting point of the LP steam jump is accurately predicted by APROS model. Furthermore, the dynamics oscillations are captured in detailed. A slight deviation in the time range between t = 100 min and t = 120 min can be noticed. During the post start-up phase, the simulated IP steam mass flow rate by Apros behaves almost the same as the measured date. The ASPEN PLUS DYNAMICS model failed to capture the developing of the LP steam mass flow rate at the beginning of the start-up phase. The steam production starts here about 15 min earlier than APROS model. In addition, the predicted steam mass flow rate in the period of time between t = 30 and t = 70 min is clearly above the measurement. From t = 75 min, a qualitative behaviour between the simulation and experiment can be observed. In the remaining time, the model accuracy is improved and a quantitative behaviour is reported.

The LP superheated temperature, illustrated in Fig. 4.24, is modelled well. The superheated temperature computed by ASPEN PLUS DYNAMICS model shows an acceptable agreement with the measured data in the start-up phase and a good agreement during the post start-up phase. Prior to the start-up, the initial temperature is 20 °C higher than the real power plant. The reason of this deviation is already explained. Approximately 18 min from the start-up, the superheated temperature increases from 155 °C to 250 °C within 15 min (thermal gradient of about 5.5 °C/min). The quick response of the numerical model to the gas turbine start-up curve is caused due to the fact that the inertia of the numerical model does not agree with the real plant. Although the temperature jump starts 20 min before the measured data, the simulated thermal gradient is close to the measurement with a relative error of about 15%. From t = 150 min, the dynamic behaviour of the simulated superheated temperature shows good agreement with the measurement. Compared to the ASPEN PLUS DYNAMICS model, the APROS model shows a quantitative agreement with real power plant during both start-up and post start-up phases.

The low-pressure drum level is reduced to approximately 2.2 m, which represents 90% the nominal setpoint. By contrast to the LP drum, the level reduction in the HP and IP drums amounts to 70% their nominal values. This is due to the fact that the low-pressure evaporator located at the end of the flue gas path and thus the risk is significant lower to have steam bubbles in the low-pressure drum during the start-up process. The LP setpoint is reset to its nominal setpoint (2.45 m) at t = 25 min. During the post start-up phase, the low-pressure drum level remains then fairly constant throughout the measurement due to the control structure. Prior to the start-up, the LP drum level in the APROS model and experiment is equal to 2.3 m, while it is 2.22 m in the ASPEN PLUS DYNAMICS model. The level of the LP drum obtained numerically by both models agrees very well with the plant data during the entire start-up procedure. However, the numerical models failed to simulate the oscillations in the drum level in the period of time between t = 20 min and t = 120 min. The ASPEN PLUS DYNAMICS model shows an almost smooth development of the LP drum level, while the APROS model predicts the level oscillations with a better accuracy. The less mass inertia of the numerical models compared to the real power plant and the optimised values of three-term controller of the drum level control circuit represent the main reasons of this discrepancy.

Hot Start-up/Shutdown

The simulation of hot start-up and shutdown procedure is performed by imposing flue gas mass flow rate, GT exhaust gas temperature and flue gas composition as boundary conditions for the HRSG models (see Fig. 4.25). The complete cycle can be divided in pre-start-up phase (minutes 0–15), start-up (minutes 15–85), load-following operation (minutes 85–245), design operation (minutes 245–480), shutdown and standby (minutes 480–600).

In the pre-start-up, the gas turbine is driven by the generator at approximately 20% of nominal rotation speed to blow about 7% of the nominal air mass flow rate (42 kg/s) through the HRSG. This measure is of relevance in order to purge the flue gas system of any residual hydrocarbons. The start-up is initiated with ignition and thus the gas turbine rapidly accelerates to its nominal rotation speed in approximately five minutes. Synchronisation is completed at minute 45 and the loading process is started. At minute 50, a sufficient steam quality is reached and the first

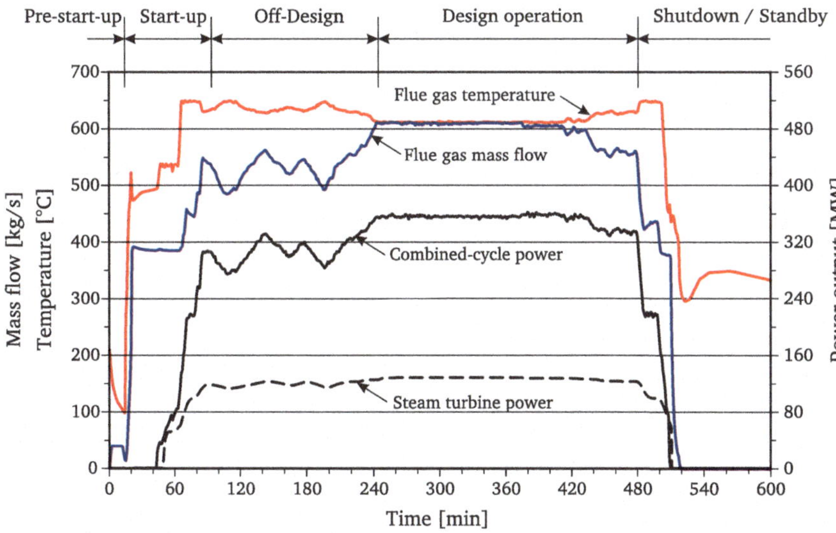

Fig. 4.25 Exhaust gas conditions, total combined-cycle power and steam turbine power measured at the generators during the hot start-up and shutdown procedure

HP superheated steam is admitted to the HP steam turbine. Then, a ramp-up to the desired load is completed when the bypass valves (HPBPCV, IPBPCV and LPBPCV) are fully closed. In the period of time between t = 85 min and t = 245 min, the combined-cycle power plant is operated in load-following operation at high part loads above 80% before switching to design operation near full load (here is 350 MW_{el}). In the shutdown procedure, the load is reduced to approximately 60%. After that, the main steam valves to steam turbine are rapidly closed and the excess steam is discharged via the bypass valves into the condenser. Finally, the gas turbine is shut-down and the steam bypasses are closed in order to maintain a fixed standby pressure for a possible HRSG restart. In the following, the response of HP, IP/RH and LP circuits to hot start-up and shutdown curve is presented only for the APROS model as an example. Dashed lines in the figures indicate simulation results and solid lines represent measurement data. Further results and discussions can be found in (Mertens et al. 2016a, b). In Fig. 4.26, the measured and simulated parameters of the HP circuit are illustrated. During pre-start-up and start-up procedures, the simulation results show qualitative agreement towards the measurement. After overnight standstill, the remaining pressure in the HP system amounts to approximately 40 bar, while the initial temperature drops to 270 °C since the HP superheater is cooled by purging air during pre-start-up. Shortly after GT ignition, the APROS model shows a temperature jump to 450 °C that precedes the measured temperature increase by an average of six minutes and is also characterised by a steeper transient (48 °C/min compared to 17 °C/min). Two reasons for this deviation can be found. On the one hand, there is an unknown distance between the measuring point at GT outlet and

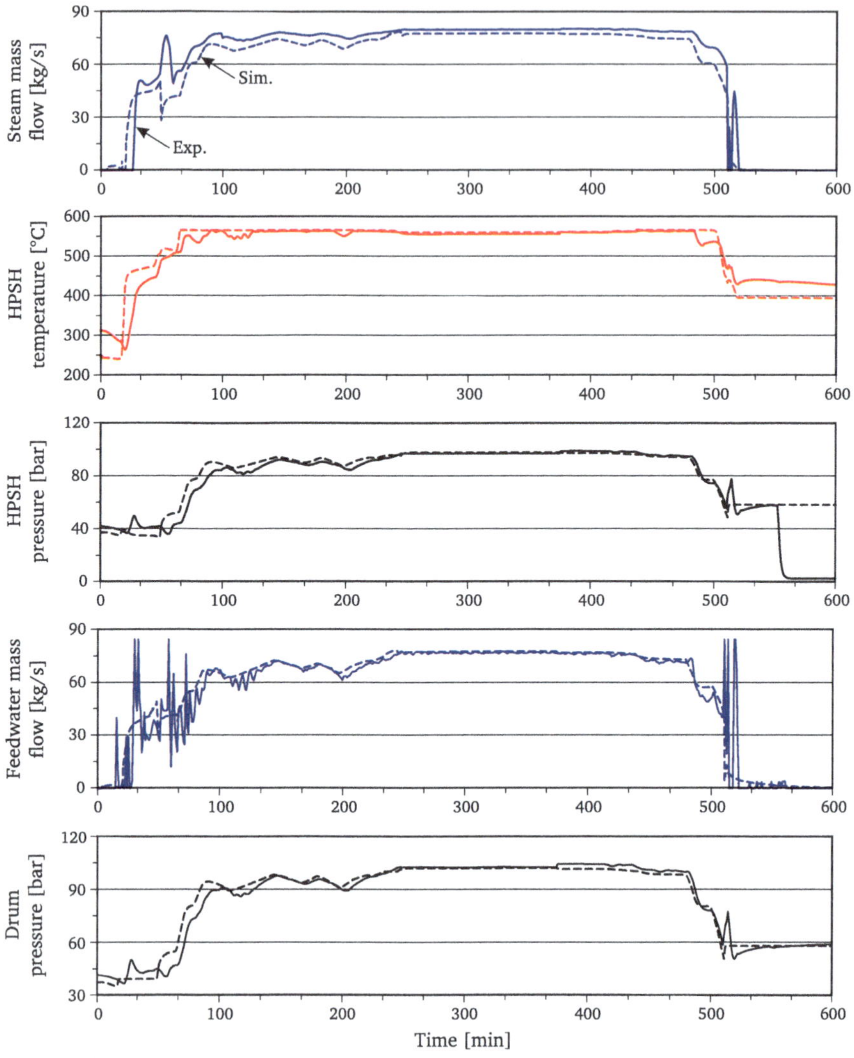

Fig. 4.26 Response of the HP circuit during the hot start-up and shutdown procedure (operation data and simulation)

the first row of superheater tubes in the HRSG. The gas turbine is modelled as ideal mass source directly upstream of the HP superheater, so that the corresponding delay of the HRSG system response is not described. On the other hand and more importantly, the model underestimates the thermal inertia of the heat recovery steam generator due to the fact that auxiliary systems (e.g. safety vents, draining systems and steam sampling lines) and supporting structures in thermal contact with the primary water/steam circuits as well as ambient losses are not considered. The tube material heats considerably faster and first steam generation in the HP evaporator is predicted six minutes earlier than observed in the real power plant. Accordingly, the calculated pressure transient precedes the measured pressure increase.

In load following operation, the calculated steam mass flow rate is in quantitative agreement with the measurement, but the steam temperature shows a discrepancy due to the attemperator control. The HP water injection keeps the simulated superheated temperature almost constant at the ideal setpoint of 567 °C. The temperature control in the real power plant tends to overshoot when engaged, causing negative temperature peaks during load changes. The steam pressure shows qualitative agreement between t = 85 min and t = 245 min, where the transients are well described by APROS model. However, the time discrepancy is passed on from the start-up phase. All HP steam conditions show good quantitative agreement (maximum deviations: −4.7%, 1.3% and −0.8% for steam flow, pressure and temperature) when the power plant operates at nominal load. The simulation results of main shutdown transients for the HP steam pressure and the mass flow rate yield good agreement with the measurement. However, the calculated steam temperature differs from the behaviour of the real power plant. In the period of time between t = 485 min and t = 505 min, the HP attemperator responds to exhaust gas temperature increase with a sudden jump of the injected water mass flow rate, so that the HP superheated temperature is reduced by 40 °C. In contrast, the idealised controller limits the water injection to maintain the design steam temperature as long as possible. During the shutdown, the superheater in the simulation is cooled more rapidly due to the lower thermal inertia, although the temperature gradients differ less compared to start-up phase (average: −9.2 °C/min calculated, −6.5 °C/min measured). Accordingly, final standby temperature is 35 °C lower and residual steam generation due to heat recovery from the tube material (indicated by the pressure and mass flow peak) is less pronounced.

At t = 552 min, the measured superheater pressure shows a sudden drop from 58 bar to near-ambient pressure level. While the data includes no physical evidence for this phenomenon, it is assumed that the live steam system was manually vented to prevent condensation in the superheater during extended standstill. The desired system pressure for the power plant restart can be re-established by auxiliary steam input; pressure level is selected in such a way that a sufficient degree of superheating with respect to the current material temperature is guaranteed.

The simulated feedwater mass flow rate agrees well with the measurement, particularly for load-following operation and at design load. In contrast, the feedwater mass flow rate during the start-up and shutdown processes in the real power plant is characterised by massive flow oscillations with maximum amplitude of 72 kg/s that could not be predicted by APROS model. When the HRSG starts generating steam, the water is discharged to the HP drum due to the decrease of average density in the HP evaporator tubes. The feedwater mass flow control circuit attempts to compensate the resulting level jump instantly and responds by closing and opening the feedwater control valve in rapid succession. The feedwater mass flow controller in the APROS model is optimised, so that a limited deviation from level setpoint is tolerated initially and then reduced over time, resulting in a smooth flow characteristic and reduced control action of the valve.

Since the drum inventory is always in saturation state, the drum temperature is a simple function of the pressure and therefore not presented. The calculated drum pressure shows good quantitative agreement for design operation and shutdown process as well as qualitative agreement in pre-start and start-up phase. Before the steam is admitted to the steam turbine at t = 50 min, the bypass control circuit aims to keep the drum pressure at approximately 40 bar. However, the detailed operating characteristic of the turbine bypass valves are not available, so that generic valves with a standard driving time (10 s) is used. This modelling error is reflected by the pressure deviations after the GT ignition, when the initial steam generation is routed through the HP bypass valve. Similar discrepancy can be observed at the end of the shutdown process.

The behaviour of the reheater system during the hot start-up procedure shows almost similar characteristic to the HP system since 90% of the reheated steam flow are constituted by the HP steam mass flow rate. Only the response of the IP steam conditions to the hot start-up and shutdown processes of the gas turbine is illustrated in Fig. 4.27. In design operation, the simulation results of IP superheated temperature and mass flow rate are in good agreement with the measurements (maximum deviations: −4.1%, −2.7% and −1.3% for steam mass flow rate, pressure and temperature), while minor discrepancies are observed for the load following operation. The hard transients and the qualitative behaviour of the IP steam conditions for the hot start-up and shutdown processes are also well reflected with two noticeable exceptions. Firstly, the simulated temperature increase after the GT ignition precedes the measurement by an average of 12 min, compared to six minutes for the HP circuit. This can be explained by the fact that the IP superheater is the smallest heat exchanger of the HRSG, so that the neglecting of auxiliary systems is more significant in relation to the tube mass. Secondly, the development of IP standby temperature cannot be predicted since the heat exchangers in the model are perfectly isolated once the flue gas mass flow and the steam mass flow rates are both zero (starting from t = 520 min).

The simulated drum pressure is in good agreement with the measurement. The feedwater mass flow rate is subjected to strong flow oscillations that are prevented in the APROS model by suitable adjustment of the controller. In contrast to the HP and LP circuits, the initial IP feedwater mass flow rate in the power plant is not zero because of the water used for fuel preheating. This pre-start procedure is enabled by the general HRSG design. The suction line of the IP boiler feedwater pump is connected to the LP drum (instead of the condensate system), so that the IP feedwater temperature is still 135 °C before restart compared to 145 °C in design operation.

The simulation results of the LP steam temperature, the pressure and the mass flow rate are compared with the operation data in Fig. 4.28, showing that the LP system response is well described by APROS model. The LP pressure is subjected to weak transients and varies in a relatively narrow range between 2.8 bar and 4 bar. The calculated temperature profile confirms the general trend observed for the IP and HP circuits. In particular, the results show good agreement during the load following and design operation (maximum deviation: −2.4%) as well as early

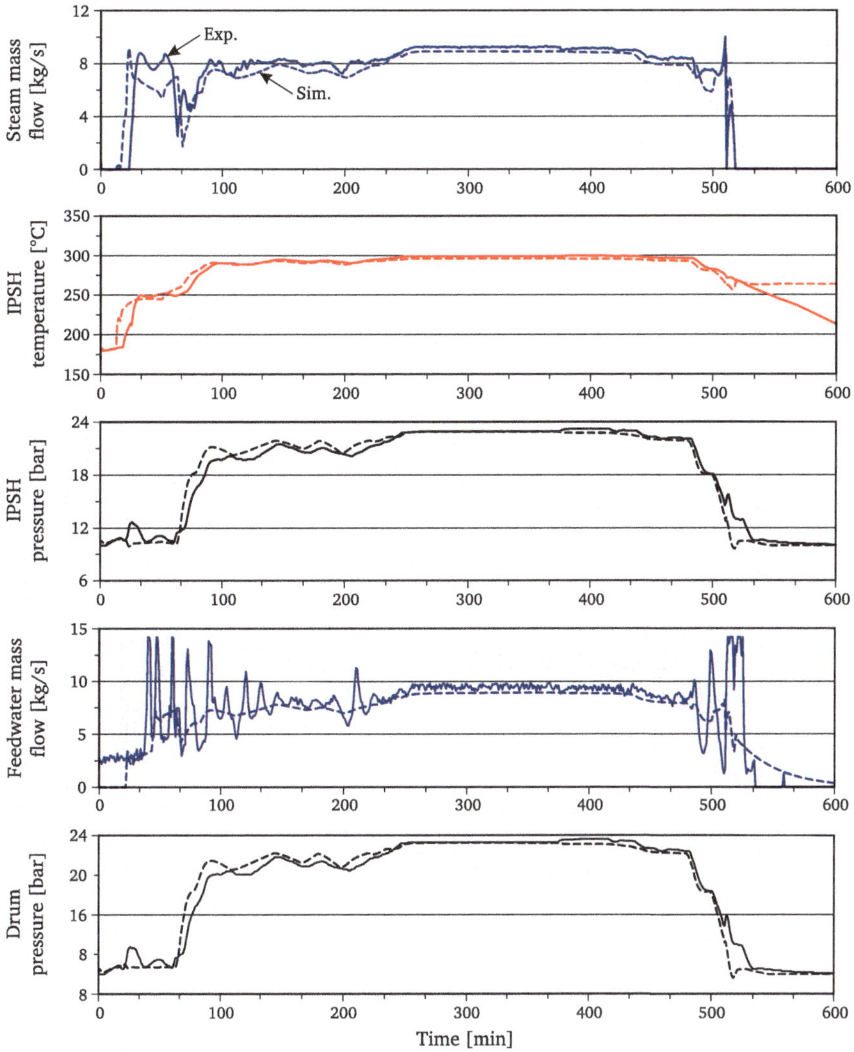

Fig. 4.27 Response of the IP circuit during the hot start-up and shutdown procedure (operation data and simulation)

temperature increase at GT ignition (average time deviation: nine minutes) and steeper temperature gradients for start-up and shutdown procedures due to the neglecting of auxiliary heat structures. Heat conduction and natural convection are not considered in APROS model and thus the continued temperature reduction post-shutdown is not reflected.

The strong transients of the LP steam mass flow rate during the hot start-up and shutdown processes are in good agreement with the measurement. However, the

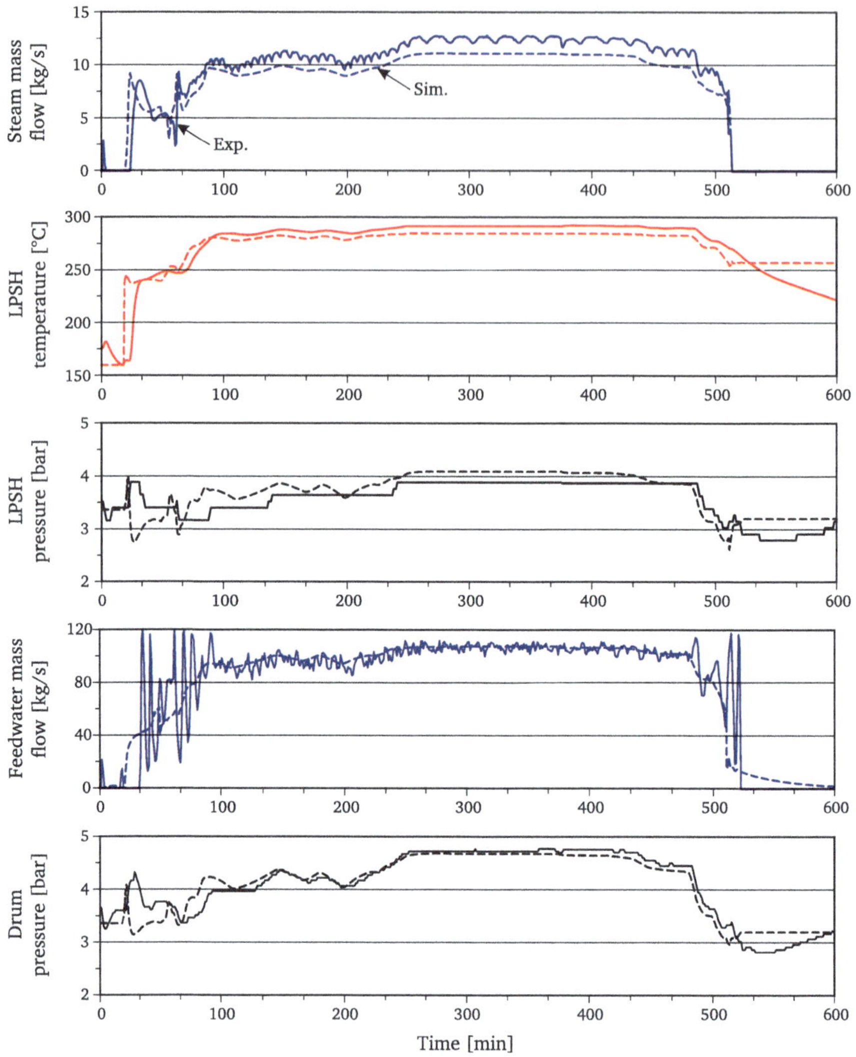

Fig. 4.28 Response of the LP circuit during the hot start-up and shutdown procedure (operation data and simulation)

steam generation in load following and design operation is underestimated by 12% on average. The model is based on steady-state design data of the HRSG, which specifies that the internal steam generation of the evaporator at full load is indeed reduced by 1.2 kg/s due to feedwater heating demand. Since the inlet conditions for the IP and HP circuits are also defined by the state of the LP drum, it is assumed that this represents a safety reserve in the design heat balance.

The calculation of the drum pressure is in quantitative agreement for load following (maximum deviation: 9.9%) and design operation (maximum deviation: 5.4%). By analogy with the HP circuit, the pressure profile shows discrepancy during the start-up procedure and in the transition from shutdown to standby. These phases are determined by control action of the LP main steam valve and the steam bypass valve (LPMSCV and LPBPCV), which is subjected to modelling error because the detailed characteristic curve of these valves is not available. The qualitative response of the LP feedwater mass flow rate is well reproduced by the HRSG model, with the exception of flow oscillations during the hot start-up and shutdown procedures, in particular that are already explained.

4.1.1.3 Process Improvements

Independent of hot, warm- or cold start-up, modern gas turbines need 7 min after ignition to achieve approximately 70% of their nominal values and additional 13 min to reach 100% of their nominal load. Heat recovery steam generators must be equipped with optimised components (e.g. advanced steam bypass systems, main and terminal attemperators) in order to enable the GT fast start-up. Furthermore, they should be constructed as natural circulation HRSG that has positive effects on capital and operational costs or as once-through boiler for achieving higher efficiency and flexibility. In the following sections, the influence of fast gas turbine start-ups on the dynamic behaviour of a supplementary-fired HRSG, once-through HRSGs with sub-critical and super-critical steam properties are presented. Furthermore, the design of an existing vertical type HRSG is enhanced by replacing the forced circulation in the evaporator systems with a natural circulation.

Supplementary-Fired HRSG

Large combined-cycle power plants with a supplementary firing of the heat recovery steam generator is receiving less attention in the research, since the interest of this technology appears to be low due to efficiency consideration. The process efficiency of a supplementary-fired combined-cycle power plant is lower than an unfired plant, but the plant flexibility to produce more power output during periods of high electricity prices represents recently an economically viable option. When electricity prices are low, the supplementary firing of the HRSG is switched off and the combined-cycle power plant operates at its nominal load, ramps to a suitable part load or shut-downs the unit. The benefit of a supplementary-fired large combined-cycle power plant is therefore a trade-off between number of hours with high and low electricity prices. Furthermore, this benefit depends on several other factors, of which the most important are the cost of fuel, the cost of plant and the demand for electricity.

The dynamic behaviour of a natural circulation heat recovery steam generator with a supplementary firing was investigated (Alobaid 2018b). The combined-cycle power plant (designed by Doosan Heavy Industries and Construction and General Electric) consists of a gas turbine connected to a horizontally-based heat recovery steam generator with an additional flue gas heating by a duct burner application. The power output of the combined-cycle power plant is 300 MW_{el} and the supplementary firing adds approximately 50 MW_{el} to the plant power output. The water/steam side of the HRSG consists of three pressure circuits with a natural circulation in the evaporator paths for HP, IP, LP and a reheater section after the high-pressure turbine, as follows:

- Natural circulation evaporator path in high, intermediate and low-pressure circuits
- Reheater section after the high-pressure turbine
- Three steam turbine stages HP, IP and LP
- Condensation system after the LP turbine

A schematic flow diagram of the combined-cycle power plant including the gas turbine, the three pressure circuits of the HRSG, the steam turbine and their steam bypass systems is illustrated in Fig. 4.29.

If electricity price is low, the supplementary firing of the heat recovery steam generator is switched off. This process lasts five minutes, wherein the fuel gas mass

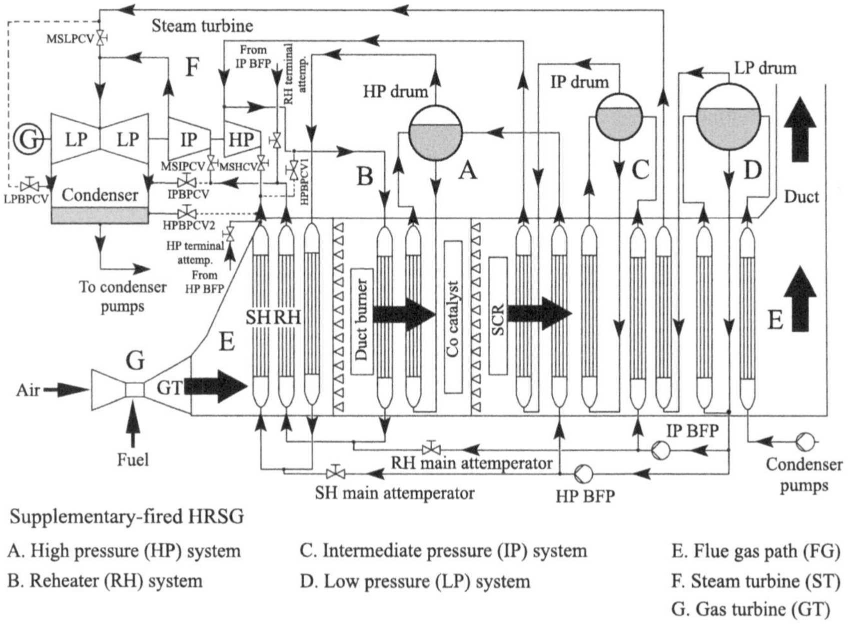

Fig. 4.29 Schematic flow diagram of the combined cycle power plant with a supplementary-fired heat recovery steam generator

Table 4.5 Technical data of the combined-cycle power plant with supplementary-fired heat recovery steam generator (duct burner is switched off)

Sub-critical heat recovery steam generator				Flue gas	ST power GT power
High-pressure system	Reheater section	Low-pressure system	Condenser inlet	$\dot{m}$ = 404 (kg/s) T_{inl} = 605 (°C) T_{out} = (72 °C)	100 (MW_{el}) 200 (MW_{el})
p = 84 (bar) T = 566 (°C) $\dot{m}$ = 50 (kg/s)	p = 20 (bar) T = 566 (°C) $\dot{m}$ = 55 (kg/s)	p = 4 (bar) T = 280 (°C) $\dot{m}$ = 7 (kg/s)	p = 56 (mbar) T = 35 (°C) $\dot{m}$ = 62 (kg/s)		

flow rate decreases from 2.8 kg/s to zero. Simultaneously, the exhaust flue gas mass flow rate of GT increases from 383 kg/s to 404 kg/s, while the flue gas temperature of GT decreases from 620 °C to 605 °C (see Table 4.5).

The heat recovery steam generator model, generated in APROS, has three pressure circuits up to 135 bar in the high-pressure system and a reheater section after the high-pressure turbine. The heat exchangers are divided into sections, as they are located in the flue gas path in order to get accurate temperature for each section. The developed model describes the flue gas path and the water/steam side in a high level of detail, while the gas turbine section is modelled as an inlet boundary condition with the characteristic parameters temperature, mass flow rate, pressure and composition. The flue gas path is described from the gas turbine exit to the HRSG outlet (see Fig. 4.30).

On the water/steam side, all bundle heat exchangers such as superheaters, reheaters, evaporators and economisers have been modelled with real geometry data. Furthermore, for each pressure circuit drum, pumps, headers, valves and

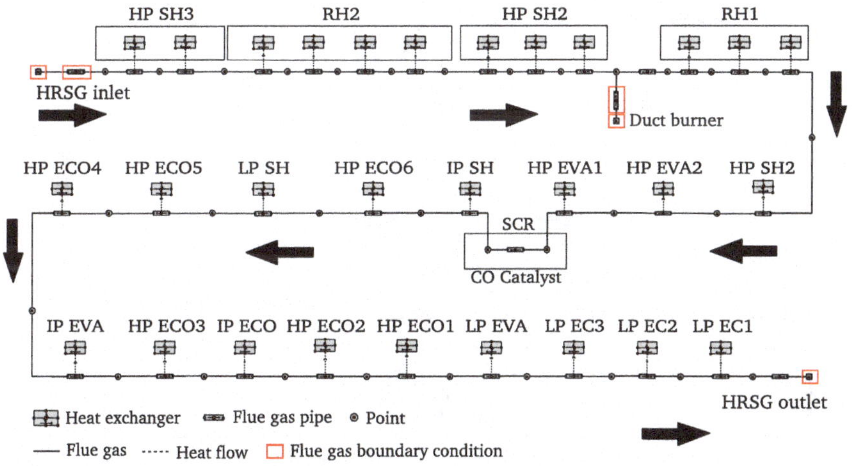

Fig. 4.30 Flue gas path of the supplementary-fired heat recovery steam generator model

turbine are considered, either. As examples for the HRSG model nets, the high-pressure and low-pressure circuits as well as the steam turbine stages are described below.

The high-pressure circuit, illustrated in Fig. 4.31, is modelled from the outlet of the high-pressure boiler feedwater pumps (HP BFP) to the outlet of superheater 3 (HP SH3). The feedwater mass flow rate streaming into the HP circuit is controlled by the HP drum level control valve (located in the LP net, see Fig. 4.32) and is delivered by the HP BFP. The feedwater flows into the HP drum via the HP economisers with the order of ECO1, ECO2, ECO3, ECO4, ECO5 and ECO6. The water in the HP drum circulates through the HP evaporators, where is heated by the flue gas and converted into water/steam mixture. Since the drum works as a separator, the water stays in the drum and the steam flows into the HP superheaters (SH1, SH2 and SH3). The HP superheated steam exits the HRSG and enters the HP steam turbine. A main attemperator is installed at the inlet of the last superheater (SH3) to control the temperature at the inlet of the HP steam turbine at a certain value (here equal to 566 °C). For reduced start-up times, also a terminal attemperator is in operation when the HP bypass system is active. The controller of the terminal attemperator maintains the steam temperature at the SH3 outlet at 530 °C for hot start-up and 370 °C for warm and cold start-ups. The main and terminal attemperators use water from the high-pressure boiler feedwater pump.

The low-pressure net consists of two circuits: main circuit and sub circuit (see Fig. 4.32). In the main circuit, the feedwater led to LP economisers are delivered by

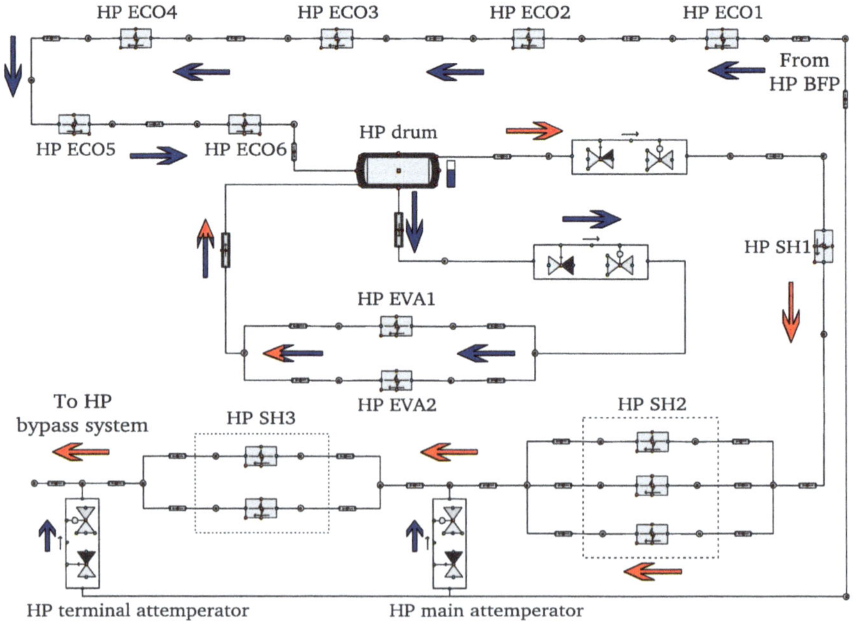

Fig. 4.31 High-pressure circuit of the supplementary-fired heat recovery steam generator model

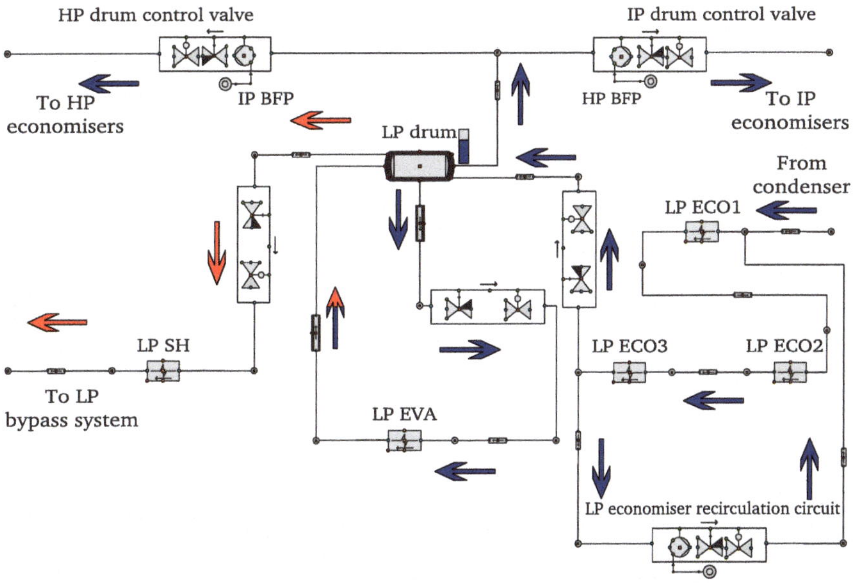

Fig. 4.32 Low-pressure circuit of the supplementary-fired heat recovery steam generator model

condensate pumps. A level control valve controls the feedwater mass flow rate to the LP drum. The water in the LP drum circulates through the tube bundles of the LP evaporator, and is heated by the gas turbine exhaust gas and converted into water/steam mixture. The dry steam exits the LP drum and flows into the LP superheater. After leaving the LP SH, the superheated steam enters the LP steam turbine. The HP BFP and the IP BFP take out the feedwater from the LP drum and deliver it into the HP and IP economisers, respectively. Beside the above-mentioned major LP path (main circuit), there is a secondary path, so-called the LP economiser recirculation system (sub circuit) that controls the temperature at the inlet of the first low-pressure economiser (LP ECO1).

The model net for the steam turbine stages, including high, intermediate and low-pressure is illustrated in Fig. 4.33. The HP superheated steam enters the HP steam turbine with high temperature and pressure of about 566 °C and 135 bar (if the supplementary firing is switched on) and to 566 °C and 84 bar (if the supplementary firing is switched off). After leaving the HP steam turbine, the cold and partly expanded HP steam mixes with the superheated steam from the IP circuit and flows into the reheater, before entering the IP steam turbine. At the outlet of the IP steam turbine, the further expanded steam is mixed with the superheated steam from the LP superheater and sent into the LP steam turbine. The fully expanded steam at the outlet of the LP turbine streams into the condenser.

The control concept of the supplementary-fired HRSG model (e.g. drum level control circuit, attemperator control circuit and economiser water bypass control circuits) is similar to the previous studies (see Sect. Control Structure). Only, the

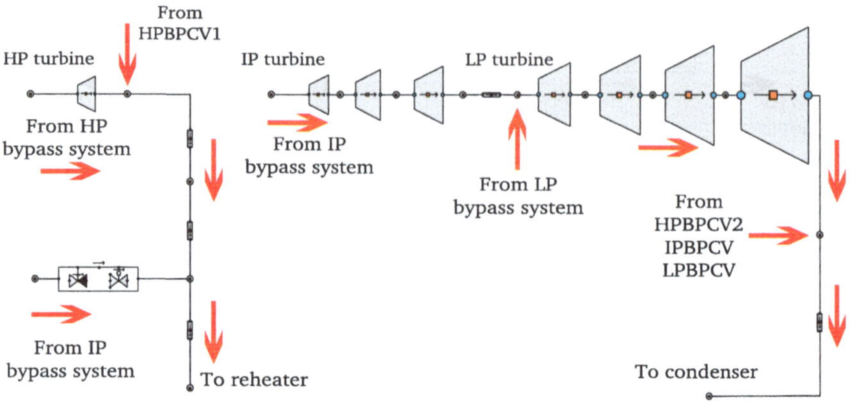

Fig. 4.33 Steam turbine model of the supplementary-fired heat recovery steam generator

control structure of the high-pressure bypass system is quite different (Fig. 4.34 compared to Fig. 4.7). The control valve of the main steam mass flow rate to the HP steam turbine (MSHPCV) is operated by a PI-controller that gets its setpoint value by a time gradient. The operation algorithm of this controller is described as follows:

- At the beginning of the start-up procedure, the MSHPCV is initially closed. Furthermore, the high-pressure bypass control valve (HPBPCV1) that regulates the mass flow rate to cold reheater is closed, too. When the steam in the HP drum reaches a certain pressure (known as fixed pressure and here equal to 49 bar), the HPBPCV1 will open first, so the steam after the terminal attemperator will be routed directly into the reheater. If the steam mass flow rate exceeds a limit value (here equal to 25.2 kg/s), the high-pressure bypass control valve (HPBPCV2) to condenser start opining, so that the difference between total steam mass flow rate and steam mass flow to the reheater (setpoint is 25.2 kg/s) should be bypassed. When the steam turbine is put in operation, the main steam to steam turbine (MSHPCV) starts opening and the bypass control valve to reheater (HPBPCV1) closes first, followed by the bypass control valve to condenser (HPBPCV2). In order to implement these sequences, the setpoint value of HPBPCV1 will be shortened by the increasing main steam mass flow rate measured behind the main steam control valve. In addition, the setpoint value of HPBPCV2 is a summation of the steam mass flow rate coming from the SH3 minus the steam mass flow rate to reheater and main steam mass flow rate to steam turbine. There is another controller in the bypass system that has the task to reduce the maximum pressure in the drum by opening the HPBPCV2 to release pressure.
- During the shutdown procedure, the MSHPCV and HPBPCV1 are closed, while MSHPCV2 controls the boiler pressure at its standing idle setpoint. Holding the pressure during the shutdown process has the advantage that the plant can reach its fixed pressure value faster during the following start-up procedure.

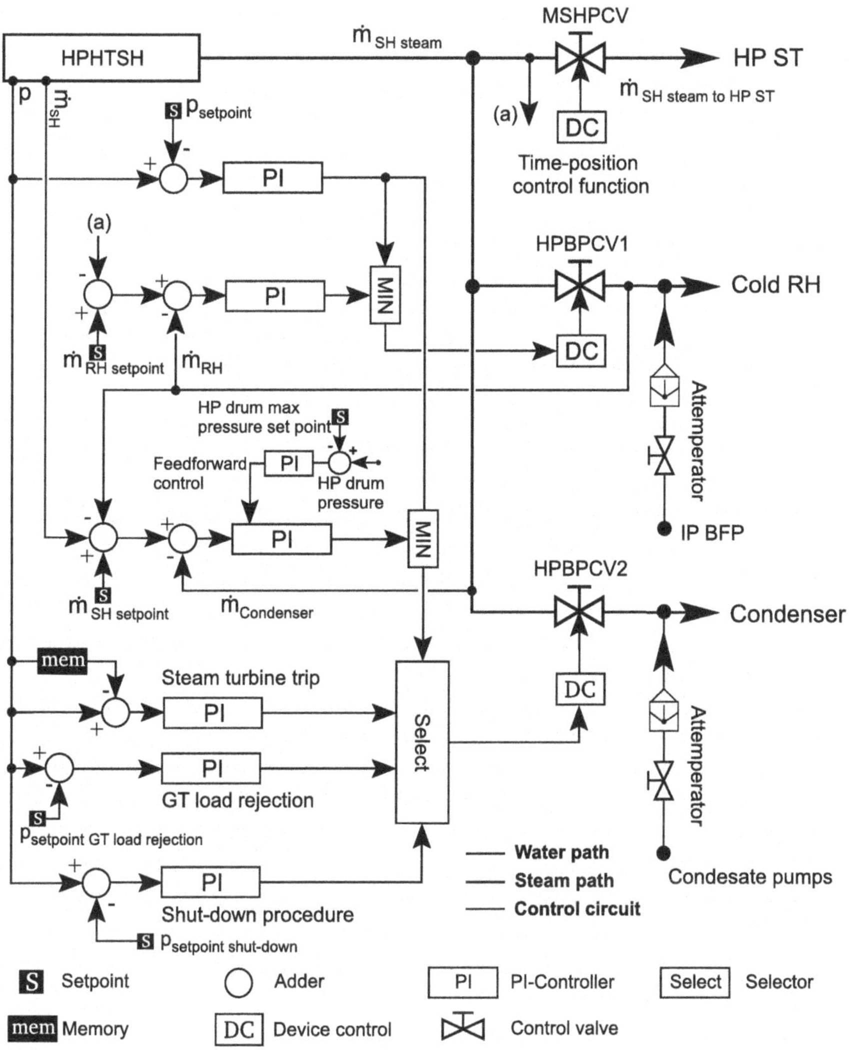

Fig. 4.34 High-pressure bypass control circuit

- In case of a steam turbine trip, the main steam to turbine valve closes immediately and the steam mass flow rate is redirected into HPBPCV2. A pressure regulation is necessary, which maintains the pressure in the boiler to reach the pressure level that was existent before the steam turbine trip took place. Using a memory component, the pressure is stored when the control concept is switched on and re-used as a setpoint for the pressure control. This makes the plant is already prepared for a hot restart.

- In case of a gas turbine load rejection, the exhaust flue gas temperature of the gas turbine drops very rapidly to ambient temperature. The HP superheated steam temperature will decrease accordingly, but the HP pressure remains at its nominal value. In order to prevent the condensing in the HP superheater, the HPBPCV2 will open discharging the pressure. The HP pressure is reduced to its saturation temperature below the exhaust flue gas temperature of the gas turbine. This allows the medium in the HP superheater to stay in the steam state and thus the undesired condensation is prevented.
- Finally, it should be mentioned here that the bypass system has two desuperheaters that cool the steam behind the control bypass valves (HPBPCV1 and HPBPCV2) down to 50 °C above the saturated steam temperature. The desuperheaters of HPBPCV1 uses water from the IP boiler feedwater pumps, while the condensate pumps feed the desuperheater of HPBPCV2.

The model analyses with design data at different processes (part loads and switching on the supplementary firing) have been performed and thus the start-up simulations with the validated model will be presented. The exhaust mass flow rate and temperature of the gas turbine are applied as dynamic boundary conditions for the heat recovery steam generator model with different start-up times for hot, warm and cold start-ups. In these simulations, the supplementary firing of the HRSG is always switched off and the flue gas composition has been assumed as constant over the entire load range of the gas turbine.

Figure 4.35-(left) shows the GT exhaust gas temperature and mass flow rate as well as the steam turbine power output, after a shutdown of the duct burner. The impact of switching off the supplementary firing on the HRSG behaviour is only discussed for the high-pressure circuit (see Fig. 4.35-(right)), as an example. After the supplementary firing is switched off, the steam temperature decreases sharply to 545 °C and then goes back to 566 °C, controlled by the HP main attemperator. Within 10 min, the steam mass flow rate and the pressure decrease from 90 kg/s to about 50 kg/s and from 135 bar to 84 bar, respectively.

Figure 4.36 shows the flue gas conditions during hot, warm and cold start-ups for different start-up times (baseline, (a), (b) and (c)). The gas turbine in cold

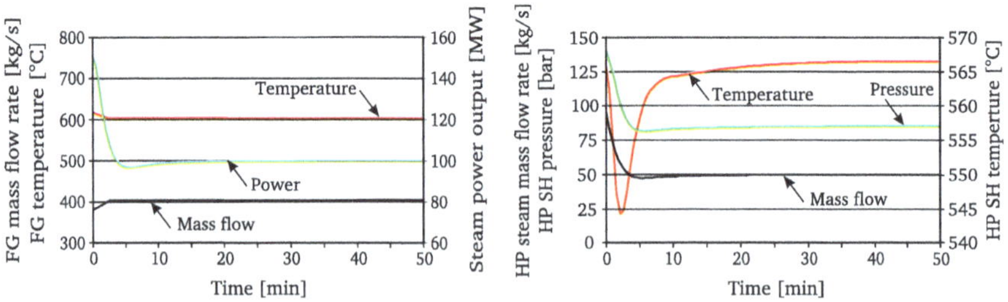

Fig. 4.35 Switching off the supplementary firing of heat recovery steam generator: (left) exhaust flue gas temperature, exhaust flue gas mass flow rate and steam turbine power output and (right) dynamic behaviour of high-pressure circuit

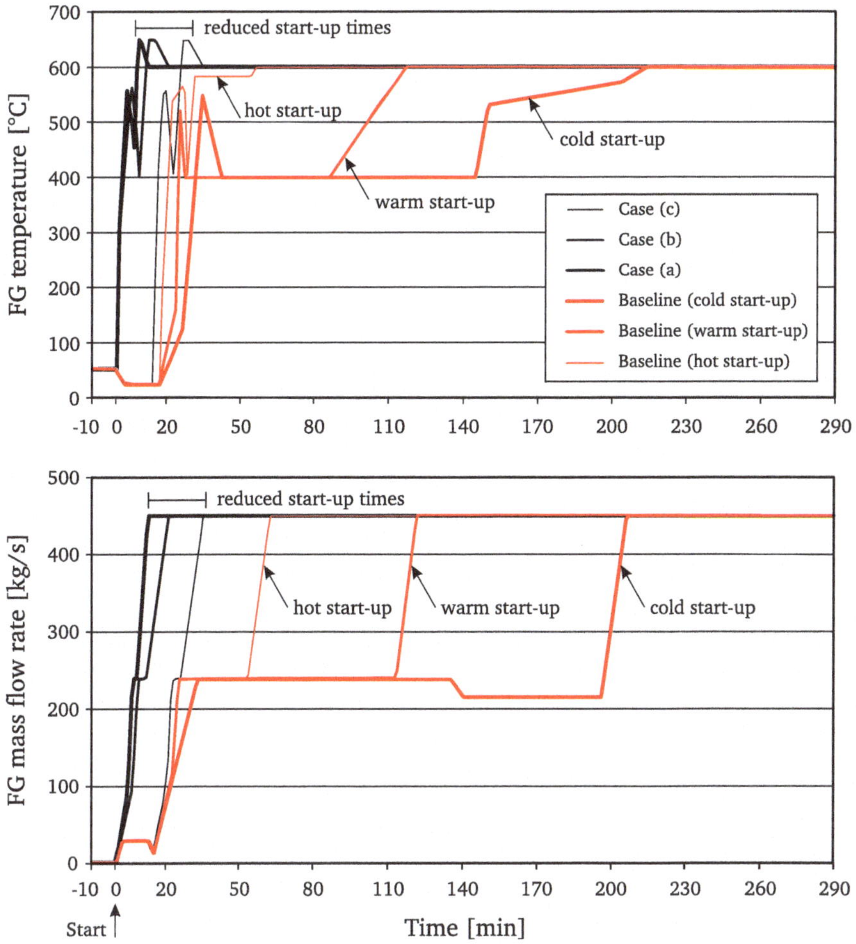

Fig. 4.36 Exhaust temperature and mass flow rate of the gas turbine during different start-up times

start-up (baseline) needs 210 min to reach its nominal operation values, while the GT runs to its full load operation in 120 min for warm start-up (baseline) and 60 min for hot start-up (baseline). In cases (a), (b) and (c), the start-up times of the gas turbine is reduced significantly. The additional time difference between case (a) and the two other cases (b) and (c) is attributed to the faster gas turbine start-up load ramp that is 16.67% per minute for case (a) instead of 8.33% per minute for the cases (b) and (c). A main difference between all cases is the purge credit (venting process) that it is completed after the shutdown procedure in cases (a) and (b), but before the start-up in cases (c) and baseline. In venting process, the GT purges the HRSG with ambient air to remove the residual combustible gases that can cause an explosion during the GT start-up, resulting in great damage in the facility.

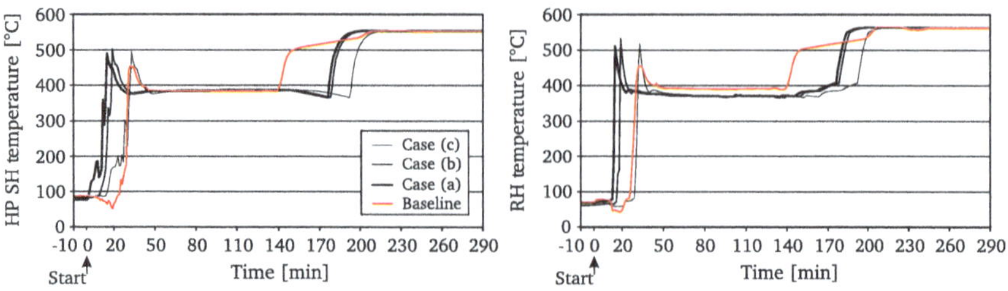

Fig. 4.37 Dynamic behaviour of the high-pressure superheated and reheated temperatures during the cold start-up procedure

In case (a), the gas turbine reaches its full load within 15 min after ignition, while it needs 24 min in case (b). With the purge credit in case (c), the gas turbine is in full load after 38 min. Finally, it should be mentioned that the reduced flue gas conditions in cases ((a), (b) and (c)), presented in Fig. 4.36, are applied to all start-ups types (hot, warm and cold).

As an example, only the dynamic behaviour of the supplementary-fired heat recovery steam generator model during the cold start-up is represented here. The temperature profiles of SH and RH during the cold start-up with different cases are displayed in Fig. 4.37. In cases (a), (b) and (c), the temperatures of SH and RH raise very fast to 500 °C within few minutes due to the flue gas mass flow rate and temperature jump, exceeding the temperature of baseline case with approximately 50 °C. In this time, the HP and RH terminal attemperators are in operation. They cool the high-pressure superheater and reheated steam mass flow rates down to around 370 °C. The HP and RH terminal attemperators are not in simulation in the base line case, but the HP and RH temperatures are decreased due to the flue gas temperature reduction and the start-up hold point. From t = 185 min, the HP and RH terminal attemperators are switched off and the SH and RH temperatures increase steadily to get their setpoint (566 °C), controlled by the HP and RH main attemperators.

The pressure in the HP circuit, displayed in the left part of Fig. 4.38, responds qualitative to the flue gas condition with different cold start-up times. In cases (a), (b) and (c), the HP pressure sharply jumps one-time to its nominal value, while in baseline case it increases gradually. Here, the HP pressure rises to 50 bar, then it remains constant before it increases again to its operation nominal value. As an example for the high-pressure bypass behaviour during the cold start-up procedure, the HP bypass trends in case (a) is presented (see the right part of Fig. 4.38).

Before start-up, the main steam to HP steam turbine control valve (MSHPCV) and both HP bypass control valves (HPBPCV1 and HPBPCV2) are initially closed. No steam will flow in the HP superheater since the HP drum pressure is less than 49 bar. When the drum pressure exceeds this setpoint, the HPBPCV1 starts discharging the HP steam to cold reheater. The HPBPCV2 is still closed as long as the discharged HP steam mass flow rate to the cold reheater is less than 25.2 kg/s. If the steam mass flow rate passes over the mass flow limitation, the HPBPCV2 starts

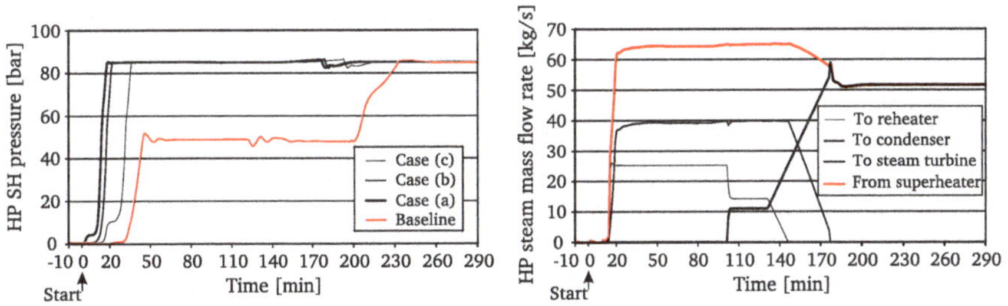

Fig. 4.38 Dynamic behaviour of the superheater pressure and the high-pressure bypass system (case (a)) during the cold start-up procedure

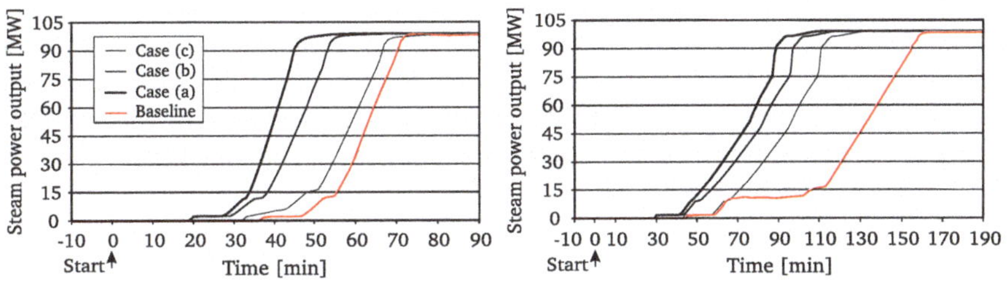

Fig. 4.39 Comparison between steam turbine power outputs for different cases during hot start-up (left) and warm start-up (right)

bypassing the remaining steam mass flow rate to the condenser. From t = 105 min, the main steam to HP steam turbine control valve (HPMSCV) starts opening. At the same time, the HPBPCV1 to reheater starts closing, followed by the HPBPCV2.

Figure 4.39 shows the steam power outputs at different hot and warm start-up cases ((a), (b), (c) and baseline). The steam turbine output in case (a) reaches its full load after 50 min, that is about 30 min earlier than baseline case and imply a time reduction of 38%. As a result, the case (a) achieved an additional energy supply of about 50 MWh per start-up, compared to baseline case.

The hot start-up time in case (b) lasts 10 min more than case (a), so that the case (b) has an additional energy supply of approximate 33 MWh towards the baseline case. The reason for these differences between both cases (a) and (b) towards the baseline case is due to the venting process at the beginning of start-up that causes a delay of 14 min as well as the long holding time of the flue gas parameters at the middle stage of the start-up curves. In cases (a) and (b), the venting process is performed at the end of the shutdown procedure. Between case (c) and baseline, a slight additional energy supply of 12 MWh to case (c) can be observed. All in all, the result shows a flexibility improvement of 38%, 25% and 9% during the hot start-up in cases (a), (b) and (c), compared to the baseline case. The steam power outputs of warm start-up cases ((a), (b), (c) and baseline) are also depicted in Fig. 4.39. Here, the main steam control valves start opening later than the hot start-up procedure. The case (a) reaches its full load after 110 min, corresponding to

31% faster than the baseline (additional energy supply power of 83 MWh per start-up) that needs 160 min to achieve its full load. The case (b) with its lower start-up gradient is about 15 min slower than case (a), but still produces 58 MWh more energy supply as baseline. Between case (c) and baseline, there is an additional energy supply of 33 MWh.

The steam power outputs during cold start-up cases ((a), (b), (c) and baseline; not illustrated in Fig. 4.39) show similar characteristic to hot and warm start-ups. The time until the main steam control valves are fully opened lasts almost 4 times longer than the hot start-up and 2 times than the warm start-up. The position of the main steam control valves is hold for 28 min in cases (a), (b) and (c) as well as for 80 min in baseline case, while the steam turbine boundaries for hot and warm start-up procedures are not hold. Due to the additional time for opening the main steam control valves, the differences decrease between cases (a), (b) and (c).

The cold start-up in case (a) is 50 min faster than baseline, corresponding to a reduction of 20%. The additional energy supply constitutes to 83 MWh for each start-up. The case (b) is only a little bit inferior as case (a) with additional energy supply of 75 MWh. Between the case (c) and baseline, an additional energy supply of 50 MWh is observed.

Vertical Natural Circulation HRSG

Horizontal heat recovery steam generator is designed with natural circulation evaporator systems. In this evaporator system, the water in the drum flows through the downcomers in direction of the evaporator. Due to the external heating of evaporator tubes by the hot flue gas, the water inside the tubes starts boiling. The density difference between the water in the downcomers and the water/steam mixture in the risers creates an uplift force that promotes the upward movement of the water/steam mixture to the drum. In the drum, the separation of the water from the water/steam mixture takes place by force of gravity using the density difference between phases. Finally, the water flows back through downcomers into the evaporator, ensuring stable and uniform circulation. This current flow does not require a circulation pump, which in turn means a saving in investment and operating costs. Vertical heat recovery steam generator is associated with forced circulation evaporator path in high, intermediate and low-pressure circuit in order to avoid two-phase flow instabilities in the evaporator systems.

The 450 MW_{el} Deir Aamar Power Station II, located in Lebanon, has three gas turbine/HRSG sets, supplying steam to one steam turbine (designed by Doosan Heavy Industries and Construction and General Electric). Each gas turbine can be operated with different fuels (heavy fuel oil (HFO), light fuel oil (LFO) and natural gas (NG)). The vertical HRSGs are unfired, forced circulation, non-reheat, two pressure units (high-pressure and low-pressure sections) and equipped with a diverter damper. When the gas turbine operates with the natural gas (fuel with low sulphur content), the LP economiser is in service with the LP economiser recirculation in order to obtain the temperature setpoint (based on fuel sulphur content) at the inlet of the LP economiser. For light oil (fuel with a higher sulphur content), the LP economiser is partly bypassed in combination with the LP economiser recirculation. For a very high sulphur content in the fuel (heavy fuel oil), the LP economiser is completely bypassed. The exhaust gas heat absorbed by the LP system decreases considerably and the inlet temperature of the condensate increases accordingly, preventing the corrosion of the LP economiser surfaces.

Based on Deir Aamar Power Station II, the design of the HRSGs has been modified for making a guideline for the natural circulation evaporator with the horizontal tubing by evaluating the flow stability in the HP and LP circuits during start-up procedures. A dynamic simulation model, based on six-equation flow model, is developed to analysis the flow stability in a vertical heat recovery steam generator with natural circulation evaporators (Alobaid 2018a). Static flow instability including Ledinegg instability, boiling crisis instability and flow pattern instability as well as dynamic flow instabilities such as density-wave oscillations, pressure-drop oscillations and thermal oscillations, in addition to flow pattern map of evaporators with the aid of Taitel and Dukler's are investigated. In order to obtain a clear understanding of flow instabilities, the obtained dynamic result in time-domain is enhanced by extending the investigation to the frequency domain. The temporal fluctuations of mass flow rate, density, pressure and other parameters are recorded with a fast Fourier transformation (FFT) that renderers the temporal fluctuations into a frequency spectrum. According to analyses conducted by the developed dynamic simulation model of Deir Aamar Power Station, the design of the vertical HRSG (thermo-hydraulic and geometry) has been modified in order to have stable behaviour during load changes and start-up procedures.

The target combined-cycle power plant model consists of 3 GTs $\times$ 3 HRSGs with a single steam turbine. For simplification, only one of GT/HRSG units is modelled using the process simulation software APROS. The gas turbine section is considered as an inlet boundary condition with the characteristic parameters temperature, mass flow, pressure and flue gas composition. The HRSG is unfired, non-reheat and dual pressure systems with a natural circulation in the evaporator path for HP and LP circuits, in detail as follows:

- Natural circulation evaporator path in the high and low-pressure circuits
- Two steam turbine stages HP and LP
- HP and LP bypass systems

A schematic flow diagram of the combined cycle power plant including the gas turbine (E), the flue gas path (C), the dual pressure circuits of the HRSG (A and B), the steam turbine (D), the feedwater and the steam bypass systems is shown in Fig. 4.40. The flue gas leaves the gas turbine and enters vertically the inlet duct of the HRSG, passing through the following heat exchangers: high-pressure superheater 2 (HP SH2), high-pressure superheater 1 (HP SH1), high-pressure evaporator (HP EVAP), high-pressure economiser (HP ECO), low-pressure superheater (LP SH), low-pressure evaporator (LP EVAP) and low-pressure economiser (LP ECO). The model of the flue gas path includes a soot blower system that utilises steam

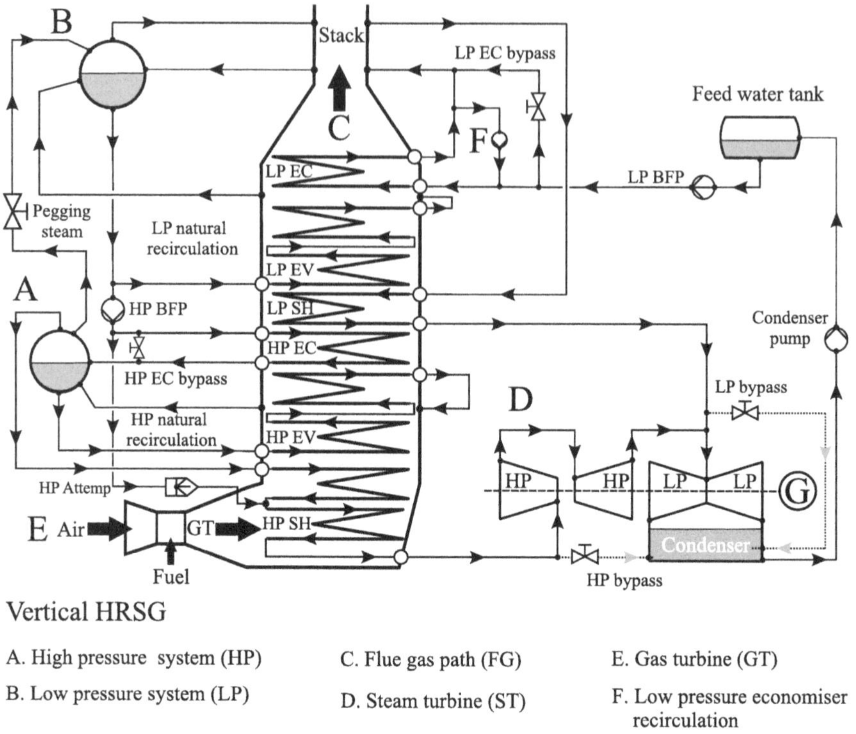

Fig. 4.40 Schematic flow diagram of the combined-cycle power plant with a vertical natural circulation heat recovery steam generator

Table 4.6 Technical data of the combined-cycle power plant with a vertical natural circulation heat recovery steam generator

Vertical natural circulation heat recovery steam generator				Flue gas	ST power GT power
High-pressure system	Reheater section	Low-pressure system	Condenser inlet	$\dot{m}$ = 417 (kg/s) T_{inl} = 530 (°C) T_{out} = 105 (°C)	150 (MW_{el}) 300 (MW_{el})
p = 63 (bar) T = 492 (°C) $\dot{m}$ = 50 (kg/s)	–	p = 5 (bar) T = 226 (°C) $\dot{m}$ = 3 (kg/s)	p = 56 (mbar) T = 35 (°C) $\dot{m}$ = 53 (kg/s)		

from the inter-stage of the HP superheater. The main task of the soot blowing system is to keep the HRSG heat transfer surfaces clean with the operating on any one of the specified fuels.

Table 4.6 includes the characteristic technical data of the real power plant for the gas path and the water/steam side.

For flow instability analyses, the heat exchangers of the HRSG is modelled as several tubes as in real plant (see Fig. 4.41 compared to Figs. 4.4 or 4.31). The heat

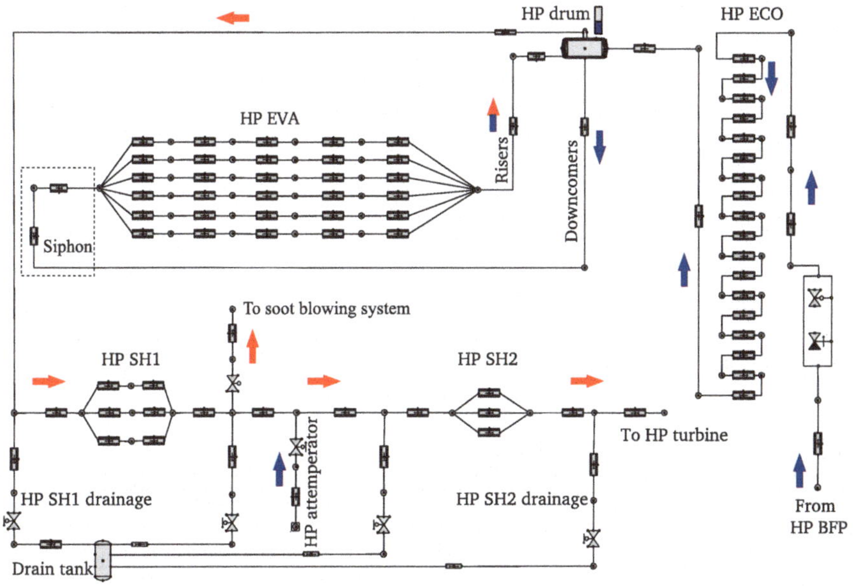

Fig. 4.41 High-pressure circuit of the vertical natural circulation HRSG model

absorbed by each tube depends on its position in the flue gas path (the closer to the gas turbine outlet, the merrier heat can be absorbed from the flue gas by these tubes). The HP feedwater mass flow rate, delivered by the HP BFP, streams into the HP drum via the high-pressure economiser (HP ECO). The water in the HP drum circulates based on the natural circulation concept, and is heated by the flue gas and converted into saturated steam in the HP drum. As the drum works as a separator, the liquid phase stays in the drum and the steam exits from the HP drum. The HP steam flows to the HP superheater, where the steam absorbs additional heat from the flue gas. Finally, the superheated steam exits the HRSG and enters the HP Turbine section. One inter-stage attemperator is installed to control the maximum temperature at the inlet of the high-pressure turbine. The attemperator uses water from the high-pressure boiler feedwater pumps. The LP feedwater mass flow rate streams to the LP drum via LP economiser (known also as condensate preheater system). The water of the LP drum circulates through the LP evaporators tube bundle, and is heated by the gas turbine exhaust gas and converted into saturated steam in the LP drum. The preheated water in the LP drum is used to feed the HP boiler feedwater pump. The dry steam exits the LP drum and flows through the LP superheater, before entering the LP turbine.

The control structures such as drum level, attemperator, economiser water bypass and steam bypass control circuits are similar to those used in Sect. Control Structure. Figure 4.42 shows the hot and warm start-up curves for GT exhaust gas mass flow rate and temperature. The type of evaporator circuit affects highly the dynamic behaviour of the steam generator. With the actual evaporator design

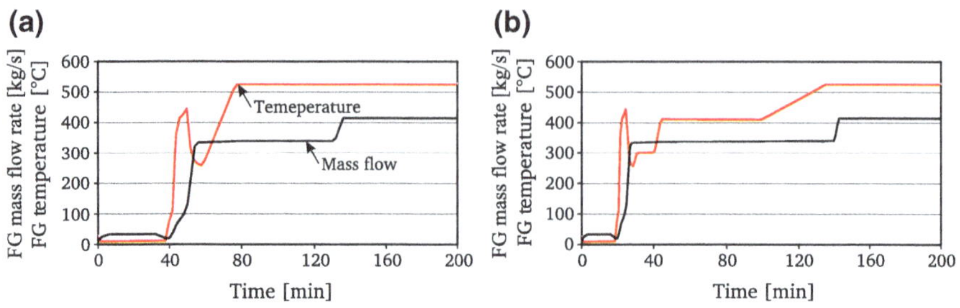

Fig. 4.42 Exhaust flue gas conditions: **a** exhaust temperature and mass flow rate of the gas turbine during hot start-up procedure and **b** exhaust temperature and mass flow rate of the gas turbine during warm start-up procedure

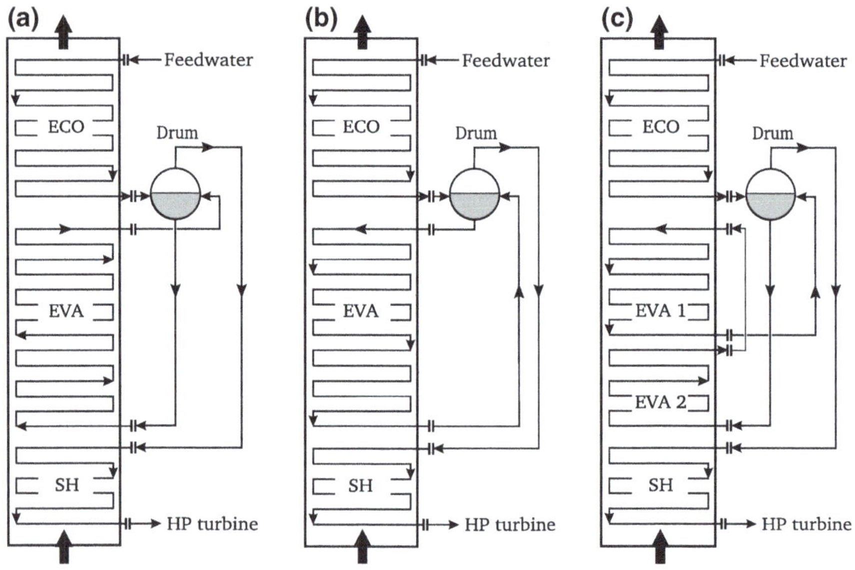

Fig. 4.43 Different configurations of the evaporator: **a** co-current flow circuit, **b** counter flow circuit and **c** co-current/counter flow circuit

(co-current, see Fig. 4.43a), instability problems has been observed during the start-up procedures. Therefore, the HP and LP evaporator circuits have been modified to counter flow circuit and co-current/counter flow circuit (see Fig. 4.43a, b).

The dynamic behaviour of the high-pressure and low-pressure circuits during the hot start-up is presented in Fig. 4.44. Here, the steam mass flow rate, the steam pressure and temperature are displayed for different configurations of the HP and LP evaporators. The results show that the co-current flow circuit of the HP and LP evaporators results in considerable flow instabilities in the LP system (fluctuations are not subsiding), while the HP system only oscillates during the last period of the

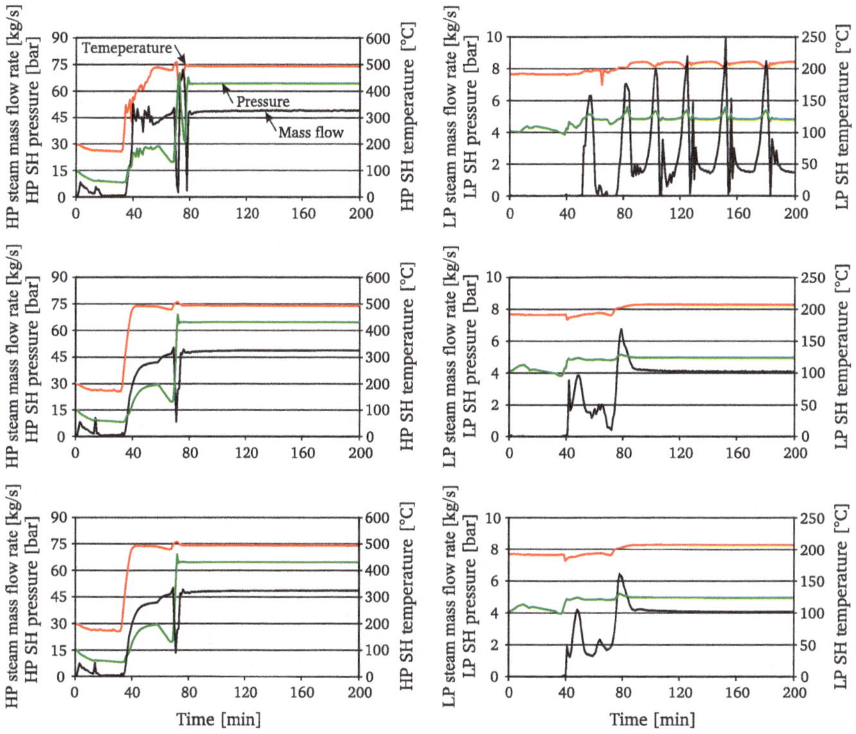

Fig. 4.44 Dynamic behaviour of the HP and LP circuits during the hot start-up procedure for different types of evaporator: (upper row) co-current flow circuit, (middle row) counter flow circuit and (lower row) co-current/counter flow circuit

hot start-up procedure. A decisive influence on fluctuations is the system pressure (an increase in the pressure leads to reduce the oscillation amplitude). This means that the LP circuit tends to produce more flow instabilities in comparison to the HP circuit. The counter flow circuit of the HP and LP evaporators under the same conditions improves the flow stabilities in the HRSG. However, the steam production in the LP system suffers from minor oscillations. The co-current/counter flow circuit of the HP and LP evaporators (the evaporator is split in two stages, connected with an inter-header) shows more favourable behaviour compared to the co-current and counter flow circuits. This is due to the fact that the heat input from the flue gas to the evaporator takes place in two stages. In the evaporator 2, the first partial evaporation of the preheated feedwater up to the saturation temperature will occur in a co-current flow circuit. In evaporator 1, the evaporation of the two-phase mixture will take place at lower flue gas temperature in a counter flow circuit. Since the feedwater mass flow rate from the drum enters the evaporator at the highest flue gas temperature, the water supply has a significant effect on the behaviour of the evaporator during load changes and start-up procedures. However, the co-current/counter flow circuit is more complex and has much higher investment costs than the

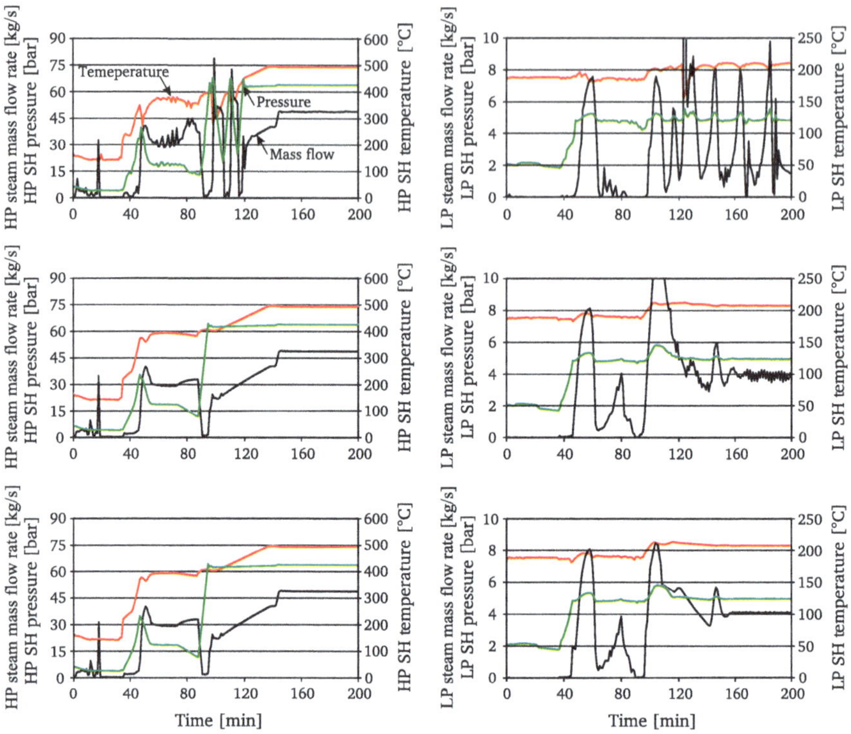

Fig. 4.45 Dynamic behaviour of the HP and LP circuits during the warm start-up procedure for different types of evaporator: (upper row) co-current flow circuit, (middle row) counter flow circuit and (lower row) co-current/counter flow circuit

co-current or the counter flow circuit, and thus is only suggested to the LP evaporator.

Figure 4.45 shows the dynamic behaviour of the high-pressure and low-pressure systems during the warm start-up for different configurations of the HP and LP evaporators. Similar to the hot start-up, the co-current flow circuit of the HP and LP evaporators leads to flow instabilities, especially in the LP circuit. Here, the oscillations of the steam mass flow rate in the period of time between t = 100 min and t = 200 min are not subsiding; even after the start-up. The HP steam mass flow rate fluctuates during the start-up, but these oscillations are subsided from t = 140 min. In case of a counter flow circuit of the HP and LP evaporators under the same conditions, both HP and LP circuits behave significantly better. At full load in the time range between t = 160 min and t = 200 min, the LP steam mass flow still face minor fluctuations with an am amplitude value of approximately 0.2 m. The co-current/counter flow circuit of the HP and LP evaporators reduces further the instabilities of the LP system. Here, the minor fluctuations in the LP steam mass flow at nominal load are decayed.

Sub-Critical Once-Through HRSG

The once-through HRSGs do not have a high-pressure drum and can follow the load change velocity of the GT with better dynamic characteristics. They are designed for daily start-stop operation and can be run with steep start-up curves owing to the absence of the high-pressure drum. The HP drum greatly limits the allowable temperature gradients by the reason of its large wall thickness. Faster plant start-ups are equivalent to fuel savings and can directly impact on CO_2 emissions per MWh. With almost unlimited numbers of fast start-ups, the once-through has the advantage against the conventional drum HRSG with limited start-ups numbers due to the thermal stresses in thick wall of the high-pressure components.

The influence of fast gas turbine start-up (within 20 min) on the dynamic behaviour of once-through heat recovery steam generator with sub-critical steam properties is investigated. The combined-cycle power plant (designed by Doosan Heavy Industries and Construction and Siemens) comprises of a gas turbine connected to an unfired, horizontally-based heat recovery steam generator.

The once-through sub-critical HRSG has three pressure systems and a reheater section after the high-pressure turbine, in detail as follows:

- Once-through evaporator path in the high-pressure circuit
- Natural circulation evaporator path in the intermediate and low-pressure circuits
- Reheater section after the high-pressure turbine
- Three steam turbine stages (HP, IP and LP)

A schematic flow diagram of the combined-cycle power plant including the gas turbine (G), the flue gas path (E), the three pressure circuits of the once-through sub-critical heat recovery steam generator (A, B, C and D), the steam turbine (F), the feedwater and steam bypass systems is shown in Fig. 4.46.

The characteristic technical data of the thermal power plant for gas and water/steam sides is presented in Table 4.7.

Using the process simulation software APROS, the once-through sub-critical heat recovery steam generator is built. The model includes the flue gas path and the water/steam side at a high level of detail, while the gas turbine section is considered as an inlet boundary condition with the characteristic parameters temperature, mass flow rate, pressure and flue gas composition. The flue gas path is modelled from the gas turbine exit to the HRSG exit. On the water/steam side, all bundle heat exchangers, namely superheaters, reheaters, evaporators, and economisers have been implemented with real geometry data. Furthermore, for each pressure circuit: turbines, drums, pumps, condenser and valves are modelled, either. In this study, the high thermal stressed heat exchangers (superheater and reheater) are divided in sections as they are located in the high temperature region of the flue gas path. The purpose of this is to get the exact temperatures of each section for their structural integrity. The overall heat absorbed was divided by the sections depending on its position in the flue gas path (the closer to the gas turbine outlet, the merrier heat can

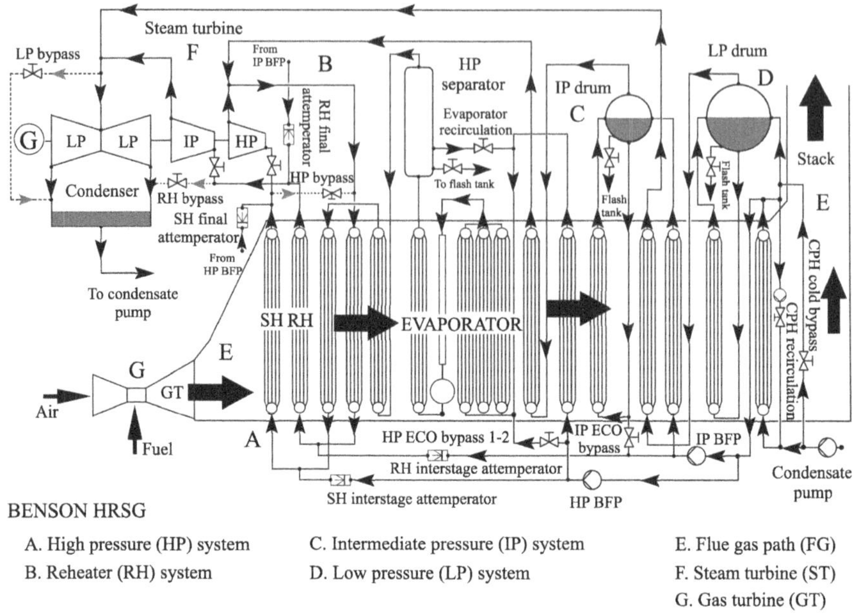

Fig. 4.46 Schematic flow diagram of the combined-cycle power plant with once-through sub-critical heat recovery steam generator

Table 4.7 Technical data of the combined-cycle power plant with once-through sub-critical HRSG

Once-through sub-critical heat recovery steam generator				Flue gas	ST power GT power
High-pressure system	Reheater section	Low-pressure system	Condenser inlet	$\dot{m}$ = 689 (kg/s) T_{inl} = 590 (°C) T_{out} = 87 (°C)	145 (MW_{el}) 235 (MW_{el})
p = 129 (bar) T = 567 (°C) $\dot{m}$ = 77 (kg/s)	p = 31 (bar) T = 565 (°C) $\dot{m}$ = 90 (kg/s)	p = 4.4 (bar) T = 235 (°C) $\dot{m}$ = 11 (kg/s)	p = 56 (mbar) T = 35 (°C) $\dot{m}$ = 101 (kg/s)		

be absorbed from the flue gas by these sections). The heat transfer coefficient between the flue gas and the heat exchangers has an important influence on the heat flux and thus on the dynamic behaviour of the heat recovery steam generator. For the finned-tubes heat exchangers, there are a lot of heat transfer correlations available such as Schmidt, VDI and traditional ESCOA and revised ESCOA. However, the application of one correlation is only possible for a definition region for Re and Pr numbers as well as a tube geometry (e.g. fin pitch, bare tube diameter, number of fins per meter). In HRSG model, the selection of the heat transfer correlation is depending on analysed operation mode as well as the heat exchanger geometry.

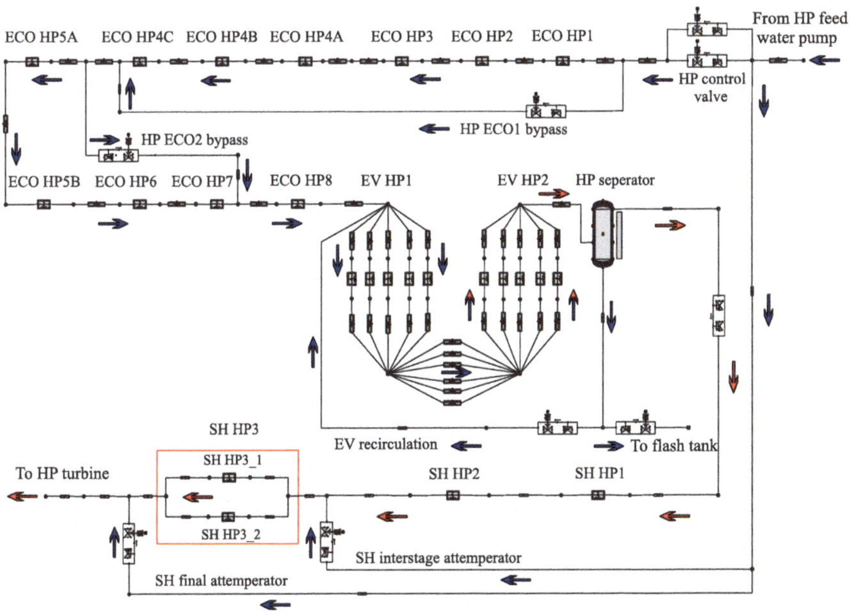

Fig. 4.47 High-pressure circuit of the once-through sub-critical heat recovery steam generator model

As an example for the once-through sub-critical HRSG model, only the high-pressure system is explained (see Fig. 4.47) since other systems are almost similar to conventional drum HRSGs, presented in Sect. 4.1.1.1. The feedwater mass flow rate, which is streaming into the high-pressure circuit, is delivered by the HP BFP. The feedwater flows into the HP evaporators via the high-pressure economisers. The water in the HP once-through evaporators is heated by the flue gas and converted into steam. Subsequently, the steam exits the high-pressure evaporators and streams into the superheater system, where the steam absorbs additional heat from the flue gas. The superheated steam exits the HRSG and enters the high-pressure turbine section. Inter-stage and final attemperators are provided at the inlet and outlet of the last superheater to control the temperature at the inlet of the HP turbine. Both attemperators will use the water from the high-pressure feedwater pumps. The feedwater mass flow rate is controlled by a feedwater control valve located in the same net. The high-pressure economiser II (ECO HP5A, ECO HP5B, ECO HP6 and ECO HP7) is partly bypassed during the part load operation in order to prevent the high-pressure economiser III (ECO HP8) from boiling. The bypassed mass flow is varied to keep the outlet temperature of the high-pressure economiser III below the boiling point, avoiding instability in the HP evaporators. During heavy oil operation, the high-pressure economiser I (ECO HP1, ECO HP2, ECO HP3, ECO HP4A, ECO HP4B and ECO HP4C) is partially bypassed so that economiser II absorbs more heat as the result of the lower inlet temperature. Due to

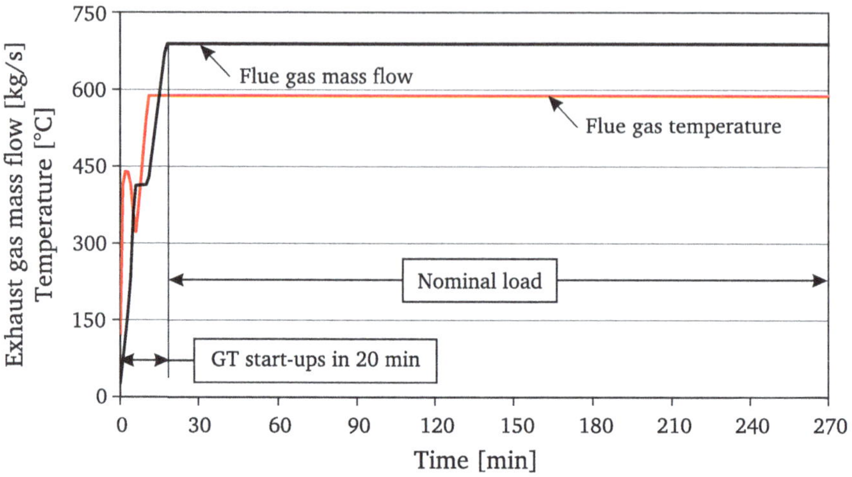

Fig. 4.48 Flue gas mass flow rate and temperature of the gas turbine during the fast start-up procedure

the decreased heat input into the IP evaporators, less amount of the IP steam is generated. Consequently, the intermediate-pressure mass flow is reduced together with the increased pressure and the decreased steam in the low-pressure drum. In view of the available overplus heat for the condensate preheater, the condensate inlet temperature therefore can be increased to the setpoint value avoiding the dew point corrosion on the low-pressure economisers' surfaces.

The HRSG model analyses for steady-state, different load changes and breakdown cases (e.g. steam turbine trip) have been performed, showing a very good agreement with the design data (Alobaid et al. 2012b; Mertens et al. 2015). For the fast start-up simulations, the exhaust flue gas mass flow and the exit temperature of gas turbine are applied as dynamic boundary conditions to the once-through sub-critical HRSG model. The exhaust flue gas composition has been assumed as constant over the entire load range of gas turbine. Independently from the hot, warm- or cold start-up, the gas turbine reaches 100% of its nominal load in approximately 20 min (see Fig. 4.48). The influence of the fast warm start-up on the high-pressure separator as well as the influence of the fast cold start-up on the physical values (temperature, pressure and the mass flow) of the high-pressure superheater is presented for once-through sub-critical HRSG. Furthermore, the result obtained from a drum-type HRSG is provided for the purpose of comparison. Figure 4.49 shows the dynamic behaviour of the HP separator. Initial system conditions are 190 °C and 12 bar.

Due to the relatively small volume of the HP separator the water inventory is quickly evaporated so that the temperature in the separator starts to rise after approximately 15 min. Consequently, the steam expands and pressure increases faster than in case of the drum-type HRSG. The steady-state temperatures in drum and separator differ from each other since part of the heat exchanger surface in the

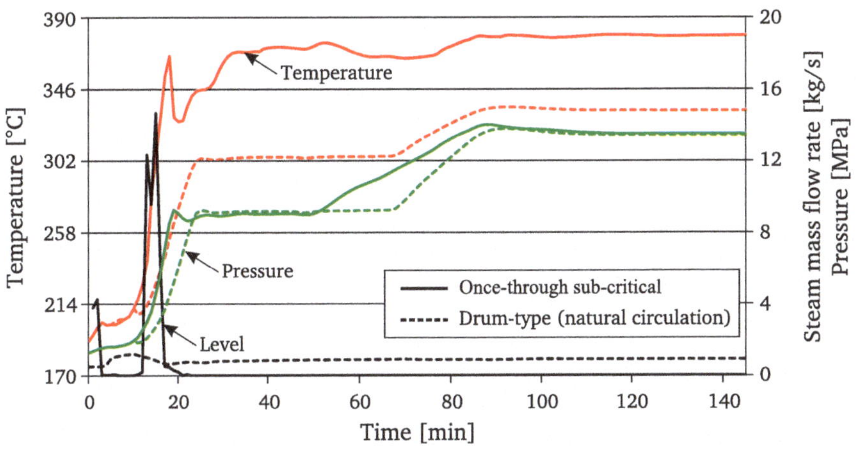

Fig. 4.49 Dynamic behaviour of separator/HP drum during the warm fast start-up procedure

once-through evaporator is already used for superheating. However, this deviation vanishes downstream due to larger HP superheater surface in the natural circulation HRSG.

In the fast cold start-up, the once-through sub-critical HRSG needs 150 min after the gas turbine start-up to reach its full load operation. In Fig. 4.50, the behaviour of the relevant parameters at the last high-pressure superheater is shown. The high-pressure increases gradually from 1 bar to the minimum pressure setpoint of 20 bar within 10 min. For circa 5 min, the high-pressure remains at the same value. The high-pressure jumps then rapidly from 20 bar to the fixed pressure value (here is 80 bar) within 10 min. In the period of time between t = 25 min and t = 135 min,

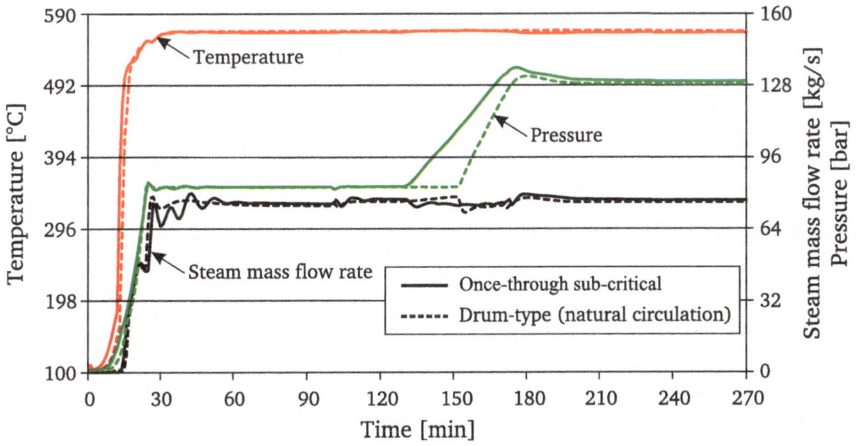

Fig. 4.50 Dynamic behaviour of high-pressure superheater during the fast cold start-up procedure

the high-pressure is kept constant equal to 80 bar. Beginning from t = 135 min until t = 180 min, the high-pressure starts increasing to reach the design value of 128 bar. The high-pressure bypass system enables a smooth build-up of the high-pressure during the fast cold start-up that matches well with the aimed pressure trend (see Fig. 4.8). The HP steam temperature rises extremely from 120 °C to 565 °C within 7 min (63 °C/min). The temperature gradient is circa 36.5% faster than the warm start-up and 52% faster than the hot start-up. After 18 min from the start-up, the high-pressure steam mass flow rate starts increasing. It needs then about 130 min to achieve its nominal value.

Super-Critical Once-Through HRSG

The once-through sub-critical HRSGs are already operated in several locations around the world. For example, a horizontal flow, once-through sub-critical HRSG has been installed at Cottam in England. The HRSG in Cottam power plant has provided an advanced concept that offers potential for future development in high efficiency combined-cycle power plants. Furthermore, Ansaldo Caldaie once-through sub-critical HRSG has achieved a time reduction in the cold start-up by 60% and a weight-saving by 5% compared to a conventional drum HRSG. The pressure in the HP circuit in these once-through HRSG is ranged between 130 bar and 180 bar. In this section, a dynamic process model of a super-critical once-through HRSG with 250 bar in the HP circuit is developed. The projected super-critical HRSG (designed by Doosan Heavy Industries and Construction) is composed of a horizontally-based gas turbine connected to a vertical gas path. The water/steam side consists of three pressure systems and a reheater section after the high-pressure turbine, as follows:

- Once-through evaporator path in high-pressure circuit with super-critical flow parameters
- Forced circulation evaporator path in the intermediate and low-pressure circuits
- Reheater section after the high-pressure turbine
- Three steam turbine stages (HP, IP and LP)

Table 4.8 displays the design technical data of the super-critical HRSG for gas and water/steam sides.

Table 4.8 Technical data of the projected combined-cycle power plant with super-critical HRSG

Super-critical heat recovery steam generator				Flue gas	ST power GT power
High-pressure system	Reheater section	Low-pressure system	Condenser inlet	$\dot{m}$ = 654 (kg/s) T_{inl} = 642 (°C) T_{out} = 72 (°C)	160 (MW_{el}) 240 (MW_{el})
p = 245 (bar) T = 566 (°C) $\dot{m}$ = 86 (kg/s)	p = 40 (bar) T = 567 (°C) $\dot{m}$ = 92 (kg/s)	p = 4.4 (bar) T = 332 (°C) $\dot{m}$ = 15 (kg/s)	p = 56 (mbar) T = 35 (°C) $\dot{m}$ = 107 (kg/s)		

The super-critical HRSG model is built in APROS on basis of a validated drum-type HRSG model (see Sects. 4.1.1.1 and 4.1.1.2). The simulation results for the dynamic behaviour of HP, IP/RH and LP circuits during a warm start-up for both super-critical HRSG and the drum-type HRSG can be found in (Alobaid et al. 2009). Only, the steam turbine power and the flue gas outlet temperature of the super-critical HRSG and the drum-type HRSG during the warm start-up is shown here (see Fig. 4.51). In the period of time between t = 0 and t = 30 min, the steam turbine power is equal to zero in both models. The steam power of the drum-type HRSG increases faster by contrast to the super-critical HRSG between t = 30 min and t = 60 min. The reason for that is the faster rising of the HP and RH steam mass flows in the drum-type HRSG comparing to the super-critical HRSG model. In the period of time between t = 60 min and t = 120 min, both models behave similar, while after 120 min, the dynamic behaviour of the super-critical HRSG shows a very good response towards the flue gas transient. Here, the outlet temperature of the flue gas remains constant around 76 °C in contrast to the drum-type HRSG, where the outlet flue gas temperature first oscillates and then fixes at 80 °C. This difference in the flue gas outlet temperature reflects directly on the steam power. Hence, the super-critical HRSG shows a net gain in the steam power of 5% compared to the drum-type HRSG. As a mathematical explanation for this gain in the steam power, Equation (2.109) and technical data of both models in Tables 4.1 and 4.8 are used. The thermal drop over the turbine section depends on the inlet enthalpy and the pressure drop ratio in the turbine. In the high-pressure turbine, the pressure drop ratio is around 0.2 in both models. The inlet enthalpy is slightly decreased with increased pressure. This means the thermal drop at the HP turbine section remains more or less the same in both models. Furthermore, in the low-pressure turbine section no difference in thermal drop can be detected.

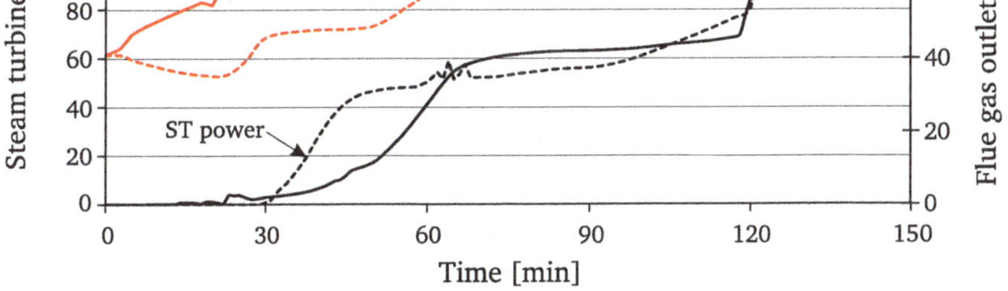

Fig. 4.51 Steam turbine power and flue gas outlet temperature during the warm start-up procedure

Actually, the steam power gain can be noticed in the intermediate-pressure turbine due to the decrease in the pressure drop ratio that is 0.1 in the super-critical HRSG and 0.2 in the drum-type HRSG.

4.1.2 Pulverised Coal-Fired Power Plant

Coal-fired power plants play a major role for the global electricity supply at present and in foreseeable future. According to World Energy Outlook, the total share worldwide electricity generation by coal-fired power plants accounted for 39% in 2015. The specific contribution varies from country to country and depends on several impact factors such as coal and gas prices, political framework, local resources and access to the world market. In Fig. 4.52, the schematic flow diagram of a modern hard-coal fired power plant is presented.

The main component of the coal-fired power plant is the steam generator, where pulverised coal entrained with the primary air flow is burned, releasing the thermal energy stored in chemical bounds. The obtained heat is transferred to the working fluid in economiser, evaporator and superheater in order to generate live steam for the Rankine cycle. Modern coal-fired power plants use single reheat and several low-pressure and high-pressure feedwater preheaters, resulting in an increase of the thermal process efficiency. The flue gas leaves the steam generator and the

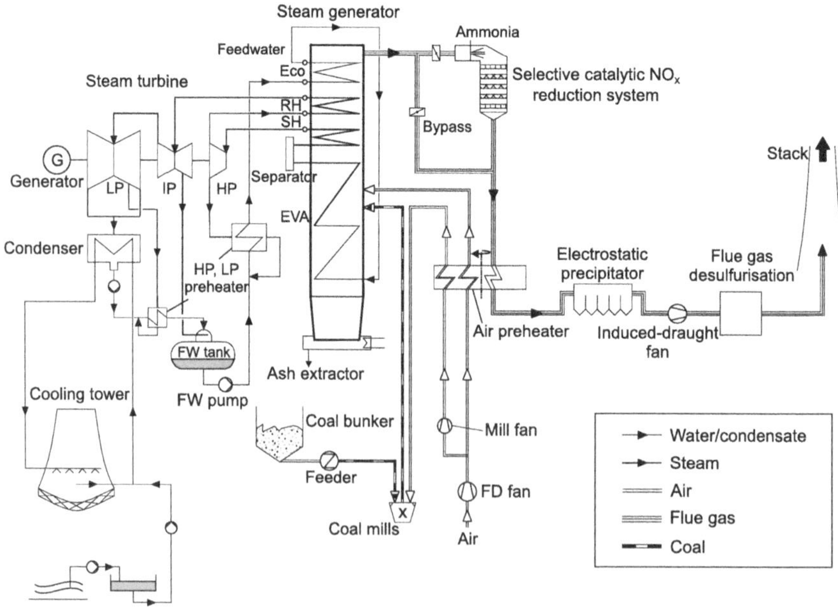

Fig. 4.52 Process flow diagram of a hard coal-fired power plant (reproduced from reference Spliethoff 2010)

remaining heat is applied to preheat the combustion air using an air preheater. The flue gas may pass different cleaning devices like selective catalytic reduction system, a particulate collector and a desulphurisation unit. The size of coal-fired power plants has a range from small industrial units to large utility power plants with up to 1300 MW_{el} per unit.

Even though all coal-fired power plants are working on the same principle, every power plant is individually engineered, which leads to different operation modes and different dynamic behaviour. The time of commissioning, fuel, control philosophy, size, environmental regulations, firing and furnace design, number of preheater and re-heater-stages and many other factors make each plant unique. The following criteria define the specific design of the coal-fired power plant:

- Coal composition: The coal composition has the most important influence on pulverisers, burners, furnace, heat exchangers, fouling and size of flue gas cleaning devices. Here, the contents of ash, sulphur, volatiles, water and fixed carbon lead to specific requirements for each component. The list of fuel categories is long and varies from country to country and reaches from meta-anthracite over bituminous and subbituminous coal to lignite.
- Fuel handling and firing concepts: Normally, pulverised coal-fired plants are equipped with a direct firing system. Here, the raw coal is pulverised in the mill and transported by primary air directly to the burners. Depending on coal composition, different mills like ball tube, ball, roller, beater mills may be used. The respond of the pulverised coal flow to a change of the raw coal feed or the primary air has a significant impact on the dynamic and transient behaviour.
- Emissions limits: The local emission regulation limits define how the flue gas cleaning devices has to be designed and if it has to applied at all.
- Water/steam circuit: While coal-fired power plants with super-critical steam parameters are always designed for once-through operation, power plants with sub-critical steam parameters may also be equipped with a natural or a forced circulation. Even though one reheater stage is state of the art, some plants are equipped with a second reheater stage.
- Reheater temperature control: The usage of water extracted from the boiler feedwater pumps in attemperators is the most common way for temperature control of coal-fired power plants. If other reheater control concepts like flue gas recirculation, tilting burners, flue gas dampers or internal heat exchangers in the water/steam circuit are applied, other controls are necessary and the dynamic behaviour may differ.

The massive growth of renewable energy in some countries changes the market conditions. In Germany for example and according to "Bundesministerium für Wirtschaft und Energie (BMWi)", the installed energy capacity of wind and photovoltaic in 2017 has reached an amount of 56.2 GW and 42.9 GW, respectively. The fluctuating wind energy can lead to extreme situations, like a sudden increase of more than 10 GW_{el} within a couple of hours, resulting in negative prices for the spot market. Traditionally, large coal-fired units were designed for an operation of

6000–8000 full load hours per year. Due to the new requirements of the market, the economic operation must be secured at 1500–3000 full load hours per year (Ketterer 2014). It is also expected that the flexibility for conventional power plants will increase even more in the near future. As a result, the power plant operators are forced to develop new technical solutions. A promising method is the use of computer aided process simulation tools. Steady-state process simulation programmes, which are limited to time independent operational points, were the preferred choice for process optimisation. For the above-mentioned fluctuating respond requirements of coal-fired power plants, a steady-state approach is no longer sufficient and dynamic process simulation becomes necessary. These dynamic models enable the researchers to test new control concepts, to modify load transient characteristic and to optimise the power plant behaviour in off-design operation conditions.

Although some dynamic simulation studies on coal-fired power plants can be found in the literature, the lack of available data for validation of the developed models is a major problem. Recently, the focus of dynamic simulation of coal-fired power plants is shifted towards carbon capture and storage, especially oxyfuel combustion. Here, the dynamic simulation tools are mostly used to evaluate the switch over from air to oxygen combustion or load changes. Since no large-scale oxyfuel-fired plants exist, measured data is not available either. In the following sections, full-scale dynamic process simulation models of a large-scale single-pass hard coal-fired power plant (Heilbronn Power Station, located in Germany), a large-scale two-pass hard coal-fired power plant (Korean Power Station, located in South Korea) and an oxyfuel single-pass hard coal-fired power plant (Jaenschwalde Power Station, located in Germany) are presented.

4.1.2.1 Single-Pass Boiler

The sub-critical, once-through steam generator of Heilbronn Power Station was designed as a single-pass boiler (tower boiler) for domestic German coals at a thermal power input of 1860 MW and an electrical output of 760 MW. Currently, imported world market coal blends are fired in the power plant. During a retrofit in 2009, an increase in the steam generator, in the pulveriser system, in the primary air system and in the steam turbines was achieved. As a consequence, the nominal steam mass flow rate to HP turbine was increased over from 606 kg/s to 659 kg/s at 195 bar and 540 °C, resulting in a sustained increase in the capacity by 40 MW_{el}. At present, the characteristic technical data of the Heilbronn Power Station for the flue gas path and water/steam sides is displayed in Table 4.9.

The schematic diagram of Heilbronn Power Station is illustrated in Fig. 4.53, including the furnace with its low NO_x tangential firing, the convective heat exchangers, mills, selective catalytic NO_x reduction system and air preheater. An overview of the air/flue gas path is shown in Fig. 4.54. The ambient conditions such as pressure, temperature and humidity are specified as a boundary condition (a). The secondary air is supplied by two parallel working forced draft fans (b), which is

Table 4.9 Technical data of the sub-critical, once-through steam generator of the Heilbronn Power Station

Heilbronn Power Station				Flue gas	ST power
High-pressure system	Reheater section	Low-pressure turbine	Condenser inlet	$\dot{m}$ = 791 (kg/s) T_{ECO} = 394 (°C) T_{Sta} = 120 (°C)	800 (MW_{el})
p = 195 (bar) T = 540 (°C) $\dot{m}$ = 659 (kg/s)	p = 42 (bar) T = 540 (°C) $\dot{m}$ = 593 (kg/s)	p = 4.5 (bar) T = 243 (°C) $\dot{m}$ = 444 (kg/s)	p = 98 (mbar) T = 45 (°C) $\dot{m}$ = 386 (kg/s)		

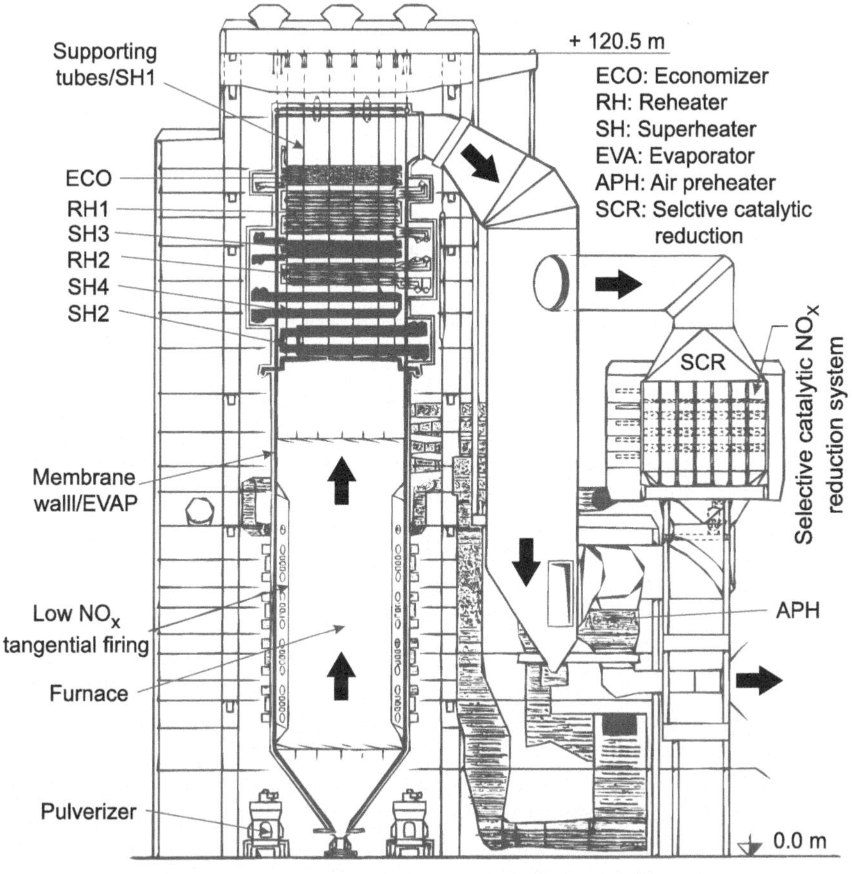

Fig. 4.53 Schematic flow diagram of the Heilbronn Power Station (Starkloff et al. 2015)

specified as an axial fan with controllable blade angle. Afterwards, the secondary air passes a steam coil air preheater (c) and the regenerative air preheater (d), before entering the furnace (e). The primary air is supplied by the speed controlled two radial primary air fans (f) and flows through the regenerative air preheater (d) into

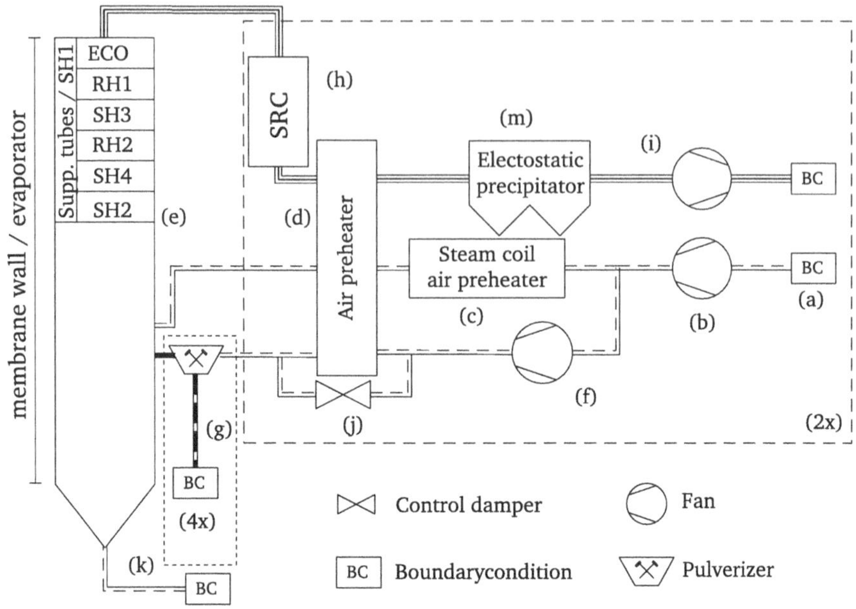

Fig. 4.54 Air/flue gas path of the hard coal-fired power plant in Heilbronn

the pulveriser (g). The now coal dust loaded air is then fed into the furnace and combusted. In order to consider the false air, which is unavoidable in a balanced-draft system, ambient air is introduced at the hopper (k) of the steam generator and adjusted for the nominal case. After exiting the steam generator, the now 350–400 °C hot flue gas flows through the selective catalytic reduction (h), the regenerative air preheater (d), the electrostatic precipitator (m) and the induced draft fan (i), which keeps the furnace at a slight negative gauge pressure.

The three different fan types are implemented according to their specifications, including a detailed characteristic maps for a reliable part load performance. Between the fans and the steam generator, the duct system is modelled according to the parameters length, cross section and elevation. Additional pressure loss due bends, branch connections or dampers is implemented via the corresponding form loss coefficients. The pneumatic transport of the coal between the mill and the burner adds another pressure loss, which is estimated for the design point and implemented by tuning the form loss coefficient. Interconnecting ducts, dampers and controls should be kept in mind when multiple gas tracks are modelled. The mill (g) is basically a mixing point of the coal and the primary air. In this mixing point, the water content of the coal is evaporated. For the dynamic behaviour of a coal pulveriser, a transfer function model is implemented, which is based on work of (Lee 1986).

Flue gas cleaning devices like the electrostatic precipitator or the selective catalytic reduction unit are modelled as pressure drops with the correct mass. In order to simulate the steam coil air preheaters or regenerative air preheater, the use of the standard APROS library components is sufficient, nevertheless for special simulation topics, e.g. regenerative air preheater trip, the model becomes more complex. A forced draft fan control was implemented, which regulates the gas pressure downstream of the air preheater at a load dependent setpoint by adjusting the blade angles. Similar to the forced draft fan, the primary air radial fans control the gas pressure in the primary air duct downstream of the air preheater by manipulation the fans speed. In a typical tangentially fired steam generator with four firing levels, the secondary air mass flow through each burner is generally controlled separately. In this model and due to one-dimensional approach, the secondary air mass flow is controlled for each level. The setpoint for the total secondary air mass flow rate in a level is defined by the feeder speed of the corresponding pulveriser and is corrected with the oxygen content in the flue gas. The primary air mass flow rate to each pulveriser is defined by the corresponding feeder speed. The induced draft fan control regulates the furnace pressure by controlling the blade angles.

In Fig. 4.55, the water/steam circuit within the steam generator is presented. At the top of the steam generator, the feedwater enters the economiser (ECO) and then is directed to the membrane wall, where the evaporation occurs. During start-up, shutdown or at lower part loads, a forced circulation is in operation. The circulation is forced by a circulation pump (CP) and the level is controlled by a circulation control valve (CCV). Furthermore, a pump protection and a sub cooling injection to prevent cavitation within the pump are implemented. The circulated mass flow

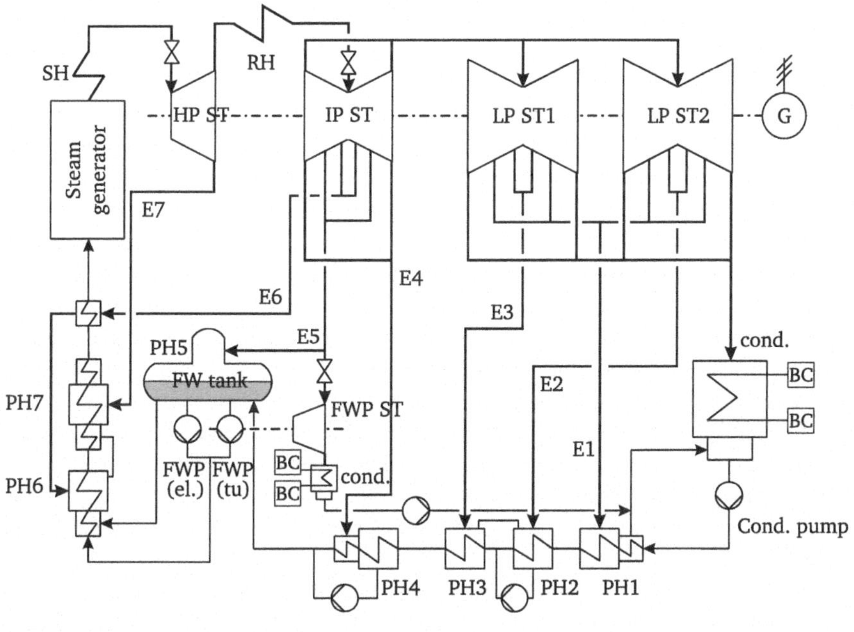

Fig. 4.55 Water/steam circuit of the hard coal-fired power plant in Heilbronn

re-enters the steam generator at the economiser inlet. In order to separate the condensate droplets within the steam flow, a cyclone is installed, before the produced steam, saturated or partly superheated, enters the steam cooled supporting tube system (superheater 1) of the convective heat exchangers. Then, the partly superheated steam flows downwards to the superheater 2 (platen superheater). After the platen superheater, the steam flows through the super heaters 3 and 4. Three attemperator stages are arranged for steam temperature control between the superheaters. The steam leaves the once-through steam generator and flows through the high-pressure turbine (HP ST) before re-entering the steam generator via the reheater system that consists of two heat exchangers and one attemperator in between. Afterwards, the steam flows through one intermediate turbine (IP ST) and two low-pressure turbines (LP ST1 and LP ST2) into the condenser. The condensed water is fed through seven feedwater preheaters back to the steam generator. For the power supply of the boiler feedwater pumps (BFP), a steam propelled turbine with a separate condenser is used. The modelling of the steam turbines is based on the given characteristics for different load cases. For accurate calculation of the heat transfer in the heat exchangers like condenser, the four low-pressure preheaters, feedwater tank and two high-pressure preheaters, the implementation of the real geometries is necessary. The condensate, feedwater and drain pumps are modelled according to their given characteristics, this is important for the speed controlled feedwater pumps. The thickness, flow area and the length interconnection piping between the components should be implemented correctly for an accurate calculation of the thermal inertia and flow speed.

The dynamic model of Heilbronn power plant includes all plant components and all required automation processes with minimum boundary conditions, based on the real design specifications, construction drawings and control schemes (Starkloff et al. 2015). Through careful development, the number of model boundary conditions is minimised. On the water/steam side, the only implemented boundary conditions are: the inlet temperatures and the mass flow rates of the cooling water into the condensers. For the flue gas path, the coal composition and the air ambient temperature are boundary conditions. Even though these boundary conditions are assumed to be constant, they may be load and time dependent. The entire model is arranged into different subsystems like air path, flue gas path, turbine, low-pressure preheaters, high-pressure preheaters, boiler feedwater pumps with turbine and steam generator. The latter is used as an example to illustrate the power plant model in APROS (see Fig. 4.56; the connection points to other modelled subsystems are marked with a circle), in addition to the low-pressure feedwater preheater system (see Fig. 4.57). The points (a) and (b) are considered as slave copies within the steam generator subsystem. Each flue gas volume in the convective area has three parallel corresponding heat exchangers on the water/steam side: a part of the membrane walls, a part of the supporting tubes and one convective heat exchanger bundle.

The feedwater control circuit plays a key role for dynamic simulation. During once-through operation, the setpoint of the feedwater mass flow rate is load dependent and corrected with enthalpy at the evaporator outlet and the HP attemperator

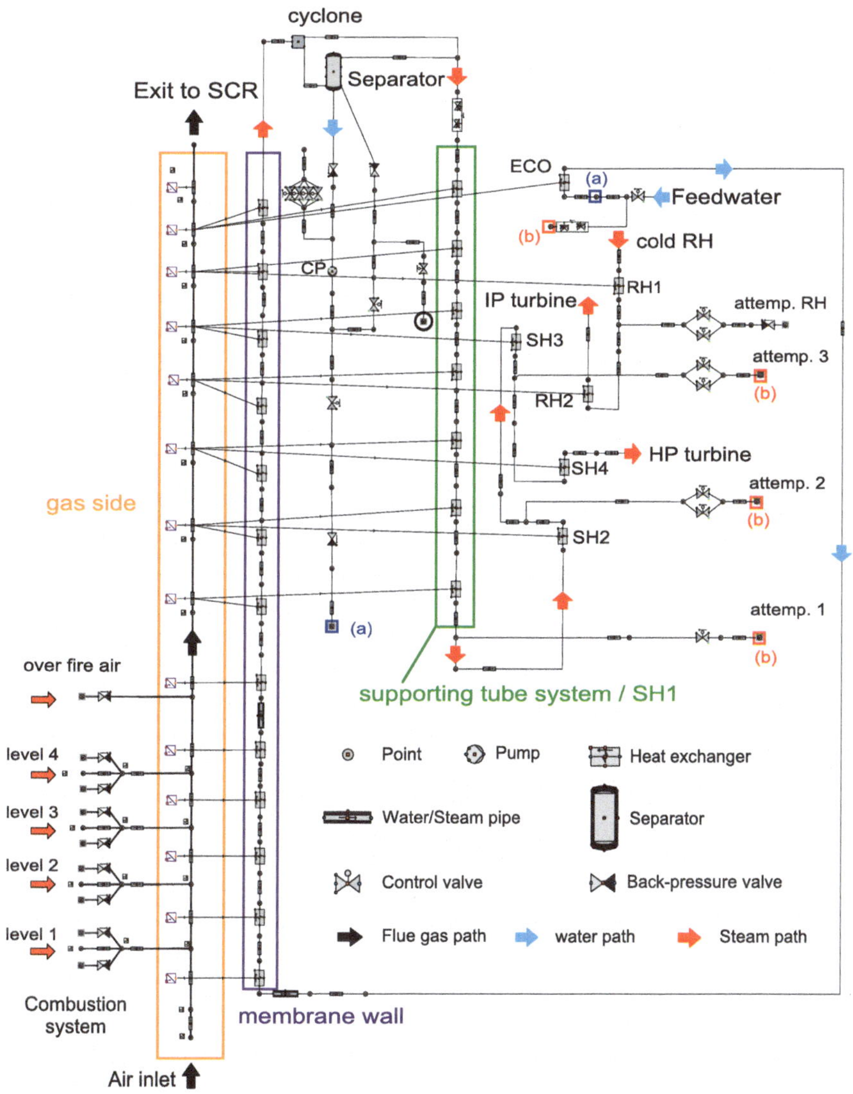

Fig. 4.56 Flue gas path and water/steam side

injection mass flow rate. During forced circulation operation, the circulation control circuit is active and keeps the evaporator mass flow rate at its minimal value. The flue gas temperature at the cold outlet of the regenerative air preheater is controlled via steam mass flow rate into the steam coil air preheater and should be kept 10–20 K above the acid dew point. Several other control systems are also implemented in the developed power plant model like pressure control in the SH and RH sections, turbine bypasses, pump protections, level control circuits etc.

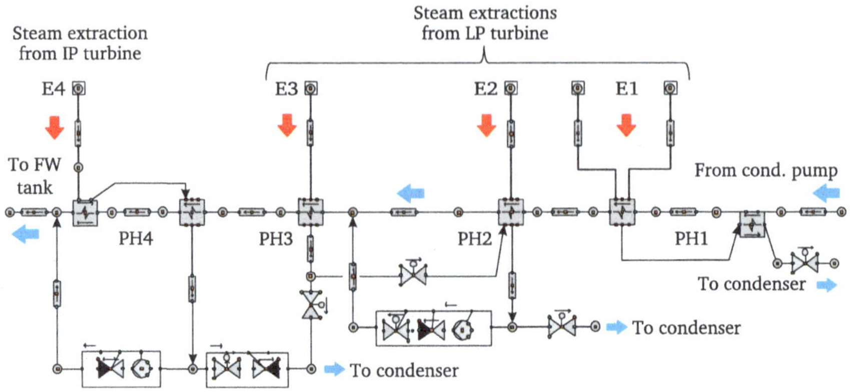

Fig. 4.57 Low-pressure feedwater preheater system

The developed model and its implemented controls must first be evaluated at different steady-state cases. In order to achieve that, a reference design point (100%) was selected and the model is tuned to this specific point. The tuned parameters have been kept constant for all following calculations. In a second step, another load point (105%) is calculated. The results show that the maximal relative deviations between the manufacturer design data and the model regarding the enthalpy and the pressure in the steam generator are approximately 0.53% and 0.17%, respectively. For the dynamic comparison with operation data, an appropriate load change scenario has to be identified. The requirements of such a scenario are: fast and wide rates of load change, change from once-through to forced circulation, start or stop of a firing level (see Fig. 4.58).

The plant operator selects a scenario, in which the load changes from 100% to 22.7%. The scenario starts with a change of the load setpoint from 100% to 67.5% after 5 min. When the setpoint is changed again from 67.5% to 42% after 110 min,

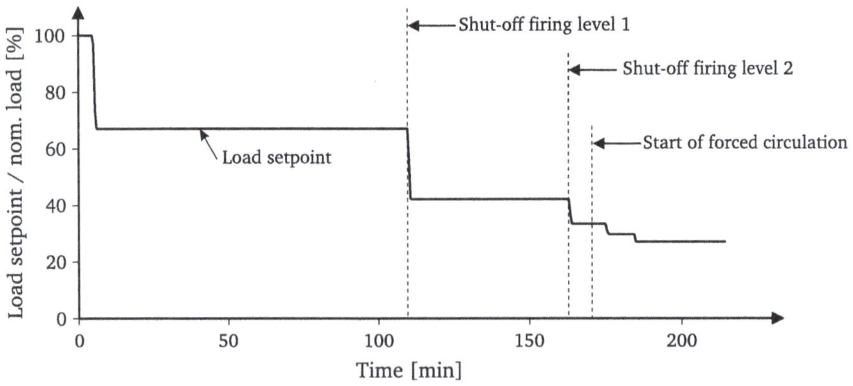

Fig. 4.58 Load change scenario

the lowest firing level is shut-down. During a pulveriser shutdown, the selected feeder speed is running down to zero, while the remaining feeders have to compensate the missing coal flow in order to keep the power plant at the desired thermal power input. Together with the next load setpoint reduction from 42% to 33.4% after 162 min, the firing level 2 is switched off. Ten minutes later the forced circulation starts and in the following the load setpoint is reduced two times more from 33.4% to 29.7% and then to 27.5% at minutes 175 and 185, respectively.

The pressure in the HP system of the steam generator is presented in Fig. 4.59. The pressure transients in the separator bottle and at the steam generator exit are plotted. The difference between the two plots represents the pressure loss in superheater system. The simulated results show a good match towards experimental data. After the first and second changes in the load setpoint from 100 to 67.5% and then to 42%, the plant moves to stationary operational points. Due the inhomogeneous coal composition of the feed in the real plants, changes in the process are unavoidable even during constant setpoint periods (between t = 60 min and t = 110 min as well as between t = 130 min and t = 170 min). One the contrary, the APROS model produces constant values until the next setpoint change occurs. Also visible is the change from sliding pressure operation to fixed pressure operation, which occurs during the third change in the load setpoint from 42 to 33.4%. Here, the valve upstream of the high-pressure turbine starts to control the pressure in the high-pressure part of the steam generator at its minimal value. A little bit earlier, i.e. after 120 min and around 55% of the nominal hot reheat pressure, the control valve before the IP turbine controls the pressure in the reheater at the minimal value. The pressure transients at inlet and outlet of the reheater system are also presented in Fig. 4.59. Similar to the pressure in the HP system, the simulated reheated pressure shows quantitative match with the measurement, suggesting that the implemented boiler feedwater pumps and its corresponding control as well as the fuel control and its dynamics are working correctly.

From Fig. 4.60, the simulated and measured mass flow rates in the high-pressure circuit of the steam generator are in a good correspondence. Three mass flow rates, including the steam to the turbine, the water at the economiser inlet and the circulated water reveal in detail the behaviour of the minimal evaporator mass flow control and the recirculation control. When the circulation pump is started, the feedwater control and the recirculation control work together to keep the minimal

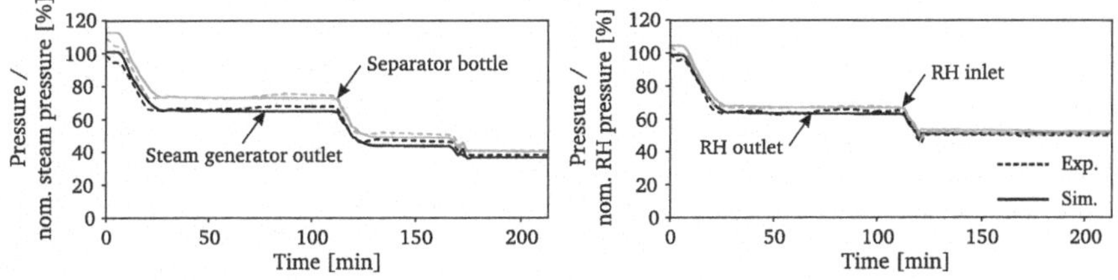

Fig. 4.59 Pressure in HP and RH sections (simulation and operation data)

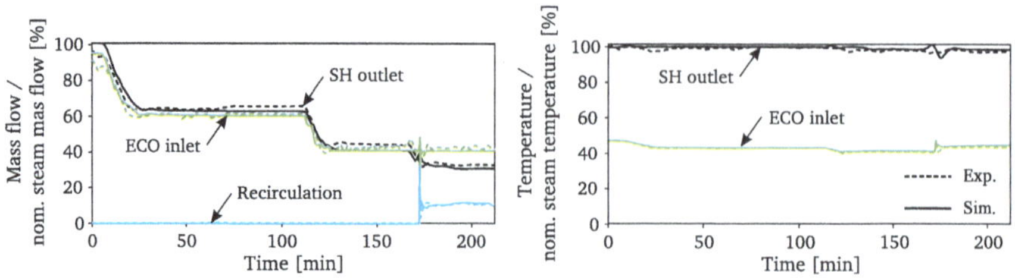

Fig. 4.60 Mass flows rate and temperatures (simulation and operation data)

evaporator flow and the level in the bottle at their limits. The inlet and outlet temperatures of the steam generator are also illustrated in the same figure. The temperature at the steam generator outlet is controlled via the complex three attemperator stages. The model can reproduce the superheated steam temperature with high accuracy. During the shut-off of the second firing level and the start-up of the recirculation system, the simulated temperature is for short periods too high due to the slow reacting of the related control circuits. The simulated water temperature at the steam generator inlet agrees very well with the experimental data and gives an impression of model quality regarding the turbine stages and the preheater section. This is due to the fact that the turbine extraction mass flow rates towards the preheaters are a result of the pressure difference between the steam turbine and the preheater condenser.

After discussing the water/steam side, simulation results and measured data from the flue gas path of the power plant during the load change scenario is presented. The flue gas system consists of two parallel working tracks, i.e. two selective catalytic reduction units, two air preheaters, two filters and two induced fans. Unlike the numerical model, where these two paths are completely identical, differences are to be expected in the real unit. Figure 4.61 contains the measured flue gas pressure after the two air preheaters and the simulated flue gas pressure after one air preheater, in addition to the oxygen content in the flue gas. The absolute gas pressure is difficult to predict since fouling or air leakage are changing during the operation. A qualitative match during the part load operation can be observed. It should be mentioned here that the good prediction of the pressure gradient

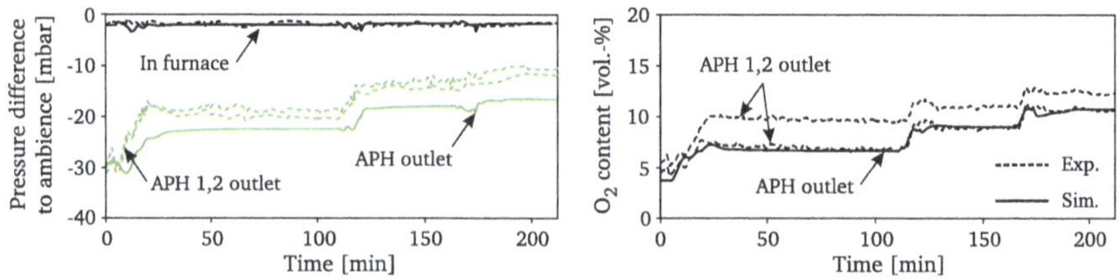

Fig. 4.61 Pressure difference to ambience and oxygen content downstream of the air preheaters (simulation and operation data)

underlines the correct control of the air mass flow rate, the air-to-fuel ratio and the accurate furnace pressure control. The air-to-fuel ratio is load dependent and is increasing with decreasing load in order to keep the gas velocities high enough to enable an effective heat transfer for the convective heat exchangers. Due to a slightly different leakage in the two air preheaters, the oxygen content at the outlets of APH 1 and APH 2 differ considerably. The trend of simulated oxygen content at the air preheater outlet follows the measurement qualitatively. It has to be kept in mind that the overall oxygen content in the flue gas of the plant is in between the two measurements, depending on the exact mass flow rates. Therefore, the good agreement of the absolute values of APH 1 outlet and the simulation should not be overrated.

4.1.2.2 Two-Pass Boiler

In single-pass boilers, the convective heat exchangers are arranged in the flue gas pass one above the other, which in turn lead to smaller differences in the temperature and the velocity of the flue gas path and water/steam side towards the top. Two-pass boilers have an off-gas deflection between the heating surfaces and this greatly increases the over-temperatures on both flue gas path and the water/ steam side, which can result in higher temperature spots (i.e. higher erosions). Furthermore, the single-pass boilers have a simple construction, rectilinear flow path with few deflections, low deposition problem, low thermal stresses and thermal expansion only in one direction. However, the construction height and construction time are much higher than the two-pass boilers.

The 1190 MW_{th} hard coal power plant in South Korea is a super-critical, once-through and two-pass steam generator. The nominal steam mass flow rate to the HP turbine and to the IP/LP turbines amounts to 445 kg/s at 260 bar and 602 °C as well as to 359 kg/s at 46 bar and 612 °C, which yields a gross electrical output of 595 MW_{el} (net generation of approximately 550 MW_{el}). Table 4.10 includes the characteristic technical data of the real power plant for the gas path and the water/ steam side.

The model, developed in APROS, describes the air/flue gas path and the water steam side in detail. The air/flue gas path, depicted in Fig. 4.62, includes the primary air supply with its pulverisers, the secondary air supply system, the air

Table 4.10 Technical data of the hard coal Power Station in South Korea

South Korean Power Station				Flue gas	ST power
High-pressure system	Reheater section	Low-pressure turbine	Condenser inlet	$\dot{m}$ = 577 (kg/s) T_{ECO} = 402 (°C) T_{Sta} = 123 (°C)	550 (MW_{el})
p = 260 (bar) T = 602 (°C) $\dot{m}$ = 445 (kg/s)	p = 46 (bar) T = 612 (°C) $\dot{m}$ = 359 (kg/s)	p = 6.8 (bar) T = 335 (°C) $\dot{m}$ = 302 (kg/s)	p = 50 (mbar) T = 32 (°C) $\dot{m}$ = 249 (kg/s)		

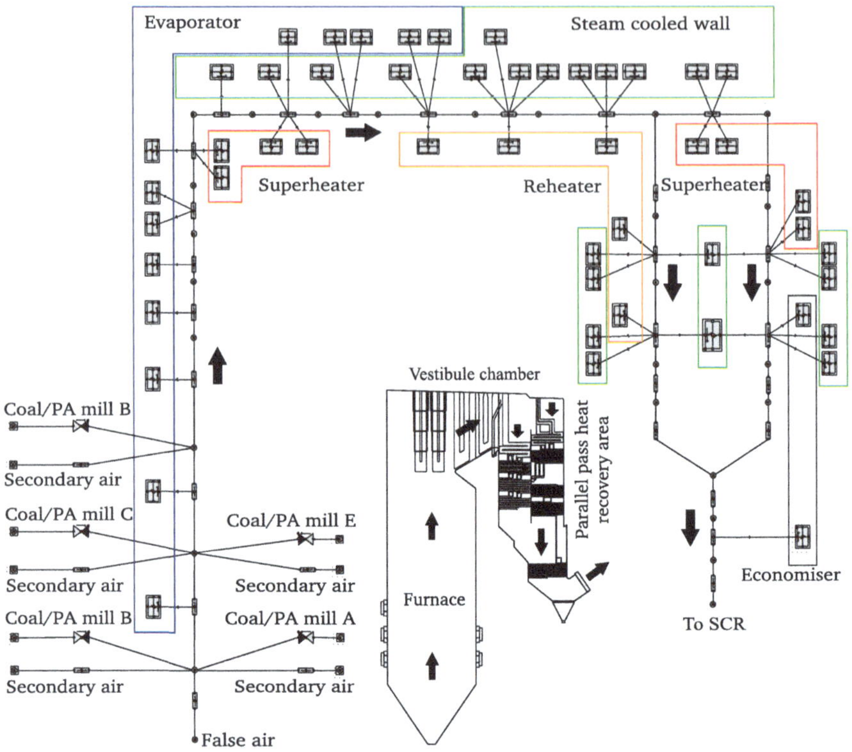

Fig. 4.62 Flue gas path of the super-critical, once-through and two-pass steam generator of the power plant (blue box for evaporator, green box for steam cooled wall, red box for superheater, orange box for reheater, black box for economiser)

preheater, the furnace chamber, the vestibule chamber and the parallel pass heat recovery area (HRA) chamber, in addition to the flue gas cleaning devices (e.g. electrostatic precipitator and selective catalytic reduction unit). The latter are considered as pressure drops with the correct mass. The primary air flows through the regenerative air preheater into the pulverisers. The coal dust loaded air and the secondary preheated air are fed into the furnace and combusted. The air supply systems and all chambers are modelled with pipes, which have the length and area of the corresponding sections. Between the pipes, there is a node, where the thermodynamic state variables are calculated.

The furnace chamber that has five firing levels starts after the tempering duct and ends in the middle section of the platen superheater. The flow area between the bottom of the furnace and the nose section is constant and equal the cross-section of the furnace. In order to account the acceleration of the flue gas due to the nose, the flow area decreases in the nose section. For that matter a mean area is used, which is calculated with the mean width of the entry and exit of the nose. The constant area in the platen superheater section is estimated with the shortest width in the nose section. Then, the flue gas stream changes its flow direction in the horizontal pipe with a constant area, followed by the horizontal vestibule chamber at the position of the final superheater and exits downstream of the final reheater. The pipe sections in the vestibule chamber represent the different heat exchanger stages. The flow area of each pipe decreases along the path by using a mean value according to the given data combined with the calculation procedure in the nose section mentioned above. Here, the height of each point is held constant. The section of the final reheater is split in two parts, since the cooling of the sidewalls is realised through different parts of the water/steam side (evaporator and steam cooled wall). The following pass of the heat recovery area is divided into a horizontal and a vertical part. The horizontal part describes the sections of the primary reheater and superheater, which the total flue gas flow passes. Afterwards this flow is directed vertically and separated into two parallel passes with equal area. In the left duct, the flue gas passes the rest of the primary reheater and in the right duct the rest of the primary superheater and the second part of the economiser. Then, the flue gas currents unite and flow through the section of the first economiser. The flue gas exists the steam generator and enters the selective catalytic reduction unit, the regenerative air preheater and the electrostatic precipitator.

The circuit of the water/steam side starts from the boiler feedwater pumps that suck the feedwater from the feedwater storage tank and supply its to the evaporator system via the high-pressure preheater system. The evaporator system of the hard coal-fired power plant consists of five parts: the economiser, the lower furnace wall, the mixing bottle, the vestibule sidewall and the upper furnace wall (see Fig. 4.63). The heat exchangers of each part are modelled via heat exchanger modules or as for the rear support case as a pipe with heat structure. Every heat exchanger is connected with the according pipe in the flue gas path. For connections between each heat exchanger as well as the inlet and outlet headers, pipes with heat structure are used. After leaving the high-pressure preheating system, the feedwater delivered by the boiler feedwater pumps enters the economiser. The latter is separated into two parts according to the different locations in the flue gas path and is modelled as a counter-current heat exchanger. Then, the water flow is directed to the heat exchangers of the lower furnace wall. These heat exchangers are designed as wall heat exchangers, which cool the furnace wall up to the lower transition header. From the lower transition header, the water/steam flow enters to the mixing bottles, where a part of the total mass flow rate streams to the vestibule sidewall. This wall heat exchanger is connected to the flue gas part to cool the sidewalls next to the first part of the final reheater. The other part of the water/steam flow is directed further to the wall heat exchangers of the upper furnace wall up to the nose section (E1). Here the water/steam flow is split in the parts evaporator upper wall furnace (E2), evaporator rear support (E3) and evaporator nose (E4). The wall heat exchangers of (E2) cool the front and sidewalls of the upper furnace wall up to the roof. The heat exchanger of (E3) that is modelled as a pipe with heat structure absorbs heat from the flue gas path. These two flows are reunited in the furnace wall upper outlet header. The wall heat exchangers of (E4) cool the rear wall of the nose section and

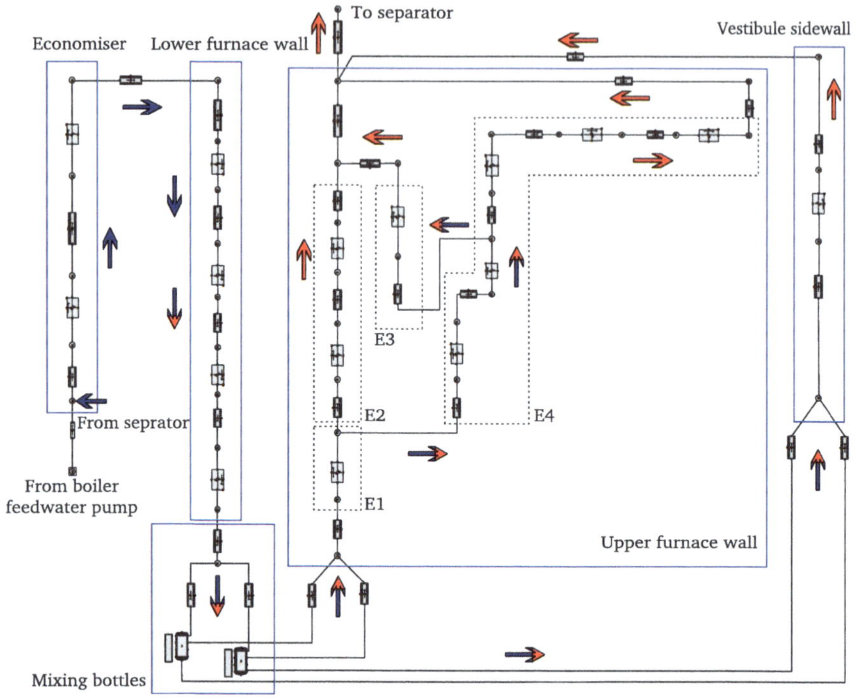

Fig. 4.63 The evaporator system of the super-critical, once-through and two-pass steam generator of the power plant

the lower wall of the vestibule chamber. All separated mass flows are joint together in the collecting header and sent to the separator.

In the separator, illustrated in Fig. 4.64, the water droplets within the steam flow is removed during part loads. At nominal load (super-critical pressure), no phase separation can take place. At low part loads or during start-up procedures, the water from water/steam mixture is separated and the collected water re-enters the steam generator at the economiser inlet. A circulation pump forces the recirculation and the water level in the separator is controlled by a circulation control valve. In order to prevent damage to components of the circulation pump due to cavitation that can dramatically shorten the pump's lifetime, a sub cooling injection system (net positive suction head (NPSH) injection) is applied.

The steam is then directed to the steam cooled wall that contains seven main heat exchanger parts. The first part is the roof of the furnace, vestibule and heat recovery area and ends downstream in the partition wall upper header. From here the total water/steam mass flow rate is separated according to the six different heat absorbing sections. These sections are the vestibule sidewall, the front wall of the heat recovery area, the left sidewall of the heat recovery area, the partition wall of the heat recovery area, the right sidewall of the heat recovery area and the rear wall of heat recovery area.

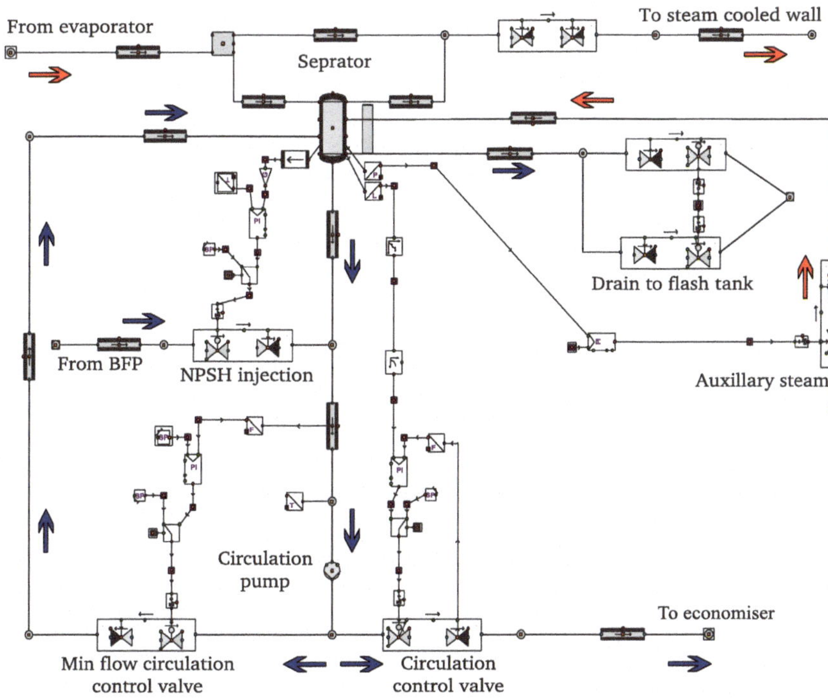

Fig. 4.64 The separator system with its control circuits

After leaving the steam cooled wall, the steam flows into the primary superheater, the platen superheater and the final superheater. The primary superheater is divided into four counter-current heat exchangers regarding the changing wall thicknesses of the heat exchanger tubes and the different arrangements in the flue gas path. Downstream of the primary superheater, an inter-stage attemperator is implemented. Here, a water mass flow rate with the conditions of the boiler feedwater pumps is injected into the steam mass flow rate exiting the primary superheater at the height of the platen superheater inlet header. The platen superheater consists of a counter-current and a co-current heat exchanger to reproduce the different steam flow directions streaming through this component. The final superheater, located downstream of the platen superheater, is modelled as two co-current heat exchangers to take into account the varying wall thicknesses and inner diameters of the heat exchanger tubes. At the inlet of the final superheater, a final attemperator that uses water from the boiler feedwater pumps is applied to control the steam temperature. The exiting steam flow with the parameters (445 kg/s at 260 bar and 602 °C) is directed further to the high-pressure turbine. The superheated steam expands in the high-pressure turbine and re-enters the steam generator via the reheater system that consists of the primary reheater and the final reheater. Due to the changing outer and inner diameters of the heat exchanger tubes and the different arrangements in the flue gas path, the primary reheater consists of three counter-current heat exchangers. Downstream of the primary reheater, the steam flow enters the final reheater that is modelled with two co-current heat exchangers with different tube lengths. An attemperator is installed at the inlet of the primary reheater to control the steam temperature at the outlet of the final reheater. The steam with the parameter (359 kg/s at 46 bar and 612 °C) is then directed to the intermediate and low-pressure turbines and finally condenses in the condenser. The condensed water flows through the feedwater preheaters back to the feedwater storage tank of the steam generator. All heat exchangers in the superheater and reheater systems are connected with the corresponding pipes in the flue gas sections. The inlet and outlet headers as well as the connecting pipes are modelled as pipes with a heat structure.

The model analyses with the design data at different loads have been performed including the load change from 100% to 75% to 100%, from 100% to 50% to 100% and from 100% to 30% to 100%, showing good agreement. With the validated model, several start-up producers (re-start, hot, warm and cold start-ups) were simulated and in this section, the hot and cold start-ups are only presented, as examples.

In Fig. 4.65, the thermal load and the net power generation during the hot start-up procedure are presented. During the pre-start-up phase, several actions take place such as start-up of the boiler feedwater pumps with 5% boiler maximum continuous rating (BMCR), start-up of the recirculation system, start-up of the induced draft fan and the forced draft fan. The oil supply starts increasing to reach 25% thermal load at t = 50 min and from this point of time starts decreasing to zero. Simultaneously, the mills (A and B) start first and follow then with mills (D and A). At t = 90 min, the high-pressure main steam to the turbine control valve

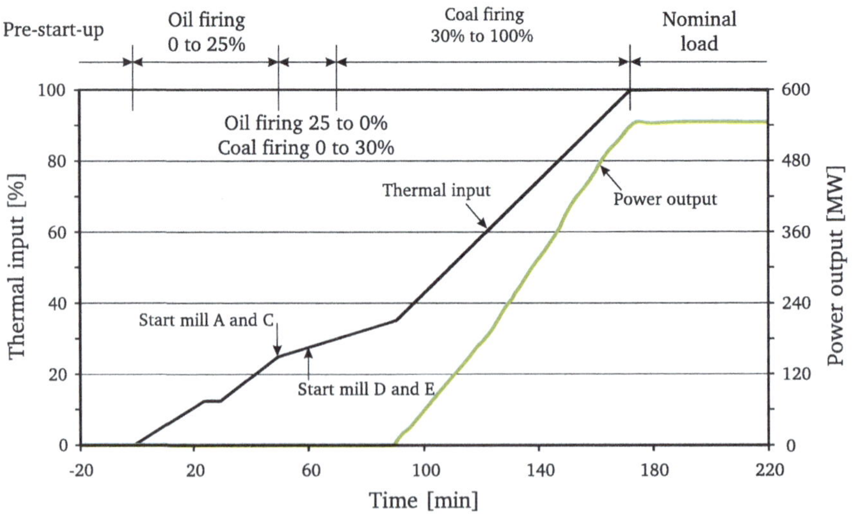

Fig. 4.65 The thermal load and the net power output during the hot start-up procedure

(HPMSCV) begins opening and the high-pressure bypass control valve closes with the same ratio. Accordingly, the power output increases and reaches its nominal load (450 MW_{el}) at t = 170 min.

The dynamic behaviour of the mass flow rates during the hot start-up procedure is displayed in Fig. 4.66. These mass flow rates are the feedwater mass flow rate to boiler (blue), the furnace mass flow rate (brown), the recirculation mass flow rate (orange), the steam mass flow rate to superheater (green) and the steam mass flow rate to high-pressure turbine (red). In the period of time between t = 0 and t = 25 min, the furnace mass flow rate is a sum of the feedwater mass flow rate to boiler and the recirculation mass flow rate. Here, the mass flow difference between the furnace mass flow rate and the recirculation mass flow rate of about 24 kg/s discharges to flash tank. After 25 min from the oil supply, the boiler starts generating steam that increases to reach 135 kg/s within 20 min. At the same time, the recirculation mass flow rate decreases and from t = 50 min the reticulation system is out of operation. At t = 90 min, the HPMSCV starts opening, allowing to the steam mass flow rate to enter the high-pressure turbine. After 170 min from the start, the power plant reaches its nominal load. The difference between the feedwater mass flow rate to boiler and the steam mass flow rate to high-pressure turbine of about 30 kg/s represents the mass flow rate that injected by the HP attemperators.

The temperature and pressure profiles of the last superheater and reheater during the hot start-up are presented in Fig. 4.67. Before the hot start-up, the temperatures and pressures of SH and RH amount to (400 °C and 70 bar) and (370 °C and 10 bar), respectively. Due to the thermal load, the SH pressure increases from its standing idle pressure to reach its fixed pressure (here is 80 bar) at t = 25 min, which in turn leads to open the high-pressure bypass valve. Accordingly, the steam mass flow rate streams into superheater and reheater sections. This is why at

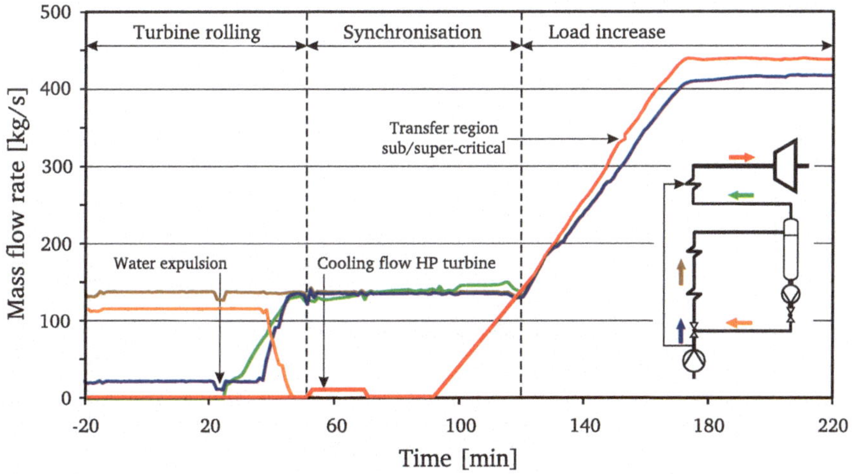

Fig. 4.66 Dynamic behaviour of the mass flow rates during the hot start-up procedure

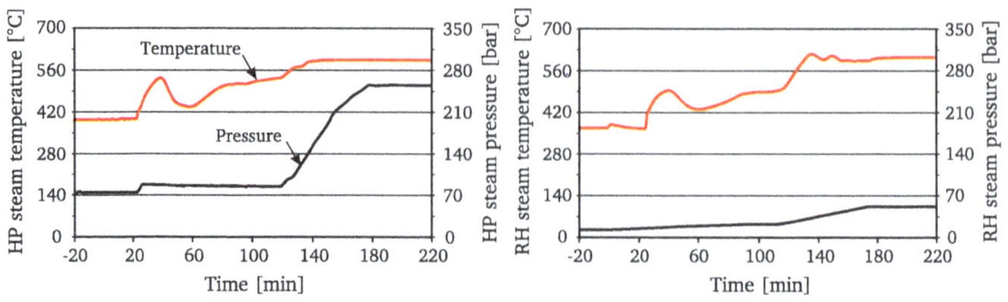

Fig. 4.67 Dynamic behaviour of the superheater and reheater sections during the hot start-up procedure

t = 25 min, the temperatures of SH and RH raise very fast to 500 °C within 15 min. From t = 25, the SH and RH pressures remain constant, before start increasing from t = 120 min to their operation nominal values (260 bar and 46 bar). In the same period of time, the SH and RH temperatures increase to get their setpoints (602 °C and 612 °C), controlled by the HP and RH attemperators.

The thermal load and the net power generation during the cold start-up procedure are shown in Fig. 4.68. The cold start-up procedure can be divided into six phases, namely the pre-start-up (20 min), the oil supply increase to 25% (105 min), the oil supply at 25% (55 min), the oil supply decrease from 25% to zero and simultaneously the coal supply increase from zero to 25% (40 min), the coal supply at 25% (50 min) and finally the coal supply increase from 25% to 100% (150 min).

In Fig. 4.69, the feedwater mass flow rate to boiler (blue), the furnace mass flow rate (brown), the recirculation mass flow rate (orange), the steam mass flow rate to superheater (green) and the steam mass flow rate to high-pressure turbine (red) are plotted during cold start-up procedure.

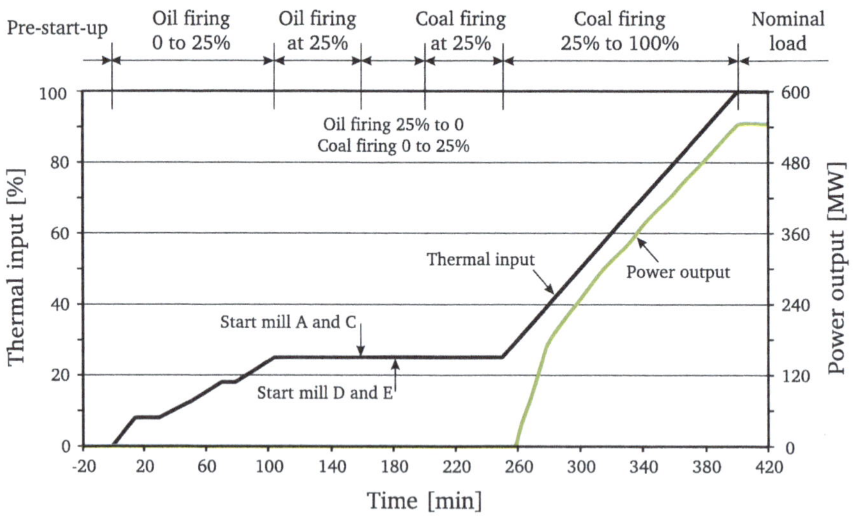

Fig. 4.68 The thermal load and the net power output during the cold start-up procedure

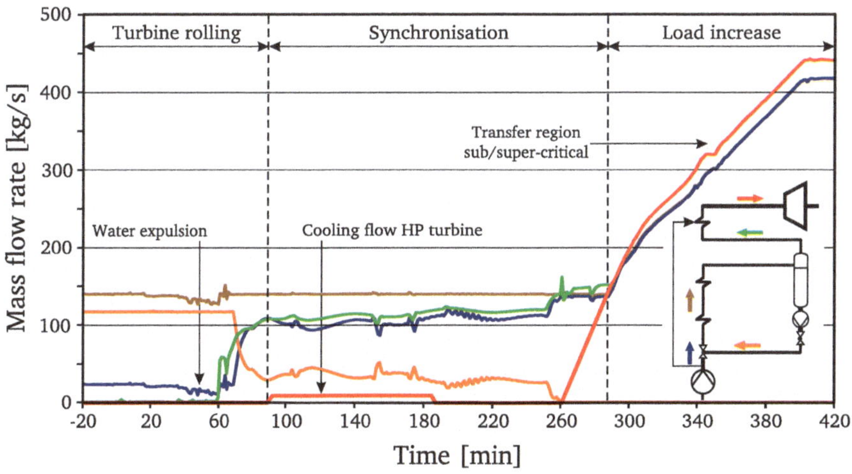

Fig. 4.69 Dynamic behaviour of the mass flow rates during the cold start-up procedure

The boiler needs approximately 60 min to start generating steam, which is about 35 min more than the hot start-up. From this point, the steam mass flow rate to superheater increases within 30 min to reach a value of 110 kg/s and this remains constant in the period of time between t = 90 min and t = 250 min. At the same time, the feedwater mass flow rate to boiler behaves almost similar to the steam mass flow rate to superheater and the recirculation mass flow rate has an average value of 35 kg/s. From t = 250 min, the thermal load starts increasing and the furnace mass flow rate is completely converted to steam that flows into the superheater. Accordingly, the recirculation mass flow rate decrease to zero and the recirculation system is out of operation at t = 260 min. Simultaneously, the high-pressure main steam to turbine control valve starts opening and the high-pressure bypass control valve starts closing. At t = 290 min, the generated steam by the boiler is completely directed to the steam turbine. After 400 min from the start, the power output reaches its nominal value of about 450 MW_{el}.

The dynamic behaviour of temperatures and pressures of the superheater and reheater during the cold start-up procedure is displayed in Fig. 4.70. During the

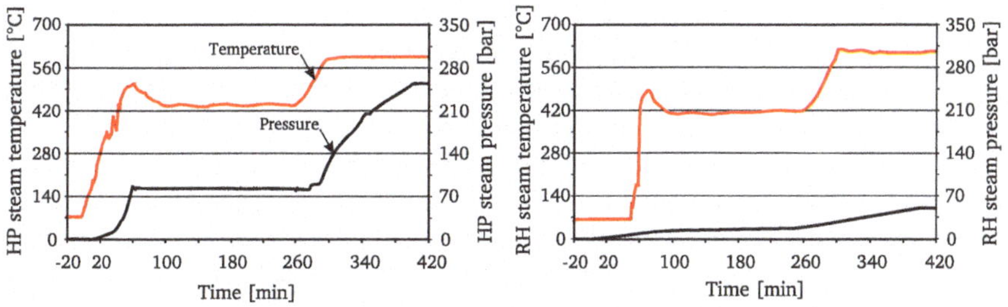

Fig. 4.70 Dynamic behaviour of the superheater and reheater sections during the cold start-up procedure

standing idle, temperatures of the SH and RH circuits decrease gradually due to the natural cooling to about 70 °C. In the period of time between t = 10 and t = 60 min, the superheated temperature increases to 520 °C with a thermal gradient of about 9 °C/min. From t = 100 min, it remains constant equal to 430 °C by the reason of the hold point of the thermal load and then jumps to 602 °C, but with lower thermal gradient (4 °C/min). Like the superheated temperature, the reheated temperature displays similar behaviour during the cold start-up procedures. The pressure in superheater section increase from 1 bar to its fixed value 80 bar within 50 min. At this point, the HP bypass valve starts opening gradually in order to prevent further increase in the pressure. From t = 290 min, the HP bypass valve is completely closed and the SH pressure increases to its nominal value of 260 bar, together with the increase of the thermal load. The pressure in the reheater section shows similar qualitative behaviour to the SH pressure.

4.1.2.3 Oxyfuel Boiler

In oxyfuel coal-fired power plants, the combustion of fuel takes place with a nitrogen-free oxidant. Air is therefore separated using an air separation unit into nitrogen and 95% pure oxygen. The oxygen enriched air is burned together with circulated flue gas in order to keep the flame temperature in the range of conventional power plants. A challenge for the design and operation of an oxyfuel coal-fired power plant is the flue gas path with its complex recirculation path, which in turn results in an undesired interaction between the involved fans. Here, the knowledge of the transient gas behaviour during malfunctions or trips are essential in order to prevent severe boiler damage or implosions. There are different designs for commercial large-scale oxyfuel pulverised coal-fired power plants. Nevertheless, the basic structure of the recirculation path is basically similar. Unfortunately, the design and operation experiences of large conventional thermal power plants can be only partly applied since the entire structure of the gas path is quite different. Furthermore, the experiences of oxyfuel pilot facilities can be only partly adopted for large-scale power plants due to test plants character and inadequate knowledge of the real process. The circulated fluidized bed boiler offers another option for a large-scale application of oxyfuel coal power plants.

A dynamic simulation model of an oxyfuel pulverised coal-fired power plant is developed in order to investigate the gas dynamics during malfunctions, e.g. the fuel trip and the blackout (Postler 2012; Starkloff et al. 2016). The oxyfuel power plant (designed by ALSTOM) is planned to be erected at the Jaenschwalde Power Station, Germany with a gross power output of 250 MW_{el} (see Table 4.11).

The tower type, sub-critical, once-through boiler has three firing levels, which are fuelled with pre-dried lignite from the local mines (lower heating value of about 20 MJ/kg). On the flue gas path, all components and their pressure losses as well as their thermal masses are considered in the developed model. The only boundary conditions are the ambient conditions at the air inlets, the oxygen coming from the air separation unit, the coal composition and the pressures at stack and the gas

Table 4.11 Technical data of the oxyfuel power plant

Jaenschwalde Power Station				Flue gas	ST power
High-pressure system	Reheater section	Low-pressure turbine	Condenser inlet	$\dot{m}$ = 319 (kg/s) T_{ECO} = 347 (°C) T_{Sta} = 130 (°C)	250 (MW_{el})
p = 167 (bar) T = 535 (°C) $\dot{m}$ = 235 (kg/s)	p = 32 (bar) T = 535 (°C) $\dot{m}$ = 217 (kg/s)	p = 4.5 (bar) T = 250 (°C) $\dot{m}$ = 190 (kg/s)	p = 90 (mbar) T = 43 (°C) $\dot{m}$ = 166 (kg/s)		

processing unit. The flue gas path and the water/steam side of Jaenschwalde Power Station is illustrated in Figs. 4.71 and 4.72, respectively. Starting at the boiler outlet (a), the flue gas enters the hot side of the gas-gas heat exchanger (GGH) and then flows through the electrostatic precipitator (ESP), the first induced draft fan (IDF 1), the heat exchanger for condensate preheating (CPH) and the wet flue gas desulphurisation unit (WFGD).

During oxygen-fired operation, the most water content in the flue gas (approximately 90%) is condensed in the flue gas condenser (FGC). The dried flue gas flows through the second induced draft fan (IDF 2) into the gas processing unit (GPU), where it is compressed and liquefied for storage or utilisation, respectively. The bypass ducts (a1 and a2) are in service for air-fired operation, during start-up/shut-down and during malfunctions. A small amount of the cleaned and dried flue gas is circulated through the primary recirculation path (b). The flue gas stream (b1) delivered by the transport gas fan (TGF) is used for the pneumatic transport of the

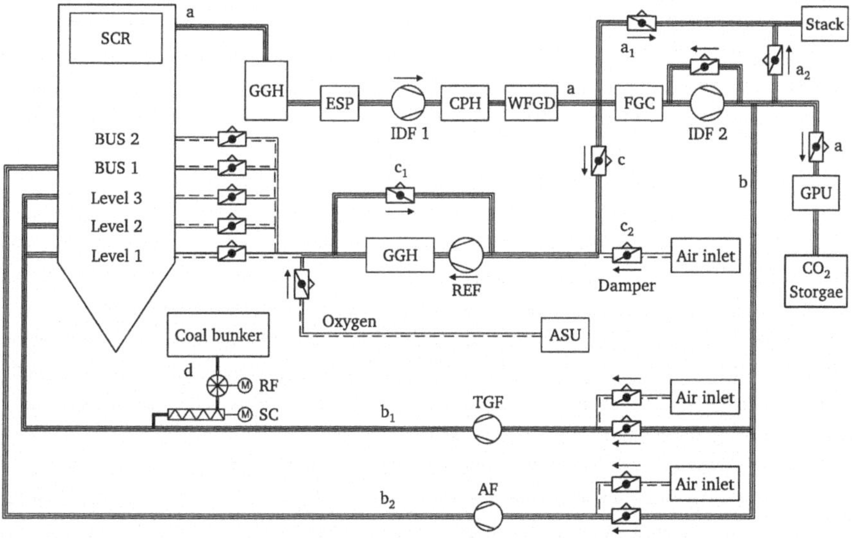

Fig. 4.71 Gas side of Jaenschwalde oxyfuel pulverised coal-fired power plant

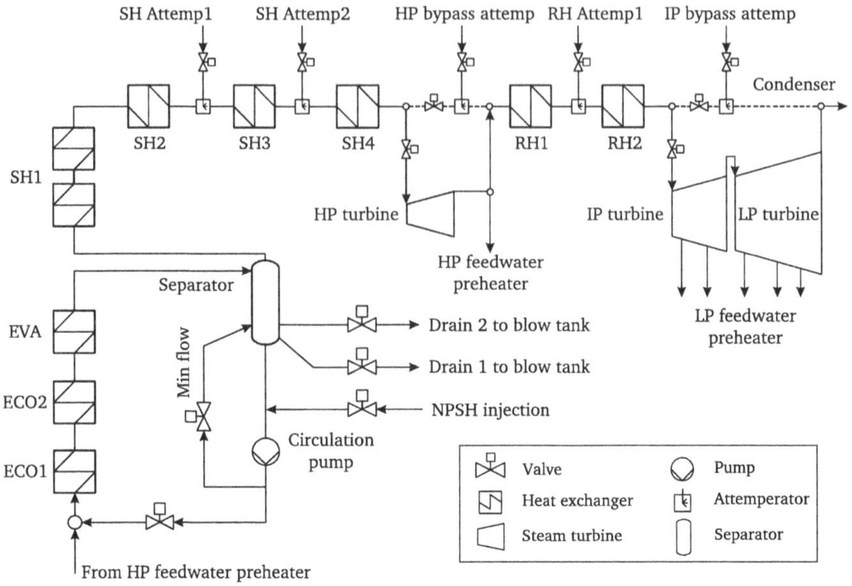

Fig. 4.72 Water/steam side of Jaenschwalde oxyfuel pulverised coal-fired power plant

pulverised coal to the furnace, while the auxiliary fan (AF) supplies the power plant with flue gas for sealing, purging or cooling purposes (b2). Around two-thirds of the flue gas is circulated by the recirculation fan (REF) through the secondary recirculation duct system (c) to the boiler. The circulated flue gas is mixed with the oxygen from the air separation unit. Afterwards, the gaseous mixture is delivered to the firing levels and the over fire burn-out supply (BUS). The secondary path has an internal recirculation (c1) for temperature control of the cold end of the main flue gas in the gas-gas heat exchanger. When the power plant is in air-fired operation, the dampers to the flue gas duct are closed and the air-inlet dampers are opened. In this case, the recirculation fan works like a conventional forced draft fan. The indirect firing system (d) contains a coal bunker for the pulverised coal, a rotary feeder (RF) and a screw conveyor (SC), which feeds the coal into the transport gas duct system. The water/steam side, presented in Fig. 4.72, has a similar structure as in the conventional coal-fired power plant and thus is not explained here.

The detail response of the numerical model to the fuel trip and blackout can be found in (Starkloff et al. 2016) and selected results are presented here. Since the projected large-scale oxyfuel-fired plant does not currently exist, operation data is not available for validation of the developed model.

In thermal power plants, the fuel trip scenario can be induced by different triggers, e.g. trip of other components, temperature limits etc. In an oxyfuel boiler, this scenario is getting even more crucial since the inlet and the outlet of the furnace are coupled via the flue gas recirculation. The most critical safety aspect during a fast fuel reduction is the negative gauge pressure in the furnace, which in turn can lead to boiler implosion or massive damages in the gas duct system. A reference fuel trip is simulated, where the power plant operates stationary at its nominal load (100% boiler maximum continuous rating (BMCR)) for the first 30 s before the fuel trip signal occurs. Accordingly, the coal mass flow rate is decreased linearly to zero within 25 s, whereby the firing levels are shut-down sequentially. Other measures that start immediately with fuel trip signal are: start of oxygen reduction, the oxidant control dampers switch to manual operation mode and remain in their positions, the recirculation isolation damper starts closing, the air inlet damper starts opening, the bypasses a1 and a2 start opening, IDF 2 shuts-off and the recirculation fan (REF) switches to manual control mode and keeps the blades at the current positions. The furnace pressure difference, the furnace temperature difference, the fuel mass flow setpoint and the false air in the furnace are illustrated in Fig. 4.73. The reduction of the coal mass flow to zero leads to an extinction of the flame. The maximum pressure decrease is equal to −28 mbar after 57 s and maximum pressure increase in the recovery phase is +13 mbar after 91 s. Since the pressure in furnace stays under the threshold of the IDF 1 shut-down (typically 30–40 mbar in conventional air-fired units), the IDF 1 is in service and the pressure in the furnace is controlled at its setpoint again. Due the fact that the value of the false air mass flow rate into the furnace is correlated with the pressure difference to the environment, an inverted behaviour can be observed during the fuel trip. This means, the false air will enter the boiler when the pressure difference between the furnace and the environment is negative, while it escapes to environment if the pressure difference between the furnace and the environment is positive. The furnace temperature is plotted as a difference to its nominal value. The temperature difference responds to the coal mass flow reduction and falls by 700 K during the observed time period.

An overview of the mass flow developments in the recirculation path relative to the nominal IDF 1 mass flow is given also in Fig. 4.73. The oxygen supply shuts-down linearly, starting at 31 s and reaches zero within 30 s. At the inlet of the recirculation fan, the flue gas is switched over into air. This process is running almost parallel and is finished at t = 62 s. After this time, the oxidant mass flow is a pure air. Similar to false air flow, the air mass flow rate at the REF inlet is supported by the large pressure difference between the furnace and the environment. The leakage in the gas-gas heat exchanger can be observed from the difference between the air and the oxidant mass flow rates. One minute after the trip signal, the recirculation fan is moved back in auto mode, which is clearly visible in the unsteadiness after 90 s.

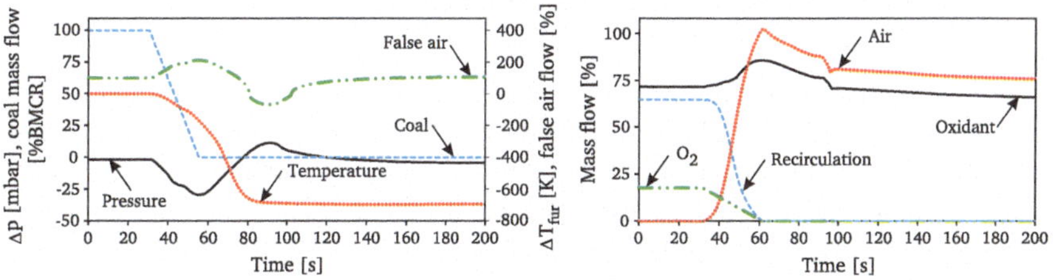

Fig. 4.73 Process parameters during the fuel trip

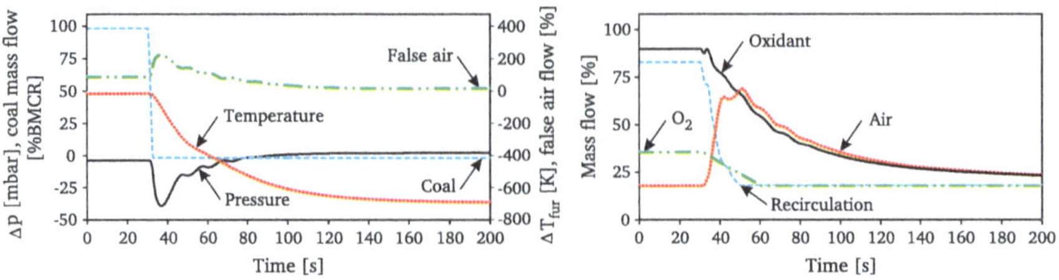

Fig. 4.74 Process parameters during the blackout

The blackout of the thermal power plants differs fundamentally from the fuel trip scenario. The reason for that is the sudden and complete lack of electricity in the power plant, which causes the motors of the fuel supply components and all fans to shut-down immediately. Furthermore, no control of the dampers is possible and they remain at their position prior to the blackout scenario. The passive safety systems have to operate the power plant during the blackout. Due their fail-safe design, a few dampers move to their safe position, e.g. the recirculation isolation damper and the air inlet damper. The scenario starts at 100% BMCR in steady-state oxyfuel operation, after 30 s the blackout occurs. Despite the extremely short fuel reduction time, the lowest negative furnace pressure is only −36 mbar (see Fig. 4.74). The mean reason for this desirable behaviour is the undeleted shut-off of the induced draft fan 1. Corresponding with the relatively moderate furnace pressure, the amount of the false air enters into the furnace stays also at moderate values. The absolute maximum mass flow rate of the false air is about two and half times the false air mass at nominal case. Due to the short reduction time of the fuel mass flow rate, the temperature drop in the furnace is quite progressively to nearly 700 K. Due to the immediate trip of the recirculation fan, the flow from the WFGD unit to the circulated path reduces sharply. Parallel, the recirculation isolation damper closes together with opening the air inlet located damper upstream of the REF. The air mass flow rate is driven by the negative gauge pressure amplitude into the furnace. Approximately at t = 200 s, the leakage flow in the gas-gas heat exchanger becomes almost zero and the air mass flow rate is equal to the oxidant mass flow rate.

4.1.3 Municipal Solid Waste Incineration

It is obvious that the suitable waste management is one of major challenges that nowadays the society should deal with. In last few decades and due to the rapid development of national economies, rapid urbanisation of rural areas and continued improvement of living standard, the solid waste output is constantly increasing. In order to effectively eliminate the municipal solid waste (MSW) from an economic and ecological point of view, different solutions have been suggested such as recycling, reduction of waste generation, energy recovery and as a last option the landfills. A proven approach to disposal large quantities of municipal solid waste is the thermal treatment of waste in grate systems (incinerator). The incineration scenario becomes a preferred option by the reason of the advantages it offers such as the cost reduction of remaining landfill disposal due to the lower volume of end products (one tenth of the original volume) and decreasing the total organic carbon (TOC) of waste, resulting in more inert residues unable to produce landfill gas. The released heat during the waste combustion (lower heating value: approximately 10 MJ/kg) can be recovered by a water/steam circuit for the supply of electricity and district heating. Accordingly, the residual waste can be used as substitute fuel for the conventional fossil fuels.

Municipal solid waste incinerators, although most pollutants are destroyed by the combustion, can emit high quantities of pollutants to atmosphere such as solid residues (particulate matter), heavy metals, acid gases and nitrogen oxides. Therefore, complex air pollution control devices are required before the cleaned flue gas passes through the stack into the environment. In order to remove the solid residues from the flue gas, separators are attached at transition regions between vertical passes and over the entire length of the horizontal pass. Furthermore, a commercial dust filter such as electrostatic precipitators or fabric filter is installed. The collected solid residues (slag and fly ash) are fed to the slag bunker and ash silos, stored temporarily and then discharged to be used as a filler material in road constructions or in cement industries. The acid gas concentrations in the flue gas, e.g. hydrogen chloride and sulphur dioxide are directly related to the chlorine and sulphur content of waste. For separation of pollutants, particularly acid gases, sulphur oxides and heavy metals, wet gas scrubbing or dry absorption with the addition of calcium compounds and/or activated carbon are generally used. The formation of nitrogen oxides is dependent on the nitrogen quantity of solid waste, combustion temperature and air/fuel ratio. Nitrogen oxides can be removed from the flue gas by employing a selective non-catalytic reduction (SNCR), in addition to conventional primary measures (e.g. staged combustion, especially air-staged supply). Finally, the cleaned flue gas passes through the stack into the environment.

In addition to its main task, volume reduction of municipal solid waste, the waste incinerator can be used to treat sewage sludge. The term "sewage sludge" refers to the residual material, resulting as a by-product during the treatment of industrial or municipal wastewater. Nowadays, approximately 10.2 million tonnes of dry sewage sludge per year are accumulated in Europe according to Eurostat (2015). The dry sewage sludge contains up to 70% organic components and the remaining 30% are composed of, among others, silicates, phosphates, phosphor and heavy metals. The sewage sludge with its valuable components can be recycled in agriculture and landscaping to improve the plant nutrition and to enhance the physical and chemical properties of the soil. Because of its higher calorific value (9 to 12 MJ/kg_{dried}), however, more than half of the sewage sludge was incinerated in Germany in 2010.

As an example, the dynamic simulation model of the municipal solid waste incinerator (60 MW_{th}) is described below.

4.1.3.1 Characterisation of Incinerator

The investigated 60 MW_{th} incinerator has been built in Finland, designed by Steinmüller Babcock Environment GmbH; a company specialised in engineering and construction of waste treatment and flue gas cleaning systems. The incinerator burns municipal solid waste with an ideal lower heating value (LHV) of 10.5 MJ/kg and a feed rate of 5.7 kg/s (492 t/d) as well as biogas with a LHV of 17.7 MJ/m^3 and a mass flow rate of 0.072 kg/s. It discharges 36.3 kg/s of flue gas at 160 °C as well as 0.7 kg/s of slag at 450 °C. The flue gas side and the water/steam side of the municipal solid waste incinerator are shown in Figs. 4.75 and 4.76, respectively.

The combustion system, illustrated in Fig. 4.75, comprises of a primary combustion zone on the grate, a post-combustion zone (also known as a secondary combustion zone), in which the secondary air is injected and a zone with auxiliary

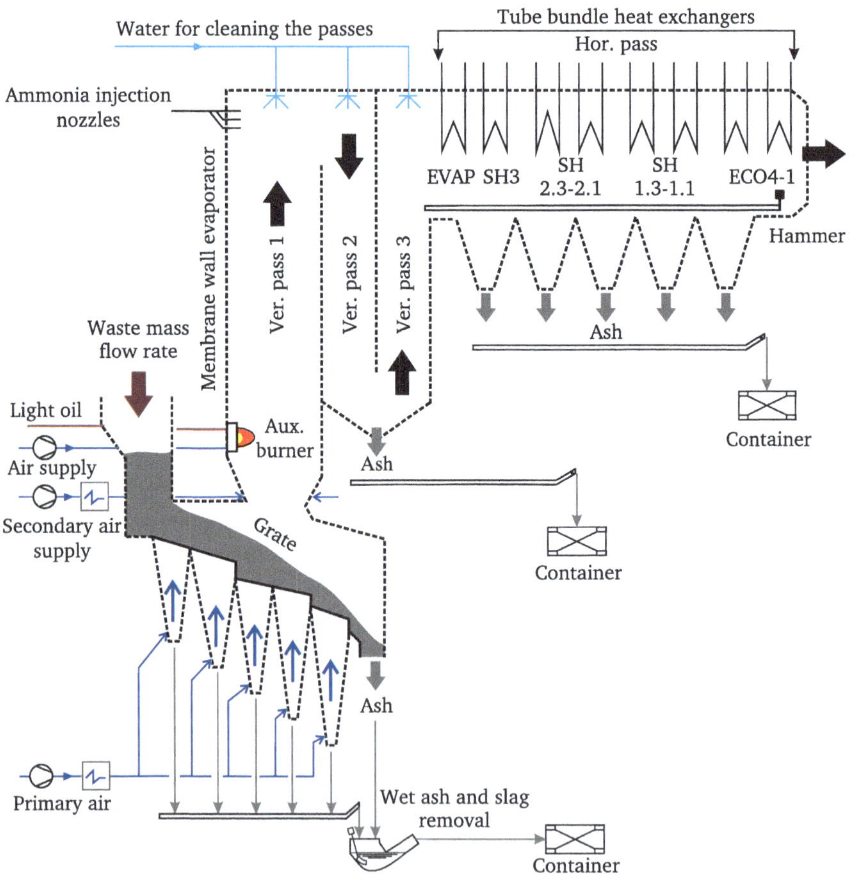

Fig. 4.75 Schematic representation of the flue gas side of the municipal solid waste incinerator in Finland

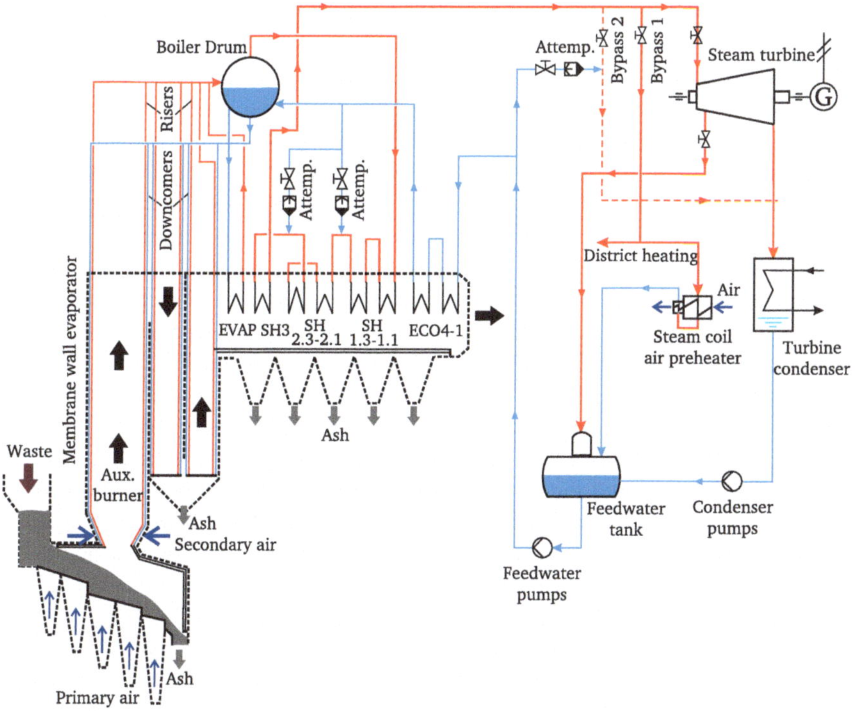

Fig. 4.76 Schematic representation of the water/steam side of the municipal solid waste incinerator in Finland

burners. The latter are required on the one hand for the plant start-up and on the other hand to control the combustion temperature, if the calorific value of the waste is insufficient to maintain the required minimum temperature and residence time.

On the grate, the incineration of solid waste takes place at different stages by means of primary air. The primary air is supplied via five hoppers below the grate, which is also used to collect and discharge ash and slag. Generally, each hopper consists of two chambers and characterises the combustion zone (drying, pyrolysis, combustion of volatile matters and burn-out of char). The processes of drying and pyrolysis occur along the first and second hoppers, while volatile matters are burnt approximately above the third hopper. Finally, the char is burned-out along the fourth and fifth hoppers. The combustion zones may merge into one another, depending on variation of the solid waste composition. However, the combustion zones can be fixed locally by adjusting the amount of solid waste dosage and the primary air mass flow rate as well as the speed of the grate. In addition to its task of providing the oxygen for the solid waste combustion (air/fuel ratio between 0.4 and 1.3), the primary air is used for grate cooling. The temperature on the grate ranges from 100 °C in the drying zone, over 800 °C in the pyrolysis zone to up to 1350 °C in the main combustion zone.

In the post-combustion zone, the remaining combustible substances are burned-out with excess secondary air that is injected via nozzles, located in the narrowest cross-section of the first flue gas pass. The arrangement and blade angle of the nozzles are set, so that a sufficient penetration depth and a high turbulence swirl flow can be established. In this post-combustion zone, the remaining combustible substances are combusted with excess air (approximately 1.5 above stoichiometric level). The exhaust gas flows into the flue gas path, including vertical passes (also known as radiation passes) and a horizontal pass (also known as a convection pass). Pass cleaning devices are installed at the highest point of first, second and third passes. At the transition region from second to third pass as well as over the entire length of the horizontal pass, slag and fly ash separator devices are used. The heat exchanger surfaces of the evaporators are located at the walls of all flue gas passes. The first heat exchanger in the horizontal pass is an evaporator with natural circulation. The next heat exchangers are superheaters and the last heat exchangers are the economisers. In order to ensure consistently good heat transfer between the flue gas and the tube bundle heat exchangers in the horizontal pass, hammer devices that generate vibrations on the pipes are used. Because of these vibrations, impurities are removed and fall into the hoppers that are attached along the horizontal flue gas pass. Finally, the flue gas leaves the incineration at approximately 160 °C and enters the flue gas cleaning devices.

The thermal energy released from the combustion of the municipal solid waste is absorbed by several shell and tube heat exchangers, converting water in superheated steam that is fed to a steam turbine. The process, illustrated in Fig. 4.76, starts with the economisers in the horizontal pass, which preheat the feedwater delivered from the feedwater tank before entering the boiler drum. The saturated water in the drum flows via downcomers to the evaporators and absorbs additional heat. The density differences at inlet and outlet of the evaporators are the main driving force for the fluid. The resultant water/steam mixture moves back via risers into the boiler drum that works as a separator. While the saturated water circulates again in the evaporator system, the saturated steam flows to the superheaters. Attemperators are installed at superheater surfaces to control the steam temperature at the inlet of the steam turbine. The attemperators use water directly from the boiler feedwater pumps. The injected mass flow rate is adjusted by a control circuit. In the steam turbine, the superheated steam expands and the enthalpy of steam is converted into mechanical energy. Finally, an alternating voltage is induced in the generator.

The steam turbine consists of high-pressure and low-pressure steam turbine sections. In between, there is a decoupling for steam extraction, which is required for the deaerator and for preheating the water in the feedwater tank. In the deaerator, dissolved oxygen, carbon dioxide and other non-condensable gases can be removed from the feedwater, preventing the formation of iron oxides (rust). There are also two bypass systems (bypass 1 and bypass 2). The first bypass is used to feed the thermal energy system of the district heating network and for the preheating of the combustion air, while the second bypass is in operation during plant start-up and malfunction cases. At the outlet of the steam turbine, the completely expanded steam is condensed before it is discharged into the feedwater tank by condenser pumps. The task of the feedwater tank is to collect the condensate and to promote the circulation to the boiler. The boiler feedwater pumps increase the pressure and transport the water into the economisers.

4.1.3.2 Incinerator Model

A dynamic simulation model of the waste incineration power plant is generated in order to investigate the influence of waste heating value variations on the plant efficiency as well as to evaluate the plant behaviour at part loads and during the start-up procedures (Alobaid et al. 2018). Table 4.12 shows the characteristic technical data of the power plant at 100% nominal load. The plant has 4 economisers, 7 superheaters and 5 evaporators with natural circulation. At full load, the superheated steam mass flow rate at turbine inlet amounts to 20.5 kg/s at 400 °C and 45 bar. The gross electrical output of the steam turbine is approximately 16 MW_{el}, which yields an electrical gross efficiency of 26.5%.

The process simulation software APROS is employed for the dynamic process modelling of the MSW incinerator. For grate modelling, the geometry of the grate and combustion chamber, e.g. the length of the pipes and cross-sectional areas of combustion zones, is required. However, it is quite difficult in process simulation to describe the grate due to the fact that no explicit distinctions can be made with respect to the combustion points (i.e. drying, pyrolysis, volatile matters combustion and burn-out of char cannot be calculated individually). Furthermore, the two-dimensional waste mass flow rate along the grate cannot be reproduced. For the purpose of simplification, the grate can be modelled in such a way that the solid waste mass flow rate, the air mass flow rates and all combustion zones are concentrated in one point. However, in this work, a detailed grate model is proposed, as presented in Fig. 4.77. Here, five main combustion points (a–e) are modelled, which represent the combustion on the real grate. Furthermore, five additional combustion points (f–k) are placed above the main combustion points (no real process component as a counterpart).

Table 4.12 Technical data of the municipal solid waste incineration power plant located in Finland

Municipal solid waste incinerator			Flue gas	ST power
High-pressure system	Reheater section	Condenser inlet	$\dot{m}$ = 36.3 (kg/s) T_{out} = 160 (°C)	16 (MW_{el})
p = 45 (bar) T = 400 (°C) $\dot{m}$ = 20.5 (kg/s)	–	p = 60 (mbar) T = 36 (°C) $\dot{m}$ = 16.9 (kg/s)		

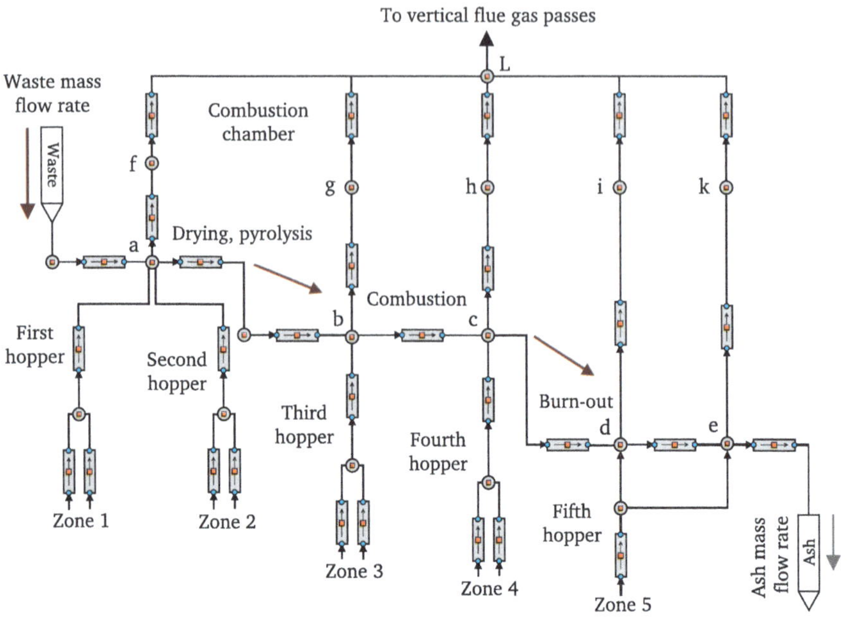

Fig. 4.77 The grate and combustion chamber

The distinction among the combustion points can be achieved by variation of the mixing factor (empirical factor) and the amount of oxygen supply. The solid waste mass flow rate enters the grate and flows to the first combustion point (a), in which the primary air mass flow rates from the first and the second hoppers are supplied. The composition of the solid waste is previously defined on a different net of the model. At point (a), the mixing factor is set to a low value in order to prevent the complete combustion of the solid waste. A small part of the solid waste and volatile matters moves upwards to combustion point (f), while most of the unburned solid waste flow continues to the next combustion point (b) that has a high mixing factor. At this point, part of the mass flow flows upwards again and a larger part flows to the next combustion point (c). This is continued until point (e) and the primary air mass flow rate is supplied at each combustion point in order to provide sufficient oxygen. Using this representation, it is possible to establish the temperature profile along the grate via the amount of primary air and the mixing factor. However, this temperature profile may have low priority in the process simulation, since all these flows will be recombined again in point (L). At this point, the flue gas parameters, including mass flow rate, temperature and composition are the important criteria for the incinerator model.

The flue gas, resulting from the solid waste combustion on the grate, flows through three vertical passes, before entering the horizontal pass (see Fig. 4.78). Modelling of the vertical flue gas passes is realized using the tube module, whose cross-sectional area, length and wall material is set based on the real specification of

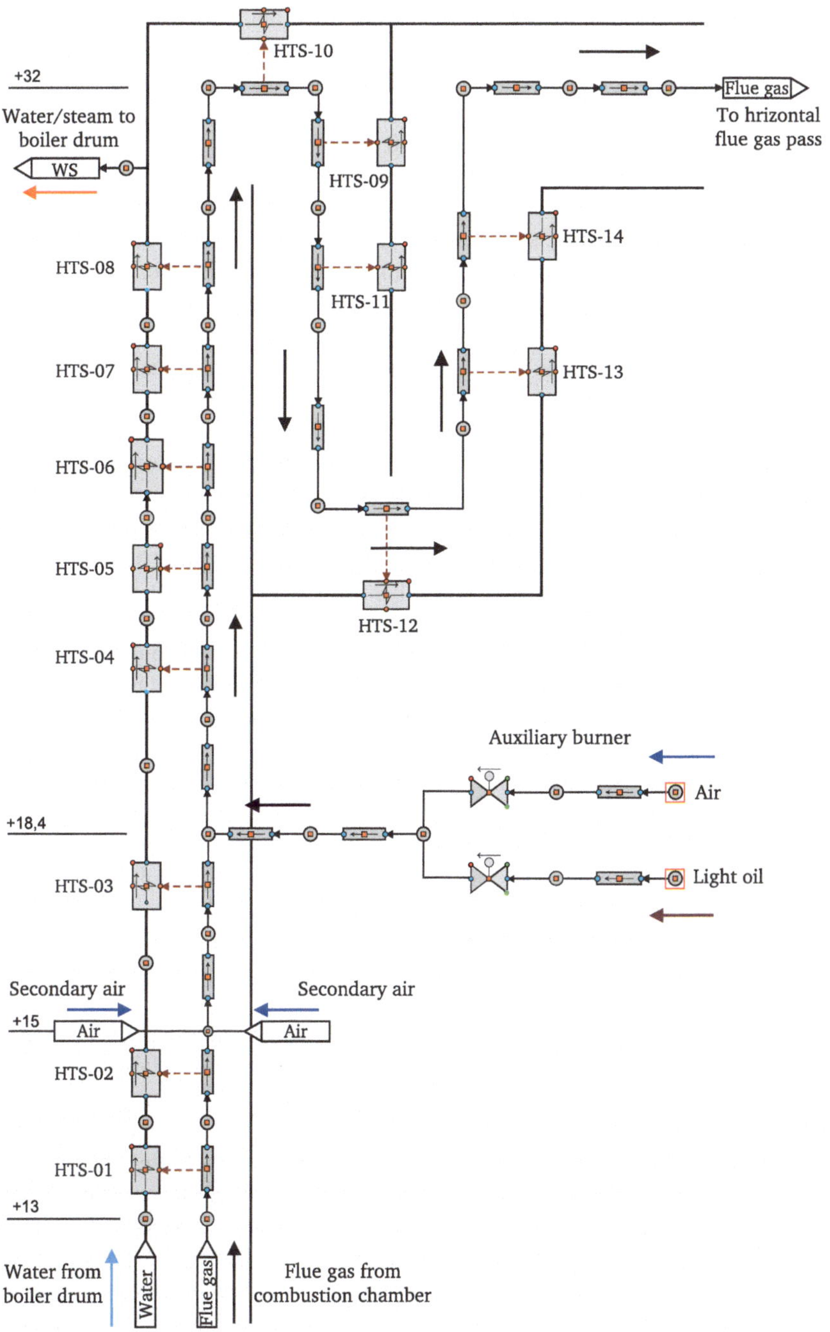

Fig. 4.78 The vertical flue gas passes

the incinerator. The first flue gas pass starts from an altitude of about 13 m above the ground (located directly above the combustion chamber). The location of the secondary air nozzles is about 14 m. The primary and secondary air mass flow rates are supplied by means of fans with the required mass flow rates. They are separately preheated in heat exchangers to about 150 °C and 230 °C using steam, which is extracted from the inlet of the high-pressure turbine. A detailed distribution of the secondary air flows in the respective nozzles is not necessary in the model. Only two nozzles are modelled, where the total mass flow rate of about 12 kg/s is injected into the flue gas stream. This simplification has no effect on the model accuracy since the pressure losses, raised from the air mass flow rates through the nozzles (in total 29 nozzles) has only an influence on the secondary air fan and its path. The auxiliary burners are located above the secondary air nozzles (about 18.4 m above the ground) and are operated with light oil and air. The main task of the auxiliary burners is to warm-up the incinerator and to support the combustion process during the start-up procedure. They are also switched on during operation at any load when the heating value of the solid waste is not sufficient to reach the required minimum temperature (850 °C) or the required residence time (2 s). For example, when they are in operation during the start-up procedure, average oil mass flow rate of about 1.2 kg/s is used, reaching a temperature of about 1300 °C. In real plants, several auxiliary burners are often installed on the sidewalls in order to reach the entire cross-sectional area of the incinerator. Since the simulation does not represent the cross-sectional area, the model is simplified by using one auxiliary burner. This simplification has no effect on the results; only the burner has to inject more fuel, if necessary.

The transition regions from the first flue gas pass to the second as well as to the third flue gas pass are also modelled using the tube module with adjustment of the pressure loss factor of the pipe in order to consider the additional pressure losses due to deflections. However, the pressure losses are relatively small because of the large cross-sectional area of the flue gas path. The dashed red arrows on the flue gas pipes in the first, second and third flue gas passes are the heat connections to the membrane wall heat exchangers. The last arrow at the third flue gas pass connects to the horizontal flue gas pass.

In the horizontal flue gas pass, the convective tube bundle heat exchangers are installed. These heat exchangers have the order of evaporator (EVAP), high temperature superheater (SH3), intermediate temperature superheater (SH2.3, SH2.2 SH2.1), low temperature superheaters (SH1.3, SH1.2 SH1.1) and economisers (ECO4, ECO3, ECO2, ECO1). In addition, the wall surfaces of the horizontal flue gas pass form a membrane wall heat exchanger that is connected to the boiler drum. The horizontal flue gas pass and the heat exchangers are modelled using the APROS tube and heat exchanger modules. Here, the number of pipes and tube rows, diameters, lengths, etc. are set according to the real geometry of the flue gas pass and the heat exchangers. The heat transfer coefficients are determined using the known heat flow.

In Fig. 4.79, the evaporator system of the 60 MW_{th} municipal solid waste incinerator is displayed. It consists of five natural circulation evaporators,

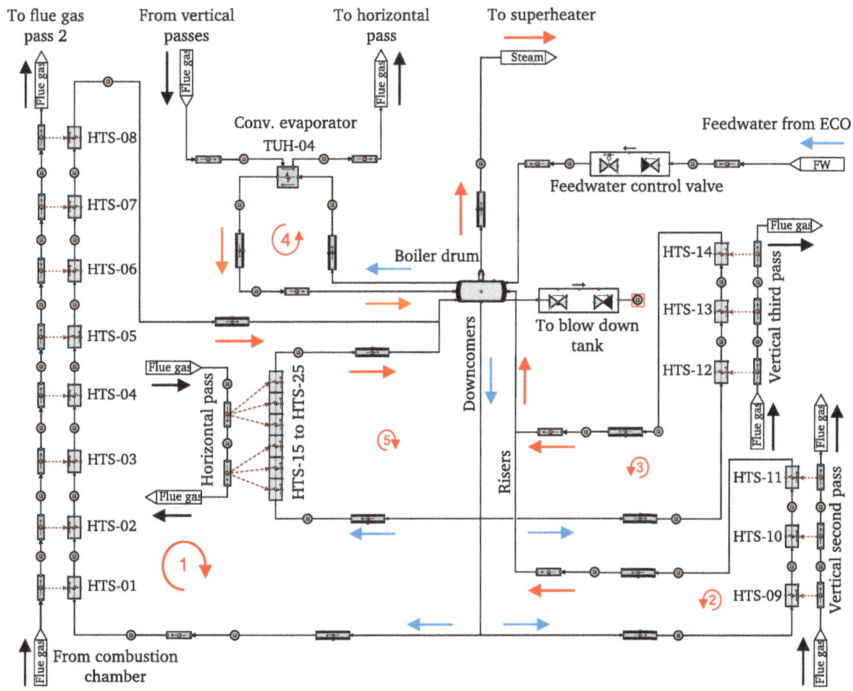

Fig. 4.79 The evaporator system with its five natural circulations

downcomers, risers and a boiler drum. The latter represents the core element of the evaporator system and has a variety of tasks. It is used as a connection between the downcomers and the risers, enabling the circulation of the working fluid through the evaporators. Furthermore, the drum separates steam from the water/steam mixture by force of gravity using the density difference between gaseous and liquid phases. The feedwater is fed to the boiler drum through economisers with a certain degree of the sub-cooling and mixed with the saturated water in the drum. A control valve regulates the feedwater mass flow rate to the boiler drum in order to maintain a certain water level in the drum. The saturated water is then directed through the downcomers to the four membrane wall evaporators and the convective evaporator, where the evaporation takes place. The first natural circulation that is located at the walls of the first vertical flue gas pass is the largest one and produces most of the steam. The second and third natural circulations are located in the second and third flue gas passes' walls. The fourth natural circulation is located at the beginning of the horizontal flue gas pass and the fifth natural circulation is installed at the walls of the horizontal flue gas pass. The resulted water/steam mixture streams back into the boiler drum via risers. While the liquid stays in the boiler drum and mixes with water coming from the economisers, the steam leaves the drum through a steam dryer that enhances the removal of liquid droplets entrained in the vapour stream. The content of salts in the pre-warmed feedwater cannot leave the drum together

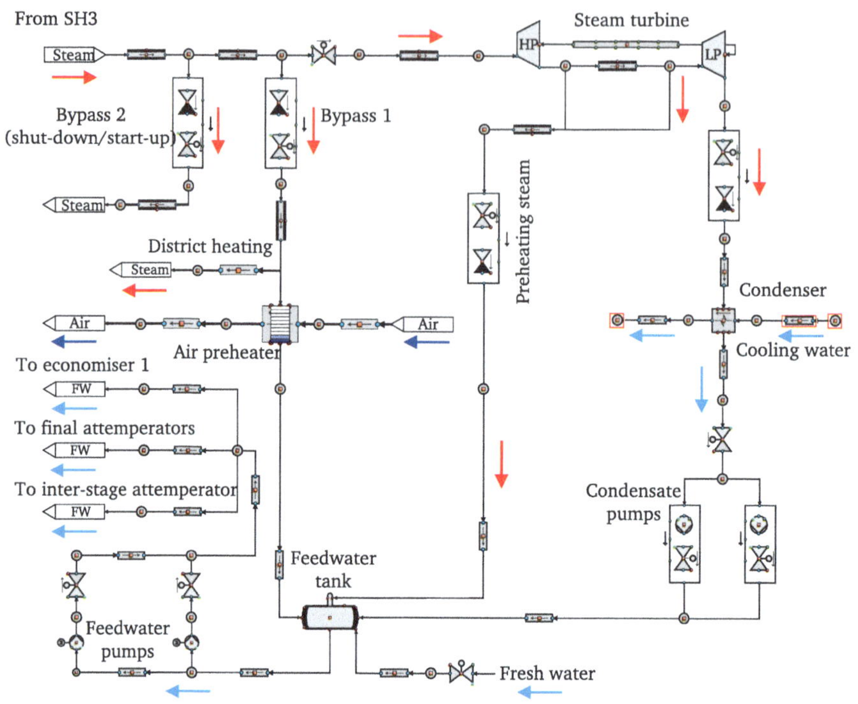

Fig. 4.80 The steam turbine, condenser and feedwater supply system

with the steam and remains in the evaporator system. Accordingly, a specific amount of water is constantly discharged from the drum to the boiler blowdown tank, limiting the concentration of salts in the evaporation system.

The steam turbine, the bypass systems, the condenser, the feedwater tank and the boiler feedwater pumps are illustrated in Fig. 4.80. The 19.6 kg/s superheated steam enter the high-pressure steam turbine at 400 °C and 45 bar, where it expands to values of 5 bar and 160 °C. In the low-pressure turbine, the steam expands further to about 0.06 bar, which is a common value in wet-cooling condenser systems; even 0.01 bar can be achieved, leading to higher power outputs in the low-pressure steam turbine. Steam is extracted between the high-pressure and low-pressure turbine for preheating the water in the feedwater tank. Two bypass systems (top part of Fig. 4.80) exist. The first bypass system is in service at normal operation of the incinerator, providing steam for preheating the primary air and the secondary air as well as for supplying steam to the district heating network. The second bypass system is in operation during malfunction and shutdown/start-up procedures. The condenser is installed downstream of the low-pressure turbine. In the condenser, the fully expanded steam condenses to water that is collected in the condenser tank. Using condensate pumps, the condensate flows into the feedwater tank at a pressure of 5 bar. The latter supplies the water to the economisers via boiler feedwater pumps and serves also as a storage tank of water, regardless of the condensate in the condenser tank. As a rule, the feedwater storage capacity of the feedwater tank is designed so that the steam can still be generated for a few minutes at full load. Furthermore, preheating and deaeration of feedwater are additional tasks of the feedwater tank, so that the feedwater tank is frequently referred to as a mixing preheater. These additional tasks are realised via steam extraction from the steam turbine with parameters of 2.64 kg/s, 160 °C and 5 bar. The two boiler feedwater pumps are installed parallel for redundant and reliable supply of feedwater to the boiler, even if one of these pumps fails. The feedwater at the outlet of the boiler feedwater pumps with the parameters of 21.27 kg/s, 140 °C and 55 bar is divided into three parts. The main part is used to supply feedwater to the boiler, while the second and third parts supply water to attemperators that are installed at superheater surfaces to control the steam temperature at the inlet of the high-pressure steam turbine.

The control concept of the waste incineration power plant (e.g. drum level control circuit, attemperator control circuit and auxiliary burner control circuit) is similar to previous sections. Only, the control structure of the steam bypass system is quite different. The steam turbine bypass system with its control circuits is illustrated in Fig. 4.81. Steam coming from the last superheater (SH3) can flow into three paths, namely the steam turbine, the air preheater and district heating network (bypass 1) and the condenser (bypass 2). Each path is equipped with a control valve that adjusts the mass flow rate and the pressure.

The working principle is described as follows:

- At normal operation, the control valve of the second bypass (BPCV2) is closed, while the main steam to the turbine valve (MSCV) is open and the mass flow rate to the air preheater and district heating network is controlled by (BPCV1), so that the difference between the total steam mass flow rate and the steam mass flow rate to the turbine (setpoint is 19.6 kg/s) should be bypassed. In the case of increased demand in district heating, the setpoint value can be lowered, which means that the steam turbine with its generator produces less electrical power.
- In case of a steam turbine trip, the main steam to turbine valve closes immediately and the steam mass flow rate is redirected into the second bypass via BPCV2. A pressure regulation is necessary, which maintains the pressure in the boiler to reach the pressure level that was existent before the steam turbine trip happened. Using a memory component, the pressure is stored when the control concept is switched on and re-used as a setpoint for the pressure control. Holding the pressure at a high level has the advantage that the plant is already prepared for a hot restart. The control valve of the first bypass (BPCV1) works similar to the case of normal operation.
- During the shutdown procedure, the MSCV and BPCV1 are closed, while the BPCV2 controls the boiler pressure at its standing idle setpoint (here equal to 10 bar in case of shutdown/hot-start-up procedure). Holding the pressure during the shutdown process has the advantage that the plant can reach its nominal operation values faster during the following start-up procedure.

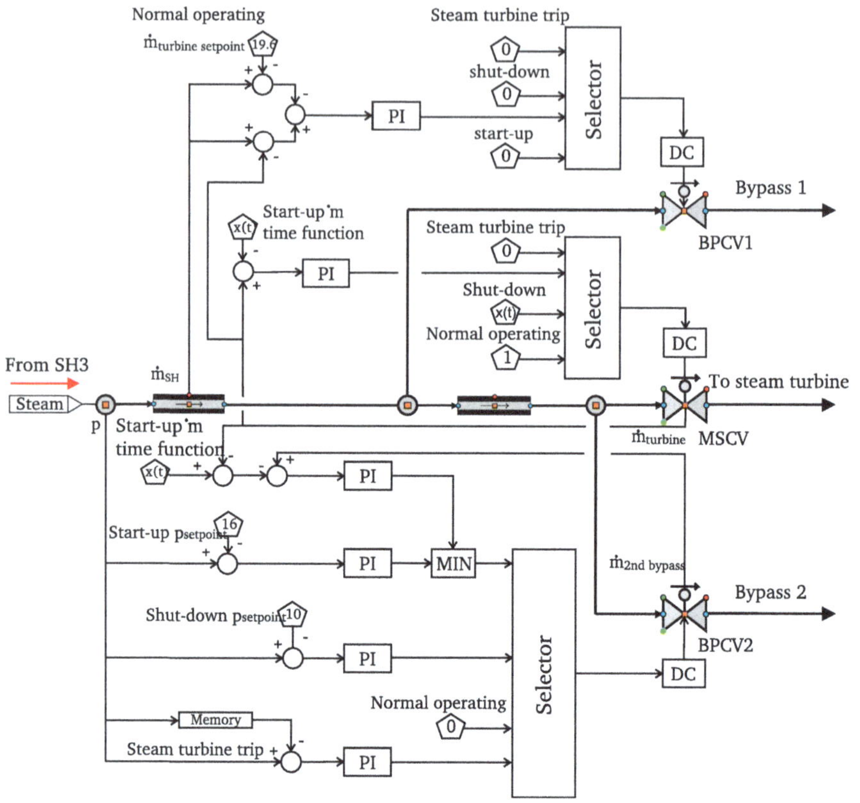

Fig. 4.81 Bypass system with its control circuits

- At the start-up procedure, the flue gas mass flow rate and the temperature increase over a certain period of time. During the first stage of the start-up, all control valves (MSCV, BPCV1 and BPCV2) are initially closed. This procedure enables the pressure in the boiler to increase rapidly. When the fixed pressure setpoint (here equal to 16 bar) is met, the second bypass valve starts opening to counteract further increase in pressure. Due to the continuous pressure rise caused by an increase in load, BPCV2 is opened forward. Once MSCV begins opening, BPCV2 closes with the same ratio. BPCV1 is in operation as soon as the steam mass flow rate to the steam turbine reaches the value of 19.6 kg/s.

For model validation, the numerical results obtained from the developed model are compared with the design data at nominal load operation. The temperature profile (design data and simulation) within the vertical passes of the waste incineration power plant at full load operation is displayed in Fig. 4.82. The x-axis represents the flue gas temperature and the y-axis is the height above the ground. The combustion temperature on the grate (approximately 1350 °C) is reduced to

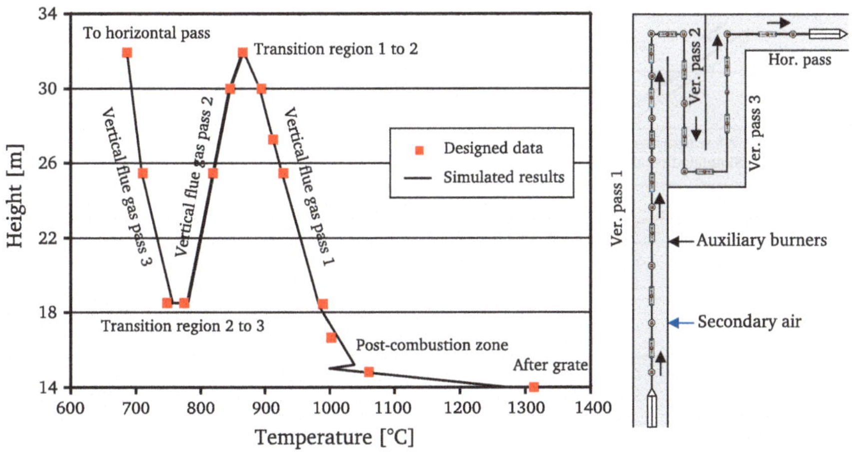

Fig. 4.82 Temperature profiles in the vertical flue gas passes at nominal load operation, comparison between simulated results and designed data

about 1271 °C at the height of 14 m due to the heat flow from the flue gas to the combustion chamber walls (water/steam side). In the height range between h = 14 m and h = 15 m, the flue gas temperature decreases rapidly by the reason of secondary air injection and the high heat flow into the evaporators HTS01 and HTS02. From h = 16 m, the heat flow to the membrane wall heat exchangers is slowed down in order to maintain the requirement of minimum temperature. The flue gas temperatures at the inlet of the second and third vertical flue gas passes are about 850 °C and 750 °C, respectively. The simulated flue gas temperature along the vertical flue gas passes shows good agreement with design data. However, a slight difference of about 50 °C at the post-combustion zone can be observed. The reason of this deviation is that the 1D process simulation model enables an ideal mixing between the secondary air and the flue gas coming from the grate, which results in rapid decrease in the flue gas temperature. In the real plant, the secondary air nozzles cannot establish a sufficient penetration depth of the secondary air into the flue gas path (i.e. a complete mixing is not possible). The 36.3 kg/s flue gas mass flow rate, including the primary and secondary air has an average velocity of 5 m/s in the vertical flue gas passes and maximum velocity of about 6 m/s at the narrowest cross-section area of the second flue gas pass. The pressure loss in the vertical flue gas passes is very small, while it amounts to 0.05 mbar in the superheaters and 0.3 mbar in the economisers, caused by the flow around the tube bundles heat exchangers.

The temperature profiles of the flue gas side path and the water/steam side in the horizontal flue gas pass are illustrated in Fig. 4.83. The location of the heat exchangers is displayed on the x-axis and the temperature is presented on the y-axis. The flue gas enters the horizontal flue gas pass with a temperature of 680 °C and exits with temperature of 160 °C as well as an average velocity of 2 m/s.

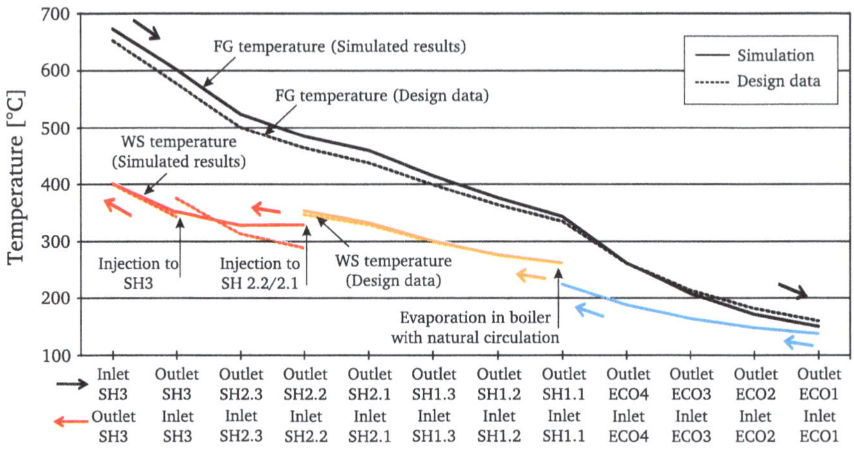

Fig. 4.83 Flue gas and water/steam temperature profiles in the horizontal flue gas pass, comparison between simulated results and designed data

The transferred heat flow to economisers, superheaters and all natural circulation evaporators are 7.8 and 13.5 and 27.2 MW, respectively. This results in a total rate of heat flow from the flue gas path to the water/steam side of 48.5 MW, while the remaining heat (approximately 11.5 MW) cannot be recovered and thus is dissipated into the atmosphere.

The incinerator model analyses with the design data at base load have been performed. Finally, the start-up simulation with the validated model can be carried out. The behaviour of the relevant parameters at the last high temperature superheater during the hot start-up procedure is presented in Fig. 4.84-(left) (steam mass flow rate (black line), pressure (green line) and steam temperature (red line)). The pressure starts increasing gradually from its standing idle pressure (10 bar) at t = 30 min to its fixed pressure setpoint of 16 bar within 15 min. In the period of time between t = 45 min and t = 110 min, the pressure is kept constant equal to 16 bar. Starting from t = 110 min, the pressure starts increasing to reach the design

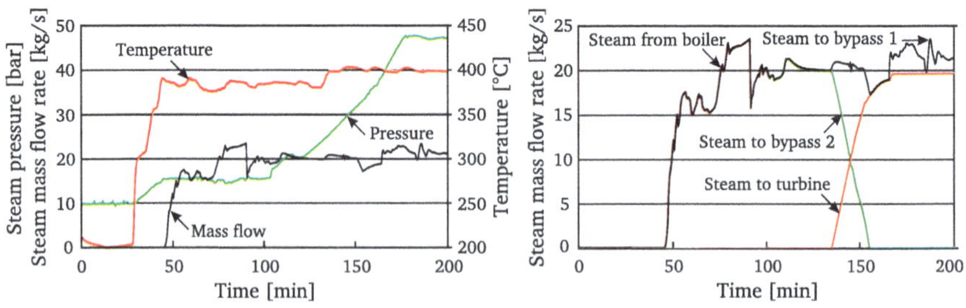

Fig. 4.84 Dynamic behaviour of the parameters during the hot start-up procedure at: (left) the last superheater and (right) the bypass system

value of 45 bar at t = 170 min. The steam temperature is around 200 °C during the first 30 min of the start-up. Then, it rises extremely from 200 °C to 390 °C within 20 min with a temperature gradient of circa 10 °C/min. After approximately 50 min from the start-up, the steam mass flow starts increasing. It requires approximately 60 min to achieve its nominal value. From t = 135 min, the second bypass control valve starts closing and the control valve to steam turbine begins opening. The oscillations in the HP steam mass flow result from auxiliary burners and switching over from BPCV2 to MSCV.

In Fig. 4.84-(right), the dynamic behaviour of the bypass system is illustrated (steam mass flow rate (black line), steam mass flow to bypass2 (green line) and steam mass flow to turbine (red line)). The main steam and the bypass valves are initially closed before the start-up in order to enable the fast increasing in the high-pressure (see Fig. 4.84-(left)). From point of time t = 45 min, the pressure reaches its fixed pressure setpoint and the BPCV2 starts opening. In sequence for this, the steam mass flow rate increases sharply. In the period of time between t = 45 min and t = 135 min, the steam streams into the condenser after it is cooled down by a desuperheater. At t = 135 min, the MSCV starts opening and the BPCV2 starts closing in the same ratio. After 155 min from the start-up, the BPCV2 is completely closed and the MSCV is completely opened. Here, it should be mentioned that the MSCV is commanded by a time function depending on the steam turbine start-up requirements. From t = 165 min, a constant steam mass flow rate of 19.6 kg/s flows into the steam turbine, while the remaining steam is directed to the first bypass (BPCV1) for the air preheating as well as for the district heating network.

4.1.4 Concentrated Solar Power Plant

In concentrated solar power (CSP) plants, solar rays are concentrated by means of mirrors, lenses or a combination of both to heat a working fluid, which then directly or indirectly drives a thermodynamic process in order to generate electric power. CSP technologies are divided into two main techniques, namely point-concentrating technology (parabolic dish and solar tower) and line-concentrating technology (parabolic trough and Fresnel reflector). Concentrating solar power plants are showing increasing interest in field of research and application, mostly as parabolic trough collectors and solar tower collectors. This is due to the fact that the CSP technology can easily be coupled with thermal energy storage and with fossil fuel combustion system, increasing the plant availability, especially during low radiation periods. However, the daily and monthly variation of the solar radiation is a main drawback. At the end of 2013, the total installed CSP capacity amounted to 3.6 GW. Spain leads the world in solar thermal energy with 2.3 GW of the cumulative CSP capacity, representing 2% of annual generated electricity. Generally, CSP plants consist of a solar field, an energy storage system (optional)

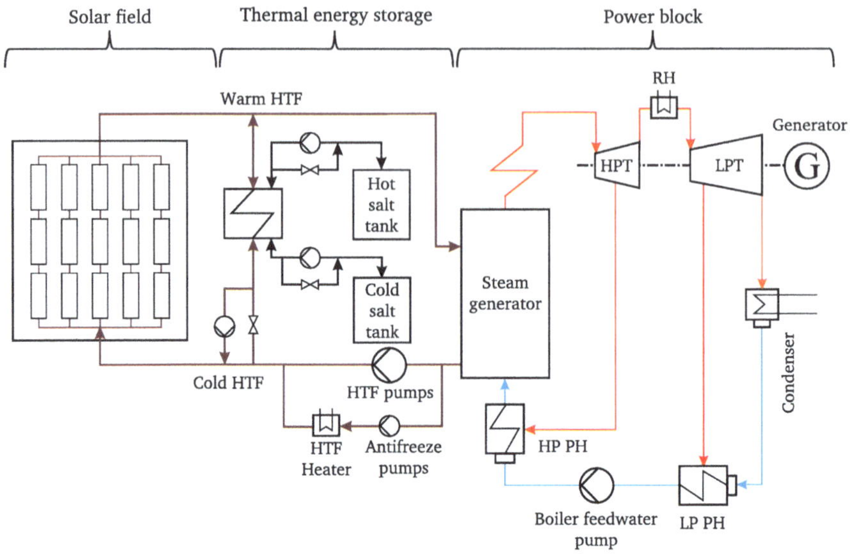

Fig. 4.85 Setup of a concentrated solar power plant with a thermal storage

and a power block (see Fig. 4.85). The overall efficiency of a concentrated solar thermal power plant is determined by the optical efficiency of the reflector, the efficiency of the receiver (absorber tube), the thermal losses during fluid transport, thermal energy storage efficiency (optional) and the efficiency of energy conversion in the power block. As an example, the dynamic simulation model of Andasol II parabolic trough power plant (50 MW_{el}) is described below.

4.1.4.1 Characterisation of Parabolic Trough Power Plant

The investigated 50 MW_{el} parabolic trough solar thermal power plant, located in Spain is designed by TSK Flagsol GmbH; a company specialised in engineering and construction of concentrated solar power plants. The plant consists of three main subsystems: the solar field, the power block and the thermal storage system in between. The solar field has 156 parallel loops of parabolic troughs. Each loop contains 4 solar collector assemblies (SCA), connected in series. One SCA consists of 12 solar collector elements (SCE). A single SCE is made of highly reflective parabolic mirrors with absorber tubes or heat collection elements (HCE) installed at the focal line of the parabolic surface. The heat transfer fluid (HTF) that flows through the heat collection elements in the solar field and heat exchangers in the power block is "Therminol VP-1". The power block consists of a steam generator, high-pressure feedwater preheaters, low-pressure feedwater preheaters, steam turbines and an electricity generator, a condenser, cooling systems (air-cooler condenser and cooling tower), auxiliary equipment and the related control structures. The steam generator has two heat exchanger trains, including economisers, evaporators, superheaters and reheaters. The water/steam side of the steam generator is a single-pressure system and includes a reheater section after the high-pressure turbine, in detail as follows:

- Forced circulation evaporator path in high-pressure circuit
- Reheater section
- High and low-pressure steam turbine stages
- Condenser, low and high-pressure feedwater preheaters and storage tank

The absorbed thermal energy in the solar field is transferred by the HTF to the water/steam side in the power block, producing steam at high parameters of approximately 384 °C and 106 bar. The superheated steam enters the high-pressure steam turbine, where it partly expands and then is reheated to approximately 383 °C at an intermediate-pressure level of about 19.4 bar, before entering the low-pressure steam turbine. Two attemperators limit the temperature of the steam supplied to the HP and LP sections of the steam turbine, respectively. The fully expanded steam leaves the turbine and enters the condenser, where it condenses and is pumped into the low-pressure feedwater preheaters with the aid of condenser pumps. The water exits the low-pressure feedwater preheaters towards the feedwater storage tank and then is pumped by the boiler feedwater pumps (BFP) to the high-pressure feedwater preheaters. Finally, the preheated feedwater enters the steam generator. During summer months, the power plant typically runs for 10–12 h a day on solar energy at full-rated electric output. The plant is also equipped with thermal energy storage, achieving rated electric output even during overcast or night periods. Furthermore, the thermal energy storage can be used to supplement the solar output during periods of low solar radiation. The storage system consists of two tanks of molten salt. The HTF passes through a heat exchanger that is used to charge/discharge the thermal energy storage system. During charge phases, the molten salt flows from the cold storage tank through the heat exchanger and enters the hot salt storage tank with a temperature of about 386 °C. When the storage system is discharged, the molten salt from the hot storage tank is sent back to cold storage tank through the heat exchanger that is used to heat the cold heat transfer fluid. The temperature of the cold molten salt is about 292 °C. Table 4.13 lists the characteristic technical data of Andasol II parabolic trough solar thermal power plant, provided at 100% nominal load.

Table 4.13 Technical data of Andasol II Power Station located in Spain

Andasol II Power Station			HTF	ST power
High-pressure system	Reheater section	Condenser inlet	$\dot{m}$ = 1600 (kg/s) T_{inl} = 393 (°C) T_{out} = 293 (°C)	50 (MW_{el})
p = 106 (bar) T = 384 (°C) $\dot{m}$ = 54 (kg/s)	p = 19 (bar) T = 383 (°C) $\dot{m}$ = 44 (kg/s)	p = 43 (mbar) T = 30 (°C) $\dot{m}$ = 34.4 (kg/s)		

4.1.4.2 Parabolic Trough Power Plant Model

A full-scale dynamic model of Andasol II parabolic trough power plant is developed using the process simulation software APROS.

The model is generated by connecting components with real geometry data and given specifications, supplied by TSK Flagsol GmbH. Process components used in the model include headers, heat exchangers, drums, pumps, turbines, pipes, valves, solar radiation modules and heat storage tanks, as well as the associated control structures and electrical systems. The developed model contains three main subsystems: the solar field (see Fig. 4.86), the thermal storage system and the power block (see Fig. 4.87).

Sets of operation data that were recorded at a time interval of about ten minutes in different months in 2010 are applied to the model validation, i.e. during clear days with little scattered clouds and days with strong cloudy periods (García et al. 2011). According to the comparison between the measured data and the simulated results (Al-Maliki et al. 2016a, b), the following points can be concluded:

- During clear days and slightly cloudy periods (on 26th till 27th June 2010 (see Fig. 4.88a) and on 13th till 14th July 2010 (results are not presented here)), the simulation results agree quantitatively with the measured data.

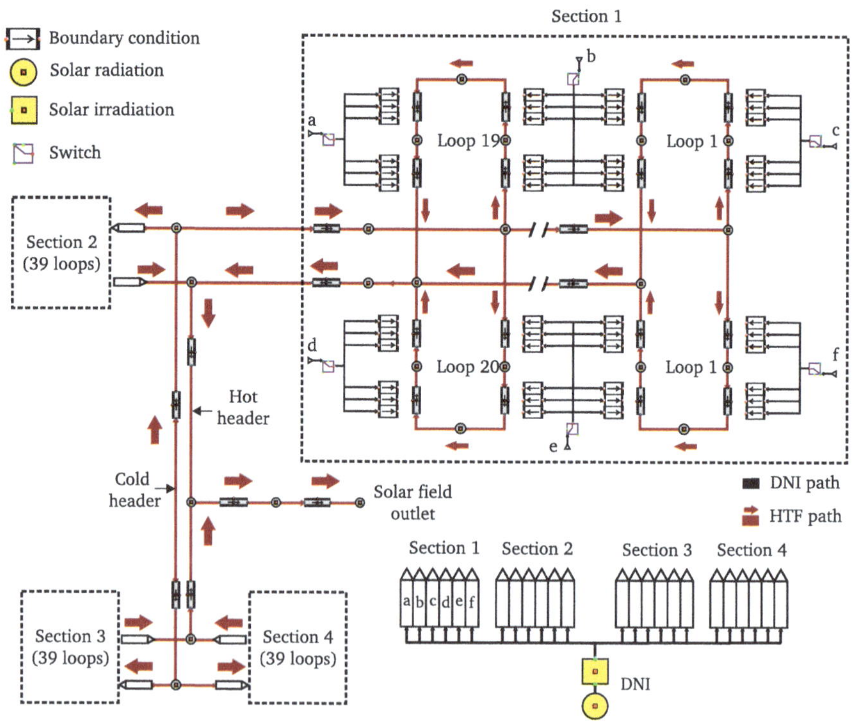

Fig. 4.86 The solar field model of the Andasol II parabolic trough power plant, located in Spain

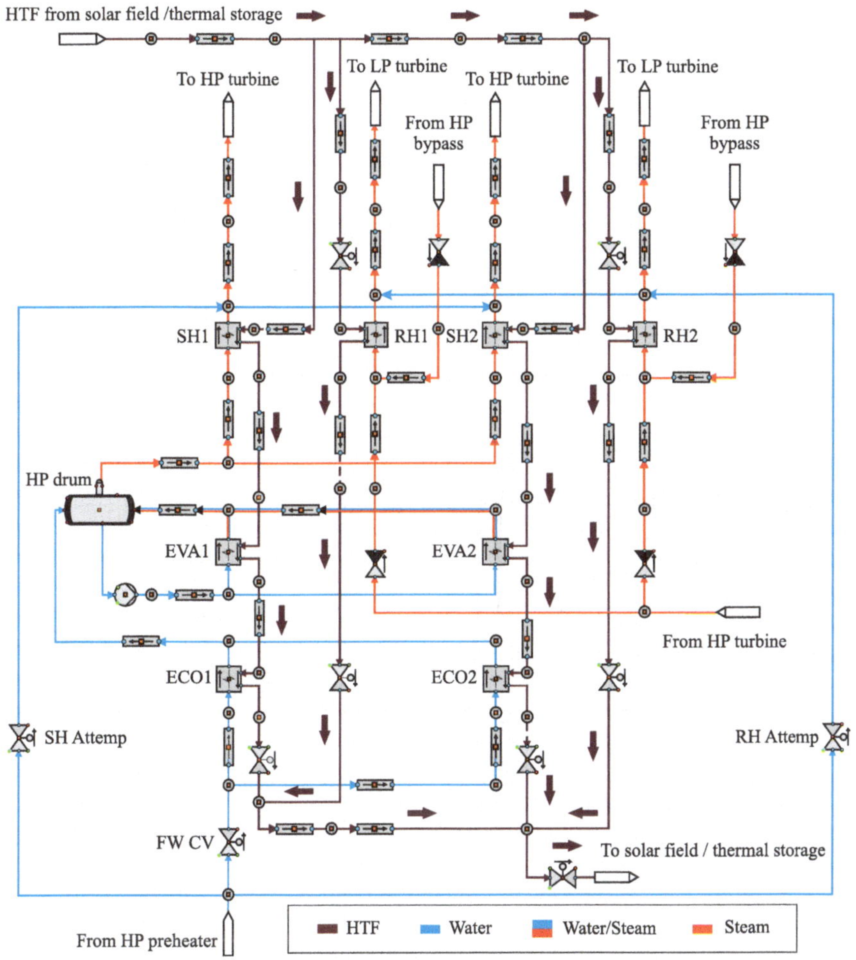

Fig. 4.87 The power block model of the Andasol II parabolic trough power plant, located in Spain

- During sunny days and after the sunset, the thermal energy storage system enables the power block to generate live steam, achieving a constant electrical power of 48 MW_{el} for approximately 7.5 h until the energy storage system is completely discharged.
- For cloudy days (e.g. 29 June 2010, see Fig. 4.88b), discrepancies can be observed. The main reason here is the unknown decisions from the operator of the power plant, e.g. for some periods, the HTF mass flow rate is redirected to the thermal storage system instead of the power block or it is completely stopped to flow within the solar field despite the fact that the solar radiation is again increased after a strongly cloudy period.

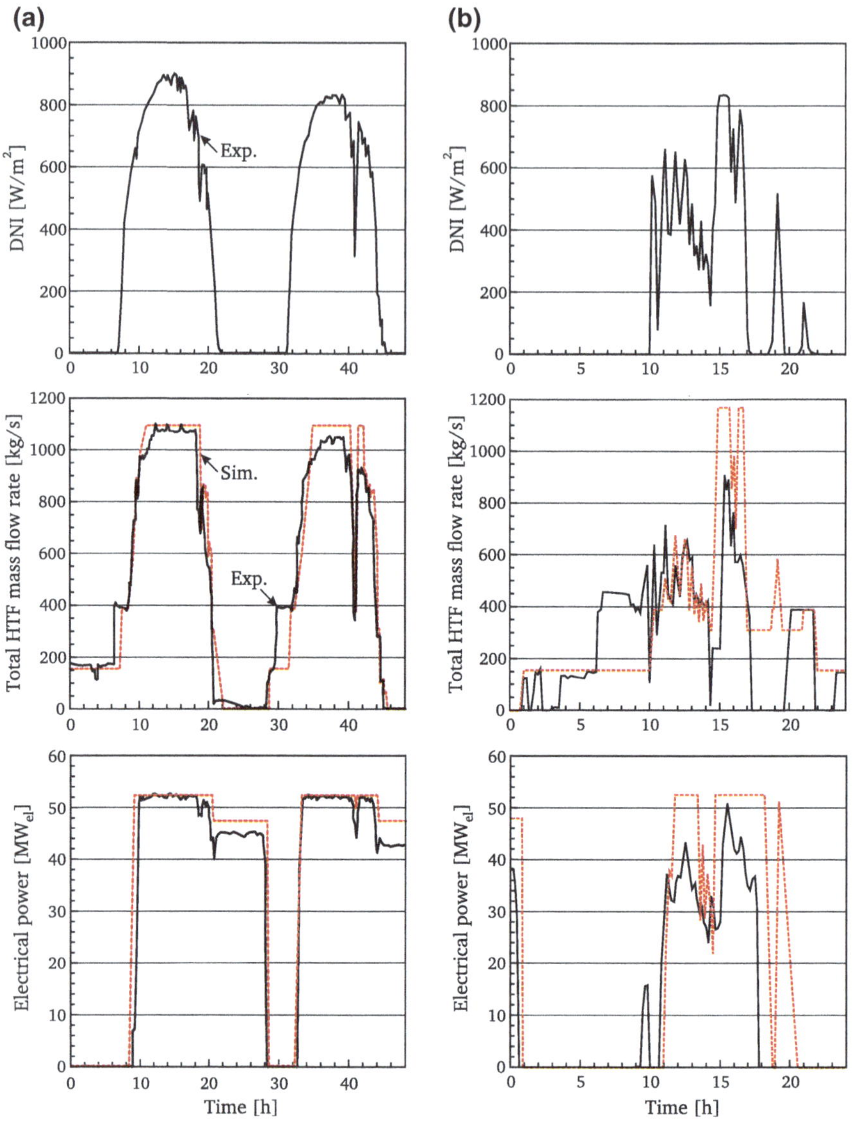

Fig. 4.88 Validation of Andasol II model: **a** during slightly day and **b** during strong cloudy day

4.2 CFD Simulation

Computational fluid dynamics can be applied to energy systems at all stages of the plant project, including design, development, commissioning and optimisation. Some advanced applications of CFD models, based on quasi-single-phase, two-fluid model and single-particle model, are presented here to various test rigs and large-scale power plants including pulverised coal-fired power plant, co-firing of biomass in coal-fired power plant, municipal waste incineration and fluidized bed systems for CO_2 capture methods. A wide set of measurements and operation data are used for the models' validations.

4.2.1 *Quasi-single-phase Model*

The quasi-single-phase model offers an efficient approach for the simulation of the thermal power plant components such as heat exchangers, combustion chambers and flue gas cleaning devices. The pollutant formation (e.g. NO_x and SO_x emissions) can be predicted under several measures for different solid fuels. Based on quasi-single-phase model, several combustion models for coal, biomass and municipal waste are developed and validated with the measurements obtained from an entrained-flow reactor, a 1 MW_{th} test facility and large-scale thermal power plants, e.g. municipal solid waste incinerator, lignite and hard coal-fired power plants.

4.2.1.1 Entrained-Flow Reactor

With the aid of an entrained-flow reactor that is erected at Technische Universität Darmstadt, pyrolysis, char burn-out and formation of corrosive gases during the combustion of solid fuels can be investigated. The central component of this reactor (combustion chamber) is a ceramic tube with an inner diameter of 70 mm and a length of 2.2 m (see Fig. 4.89). The entrained-flow reactor consists of a preheating step for the reaction gas, a reaction zone and a water-cooled quenching section. It can be operated at temperature and pressure ranges between 900 °C and 1600 °C and 1–20 bar, respectively. The preheating stage and reaction zone are heated-up by electrical heating elements that are assembled around the outer wall of the reactor to maintain the required wall temperature during operation. Fuel together with primary air is fed into the reaction tube through a water-cooled injector from the top, where the air acts as fuel carrier gas. Preheated secondary air is provided to the combustion chamber, where it is blended with primary air and fuel. After entering the reaction zone, the fuel conversion processes (pyrolysis, homogeneous reactions and char combustion) will take place, besides the chemical reactions of chlorine and sulphur-containing species.

The concentrations of main species such as CO, CO_2, CH_4, SO_2, and O_2 can be measured through several connections on the sidewall of the entrained-flow reactor. Different corrosive substances as for example HCI, H_2S and COS can also be determined through a combination of gas chromatography features with the mass spectrometry technique. Solid samples may be extracted from the reactor using the cooled movable lance at various points in the reaction zone. The sudden cooling of

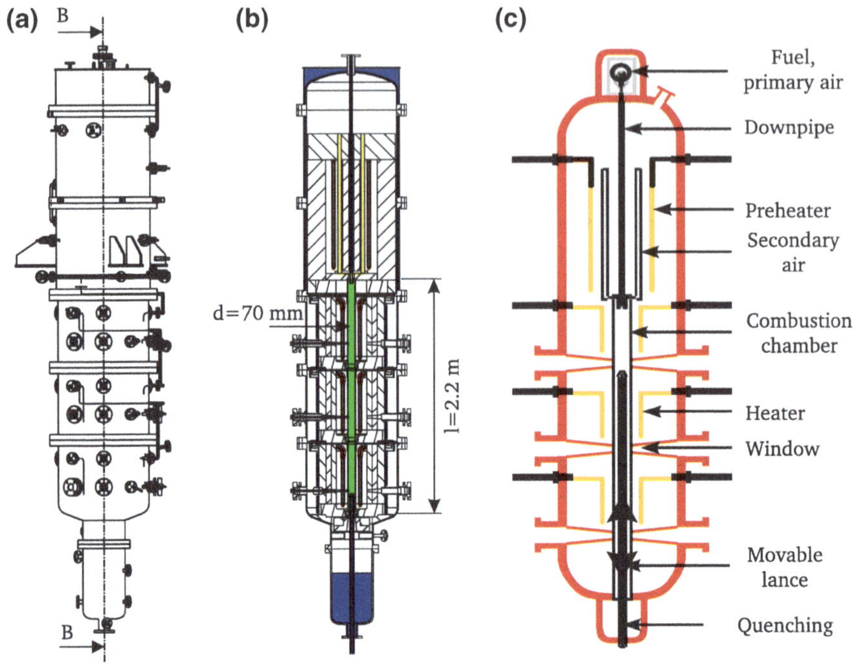

Fig. 4.89 Structure of the entrained-flow reactor: **a** 3D representation, **b** detailed cross section and **c** simplified cross section

the movable lance prevents further conversion of particles and therefore the burn-out behaviour of the solid fuel can be examined at pre-defined reaction time.

In Fig. 4.90, the 2D numerical grid of the top part of the entrained-flow reactor is illustrated.

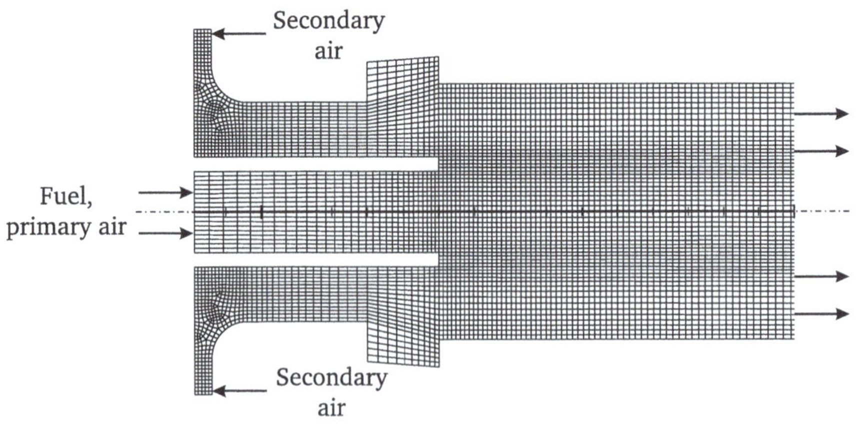

Fig. 4.90 Numerical grid of the entrained-flow reactor (top part is only illustrated)

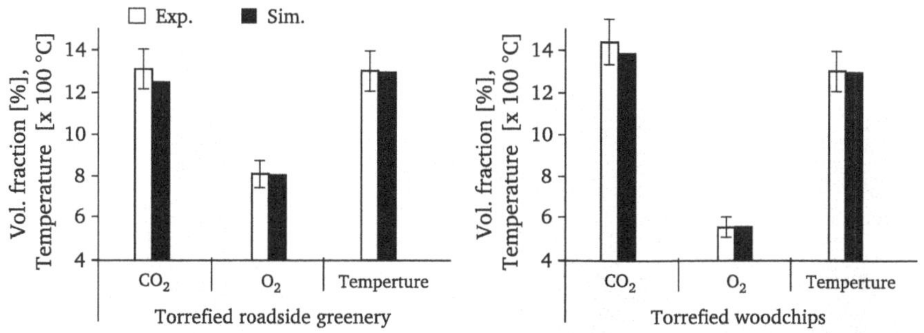

Fig. 4.91 Experimental data and numerical predictions for torrefied roadside greenery and torrefied woodchip at a height of approximately 1.8 m

Different developed combustion models based on quasi-single-phase model for coal, biomass as well as fuel blends are validated towards the measurement obtained from the entrained-flow reactor. For example, the measured data and the numerical results of temperature and volume fractions (CO_2 and O_2) are presented for torrefied roadside greenery and woodchip. In torrefaction process, the raw biomass is heated-up in the absence of oxygen to a temperature range of 200–400 ° C for a residence time of 15–90 min. Water and low calorific gaseous species are released and thus the weight related calorific value increases by approximately factor of 1.3. Torrefied biomass is a solid brittle product, which has a number of positive characteristics in comparison to raw biomass. Most relevant advantages include improved grindability and less capacity needed for transport and storage.

From Fig. 4.91, it can be observed that the developed combustion models can predict the volume fractions of CO_2 and O_2 with good accuracy (a maximum relative error of about 5%). Furthermore, the simulated temperature shows good agreement towards the measured one. The systematic measuring error is represented by error bars. Detail information on the combustion models and their applications to different solid fuels can be found in (Alobaid et al. 2012a; Epple et al. 2014).

4.2.1.2 1 MW_{th} Pilot Combustion Chamber

The 1 MW_{th} test facility at Technische Universität Darmstadt has an inner diameter of about 1 m and a height of 7 m. The combustion chamber is equipped with a top swirl burner and has a two-pass design. While the top part of the combustion chamber is constructed as regular hexagon and cooled by an evaporator tube heat exchanger, the bottom part has a cylindrical shape (see Fig. 4.92).

A schematic process diagram of the 1 MW_{th} test facility is illustrated in Fig. 4.93. The concentrations of the main species O_2, CO, CO_2, and CH_4 in addition to NO_x and SO_x emissions in the exhaust gas can be determined by a gas measuring system installed permanently. With the help of several connections on the sidewall of the furnace at different positions and the application of various

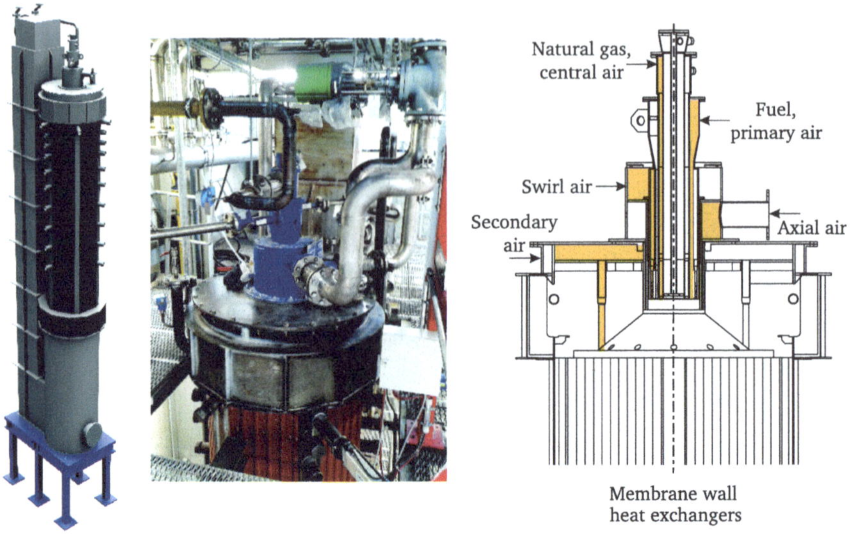

Fig. 4.92 1 MW_{th} combustion chamber facility

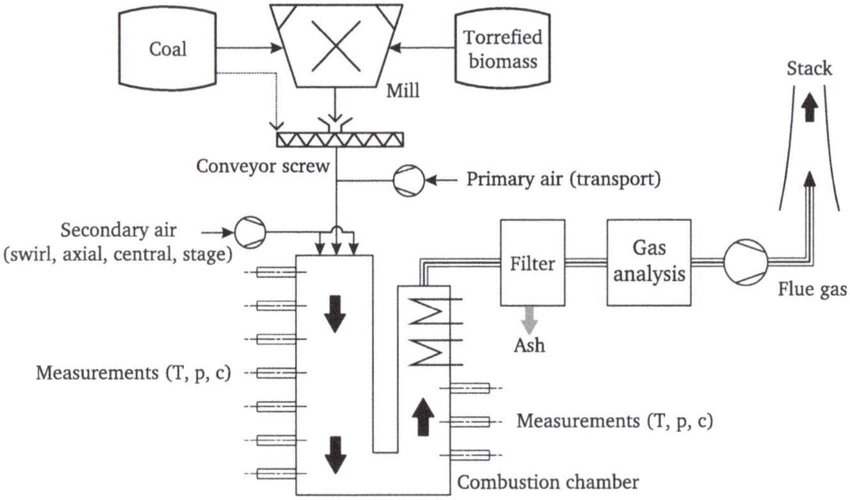

Fig. 4.93 Schematic flow diagram of the 1 MW_{th} pilot facility

measuring apertures, additional flue gas samples may be extracted and analysed. Furthermore, the solid samples at various points along the combustion chamber can also be studied for their mineral composition. Different solid fuels are fired in the 1 MW_{th} test facility, e.g. pure coal and torrefied biomass in coal (Stroh et al. 2015).

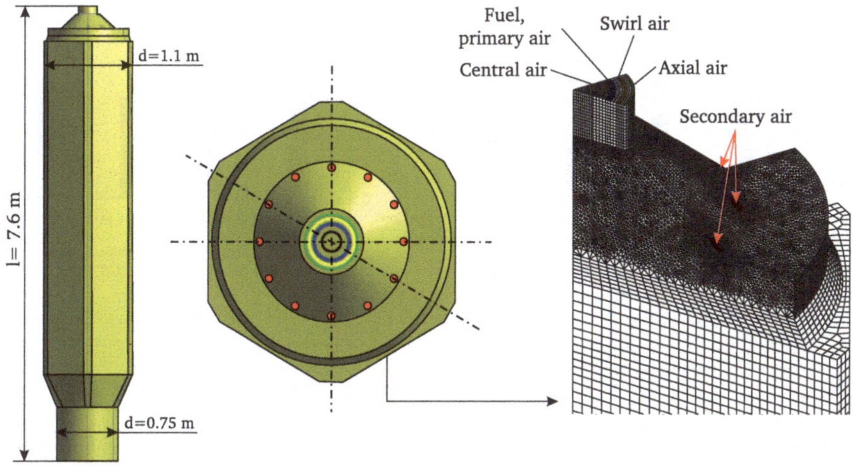

Fig. 4.94 Geometry and numerical grid of the test facility of the 1 MW_{th} test facility (top part is only illustrated)

The air inlets for primary (transport air) and secondary air (stage air, axial air, swirl air, central air) together with the top part of numerical grid of the 1 MW_{th} test facility are presented in Fig. 4.94.

The actual full-scale geometry is reduced to one sixth in order to reduce the computational time and to improve the convergence behaviour. The applied mesh has approximately 400,000 structured hexahedral and unstructured tetrahedral cells; however, the largest cell density is located at the inlet and the combustion zone. Several combustion models for solid fuels are developed and validated with the measurement data obtained from the 1 MW_{th} test facility. Here, the comparison between the coal/biomass combustion model and the test facility is presented for 9.14% torrefied sawdust in coal (thermal basis), as an example.

A highly volatile hard coal with a lower heating value of 26.5 MJ/kg, which has the origin in Columbia from the mine Calenturitas is selected due to its good combustion behaviour. Such types of coals are also often used in German power plants that also favoured the decision for this coal. The torrefied sawdust has a lower heating value of about 20.6 MJ/kg. The experimental data is measured at two different reactor heights, namely z = 1.073 m as well as z = 1.337 m and two different radial positions r = 0.1 m as well as r = 0.2 m from the reactor wall, respectively. The measurements and predictions are shown Fig. 4.95.

The numerical results of volume fractions of CO_2 and O_2 agree well with the experimental data. The average deviation for carbon dioxide for all measured points amounts to maximum 5%. The agreement for the oxygen concentration is satisfactory for the radial position at 0.1 m from the wall, while the agreement at 0.2 m is not precise. The measurement of the gas concentrations is afflicted with errors, which depend on the range of the measured gas concentration. The systematic measuring error is indicated by error bars. It is necessary to point out that the

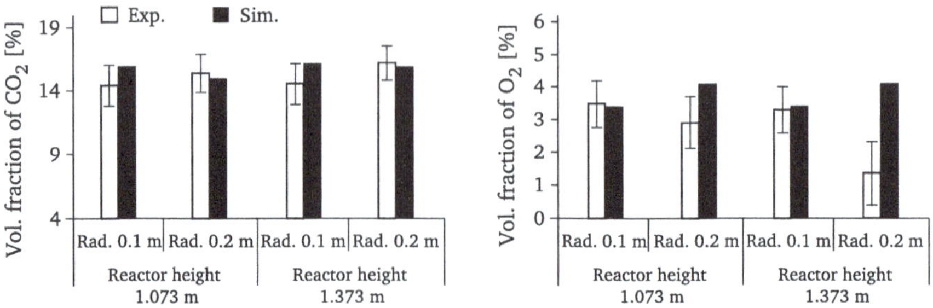

Fig. 4.95 Experimental data and numerical predictions for 9.14% torrefied sawdust in coal (experimental errors and uncertainty are bigger for low gas concentrations because of the wide range setting of the measuring devices)

systematic error is bigger for low gas concentrations due to the wide measurement range in the 1 MW_{th} test facility, which was defined to 0–30 vol% in the beginning of the experiment. This explains the considerable error for oxygen at reactor height 1.373 m and radial position 0.2 m. The $k - w - sst$ turbulence model together with the eddy dissipation model (EDM) is used in order to calculate the gas species reactions. The $k - w - sst$ model combines the advantages of two turbulence models: the original Wilcox $k - w$ model that used in wall regions and the free stream independent $k - \varepsilon$ model that used in the far field. The EDM reaction mechanism has the advantage of low computational time, but the disadvantage in prediction capabilities. From the numerical point of view, it is possible to consider more sophisticated and detailed reaction mechanism for the gas species, in which the reaction rates are determined using the reaction kinetics. In such models, the sulphur and chlorine chemistry can be included in order to estimate the corrosion process in thermal power plants.

The combustion flame and the simulated temperature distribution for the fuel blend (9.14% torrefied sawdust in coal) are shown in Fig. 4.96. For illustrative purposes, only the top part of the combustion chamber is presented. The maximum

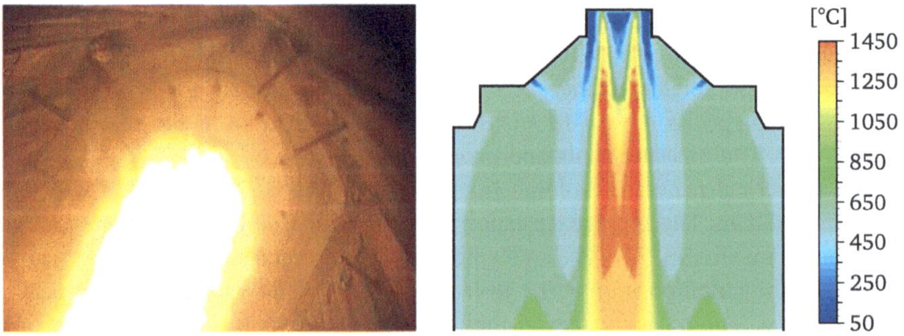

Fig. 4.96 Combustion flame in the 1 MW_{th} test facility and the simulated temperature contour for 9.14% torrefied sawdust in coal

gas temperature that is located in the vicinity of the fuel inlet reaches about 1400 °C. Due to the evaporator tube heat exchanger, the temperature decreases to about 1000 °C at a height of approximately 2 m and to about 600 °C at the outlet of the combustion chamber, respectively.

4.2.1.3 Large-Scale Power Plant

The validation of the combustion models towards the entrained-flow reactor and the 1 MW_{th} pilot combustion chamber allows reliable simulation of thermal power plants and their components, e.g. waste incineration and pulverised coal-fired power plants. Using these validated models, the fields of temperature, velocity and concentrations can be predicted with high accuracy. Furthermore, the corrosion behaviour, the mineral transformations of ash and the ash deposition on boiler walls can also be calculated for different solid fuels (e.g. hard coal, lignite coal and biomass). As an example, selected results from the simulation of municipal solid waste incinerator, lignite and hard coal-fired power plants using ANSYS-CFX and ANSYS-FLUENT are presented in the following sections.

Municipal Solid Waste Incinerator

The investigated 60 MW_{th} municipal solid waste incinerator has been built in Finland (designed by Steinmüller Babcock Environment GmbH). The incinerator burns 5.7 kg/s municipal solid waste (492 t/d) with an ideal lower heating value of 10.5 MJ/kg (see Sect. 4.1.3 for further information regarding the incinerator characterisation). The geometry and numerical grid of the incinerator are shown in Fig. 4.97. Two different numerical grid were generated, one with a coarser mesh (circa one million cells) and a second mesh with higher resolution (circa 3.5 million cells). On the post-combustion zone (the secondary combustion zone), a finer inner mesh was also generated, taking into consideration the high air injection velocity. Here, a total mass flow rate of about 12 kg/s at temperature of 230 °C is injected into the flue gas stream through 29 nozzles, distributed at both sidewalls of the first vertical pass. The primary air (19.3 kg/s) at temperature of about 150 °C is supplied via five hoppers below the grate (see Table 4.14).

High requirements in terms of wear, corrosion and fouling are placed on the wall materials of the combustion chamber and vertical and horizontal passes, so that a defined heat transfer between the flue gas and the water/steam side can take place. The heat flow from the flue gas to the membrane wall heat exchangers (evaporators) depends largely on the thermal conductivity coefficient of the wall materials. In the combustion chamber, most of the heat that is generated must be transferred to the water/steam side and thus an appropriate material is attached accordingly. Furthermore, the material has to withstand the high temperature, so that a refractory concrete or an andalusite with Al_2O_3 base is usually applied. Generally, this type of material has a low thermal conductivity coefficient and therefore the temperature of

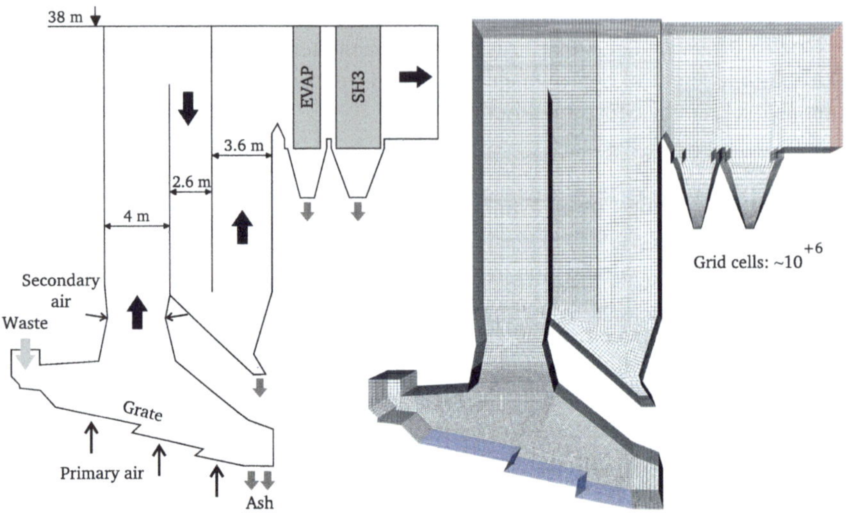

Fig. 4.97 Geometry and numerical grid of the 60 MW_{th} municipal solid waste incinerator in Finland

Table 4.14 Primary and secondary air mass flow rate distributions

Primary air					
Zone	1	2	3	4	5
Percentage (%)	9	34	26	21	10
Mass flow rate (kg/s)	1.74	6.56	5.02	4.05	1.93
Secondary air					
Nozzles (rear)	Percentage (%)	Mass flow rate (kg/s)	Nozzles (front)	Percentage (%)	Mass flow rate (kg/s)
1	2.5	0.3	1	4.3	0.52
2	2.8	0.33	2	4.1	0.49
3	3.0	0.36	3	3.8	0.46
4	3.2	0.38	4	3.6	0.43
5	3.5	0.42	5	3.3	0.39
6	3.7	0.44	6	3.0	0.36
7	4.0	0.48	7	2.8	0.33
8	4.7	0.56	8	2.8	0.34
9	4.2	0.5	9	3.0	0.36
10	3.7	0.44	10	3.3	0.39
11	3.5	0.42	11	3.6	0.43
12	3.2	0.38	12	3.9	0.47
13	2.9	0.35	13	4.1	0.49
14	2.8	0.33	14	4.5	0.54
15	2.6	0.31			

the grate is decreased by supplying a sufficient amount of primary air. Particular attention must be given to the wall materials of the vertical passes, since they have different thermal and mechanical properties. For the secondary combustion zone, silicon carbide (SiC) materials are used, which offer high resistance to mechanical and thermal stresses as well as different thermal conductivities depending on the design. The reason for using different constructions in order to achieve different thermal conductivities is the required minimum temperature of 850 °C that must be maintained over 2 s in this area of the first flue gas pass. Here, it is not allowed that too much heat is absorbed by the water/steam side in order to meet this requirement. Therefore, ventilated SiC plates are installed to reduce the thermal conductivity.

For the simulations, the grate was not modelled and the mass fractions of the substances were given over the primary combustion zones (see Table 4.15). A two-step reaction mechanism of methane combustion (methane and carbon monoxide oxidations) is used:

$$\begin{aligned} &\text{R1: } CH_4 + {}^3/_2O_2 \rightarrow CO + 2H_2O \\ &\text{R2: } CO + {}^1/_2O_2 \rightarrow CO_2 \end{aligned} \tag{4.1}$$

The reaction rates were computed by the eddy dissipation model (EDM) of Magnussen and Hjertager (Magnussen and Hjertager 1977). The $k - \varepsilon$ turbulence model was used and the discrete ordinates radiation model (DO) was applied to simulate radiation heat transfer. The heat flows over incinerator walls and convective heat exchangers were determined using the process simulation model, presented in Sect. 4.1.3. The governing equations were calculated with SIMPLE algorithm in association with a second-order upwind scheme.

In Fig. 4.98, the velocity distribution in the waste incineration is displayed. At the primary air zone, the gas velocity is relatively low (1 m/s), while a maximum value of 20 m/s can be observed in the secondary air region. In the vertical passes and the horizontal pass, the velocity is in the range between 8 m/s and 12 m/s. At the outlet, the average flue gas velocity is approximately 6 m/s. The temperature distribution of the flue gas in the waste incineration is illustrated in Fig. 4.99.

At the primary air inlets, the flue gas temperature is about 150 °C. Due to the combustion, the temperature rises and reaches a maximum value of approximately 1100 °C, before starts decreasing by the reason of membrane wall heat exchanger. From Figs. 4.98 and 4.99, one can conclude that ANSYS-CFX and

Table 4.15 Mass fractions of the substances over the primary combustion zones (Wolf 2005)

Zone	H_2O (wt%)	CO (wt%)	CO_2 (wt%)	O_2 (wt%)	CH_4 (wt%)	N_2 (wt%)
1	0.1498	0	0	0.1936	0	0.6566
2	0.1328	0.0826	0.2109	0	0.0473	0.5265
3	0.0123	0.0096	0.1503	0	0.0055	0.8224
4	0.0125	0	0.0153	0.23	0	0.7389
5	0.0126	0	0	0.2346	0	0.7529

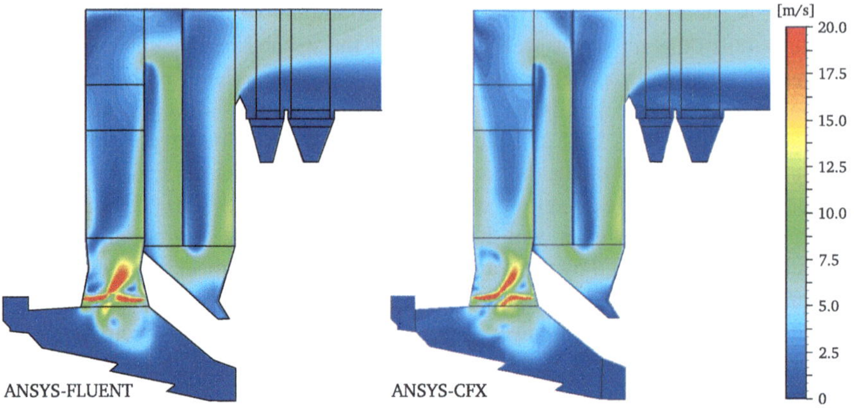

Fig. 4.98 Velocity field (longitudinal cross section of the waste incineration)

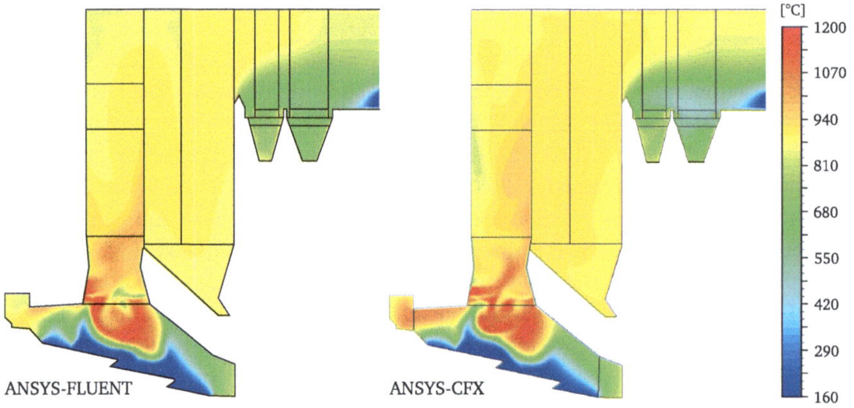

Fig. 4.99 Temperature field (longitudinal cross section of the waste incineration)

ANSYS-FLUENT show almost a similar velocity and temperature fields. It should be mentioned here that the numerical results, although the boundary conditions are completely symmetric, do not have to be symmetric. The main reason of these is the chemical fluctuations. In the real thermal power plant, these fluctuations are a measure of the combustion stability. Generally, the CFD simulations of the coal-fired power plants are computed as a half model with a symmetry plane in order to reduce the computational cost. These symmetric simulations cannot, however, give any information regarding the combustion stability. In municipal waste incinerations, variations in the waste feeding and composition are expected. The influence of the heating value changes on the combustion stability is of high relevance and thus the CFD simulations of waste incinerations should consider the entire geometry without using symmetry planes.

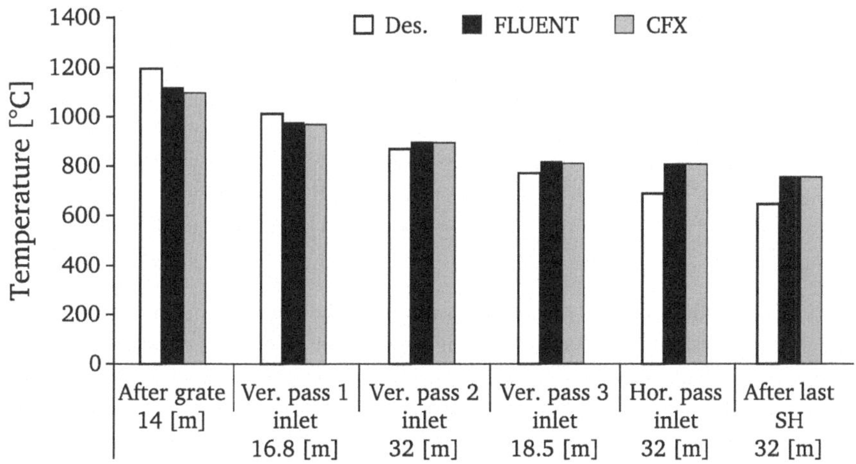

Fig. 4.100 Temperature within the vertical and horizontal passes (simulation and design data)

The temperature within the waste incineration power plant at nominal load is depicted in Fig. 4.100. Here, the numerical results obtained from ANSYS-CFX and ANSYS-FLUENT are compared with the design data, provided by Steinmüller Babcock Environment GmbH, showing a good agreement.

Lignite-Fired Power Plant

Brown coals have a very high water content and this value can reach as high as 60% in case of German origin brown coal. The raw brown coal is generally pre-dried during the grinding process using special mills (e.g. the beater wheel mill) with a huge amount of the recirculated hot flue gas at temperature of about 1000 °C. This has a major negative consequence on the plant efficiency since the flue gas with a very high temperature level is used to evaporate the water of the coal with a temperature of 100 °C. From an exergetic point of view, it is favourable to dry the brown coal using a low calorific value medium or one of the external pre-drying methods, summarised for example in (Klutz et al. 2006). The brown coal is used almost exclusively in power generation and is an important indigenous fuel source in several European countries.

The investigated 2500 MW_{th} brown coal-fired power plant (designed by ALSTOM) burns pre-dried lignite (123.6 kg/s) with a lower heating value of 19.82 MJ/kg and generates 829 kg/s of steam. The raw brown coal is treated in a preliminary process to dry the brown coal with a residual moisture content of 12% by weight. The proximate and ultimate analyses of the raw and dried lignite are presented in Table 4.16. The steam generator is designed as a single-pass boiler with 20% flue gas recirculation from the inlet of the induced-draft fan (flue gas with a low temperature). The 1258 kg/s flue gas flows in the combustion chamber that

Table 4.16 Proximate and ultimate analyses of the raw and dried lignite

Raw brown coal		Pre-dried brown coal			
C	68.19 $(wt\%)_{waf}$	C	53.24 $(wt\%)_{raw}$	fixed carbon	24.1 $(wt\%)_{raw}$
H	4.84 $(wt\%)_{waf}$	H	3.78 $(wt\%)_{raw}$	volatile	54 $(wt\%)_{raw}$
N	1.02 $(wt\%)_{waf}$	N	0.795 $(wt\%)_{raw}$	Ash	9.9 $(wt\%)_{raw}$
O	25.19 $(wt\%)_{waf}$	O	19.67 $(wt\%)_{raw}$	Moisture	12 $(wt\%)_{raw}$
S	0.76 $(wt\%)_{waf}$	S	0.596 $(wt\%)_{raw}$	LHV	19.82 (MJ/kg)

has a cross-section area of (26 m × 26 m) and a height of about 84 m. The tangential firing system consists of six firing levels with 24 swirl burners; each burner is provided with an additional wall air nozzle. The flue gas temperature at the end of the combustion chamber is of about 1030 °C and then is cooled in the convective heat exchangers to approximately 357 °C at the outlet of the economiser.

The geometry and the numerical grid of the lignite-fired power plant, including the swirl burners, the cold off-gas and the burnt-out air nozzles are presented in Fig. 4.101 (the convective heat exchangers are not illustrated).

For the simulation of the combustion chamber, the gas and solid phases are assumed to be in mechanical and thermal equilibriums (the velocities and the temperatures of both phases are locally the same). As a result, a single set of momentum and energy transport equations is solved, where the solid phase appears as a component of the mixture. The turbulence is modelled by the turbulent viscosity approach (the $k-\varepsilon$ turbulence model with standard wall functions). In the energy equation, the kinetic energy, the pressure work and viscous dissipation terms are neglected. The radiative heat transfer is modelled using the P1 radiation model with a mixture absorption coefficient as a sum of gas and particle contributions (Epple et al. 2012). The governing equations were calculated with SIMPLE algorithm in association with a first-order upwind scheme.

A global five-step reaction mechanism with three heterogeneous reactions (pyrolysis, char oxidation and Boudouard gasification reaction) and two homogeneous reactions (methane and carbon monoxide oxidations) was implemented in ANSYS-FLUENT using User-Defined-Functions:

$$\begin{aligned}
&\mathrm{R1}: \nu_{1,\mathrm{RC}}\mathrm{RC} \Rightarrow \nu_{1,C}\mathrm{C}_j + \nu_{1,C_xH_y}\mathrm{C}_x\mathrm{H}_y + \nu_{1,CO}\mathrm{CO} + \nu_{1,H_2O}\mathrm{H_2O} + \nu_{1,N_2}\mathrm{N_2} \\
&\mathrm{R2}: \nu_{2,\mathrm{C}}\mathrm{C}_j + \nu_{2,\mathrm{O_2}}\mathrm{O_2} \rightarrow \nu_{2,\mathrm{CO}}\mathrm{CO} \\
&\mathrm{R3}: \nu_{3,\mathrm{C}}\mathrm{C}_j + \nu_{3,\mathrm{CO_2}}\mathrm{CO_2} \rightarrow \nu_{3,\mathrm{CO}}\mathrm{CO} \\
&\mathrm{R4}: \nu_{4,\mathrm{C}_x\mathrm{H}_y}\mathrm{C}_x\mathrm{H}_y + \nu_{4,\mathrm{O_2}}\mathrm{O_2} \rightarrow \nu_{4,\mathrm{CO}}\mathrm{CO} + \nu_{4,\mathrm{H_2O}}\mathrm{H_2O} \\
&\mathrm{R5}: \nu_{5,\mathrm{CO}}\mathrm{CO} + \nu_{5,\mathrm{O_2}}\mathrm{O_2} \rightarrow \nu_{5,\mathrm{CO_2}}\mathrm{CO_2}
\end{aligned} \tag{4.2}$$

Here, the subscript $\nu_{i,k}$ represents the stoichiometric coefficient of the species k in the reaction i. During the pyrolysis, the raw coal (RC) produces a volatile matter (C_xH_y, CO, H_2O and N_2) and a char (C). In Eq. (4.2), the species (C_xH_y) is a general hydrocarbon (methane is generally used) and C_j denotes the char in the size

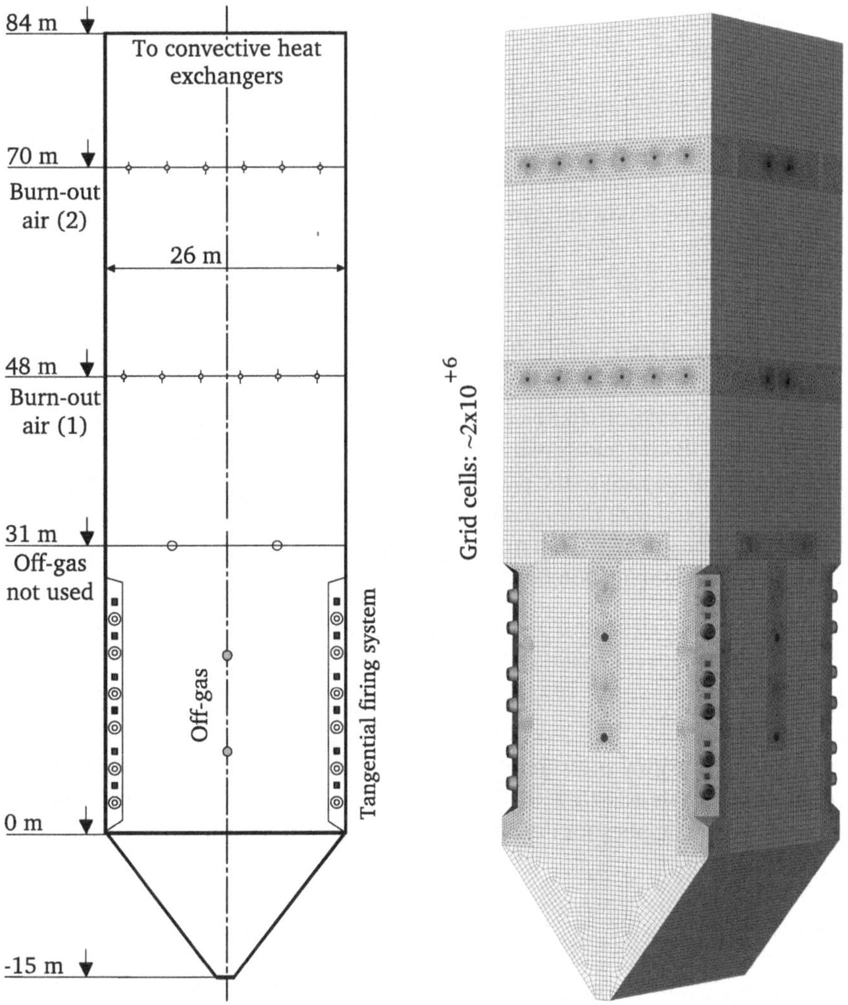

Fig. 4.101 Geometry and numerical grid of the combustion chamber of the 2400 MW_{th} lignite-fired power plant

class *j*. The combustion rates of char particles depend on the particle size distribution that is taken into account by discretising the char particles into several size classes; each of these classes has a constant particle size (three or four particle size classes provide a sufficient accuracy for the simulation of pulverised coal-fired power plants). The pyrolysis rate is not size dependent and the consideration of different size classes of raw coal is not necessary.

For determining the pyrolysis rate (R1), several models are available with different levels of sophistication. Here, the model of Badzioch and Hawskley is used

(Badzioch and Hawksley 1970), describing the pyrolysis rate by a single irreversible reaction as:

$$\frac{dm_{rc}}{dt} = -k_{pyr}^{rc} m_{rc}, \tag{4.3}$$

where m_{rc} is the raw coal mass fraction and the pyrolysis reaction rate constant k_{pyr} is modelled by the following Arrhenius rate expression:

$$k_{pyr}^{rc} = A_{pyr}^{rc} \mathrm{e}^{\frac{E_{pyr}^{rc}}{RT}}. \tag{4.4}$$

According to (Epple et al. 2012), the kinetic parameters of the brown coal pyrolysis have the values of $A_{pyr}^{rc} = 3.15 \times 10^5$ (1/s) and $E_{pyr}^{rc} = 74 \times 10^6$ (J/kmol).

The oxidation rate of char to carbon monoxide (R2) is modelled based on the assumptions of Baum and Street (Baum and Street 1971). Here, the kinetic and diffusion rates of oxygen to the particle surface determine the effective burning rate of char. If one assumes that a complete formation of CO because of the heterogeneous reaction of char with oxygen on the particle surface is achieved, the effective mean char oxidation rate for the particle size class j per surface area can be written as:

$$k_{ox;j}^{cha} = \frac{1}{\left(1/k_{kin}^{cha} + 1/k_{dif}^{cha}\right)} p_{O_2}, \tag{4.5}$$

where p_{O_2} is the partial pressure of oxygen. For modelling the kinetic reaction rate constant, Eq. (4.4) is used ($A_{ox}^{cha} = 93$ (kg/(m^2 s bar) and $E_{ox}^{cha} = 67.8 \times 10^6$ (J/kmol)) and the following applies for the diffusion reaction rate constant:

$$k_{dif}^{cha} = \frac{48 D_{dif}^{O_2,N_2}}{RT_0} 10^5 \frac{T_P^{0.75}}{d_{P,j}^{cha}}. \tag{4.6}$$

The constant $D_{dif}^{O_2,N_2}$ denotes the binary diffusion coefficient for O_2 and N_2 within the boundary layer surrounding the char particle at reference state temperature. The values suggested by (British Coal Utilisation Research and Field, 1967) have been used ($D_{dif}^{O_2,N_2} = 3.49 \times 10^{-4}$ (m^2/s) at $T_0 = 1600$ (K)). The term $d_{P,j}^{cha}$ represents the diameter of the char particle that is assumed to be spherical in the size class j.

The heterogeneous, endothermic Boudouard reaction (R3) describes the char gasification by CO_2 to CO. Corresponding to the description of the char oxidation, the Arrhenius parameters and the binary diffusion coefficient for CO_2 and N_2 must be determined. According to (Golovina and Klimov 1999), the binary diffusion coefficient $D_{dif}^{CO_2,N_2}$ has a value of 3.31×10^{-4} (m^2/s), the pre-exponential factor and the activation energy are set equal to $A_{Boud}^{cha} = 10^{12}$ (kg/(m^2 s bar)) and $E_{ox}^{cha} = 381 \times 10^6$ (J/kmol).

For obtaining the local particle area per mixture volume, the shadow method of Spalding (Spalding 1982) is applied. In this method, the hypothetical mass fraction of the particles for the hypothetical case of no-combustion is assumed the shadow mass fraction. As a result, the burn-out of the particle phase is determined from the ratio of the local mass fraction of the particles to their shadow mass fraction. For estimating the shadow mass fraction, an inert variable mixture fraction (f) is utilised (i.e. governed by a source-free transport equation and has the value of unity at the primary inlet (fuel) and the value of zero at the secondary air inlets of the combustion chamber (Benim et al. 2005)).

The homogeneous reactions (gas phase reactions) include methane and carbon monoxide oxidations (R4 and R5). The combustions rates of the gas phase reactions are modelled by utilising the turbulent-chemistry eddy dissipation model (EDM) of Magnussen and Hjertager (1977). A reaction mechanism that includes detailed hydrocarbon, nitrogen, sulphur and chlorine chemistry can be considered. In this case, the turbulence-chemistry interaction is modelled with the eddy dissipation concept (EDC) that is an extension of the eddy dissipation model to include detailed chemical mechanism in the turbulent flows. However, the computational cost increases considerably due to the high number of species and the gas phase reactions.

The mass fractions of carbon dioxide, oxygen, nitrogen and steam measured at the outlet of the combustion chamber were applied for the model validation, in addition to the flue gas temperature. The mean values of the numerical results, displayed in Table 4.17, are in good agreement with the measurements.

In Fig. 4.102, as an example, the spatial distributions of temperature, velocity, gaseous and solid species are also presented. The right snapshots show the calculated temperature and the calculated mass fractions of the species on a horizontal plane in the burner level 3 of the tangential firing system. It can be clearly observed that the flow entering from the corners is defected (clockwise) by the tangential vortex. The special wall airstreams (injected at a low angle to the wall) supply an oxygen-rich atmosphere along the wall, which in turn result in keeping the corrosive gases away from the combustion chamber walls (indication of these based on CO mass fraction). The left snapshots show the temperature at the diagonal section of the combustion chamber. Here, the mean temperature at the outlet of the combustion chamber amounts to about 1000 °C, before the flow gas enters the convective heat exchanger surfaces (i.e. SH, RH and ECO).

Table 4.17 Mass fractions of species and temperature at the combustion chamber outlet (operation data and simulation)

Parameter	Unit	Experiment	Simulation	Absolute deviation
CO_2 (outlet)	(wt%)	22.87	22.9	+0.03
O_2 (outlet)	(wt%)	2.68	2.45	−0.23
N_2 (outlet)	(wt%)	68.55	68.72	+0.17
H_2O (outlet)	(wt%)	5.90	5.93	+0.03
Temperature (outlet)	(°C)	1030	1005	−2

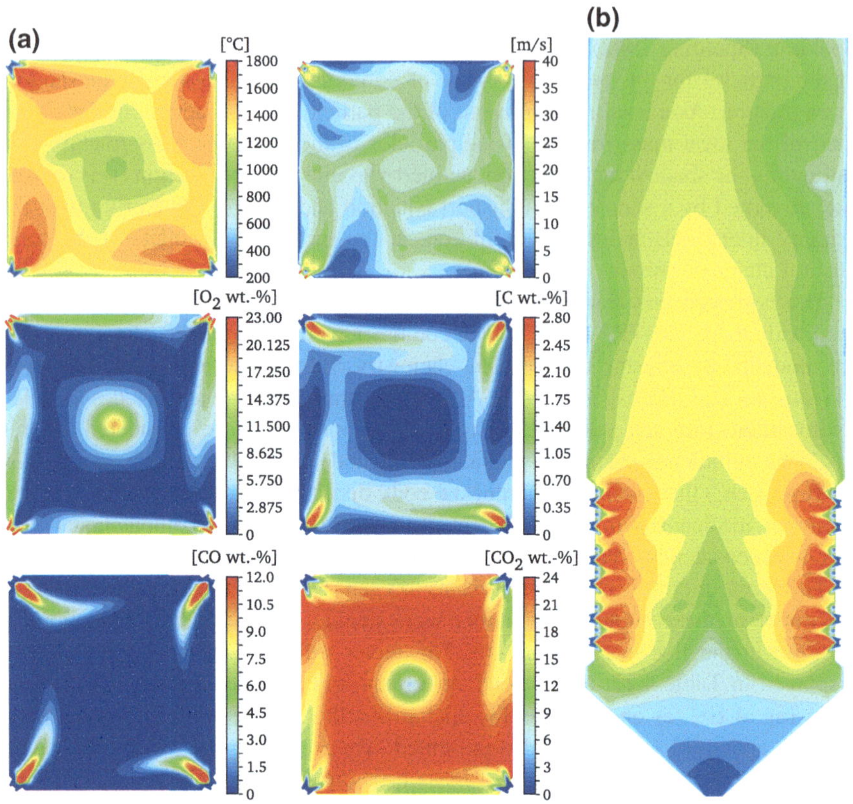

Fig. 4.102 **a** spatial distributions of the temperature, the velocity and the species at the third firing level and **b** spatial distribution of the temperature at the diagonal section of the combustion chamber

Hard Coal-Fired Power Plant

Hard coals, known as bituminous coals, contain low water content and have higher calorific values compared to raw brown coals. For example, a brown coal-fired power plant with a capacity of 330 MW_{el} can have similar dimensions to a hard coal-fired power plant with a capacity of 750 MW_{el} (Epple et al. 2012).

The investigated 2300 MW_{th} pulverised coal-fired power plant (designed by Doosan Heavy Industries and Construction) burns hard coal (mass flow rate of 97.2 kg/s) with a lower heating value of 23.93 MJ/kg. The steam generator is designed as a two-pass boiler and the combustion chamber has a cross-section area of 14.5 m × 16 m and a height of about 58 m. The firing system consists of three firing levels with 24 swirl burners; each has a centred fuel feed pipe that is surrounded by annular pipes with swirling vanes for air supply. In Table 4.18, the proximate and ultimate analyses of the used fuel are presented. The geometry and

Table 4.18 Proximate and ultimate analyses of the hard coal

Component	Value	Unit	Element	Value	Unit
Fixed carbon	53.4	$(wt\%)_{raw}$	C	62.30	$(wt\%)_{raw}$
Volatile matter	21.55	$(wt\%)_{raw}$	H	3.56	$(wt\%)_{raw}$
Ash	14.05	$(wt\%)_{raw}$	O	7.03	$(wt\%)_{raw}$
Moisture	11	$(wt\%)_{raw}$	N	1.54	$(wt\%)_{raw}$
LHV	23.93	(MJ/kg)	S	0.52	$(wt\%)_{raw}$

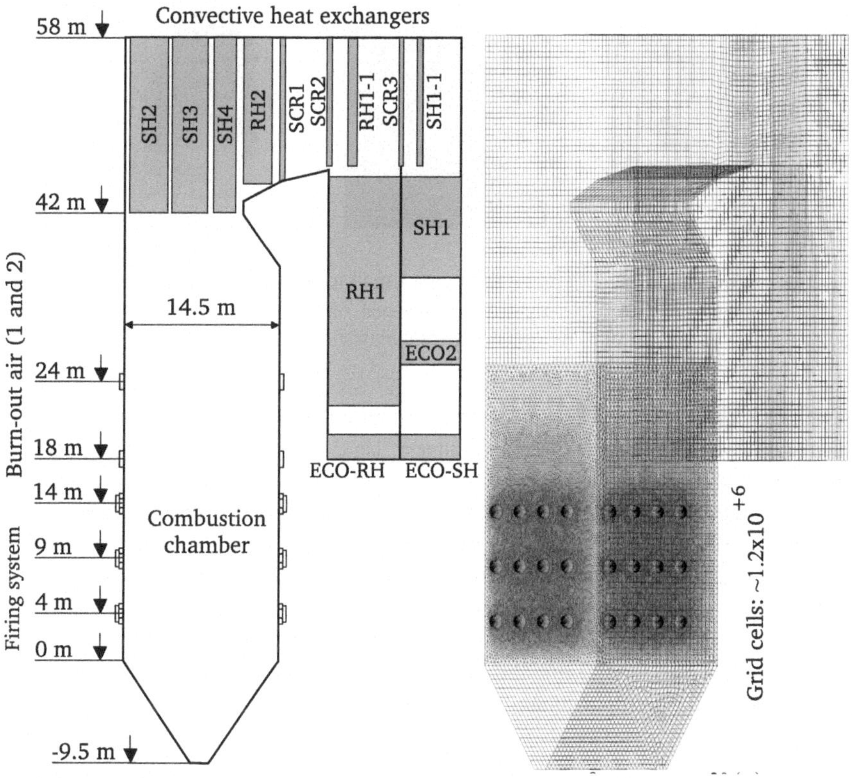

Fig. 4.103 Geometry and numerical grid of the combustion chamber of the 2300 MW_{th} hard coal-fired power plant

the mesh (approximately 1.2×10^6 grid cells) of the hard coal-fired power plant are displayed in Fig. 4.103.

The numerical model is based on the solution of the momentum components of the mixture, the turbulent kinetic energy k and its dissipation rate ε, the transport equations for chemical species and enthalpy. The heat transferred by radiation is calculated using the P1 radiation model with a mixture absorption coefficient as a sum of gas and particle contributions. The heat exchangers in the boiler are

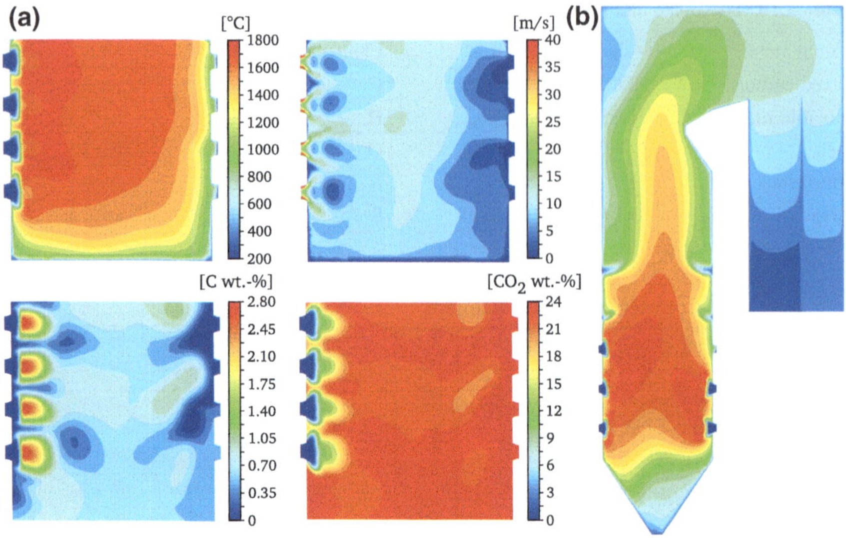

Fig. 4.104 **a** spatial distributions of the temperature, the velocity and the species at the third firing level (the burners of the right sidewall are switched off) and **b** spatial distribution of the temperature at the vertical plane (y = 5 m) of the combustion chamber

modelled with the porous media approach to provide an appropriate pressure drop calculation over tube bundles heat exchangers. Here, the volume heat sink method is applied to calculate the heat extraction from the flue gas path to the water/steam side. The governing equations were calculated with SIMPLE algorithm in association with a first-order upwind scheme. A global five-step reaction mechanism with three heterogeneous reactions and two homogeneous reactions was implemented in ANSYS-FLUENT using User-Defined-Functions (see Sect. Lignite-Fired Power Plant).

In Fig. 4.104, the simulated contour of temperature and mass fractions of the species at the firing level 3 are shown, in addition to the spatial distribution of the temperature at the vertical plane (y = 5 m) that is passed through the first burner (front wall).

4.2.2 *Two-Fluid Model*

For the numerical simulation of gas-solid flows with smaller particle size (Geldart group A and C) and high solid loading, e.g. fluidized bed reactors, the two-fluid model is particularly a desirable approach. Fluidized beds can be found in various industrial applications such as mineral processes, coating processes, CO_2 capture in the thermal power plants by chemical or carbonate-looping process as well as the solid fuels conversion including gasification and combustion. In the following sections, the application of the two-fluid model to a cold-flow pseudo-2D fluidized bed and a 3D fluidized bed under the consideration of thermochemical reactions are presented. The commercial CFD package ANSYS-FLUENT, based on the two-fluid formulation, is employed. Here, the two-phase flow is treated as interpenetrating continua with the aid of the volume fraction, which denotes to the space occupied by each phase. The volume fraction is a function of space and time and its sum in each grid cell is equal to unity. The mass, momentum and energy conservation equations are applied to each phase individually. However, additional constitutive relations are required in order to close the governing conservation equations. This can be achieved with the aid of the kinetic theory of granular flows that describes the viscous forces and the pressure of the solid phase as a function of the granular temperature.

4.2.2.1 Quasi-2D Fluidized Bed

A lab-scale cold-flow spouted fluidized bed model is constructed and built using acrylic glass plates (Plexiglasr), which allows the visual observation of the solid motion inside the bed. The geometrical details of the test rig model are illustrated in Fig. 4.105.

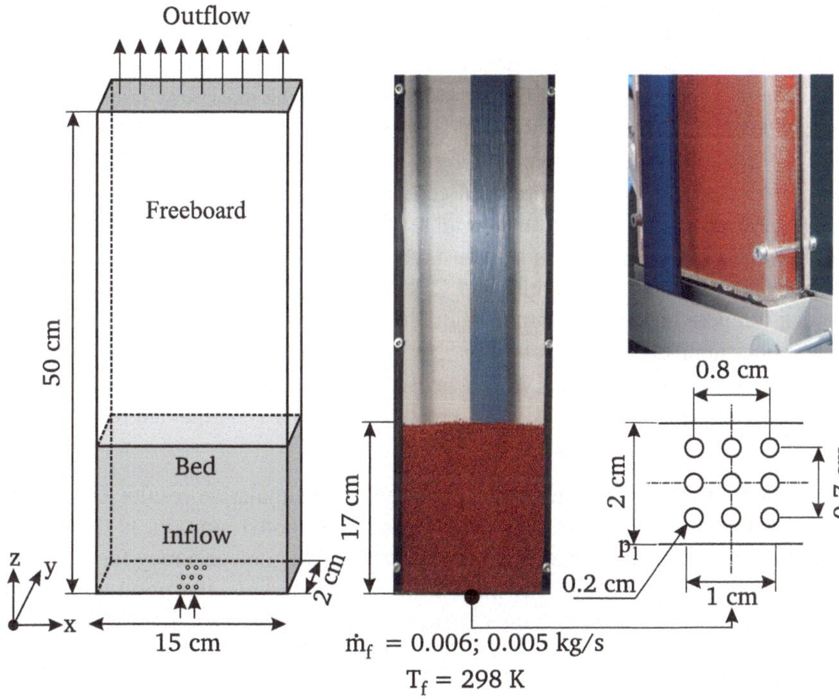

Fig. 4.105 Test rig of the quasi-2D fluidized bed

The test rig has a height of 50 cm, a width of 15 cm and a depth of 2 cm. The fluidizing medium (air with humidity of about 70%) is injected into the bed through centrally located nine holes with 0.2 cm inner diameter. The outlet is completely opened (15 cm (W) × 2 cm (D)). The solid phase consists of spherical glass beads (filled inventory of about 0.7 kg) with an average diameter of 2.5 cm and a density of 2500 kg/m^3. A coarse grid is mounted at the outlet, preventing the particles from leaving the system, but it causes a negligible pressure drop. The experiments are carried out at room temperature and the freeboard pressure is atmospheric. The depth of the test rig model is sufficiently small to achieve a pseudo-2D behaviour, which is necessary for recording the particle flow and the bubble behaviour by a high temporal resolution camera. The fluidizing air mass flow rates are 0.006 kg/s for case 1 and 0.005 kg/s for case 2. The total measuring time is 500 ms for each case.

The model geometry and the computational mesh have been generated in ANSYS-FLUENT. Unlike the experiment, the inlet in ANSYS-FLUENT is modelled as a set of nine squares instead of circles for simplification purposes. In this case, a block-structured mesh can be generated since the modelling of the inlet with 9 circular holes results in an unstructured grid. The number of grid cells is equal to 380,000 cells. In the simulation model, the initial velocity and pressure fields of the gas phase are zero. The bed inventory is about 0.75 kg and the particles in the spouted bed have zero initial kinetic energies with a maximum packing limit of 0.6. The initial bed height is set identical to the measurement (approximately 17 cm). At the inlet, the pressurized air enters the fluidized bed at mass flow rates 0.006 and 0.005 kg/s, while the solid mass flow rate is set equal to zero. The outlet is defined as a continuous outflow for the gas and solid phase. At the mass flow rates used, the vertical position of the particles should not be higher than 50 cm. If this occurs however, these particles are removed from the simulation. The numerical calculations are carried out using commercial computers with multi-core processors.

In order to identify significant simulation settings for the two-fluid model, a parameter study was carried out. The investigated input parameters include the gas-particle drag models (Syamlal and O'Brien, Gidaspow and Wen and Yu), the granular temperature approach (partial differential approach and algebraic granular temperature approach), the solid-phase wall boundary conditions in terms of specularity coefficient (0, 0.2, 0.5, 0.7 and 1.0) and the particle-particle restitution coefficient (0.6, 0.7, 0.8, 0,9) (Almohammed et al. 2014). The obtained results show the following points:

- The drag model has a significant influence on the simulation results. Compared to other drag models, Syamlal and O'Brien drag model reproduces the hydrodynamics of the present gas-solid fluidized bed with good accuracy for gas mass flow rate 0.005 kg/s. For gas mass flow rate 0.006 kg/s, the good agreement between the simulation and the experiment is, however, decreased. This might be attributed to the fact that the selection of an appropriate drag model alone can probably not improve the simulation results. Other contributions such as the particle-particle interactions, the particle rotation and the frictional viscosity model that are not appropriately modeled or neglected in the framework of the two-fluid model may have an important role.

- The algebraic granular temperature approach reproduces the flow structure reasonably for both mass flow rates. Furthermore, the algebraic granular temperature approach is more efficient compared to the partial differential one.
- The specularity coefficient has a significant effect on the flow behaviour of the spouted fluidized bed regarding the bed height and the bubble size. A value of 0.5 reproduces the flow patterns with good accuracy, suggesting that the use of a suitable specularity coefficient is critical for improving the predictions.
- In the range of values considered for the particle-particle restitution coefficient, a value of 0.6 gives the best predictions.

The numerical results at the mass flow rate of 0.005 kg/s over the time range 0–500 ms is depicted in Fig. 4.106, as an example. Due to the aerodynamic forces, a small gas bubble is formed in the jet region. The bubble expands further and the particles in the spout region rise upwards. In the period of time between t = 50 ms and t = 150 ms, the bed height and the bubble area in the two-fluid model show a good agreement with measurement. At t = 150 ms, the bed height in the two-fluid model with about 20 cm has almost an identical value to the test rig. In the period of time between t = 175 ms and t = 275 ms, the particle layer over the gas bubble accelerates and becomes thinner. At t = 300 ms, a new small bubble is built in the jet region besides to the main bubble. From time point t = 350 ms, the kinetic energy of the particles in the upper region of the bubble is almost dissipated, so that the particles fall back into the sidewalls. At t = 430 ms, a maximum measured bed height of about 25 cm is reached. The numerical model underestimates the measured bed height with a relative error of about 4%. After the bubble collapses at t = 475 ms, the bed height obtained from two-fluid model seems to be clearly below the average measured bed height with a relative error of 7%. The asymmetry in the solid distribution and the irregular bubble shape can be observed in the test rig. The simulated bubble flow, by contrast, has a regular form and the solid is symmetrical distributed relating to y-z midplane.

4.2.2.2 1 MW_{th} Pilot Plant

The validation study of the two-fluid model towards the quasi-2D fluidized bed is carried out, showing a good agreement. The relevant next step is analysing the model capability to reproduce the hydrodynamics behaviour of a warm-flow 3D fluidized bed with chemical reactions. This requires the consideration of kinetic models for pyrolysis, char-oxidation and heterogeneous reactions as well as the associated heat and mass transfer between gas and solid phases. In this work, three-dimensional, transient two-fluid models for gasification or combustion of solid fuels (coal, biomass or fuel blends) as well as the CO_2 capture by chemical or carbonate-looping process are developed. A set of experimental data obtained from a 1 MW_{th} pilot plant erected at Technische Universität Darmstadt are applied to

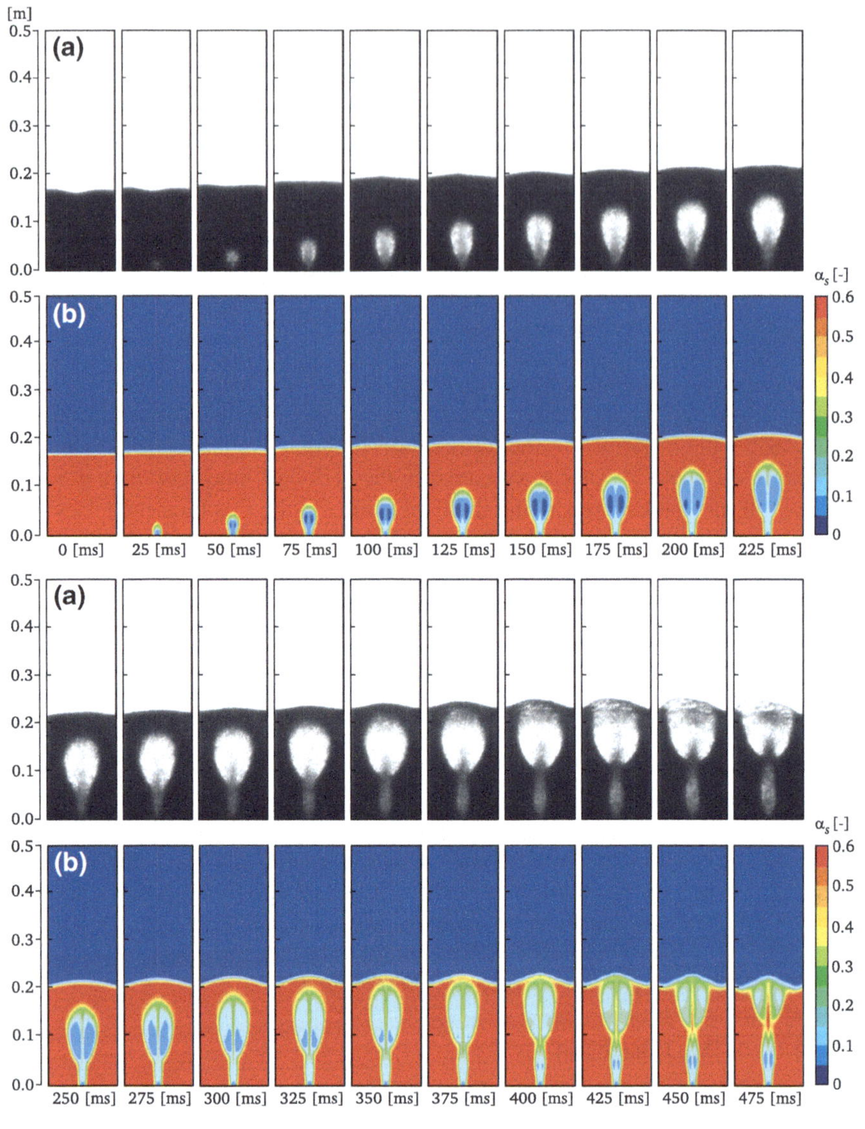

Fig. 4.106 Snapshots for the spatial distribution of solid (mass flow rate 0.006 kg/s; time range between 0 ms and 475 ms): **a** experiment and **b** two-fluid model (legend: solid volume fraction)

validate the models. The pilot plant consists of two interconnected circulating fluidized bed (CFB) reactors and a pulverised coal combustion chamber (explained in Sect. 4.2.1.2). The pilot plant is capable of running in different operation modes. For the gasification or combustion of solid fuels, one of the interconnected circulating fluidized reactors is in service. In case of the calcium carbonate-looping (CCL) process, the two fluidized bed reactors acting as carbonator and calciner, in addition to the pulverised coal combustion chamber are in operation. The chemical-looping combustion (CLC) uses two-fluidzed bed reactors acting as air and fuel reactors. As an example, the developed models of air reactor and fuel reactor of the CLC are presented here. For the numerical simulation, the ANSYS-FLUENT software is combined with an in-house code for consideration the thermochemical reactions. The results obtained from the two-fluid model/thermochemical reaction model are compared towards the measurements delivered from the 1 MW_{th} CLC pilot plant, (Alobaid et al. 2015a; May et al. 2017a, b; Sharma et al. 2017).

The schematic flow diagram of the 1 MW_{th} chemical-looping combustion plant is shown in Fig. 4.107. At the bottom of the air reactor, preheated air is supplied to fluidize and oxidize the ilmenite. The oxidation takes place at a temperature of approximately 1050 °C. In the subsequent cyclone, the mixture of gas and solid is separated. Afterwards the gas of low oxygen content is cooled down, filtered and released to the atmosphere. The particles enter a loop seal after the cyclone (LS 1). Some particles flow into the air reactor and the other particles reach the fuel reactor passing through an L-valve. In the air reactor, preheated steam fluidizes the particles and coal is provided as fuel. The average temperature in the fuel reactor is 950 °C. After the fuel reactor, the mixture of gas and solid is separated in a cyclone and the particles circulate into the fuel reactor by passing two loop seals (LS 4 and LS 5). The gas enters a post oxidation chamber (POC), after the cyclone, so that the unconverted gases are combusted with pure oxygen. Afterwards, the gas is cooled, filtered and released to the atmosphere by means of an induced draft. All loop seals and the L-valve are fluidized with nitrogen and the propane burners that were out of use during the auto-thermal operation are flushed through with nitrogen. The dashed lines illustrate alternative configurations of the pilot plant. Here, the effect of a carbon stripper that is installed between air and fuel reactors can be evaluated.

Six hours of worldwide first auto-thermal tests of the CLC were performed at the pilot plant. Different operational conditions were realised such as additional propane feed to the air reactor or various fluidization gas compositions for the fuel reactor. The maximum coal feed was up to 115 kg/h, which corresponds to a thermal capacity of 820 kW_{th}. The mass-weighted mean diameter of the coal particles was 3.1 mm. A detailed description of the experiments is given in (Ohlemüller et al. 2016). The time averaged results of 1.5 h stable operation are selected to determine the boundary conditions for the simulation. The temperature, the pressure profile of air and fuel reactors and the gas composition at the exit of the reactors, downstream the cyclones, were measured and serve for the validation of the simulation results. The mass circulation between the reactors was not measured; but, a heat and energy balance performed in ASPEN PLUS gives a rough estimation of the circulation rate (Ohlemüller et al. 2017).

For the simulation of the air reactor of the chemical-looping combustion, the geometry of the air reactor with a height of 8.66 m and a diameter of 0.59 m is built and a block-structured grid with a resolution of approximately 150,000 cells is generated. Figure 4.108 shows the boundaries of the domain: Preheated air enters the reactor from the bottom boundary "Distributor inlet" in order to fluidize the particles and to oxidize the ilmenite. The right entry "Ilmenite inlet" is the

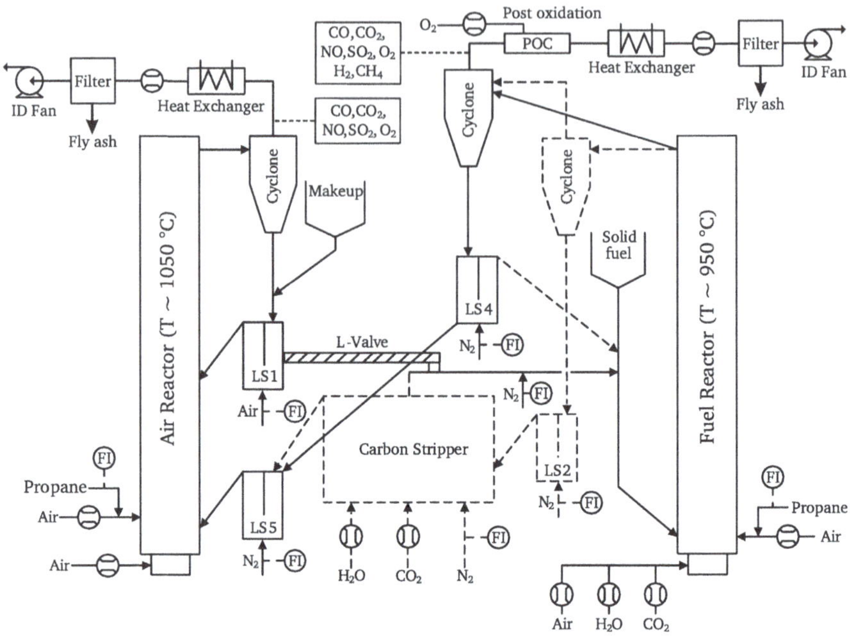

Fig. 4.107 Schematic flow diagram of the 1 MW_{th} chemical-looping combustion plant at Technische Universität Darmstadt

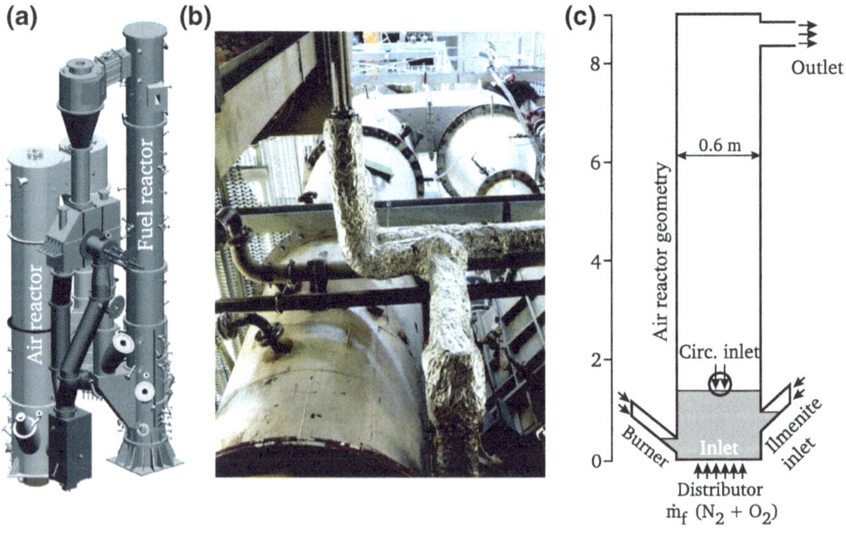

Fig. 4.108 The pilot plant: **a** 1 MW_{th} CLC, **b** test rig of the CLC air reactor, **c** geometry and simulation set-up of the CLC air reactor

connection to the fuel reactor and the boundary "Circ. inlet" is the circulation entry of particles. In this simulation, the boundary "Circ. Inlet" is not used and it is assumed that all particles enter the "Ilmenite inlet" boundary, whereby the influence of the solid circulation on the composition of the incoming ilmenite is taken into consideration. The solid at the outlet circulates into the air reactor via the "Ilmenite inlet" boundary in order to keep the mass inventory constant. Due to the incomplete combustion of coal particles in the fuel reactor, a slight amount of char gets in the air reactor. This mass flow rate of the incoming char is also considered in the boundary "Ilmenite inlet". The left entry "Burner inlet" is a start-up burner, which was not in operation during the experiments but was flushed through with air. Two phases are defined. Phase 1 contains all gaseous species: carbon dioxide, carbon monoxide, hydrogen, nitrogen and oxygen. Furthermore, char and ash are also defined as components of the phase 1, which are entrained by the gas flow. The second phase consists of the reduced and oxidized forms of ilmenite. The simulation is initialized with 214 kg of ilmenite (49.5% Fe_2O_3, 49.5% TiO_2 and 1% Fe TiO_3), which corresponds to a bed height of 0.32 m with a volume fraction of 53%. The gas phase is air (77% Nitrogen, 23% Oxygen) and the initial temperature of both phases is set to a value of 1150 °C. A time step size of 0.005 s was selected and 10,000 time steps were calculated (total simulation time 50 s). The boundary conditions of phase 1 and phase 2 were set equal to the boundary conditions in the experiments (see Table 4.19).

In the air reactor, two reactions take place, namely the oxidation of oxygen carrier and the combustion of unburned char, coming from the fuel reactor. The mechanism for char-oxidation is based on the assumption that the R1 and R2:

$$\begin{aligned} &R1: 2C + O_2 \rightarrow 2CO \\ &R2: 2CO + O_2 \rightarrow 2CO_2 \end{aligned} \tag{4.7}$$

Table 4.19 Boundary conditions of the air reactor model

Location	Temperature (°C)	Mass flow rate (kg/s)	Species mass fraction (wt%)	
Distributor inlet	395	0.24	O_2	23
			N_2	77
Burner inlet	35	0.018	O_2	23
			N_2	77
			Propane	0
Circ. Inlet (L S1 inlet)	–	–	–	–
Ilmenite inlet (L S5 inlet)	35	0.012	N_2	50
		0.012	Char	50
	1050	User-Defined-Function	Fe_2O_3	45
			TiO_2	45
			$FeTiO_3$	10
Pressure outlet	1050	0		

can be combined to form the following equation:

$$C + O_2 \rightarrow CO_2 \tag{4.8}$$

The reaction rate of the char-oxidation is expressed by:

$$r_{ox}^{cha} = \frac{1}{\left(\frac{1}{k_{kin}^{cha}} + \frac{1}{k_{dif}^{cha}}\right)} \frac{a_{cha}}{M_{cha}} p_{O_2}. \tag{4.9}$$

Here, the symbol p_{O_2} is the partial pressure of oxygen, M_{cha} represents the molar mass of char, k_{kin}^{cha} and k_{dif}^{cha} are the kinetics and diffusion rate reaction constants, which are given by:

$$\begin{aligned} k_{kin}^{cha} &= A_{pre}^{cha} e^{-\frac{E_{act}^{cha}}{RT}}, \\ k_{dif}^{cha} &= \frac{48 D_{dif}^{cha}}{RT_0} 10^5 \frac{T_P^{0.75}}{d_{0,cha}}. \end{aligned} \tag{4.10}$$

The local specific particle surface area per mixture volume a_{cha} is calculated as:

$$a_{cha} = \frac{6 \rho_g m_{cha}}{\rho_{0,cha} d_{0,cha}}, \tag{4.11}$$

where m_{cha} is the mass of char, ρ_g represents the density of the gas phase, $\rho_{0,cha}$ is the initial material density of char and $d_{0,cha}$ denotes the initial diameter of char. All kinetic parameters are listed in Table 4.20.

Ilmenite ($FeTiO_3$) was used in the experiments as oxygen carrier. It is a well-studied and cheap oxygen carrier with high availability as well as low environmental and health impact. In the oxidized form, it is a mixture of iron oxide (Fe_2O_3) and titanium oxide (TiO_2), while the reduced form is $FeTiO_3$. The oxidation reaction of the ilmenite takes place in the air reactor as follows:

$$4\,FeTiO_3 + O_2 \rightarrow 2\,Fe_2O_3 + 4\,TiO_2 \tag{4.12}$$

The reaction rate of the ilmenite-oxidation is given by (Abad et al., 2011; Ohlemüller et al. 2015, 2017):

$$r_{ox}^{OC} = \frac{dX_{ox}^{OC}}{dt} = R^{OC} \frac{3}{\tau_{ox}^{OC}} \left(1 - X_{ox}^{OC}\right)^{\frac{2}{3}}, \tag{4.13}$$

Table 4.20 Kinetic parameters for char-oxidation (Epple et al. 2012)

Parameter	Value	Unit	Parameter	Value	Unit
E_{act}^{cha}	67.8×10^3	(kJ/kmol)	T_0	1600	(K)
A_{pre}^{cha}	93	(1/s)	$d_{0,cha}$	40×10^{-6}	(m)
D_{dif}^{cha}	3.94×10^{-4}	(m^2/s)	$\rho_{0,cha}$	1000	(kg/m^3)

where τ_{ox}^{OC} is the time for the complete conversion of the oxygen carrier, expressed by:

$$\tau_{ox}^{OC} = \frac{\rho_{mol}^{OC} r_{gra}}{b_{ox} k_{ox} c_{ox}^{n}}, \tag{4.14}$$

with the molar density of ilmenite ρ_{mol}^{OC}, the grain radius r_{gra}, the average stoichiometric coefficient for the reaction of solid with oxygen b_{ox}, the oxygen concentration c_{ox}, the reaction order n and the kinetic rate reaction constant k_{ox} (see Table 4.21):

$$k_{ox} = A_{pre}^{ox} e^{-\frac{E_{act}^{ox}}{RT}}. \tag{4.15}$$

The conversion of the oxygen carrier X_{ox}^{OC} is defined using the following relation:

$$X_{ox}^{OC} = \frac{m_{ox}^{OC} - m_{tot}^{OC}\left(1 - \frac{R^{OC}}{R_{the}^{OC}}\right)}{m_{tot}^{OC} - m_{tot}^{OC}\left(1 - \frac{R^{OC}}{R_{the}^{OC}}\right)}, \tag{4.16}$$

where R^{OC} and R_{the}^{OC} represent the practical oxygen and the theoretical oxygen transport capacities of ilmenite, respectively. The oxidized mass of ilmenite m_{ox}^{OC} and the total mass of ilmenite m_{tot}^{OC} are given by:

$$\begin{aligned} m_{ox}^{OC} &= \frac{m_{TiO_2} + m_{Fe_2O_3}}{1 + R_{the}^{OC}}, \\ m_{tot}^{OC} &= \frac{m_{TiO_2} + m_{Fe_2O_3}}{1 + R_{the}^{OC}} + m_{FeTiO_3}. \end{aligned} \tag{4.17}$$

Here, the masses m_{TiO_2}, $m_{Fe_2O_3}$ and m_{FeTiO_3} are the mass of TiO_2, Fe_2O_3 and $FeTiO_3$, respectively.

Table 4.22 shows the compression between measured and simulated pressures. At height of 8.03 m the experimental value is 2 mbar higher than the simulation and at height of 4.76 m the deviation between both values is approximately 8 mbar. The pressure profiles at the bottom of the air reactor display deviations in the range of 13–22 mbar. This is an indication of a wrong particle distribution in the simulation, which may relate due to the Syamlal and O'Brien drag model and/or to the wrong initial value for the ilmenite mass in the air reactor. The dry concentrations

Table 4.21 Kinetic parameters for ilmenite-oxidation

Parameter	Value	Unit	Parameter	Value	Unit
R^{OC}	0.04	[–]	b_{ox}	4	[–]
R_{the}^{OC}	0.0526	[–]	n	1	[–]
ρ_{mol}^{OC}	31,100	(mol/m^3)	A_{pre}^{ox}	1.9×10^{-3}	(mol^{1-n} m^{3n-2}/s)
r_{gra}	1.2×10^{-6}	(m)	E_{act}^{ox}	25.5	(kJ/mol)

Table 4.22 Pressure along the fuel reactor height (experiment and simulation)

Height	Unit	Experiment	Simulation	Absolute deviation
0.265	(mbar)	87.59	66.61	−20.98
0.605	(mbar)	36.43	58.84	+22.4
2.12	(mbar)	14.21	37.52	+23.31
4.76	(mbar)	7.91	15.88	+7.97
8.03	(mbar)	4.65	2.72	−1.93

Table 4.23 Temperature and gas concentrations at the air reactor outlet (experiment and simulation)

Parameter	Unit	Experiment	Simulation	Absolute deviation
CO_2 (outlet)	(vol%)	10.1	10.3	+0.2
O_2 (outlet)	(vol%)	3.2	7.9	+4.7
N_2 (outlet)	(vol%)	86.7	81.8	-4.9
Temperature (outlet)	(°C)	1053	1061	+8

of carbon dioxide, oxygen and nitrogen were also measured during the experiment over a period of 1.5 h. A comparison between the mean values of the numerical results and experimental data (see Table 4.23) displays a good agreement for the carbon dioxide concentration at the outlet. The concentration of oxygen is too high in the simulation and a discrepancy of 4.7% points can be seen. An explanation for this difference could be the wrong particle distribution in the air reactor due to the used Syamlal and O'Brien drag model, which in turn results in an insufficient contact between gas and particles. An indicator for an inaccurate particle distribution is observed by the analysis of the pressure profile. During the experiments, the temperature was measured at the top of the reactor where the mixture of gas and particle phases leaves the reactor to enter the cyclone. The mean measured temperature was 1053 °C. The calculated mean temperature at the same height in the period from 10 s to 50 s is 1061 °C. The distribution of the species at time step 50 s is shown in Fig. 4.109, as example.

The success of the CLC system for solid fuels depends largely on the behaviour of the fuel reactor. The amount of unburnt gasses leaving the fuel reactor and the char escaping out along with the metal oxide particles affect the combustion efficiency considerably. CFD simulation can be helpful for the design, optimisation (in order to obtain a high solid fuel conversion) and scale-up of the process. For the simulation of the fuel reactor of the chemical-looping combustion using the two-fluid model, two phases are defined. The phase 1 contains all gaseous species: methane, carbon monoxide, hydrogen, carbon dioxide, steam and nitrogen. Furthermore, coal, char and ash are also defined as components of phase 1. This assumption simplifies the problem complexity, otherwise three phases are required (gaseous species, coal and char, reduced and oxidized forms of ilmenite). The second phase consists of the reduced and oxidized forms of ilmenite.

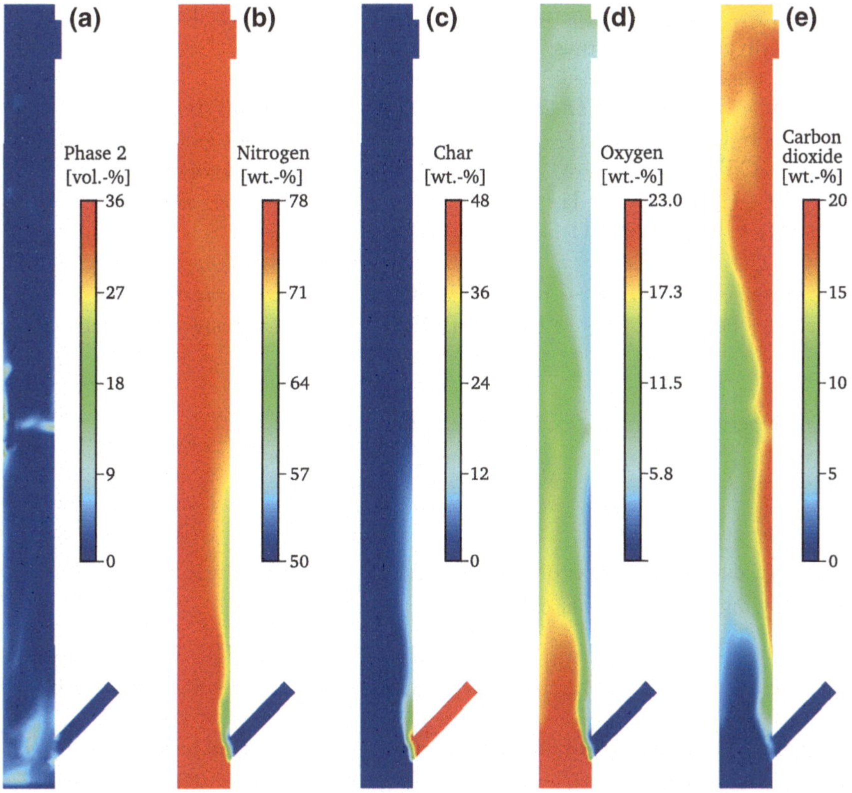

Fig. 4.109 Snapshots for the spatial distributions of gaseous species and solid at time step 50 s: **a** concentration of phase 2, **b** mass fraction of nitrogen, **c** mass fraction of char, **d** mass fraction of oxygen and **e** mass fraction of carbon dioxide

Table 4.24 Proximate and ultimate analyses of the Taldinsky hard coal

Component	Value	Unit	Element	Value	Unit
Fixed carbon	52.1	$(wt\%)_{raw}$	C	65.8	$(wt\%)_{raw}$
Volatile	32.0	$(wt\%)_{raw}$	H	4.0	$(wt\%)_{raw}$
Ash	10.1	$(wt\%)_{raw}$	O	11.8	$(wt\%)_{raw}$
Moisture	5.8	$(wt\%)_{raw}$	N	2.0	$(wt\%)_{raw}$
LHV	26.6	(MJ/kg)	S	0.5	$(wt\%)_{raw}$

Russian hard coal "Taldinsky" with a lower heating value of 26.6 MJ/kg and a mean diameter of 3.1 mm used as fuel. The ultimate and proximate analyses are listed in Table 4.24.

The thermochemical reaction model is represented in Fig. 4.110 and the reactions involved are as follows:

$$\begin{array}{ll}
\mathrm{R1:\ RC \Rightarrow C + CH_4 + CO + CO_2 + H_2 + H_2O + N_2} & \textit{Pyrolysis} \\
\mathrm{R2: C + H_2O \rightarrow H_2 + \ CO} & \textit{Steam gasification} \\
\mathrm{R3: C + CO_2 \rightarrow 2CO} & \textit{Boudouard reaction} \\
\mathrm{R4: CO + H_2O \rightarrow CO_2 + H_2} & \textit{Water-gas shift reaction} \\
\mathrm{R5: CH_4 + 4Fe_2O_3 + 8TiO_2 \rightarrow} & \\
\mathrm{2H_2O + CO_2 + 8FeTiO_3} & \textit{Methane oxidation} \\
\mathrm{R6: H_2 + Fe_2O_3 + 2TiO_2 \rightarrow H_2O + 2FeTiO_3} & \textit{Methane oxidation} \\
\mathrm{R7: CO + Fe_2O_3 + 2TiO_2 \rightarrow CO_2 + 2FeTiO_3} & \textit{Carbon monoxide oxidation}
\end{array} \tag{4.18}$$

The reaction rate of the coal pyrolysis is defined as:

$$\frac{dm_{rc}}{dt} = -k_{pyr}^{rc} m_{rc}, \tag{4.19}$$

where m_{rc} denotes the mass of raw coal and k_{pyr} is the pyrolysis rate expressed by the following Arrhenius rate expression:

$$k_{pyr}^{rc} = A_{pyr}^{rc} e^{\frac{E_{pyr}^{rc}}{RT}}. \tag{4.20}$$

The kinetic parameters of the pyrolysis reaction are set equal to $A_{pyr}^{rc} = 1.5 \times 10^5$ (1/s) and $E_{pyr}^{rc} = 74 \times 10^6$ (J/kmol) (Epple et al. 2012). The products from

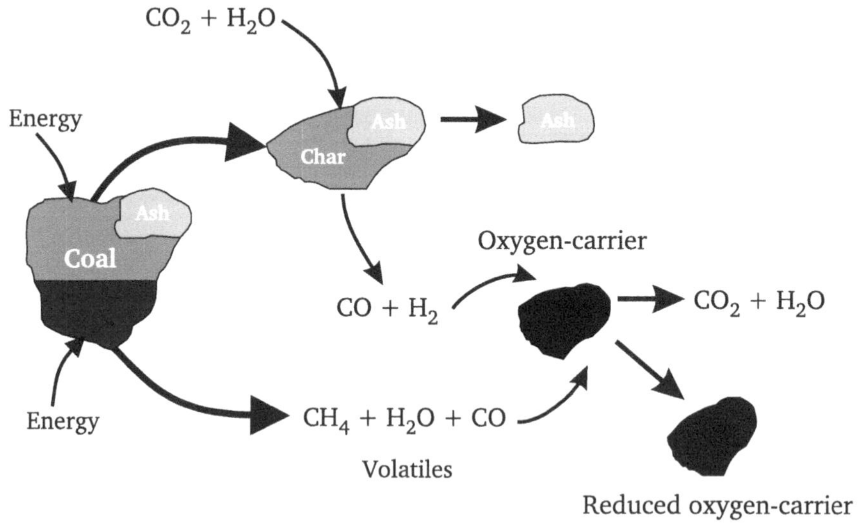

Fig. 4.110 Representation of the multi-step reaction mechanism for coal chemical-looping combustion

devolatilisation of the Taldinsky hard coal are summarised in Table 4.25. Tar and volatile organic compounds are considered to be transformed to methane and reformed to carbon monoxide and hydrogen by steam and carbon dioxide.

The Boudouard reaction is modelled with the following reaction rate:

$$r_{CO_2}^{cha} = \frac{dX_{CO_2}^{cha}}{dt} = \frac{k_{CO_2}^{cha} p_{CO_2}}{1 + K_{CO_2}^{cha} p_{CO_2} + K_{CO}^{cha} p_{CO}} \cancel{\left(1 - X_{CO_2}^{cha}\right)}. \tag{4.21}$$

Here, $X_{\mathrm{CO}_2}^{cha}$ denotes the conversion of char due to Boudouard reaction, p_{CO_2} and p_{CO} are the partial pressures of carbon dioxide and carbon monoxide, respectively. For reasons of simplification, the actual conversion of char (term in parenthesis) is neglected and therefore the reaction rate $r_{\mathrm{CO}_2}^{cha}$ is slightly higher. The kinetic reaction rate constant k_i^{cha} and the adsorption reaction rate constant K_i^{cha} are calculated using Arrhenius equation as:

$$\begin{aligned} k_i^{cha} &= A_{pre,i}^{cha} e^{\frac{-E_{act,i}^{cha}}{RT}}, \\ K_i^{cha} &= k_{pre,i}^{cha} e^{\frac{-\Delta H_i^{cha}}{RT}}, \end{aligned} \tag{4.22}$$

where A_{pre}^{cha} and k_{pre}^{cha} are the pre-exponential factors, E_{act}^{cha} and ΔH^{cha} represent the activation energies. The values of these parameters are listed in Table 4.26.

The gasification of char with steam and the water-gas shift reaction are combined to form the following equation:

$$\mathrm{C} + 1.25\,\mathrm{H_2O} \rightarrow 0.75\,\mathrm{CO} + 0.25\,\mathrm{CO_2} + 1.25\,\mathrm{H_2} \tag{4.23}$$

Table 4.25 Pressure along the fuel reactor height (experiment and simulation)

Product	Mass fraction	Unit
Char	46.28	(wt%)
CH_4	9.18	(wt%)
CO	23.16	(wt%)
CO_2	1.33	(wt%)
H_2	2.68	(wt%)
H_2O	5.78	(wt%)
N_2	2.41	(wt%)

Table 4.26 Kinetic parameters for Boudouard reaction (Ohlemüller et al. 2017)

Parameter	Value	Unit	Parameter	Value	Unit
A_{pre,CO_2}^{cha}	1.3×10^3	(1/bar s)	$\Delta H_{CO_2}^{cha}$	−112.7	(kJ/kmol)
E_{pre,CO_2}^{cha}	141.8	(kJ/kmol)	$k_{pre,CO}^{cha}$	2.9×10^{-6}	(1/bar)
k_{pre,CO_2}^{cha}	3.9×10^{-5}	(1/bar)	ΔH_{CO}^{cha}	−154.7	(kJ/kmol)

and the total reaction rate is determined as:

$$r^{cha}_{H_2O} = \frac{dX^{cha}_{H_2O}}{dt} = \frac{k^{cha}_{H_2O} p_{H_2O}}{1 + K^{cha}_{H_2O} p_{H_2O} + K^{cha}_{H_2} p_{H_2}} \cancel{\left(1 - X^{cha}_{H_2O}\right)}, \tag{4.24}$$

Here, the symbol $X^{cha}_{\mathrm{H_2O}}$ is the conversion of char due to steam and water-gas shift reactions, $p_{\mathrm{H_2O}}$ is the partial pressure of steam and $p_{\mathrm{H_2}}$ is the partial pressure of hydrogen. The kinetic rate reaction constants are calculated by the Arrhenius equation with the values, given in Table 4.27 according to (Ohlemüller et al. 2017).

The reaction rates of oxygen carrier (R5, R6 and R7) with reductive gases i are calculated using the transport capacity of the oxygen carrier R^{OC}, time for complete solid conversion $\tau^{\mathrm{OC}}_{red,i}$ and the conversion of the oxygen carrier X^{OC}_{red}:

$$r^{\mathrm{OC}}_{red,i} = \frac{dX^{\mathrm{OC}}_{red,i}}{dt} = R^{\mathrm{OC}} \frac{3}{\tau^{OC}_{red,i}} \left(1 - X^{\mathrm{OC}}_{red}\right)^{\frac{2}{3}}, \tag{4.25}$$

The time for the complete solid conversion $\tau^{\mathrm{OC}}_{red,i}$ is determined by the molar density ρ^{OC}_{mol}, the grain radius r_{gra}, the stoichiometric coefficient for metal oxide conversion $b_{red,i}$, the kinetic rate reaction constants $k_{red,i}$, the concentration of the reductive gases $c^n_{red,i}$ and the reaction order of each component using the following relation:

$$\tau^{\mathrm{OC}}_{red,i} = \frac{\rho^{\mathrm{OC}}_{mol} r_{gra}}{b_{red,i} k_{red,i} c^n_{red,i}}, \tag{4.26}$$

with the kinetic constants of reductive gases, listed in Table 4.28.

Table 4.27 Kinetic parameters of the combined reaction of steam gasification and water-gas shift reaction

Parameter	Value	Unit	Parameter	Value	Unit
A^{cha}_{pre,H_2O}	1.1×10^3	(1/bar s)	$\Delta H^{cha}_{H_2O}$	−88.3	(kJ/kmol)
E^{cha}_{pre,H_2O}	120.4	(kJ/kmol)	k^{cha}_{pre,H_2}	1.0×10^{-5}	(1/bar)
k^{cha}_{pre,H_2O}	4.1×10^{-4}	(1/bar)	$\Delta H^{cha}_{H_2}$	−151.9	(kJ/kmol)

Table 4.28 Kinetic constants of reductive gases (Abad et al. 2011)

Parameter	CO	H_2	CH_4	Unit
$b_{red,i}$	1.45	1.45	5.78	[–]
n	0.8	1	1	[–]
$A^{red,i}_{pre}$	0.1	0.062	9.8	$(\mathrm{mol}^{1-n}\ \mathrm{m}^{3n-2}/\mathrm{s})$
$E^{red,i}_{act}$	81	65	135	(kJ/kmol)
ρ^{OC}_{mol}	13,590	13,590	13,590	$(\mathrm{mol/m^3})$
r_{gra}	1.25×10^{-6}	1.25×10^{-6}	1.25×10^{-6}	(m)

The conversion of the oxygen carrier $X_{red,i}^{OC}$ is calculate using the following equation:

$$X_{red}^{OC} = \frac{m_{ox}^{OC} - m^{OC}}{R^{OC} m_{red}^{OC}}, \tag{4.27}$$

where m^{OC} represents the actual mass of the oxygen carrier, R^{OC} denotes the oxygen transfer capacity, m_{ox}^{OC} and m_{red}^{OC} are the masses of the complete oxidized and reduced oxygen carrier, respectively.

A mesh (about 100,000 grid cells) with an unstructured part at the bottom and a structured part at the upper part is generated. Figure 4.111 shows the test rig of the fuel reactor and the boundaries of the domain. Steam enters the fuel reactor from the bottom boundary "Distributor inlet". The left entry "Burner inlet" is for the start-up burner that was not in operation during the experiments, but it is required for cooling through nitrogen flushing. From the right boundary "Ilmenite inlet" nitrogen from the loop-seals, coal and ilmenite enter the fuel reactor. At the top of the bed, the outlet is defined as a continuous outflow for the gas and solid phases, and the pressure is set at an ambient atmosphere (boundary conditions are summarised in Table 4.29). For the tangential velocity and the granular temperature of the solid phase at the wall, the Johnson and Jackson model is applied (Johnson and Jackson 1987). Phase 1 has a no-slip condition at the "wall" boundary and the specularity coefficient of phase 2 was set to a value of 0.01. The turbulence is

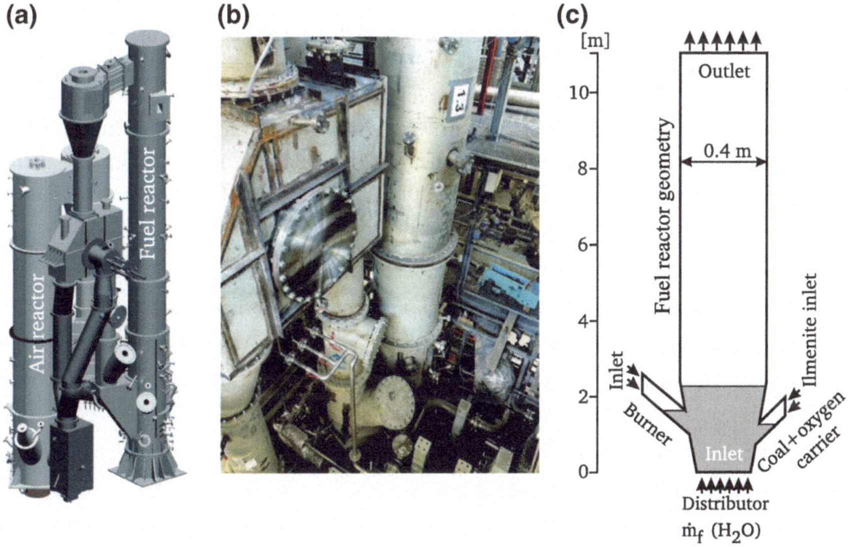

Fig. 4.111 The pilot plant: **a** 1 MW_{th} CLC, **b** test rig of CLC fuel reactor and **c** geometry and simulation set-up of CLC fuel reactor

Table 4.29 Boundary conditions of the fuel air reactor model

Location	Temperature (°C)	Mass flow rate (kg/s)	Species mass fraction (wt%)	
Distributor inlet	365	0.06	H_2O	100
Burner inlet	35	0.014	N_2	100
			Propane	0
Ilmenite inlet	35	0.032	RC	53
		0.028	N_2	47
	900	User-Defined-Function	Fe_2O_3	50
			TiO_2	50
Pressure outlet	950	0		

specified at the inlets by a turbulent intensity of 10% and the respective hydraulic diameter. The simulation is initialized with 67.3 kg of ilmenite (50% Fe_2O_3 and 50% TiO_2 at temperature of 900 °C), which corresponds to a bed height of approximately 0.4 m with a solids volume fraction of 53%. The initial velocity field and pressure of the gas and solid phases are set to zero. The time step size was set to a value of 0.005 s and 26,000 time steps (total simulation time 130 s) were calculated.

The numerical results obtained from the developed two-fluid model/ thermochemical reaction mechanism are compared with the measurement data obtained from the 1 MW_{th} chemical-looping pilot plant. In Table 4.30, the pressure values of experiment and simulation are presented. The pressures were measured at seven different heights and have averaged over the duration of the experiment. The simulated pressure shows good agreement over the whole fuel reactor height. At a height of 0.085 m, the calculated value is close to the measured pressure. The second value at a height of 0.243 m is approximately 10 mbar higher in the experiment than in the simulation. This is an indication of a slightly different particle distribution in the simulation. The next five values show again a good agreement, while the last value differs about 10 mbar from the measurement. Reason for this difference is the "outlet" boundary condition of zero gauge pressure in the simulation. In real plant, a small pressure loss exists because of the following cyclone of the fuel reactor.

Table 4.30 Pressure along the fuel reactor height (experiment and simulation)

Height	Unit	Experiment	Simulation	Absolute deviation
0.085 (m)	(mbar)	63.55	60.76	−2.79
0.243 (m)	(mbar)	59.65	48.3	−11.35
0.805 (m)	(mbar)	38.72	36.97	−1.75
2.140 (m)	(mbar)	30.71	28.05	−2.66
4.285 (m)	(mbar)	22.44	19.34	−3.1
6.73 (m)	(mbar)	18.88	11.17	−7.71
9.8 (m)	(mbar)	12.8	2.79	−10.01

A comparison between the mean values of the numerical results and experimental data regarding the gas concentrations of carbon dioxide, carbon monoxide, methane, hydrogen and nitrogen in the period of time from t = 100 s to t = 130 s at the outlet of the fuel reactor is shown in Table 4.31. The difference between the measured concentration of carbon dioxide in the experiment and the simulation is approximately eleven percentage points. This is an indication that the reaction between carbon monoxide and methane with ilmenite is not running properly. The model overestimates the carbon monoxide and methane concentrations at the outlet. For carbon monoxide, there is a discrepancy of approximately nine percentage points and for methane the difference is about two percentage points. The simulated values of hydrogen and nitrogen also show considerable deviation of approximately nine percentage points. During the experiments, the temperature was measured at the top of the fuel reactor, where the mixture of gas and particle phases leaves the reactor to enter the cyclone. The mean temperature was 948 °C in the test rig. The mean temperature in the simulation at the same height in the period of time between t = 100 s to t = 130 s is 805 °C and therefore approximately 140 °C lower than in the experiment. The low temperature slows down the reaction rates as they depend on the temperature. Furthermore, the released reaction enthalpy falls with the decreasing reaction rate. The reason for the insufficient reaction enthalpy is the poor contact between gas and solid phases due to the Syamlal and O'Brien drag model.

One reason for this discrepancy between the simulation and the experiment could be due to the wrong distribution of the species in the fuel reactor. Figure 4.112 shows the mass fractions of char (a), carbon monoxide (b), carbon dioxide (c), hydrogen (d) and methane (e) at time step t =100 s. As char was defined to be a component of phase 1, it does not fall into the bottom of the fuel reactor (dense bed), but follows the gas flow. However, the coal in the experiment had a mean diameter of 3.1 mm and therefore the diameter of the char particles after the pyrolysis was only slightly smaller. An immediate entrainment of the particles with the gaseous species is rather unlikely. Due to the coarse size of the coal particles, it is expected that the char falls into the dense bed and reacts with the incoming steam to form carbon monoxide, carbon dioxide and hydrogen. Afterwards, these gases form carbon dioxide and water when they react with ilmenite, particularly in the dense bed. This behaviour of char is also the reason for the formation of carbon dioxide, carbon monoxide and hydrogen in the upper part of the fuel reactor. The gasification reactions of char with water and carbon dioxide take place here mainly. Two solutions are suggested to correct the behaviour of char in the simulation. The

Table 4.31 Gas concentrations at the fuel reactor outlet (experiment and simulation)

Parameter	Unit	Experiment	Simulation	Absolute deviation
CO_2 (outlet)	(vol%)	21.7	10.6	−11.1
CO (outlet)	(vol%)	3.9	12.7	+8.8
CH_4 (outlet)	(vol%)	2.5	4.8	+2.3
H_2 (outlet)	(vol%)	5.3	14.2	+8.9
N_2 (outlet)	(vol%)	66.6	57.7	−8.9
Temperature (outlet)	(°C)	948	805	−143

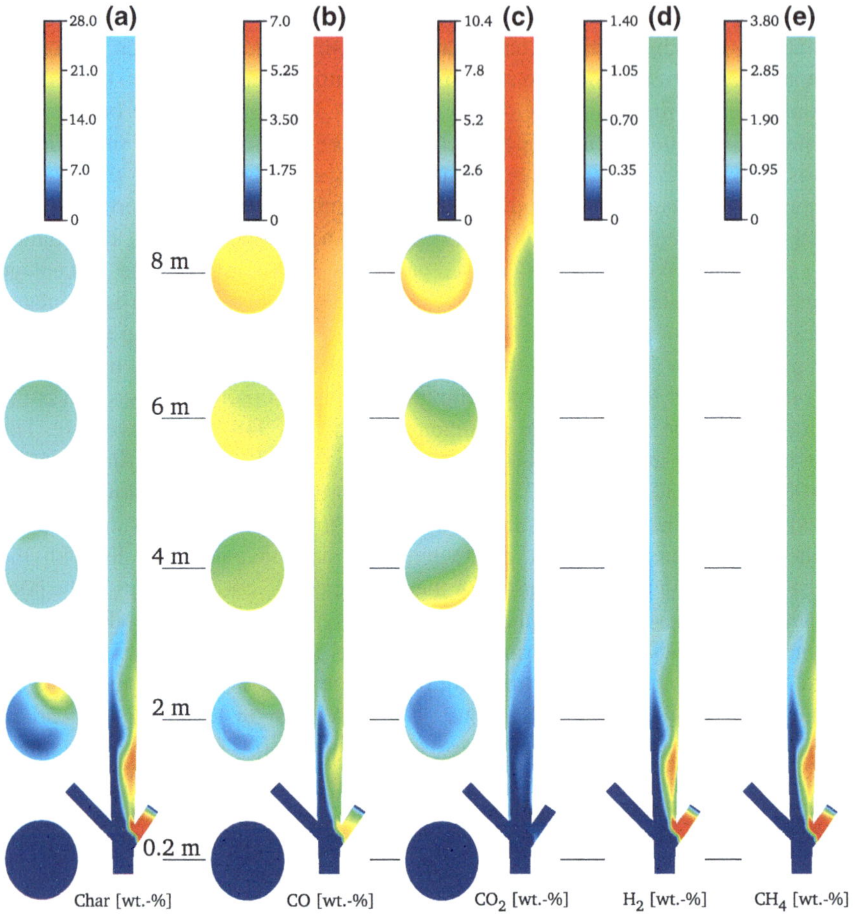

Fig. 4.112 Snapshots for the spatial distributions of gaseous species and solid at time step 100 s: **a** mass fraction of char, **b** mass fraction of carbon monoxide, **c** mass fraction of carbon dioxide, **d** mass fraction of hydrogen and **e** mass fraction of methane

first possibility is the definition of a third phase for the char particles. This has the advantage that an individual mean diameter can be defined. Nevertheless, a third phase complicates the numerical solution of the problem. An easier approach is the definition of char as part of phase 2. This would correct the atypical behaviour of char during the simulation and does not complicate the modelling and solution of the problem. However, these suggestions have no influence on the distribution of methane in the simulation as it results only from the pyrolysis of coal. Even though the simulated pressure profile fits well with the measured one, an explanation for the difference could be the wrong particle distribution due to the applied Syamlal and O'Brien drag model resulting in an insufficient contact between gas and particle phases. The low temperature in the fuel reactor is also a reason for the low conversion of methane, hydrogen and carbon monoxide.

According to the numerical result obtained and the comparison with the measurement, the following points are proposed in order to improve the model predictions:

- The gasification reaction rates are calculated by neglecting the actual conversion of char. This simplification leads to a slightly increased reaction rates. In further simulations, a mean conversion over the whole fuel reactor will be calculated.
- The distribution of species showed that the char does not fall into the dense bed of the fuel reactor. Reason for this unwanted behaviour is the definition of char as part of the gaseous phase. Since the mean diameter of raw coal was approximately 3 mm during the experiments, different behaviour of char is expected. The char particles should fall into the dense bed, reaction with steam in order to form carbon monoxide, carbon dioxide and hydrogen. After the gasification, products react with ilmenite also mostly in the dense bed. In order to avoid this behaviour of char, it has to define the char as a component of the second phase or as a third phase.
- The outgoing solid mass flow rate has a big influence on the available fresh material in the fuel reactor. A comparison of the simulated outgoing mass flow rate of solid with the mass flow rate calculated from a process model (heat and energy balance) performed in ASPEN PLUS showed deviations. Here, the drag models are the most important factor regarding the characteristic of gas solid flows. Conventional drag models (e.g. Syamlal and O'Brien, Wen and Yu) usually over-predict drag forces and under-predict slip velocities. Newly developed drag models, which are capable of describing the momentum between gas and solid phase more precisely have been developed recently. A parametric study of different drag models should be carried out, in addition to the variation of restitution coefficients and specularity coefficients in order to show their influences on the outgoing mass flow rate of solid.
- A comparison of the temperature in the simulation and the experiment showed that the calculated temperature is too low. An adjustment of the initial conditions and the definition of char as a part of the second phase, combined with an advanced drag model (e.g. three-zone, four-zone and energy minimization multi-scale drag model) will improve the results.
- The modelling of turbulence in dispersed flows is complex and usually is neglected. This assumption is justified for dense gas-solid flows as the effect of turbulence on the velocity is small. However in circulating fluidized beds, zones with low volumetric void fractions may be existent, which in turn means that the modelling of turbulence is of importance. In this work, the statistical mixture realisable $k - \varepsilon$ turbulence model was used to consider turbulent fluctuations. In order to improve the modelling of turbulence in gas-solid flows, further research in terms of experimental and numerical investigation has to be performed.

4.2.3 Single-Particle Model

The single-particle method is suitable, if detailed information on the physical and chemical properties of solid phase is required. Compared to continuum methods

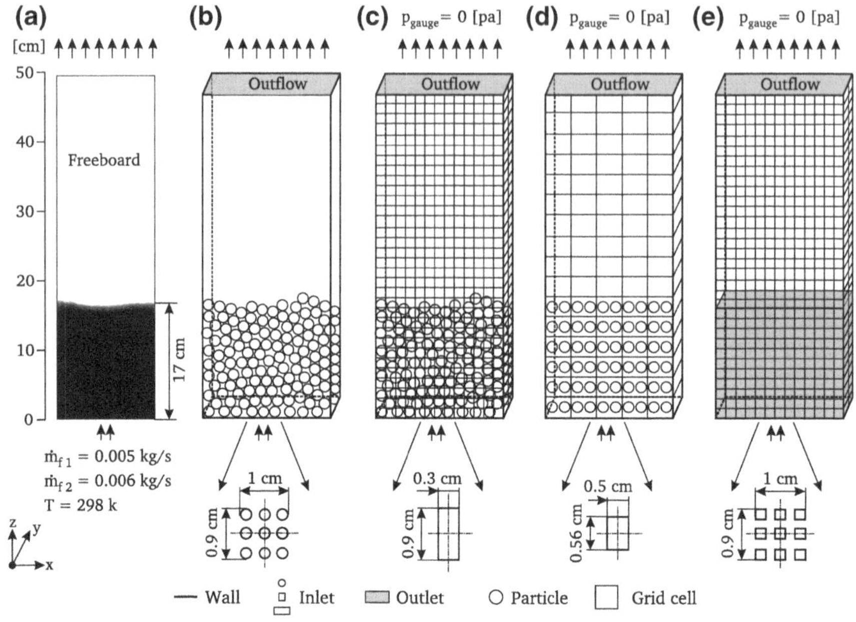

Fig. 4.113 Quasi-2D fluidized bed: (**a** and **b**) geometry and schematic diagram of test rig, **c** boundary and initial conditions for the DEM model, **d** boundary and initial conditions for the MP-PIC model and **e** boundary and initial conditions for the two-fluid model

(quasi-single-phase and two-fluid model), the single-particle model shows better agreement towards the measurement, but it is computationally expensive, especially when the number of particles increases to a few millions. The Euler-Lagrange/MP-PIC and Euler-Lagrange/DEM models are applied to simulate the lab-scale cold-flow spouted fluidized bed (see Fig. 4.105). In the Fig. 4.113, the schematic boundary and initial conditions of both models are displayed.

4.2.3.1 Euler-Lagrange/MP-PIC Model

The commercial programme CPFD-BARRACUDA®, based on the Euler-Lagrange/MP-PIC formulation, is applied to predict the hydrodynamics behaviour of the gas spouted fluidized bed (Sect. 4.2.2.1). In MP-PIC model, the Eulerian description of the fluid phase and the Lagrangian representation of the solid phase are combined. The collision between particles can be determined by an isotropic inter-particle stress tensor (see Sect. 3.2.4). The model geometry and the computational mesh have been generated using CPFD-BARRACUDA programme. The block-structured grid is applied and the number of grid cells is 25,000. In the CPFD-BARRACUDA, the grid cell volume should be much bigger than the particle size, which in turn results in coarse grid. The gas inlet is modelled as a centrally placed nozzle. The solid phase (bed inventory of about 0.75 kg) consists of monodisperse spherical particles with an average diameter of 2.5 mm. The static bed height is 17 cm and the bulk solid fraction is ca. 0.6. The initial velocity and pressure fields of the gas phase are zero. Two different fluidization mass flow rates (0.005 kg/s and 0.006 kg/s) are considered to analyse the ability of the used simulation approach to predict the hydrodynamics behaviour of the gas spouted fluidized bed. For comparison purposes with Euler-Lagrange/DEM and two-fluid model, the numerical results of the MP-PIC model are presented in Sect. 4.2.3.2.

4.2.3.2 Euler-Lagrange/DEM Model

In the DEM model, the change of the translational and angular velocities of each particle is computed by forces and momentums balances. By integration the Newtonian equations of motion, the particle trajectory and the angular displacement can be determined. In the following sections, the offset-method, the particle-grid method and the particle search-grid method are presented in addition to different application examples using the DEM model.

DEMEST Code Validation

The DEM programme "DEMEST" was first developed by Götz (2006) and further extended to model reactive gas-solid flows by Alobaid (2013). In the DEM programme, the discretisation of the non-linear, coupled and partial differential equations is carried out with the finite volume method. For the solid phase, the single-particle method combined with deterministic collision detection model is applied. The parallel simulation can be performed by special computer clusters as well as by commercial computers with multi-core processors. With the aid of a 10-gigabit network, the workstation clusters are distributed and communicated with one another via the TCP/IP protocol. The data exchange between the processors is settled through the message passing interface (MPI). The visualisation of calculated results is achieved by separate codes that are programmed by means of MATLAB functions and libraries. The programme has the following features:

- Euler-Lagrange/DEM model (4-way coupling) and thermochemical reaction mechanism
- Mono and polydisperse spherical particles
- Fine fluid grid in combination with multi-grid method for fluid phase calculation
- Particle-grid for solid phase calculation
- Particle search-grid (advanced algorithms for collision detection)
- Offset method for calculation of volumetric void fraction, momentum and heat transfers

The tool and software used in developing the DEMEST code include Microsoft Visual C++®, MPICH2, MATLAB® and Windows®. The flow diagram of the DEMEST includes the following steps: the continuity, Navier-Stokes, energy and species transport equations of the fluid phase are first discretised. The generated linear equations are solved using an iterative method. The obtained velocity, pressure and temperature are stored for each cell in the fluid field. By transferring the fluid physical variables from the fluid grid to the particle-grid and thus to the particle phase, the aerodynamic surface forces can be determined. Employing DEM model, the short-range forces between particles and the volume forces are calculated. The translational and angular velocities of particles can be solved by integrating the Newton's equations of motion and the angular momentum, respectively. The particle temperature is obtained by solving the energy equation for each particle. Subsequently, the particle-fluid interaction namely the volumetric void fraction, the momentum and heat transfers are determined in the particle-grid and transferred to the fluid cells.

For the model validation, the quasi-2D fluidized bed test rig (see Sect. 4.2.2.1) is simulated employing the DEMEST programme. The forces acting on the particles are the particle volume forces (gravitational and buoyancy forces), the aerodynamic surface forces (drag, pressure, Saffman and Magnus forces) and the short-range surface forces (contact force). As mentioned above, at the mass flow rates used (0.006 kg/s and 0.005 kg/s), the vertical position of particles should not be higher than 50 cm. If this occurs however, these particles are removed from the subsequent time steps. Besides the block-structured fine fluid grid with number of nodes: 72 (W) $\times$ 12 (D) $\times$ 440 (H), a total of 380,160, three increasingly coarser multi-grids are generated in order to accelerate the fluid calculation. With the first level coarse grid, the number of cells is reduced from 380,160 to 47,520 (number of nodes: 36 (W) $\times$ 6 (D) $\times$ 220 (H)). For the second and third level coarse grids, the numbers of cells are further decreased to 11,880 with number of nodes: 18 (W) $\times$ 6 (D) $\times$ 110 (H) and 2970 with number of nodes: 18 (W) $\times$ 3 (D) $\times$ 55 (H), respectively. The computation of the solid phase is performed on the particle-grid with a resolution of 47,520 cells (number of nodes: 36 (W) $\times$ 6 (D) $\times$ 220 (H)). The deterministic detection of particle-particle/wall collisions is performed on the particle search-grid (number of nodes: 40 (W) $\times$ 4 (D) $\times$ 160 (H), a total of 25,600 cells. In order to accelerate the calculations, the computational domain is decomposed into eight identical sub-blocks (B1 B2 B3 B4 and B6 B7 B8 B9) and a central narrow block (B5) for the inlet (see Fig. 4.114).

The numerical calculations are carried out with commercial computers with multi-core processors. In the DEMEST programme, particles with 2.5 mm diameter are generated and distributed uniformly in the entire computational domain. The initially positioned particles have zero initial kinetic energies. The particles are then left to drop freely into the test rig. The particles experience the iterative processes of move and drop under gravity. During this phase, the mass flow rate is set equal to zero, i.e. the particles undergo only to volume and short-range forces including the contact force. The aerodynamic forces that result from the interaction with the fluid

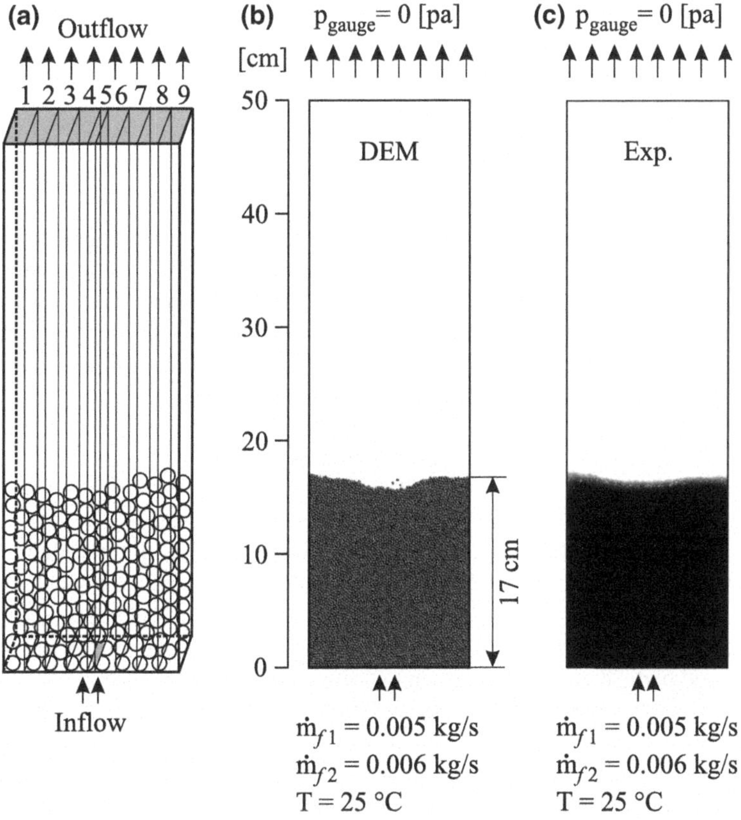

Fig. 4.114 Numeric model of the quasi-2D fluidized bed: **a** geometry divided in 9 sub-blocks, **b** the solid distribution at start point in simulation and **c** the solid distribution at start point in the test rig

phase are not considered in this stage. The final static bed height is 17 cm and the bulk solid fraction is ca. 0.6. For air mass flow rates (0.006 and 0.005 kg/s), the numerical result of the DEM model is compared against 500 ms operation of a Plexiglas spouted fluidized bed. Furthermore, the numerical simulation results obtained by ANSYS-FLUENT that is based on the two-fluid formulation and by CPFD-BARRACUDA that is based on the Euler-Lagrange/MP-PIC formulation are presented for the purpose of comparison. The comparison is achieved with the aid of the visual observation of solid distribution, the absolute bed height and the equivalent bubble diameter. Furthermore, the quantitative comparison of vertical particle velocities is evaluated.

First, the numerical results using a mass flow rate of 0.006 kg/s is discussed, followed by a comparison of the numerical predictions for the mass flow rate of 0.005 kg/s. Besides the visual observation of solid distribution, the bed behaviour during the fluidization is quantitatively analysed in terms of bed expansion (absolute bed height h_{bed}) and bubble size (equivalent bubble diameter d_{bub}^{equ}). The circular equivalent bubble diameter will be used to analyse the bubble growth over a time range of 500 ms:

$$d_{bub}^{equ} = \sqrt{\frac{4}{\pi} A_{bub}}, \tag{4.28}$$

The absolute bed height and bubble area A_{bub} for the computational snapshots are measured using the open source software GIMP. Additionally, the depth-averaged particle velocity along the bed width is discussed. The measurement errors for the bed height and the bubble area are assumed to be $\Delta h_{bub} = \pm 0.2$ cm and $\Delta A_{bub} = \pm 2$ cm^2, respectively. The uncertainties by measuring the equivalent bubble diameter can then be calculated using Eq. (4.28) as:

$$\Delta d_{bub}^{equ} = \frac{\Delta A_{bub}}{\sqrt{\pi A_{bub}}}. \tag{4.29}$$

For reasons of simplification, the calculated uncertainties will not be included in the plotting graphs in order to facilitate the results evaluations. In Figs. 4.115 and 4.116, snapshots for the spatial distribution of solid in the Euler-Lagrange/DEM model and Euler-Lagrange/MP-PIC model as well as the contours of vertical particle velocity and particle volume fraction in two-fluid model are presented in the period of time 0–475 ms for the mass flow rate of 0.006 kg/s. For the subpanels (b), (c), and (d), red indicates a particle velocity of +1 m/s, while blue represents a negative velocity of −1 m/s. The two-fluid model results shown in subpicture (e) indicate a volume fraction of 0.6 using red and a volume fraction of zero using blue. At the start point t = 0, the static bed height of three models is 17 cm. In contrast to the Euler-Lagrange/DEM model, the two-fluid model and Euler-Lagrange/MP-PIC model is not able to build the bed surface profile. This is due to the fact that the solid phase is assumed to be like a fluid in the two-fluid model. In the MP-PIC approach, the particles are generated in each grid cell with the assumption that no contacts occur between the particles by the reason of collision detection. In the period of time between t = 25 ms and t = 75 ms, a bubble is slowly formed near the gas inlet and accordingly the particles in the bed start accelerating. The bubble area and the bed height are almost the same in all models. From time point t = 75 ms, the good agreement decreases. The bubble expands very fast in the Euler-Lagrange/MP-PIC model and thus the bed height increases to 33 cm at t = 150 ms with a relative error of 25%. On the other hand, the two-fluid model shows an opposite effect. Here, the bubble grows slowly and the bed height reaches about 23 cm at t = 150 ms with a relative error of 8%. In the time period t = 150 ms and t = 300 ms, the numerical predictions of the Euler-Lagrange/MP-PIC and two-fluid approaches agree with the DEM model and accordingly with the experiment. However, there is a considerable deviation in the bubble size and the bed height. It can be noticed that the bubble area increases sharply in the Euler-Lagrange/MP-PIC model, while the two-fluid model underestimates the measured data.

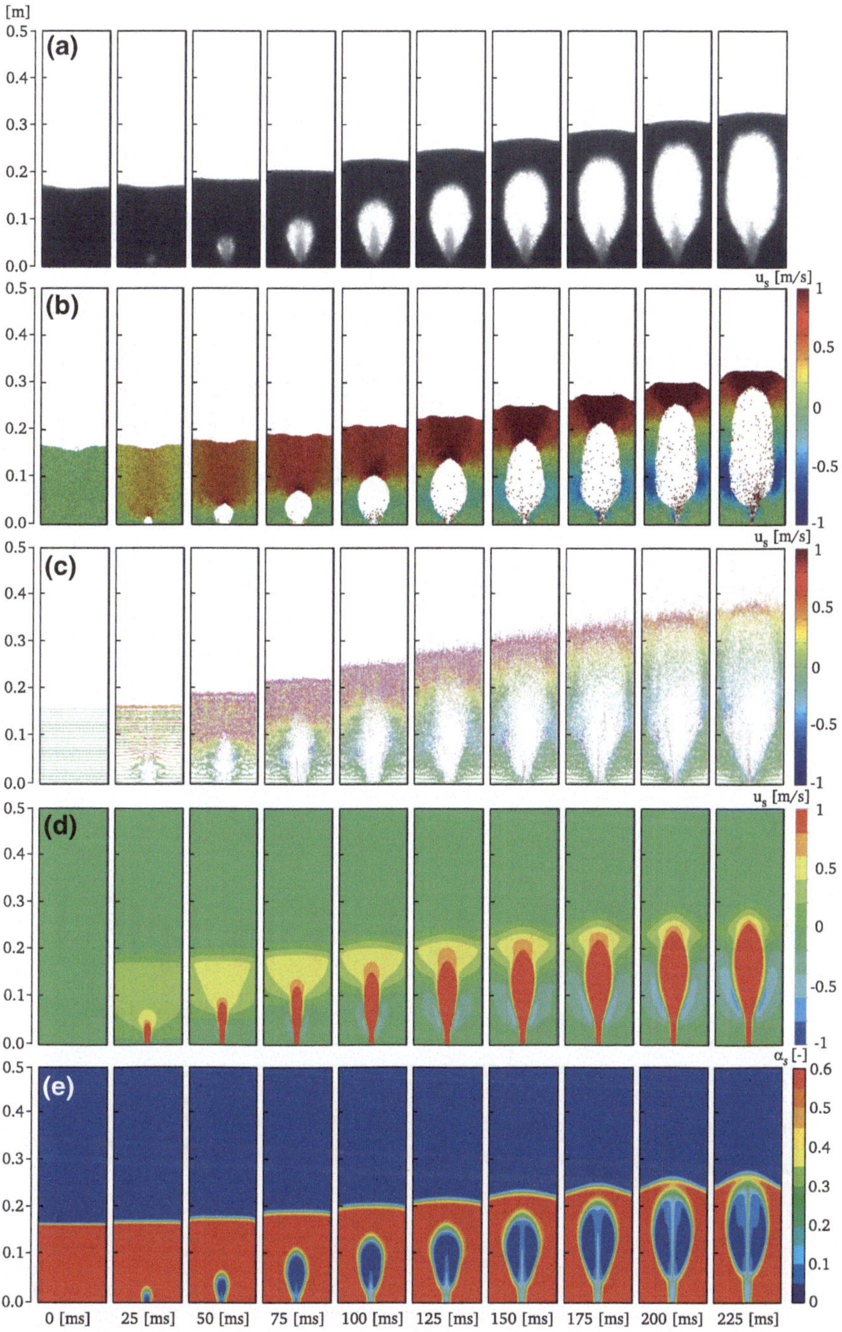

Fig. 4.115 Snapshots for the spatial distribution of solid (mass flow rate 0.006 kg/s; time range between 0 and 225 ms): **a** experiment, **b** Euler-Lagrange/DEM model (legend: vertical particle velocity), **c** Euler-Lagrange/MP-PIC model (legend: vertical particle velocity), **d** two-fluid model (legend: vertical particle velocity) and **e** two-fluid model (legend: particle volume fraction)

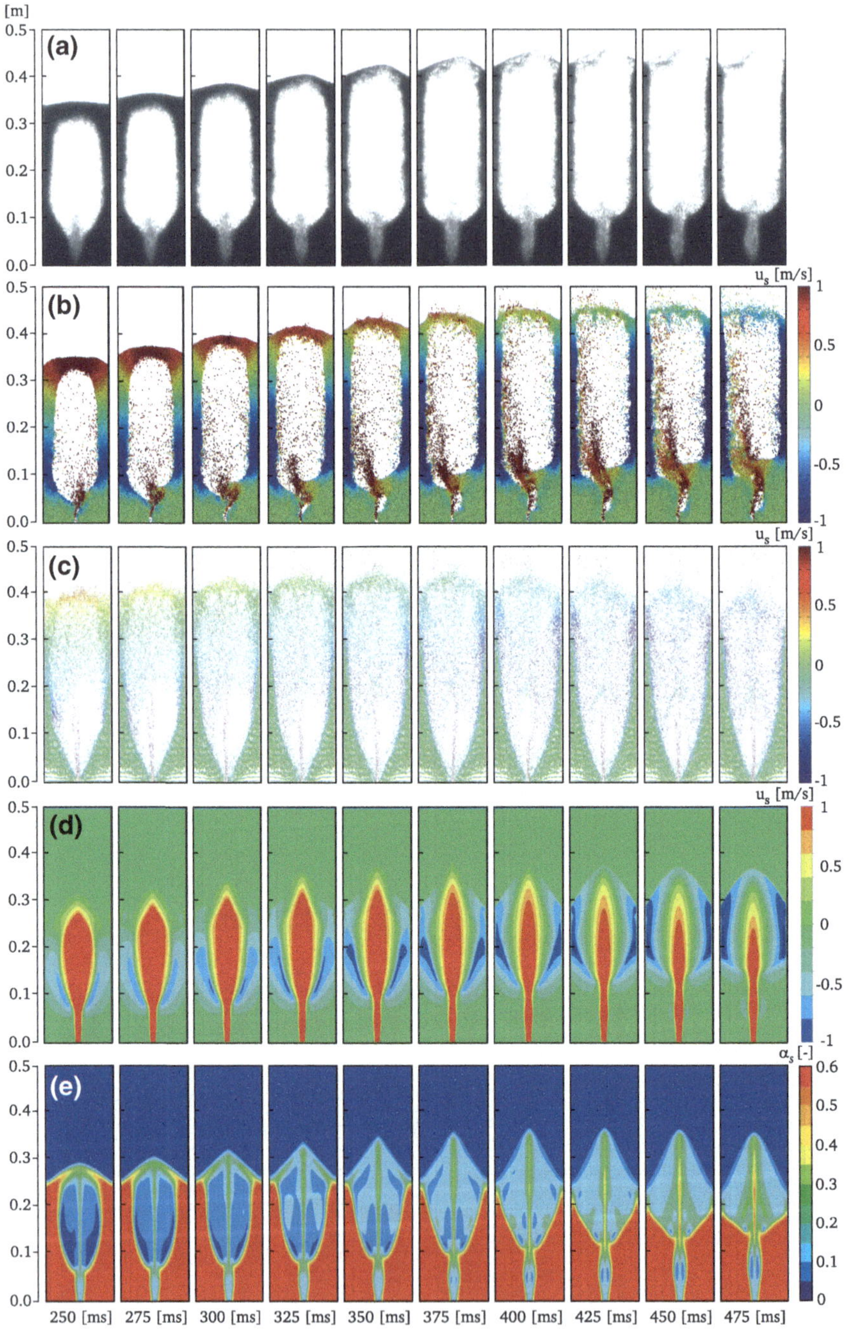

Fig. 4.116 Snapshots for the spatial distribution of solid (mass flow rate 0.006 kg/s; time range between 250 and 475 ms): **a** experiment, **b** Euler-Lagrange/DEM model (legend: vertical particle velocity), **c** Euler-Lagrange/MP-PIC model (legend: vertical particle velocity), **d** two-fluid model (legend: vertical particle velocity) and **e** two-fluid model (legend: particle volume fraction)

At time point t = 325 ms, the bed height in the Euler-Lagrange/MP-PIC model reaches a maximum value of about 46 cm. The two-fluid model, by contrast, seems to be clearly below the average measured bed height with a relative error of 20%. The asymmetry in the solid distribution can be seen in the Euler-Lagrange/DEM model, especially in the nozzle region. In the Euler-Lagrange/MP-PIC and two-fluid approaches, the spatial distribution of solid and the contours of particle volume fraction are axially symmetrical relating to the y-z midplane. At t = 350 ms, the typical flow regime of a spouted bed can be clearly observed. The gas penetrates the entire bed, dividing the bed into three regions, namely the spout in the centre, the fountain above the initial bed surface and the annulus between the spout and the walls. In the inlet region, a dilute flow is existent and the particles move upward, while in the annular flow region the particles move downward to the core again. The bed height and the bubble area in the Euler-Lagrange/MP-PIC model starts decreasing from t = 350 ms until the bubble collapses completely. In general, the bubble area is relatively larger than the Euler-Lagrange/DEM model. A considerable discrepancy between the calculated contours of particle volume fraction in the two-fluid method and the DEM models is clearly noticed, but the predicted annular flow region is still reasonable and agrees with the experiment. At t = 475 ms, the bed height in the Euler-Lagrange/MP-PIC model reaches approximately 40 cm with a relative error of about 11%, while the bed height in two-fluid model is about 34 cm with a relative error of 30%.

Figure 4.117 displays the quantitative comparison of the experimentally observed and simulated bed expansions including the bed height and equivalent bubble diameter as a function of time. During the first stage of the bubble formation (t = 0 and t = 100 ms), a good agreement between two-fluid model and the experimental data can be shown. The Euler-Lagrange/MP-PIC model, by contrast, exceeds the measured bed height and the bubble size. In the period of time between t = 100 ms and t = 300 ms, the predicted bed height and bubble area in the two-fluid model fall below the experimentally observed ones, while the Euler-Lagrange/MP-PIC model overestimates the measured bed expansion. The maximum bed height of about 46 cm is achieved after 440 ms in the experiment as well as in the

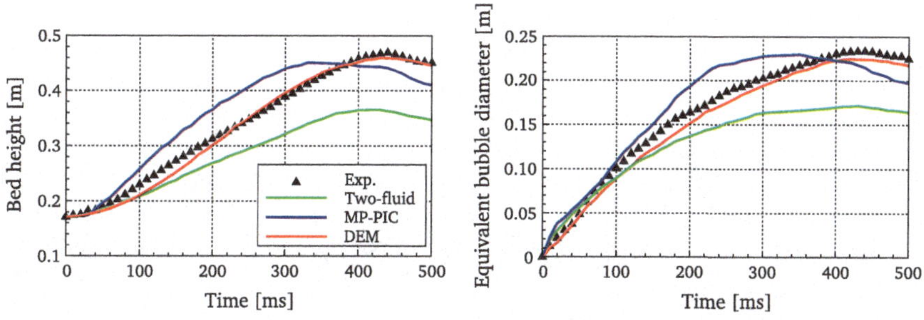

Fig. 4.117 Bed expansion behaviour for Euler-Lagrange/DEM model, Euler-Lagrange/MP-PIC model and two-fluid model at gas mass flow rate 0.006 kg/s, (left) bed height and (right) equivalent bubble diameter

Euler-Lagrange/DEM model. The Euler-Lagrange/MP-PIC model reaches the maximum bed height (46 cm) and the maximum bubble size faster than the measured data at t = 325 ms. From t = 350 ms, the bubble collapses and thus the bed height and the bubble size decrease accordingly. On the other hand, the two-fluid model reaches the maximum bed height without delayed in time similar to the experiment. However, this model predicts smaller expansion with a maximum relative error of about 22%. In the final stage of the bubble formation, this error becomes larger and reaches approximately 25%. The deviation in the bubble size increasers gradually from 3% at t = 150 ms and reaches a maximum value of about 30% at t = 480 ms. The vertical particle velocity profiles on the midline at different locations of the bed height (z = 15 cm and z = 25 cm) are presented in Figs. 4.118 and 4.119. The heights of the observation lines are selected, so that the flow regime of the spouted bed including the main three regions (i.e. spout, annulus and fountain) can be covered during the fluidization.

Figure 4.118 displays the vertical particle velocity profiles at z = 15 cm for the Euler-Lagrange/DEM, Euler-Lagrange/MP-PIC and two-fluid models at gas mass flow rate of 0.006 kg/s at different time steps (150, 250, 350 and 450 ms). In the spout region, the computed vertical particle velocities in the Euler-Lagrange/MP-PIC and two-fluid models show good agreement with the Euler-Lagrange/DEM model. In the period of time between t = 250 ms and t = 475 ms, the negative particle velocity can be noticed in the annulus. These negative velocities of the particles indicate that the particles move downwards in the annular region in the direction of the inlet. Here, the two-fluid model underestimates the magnitude of

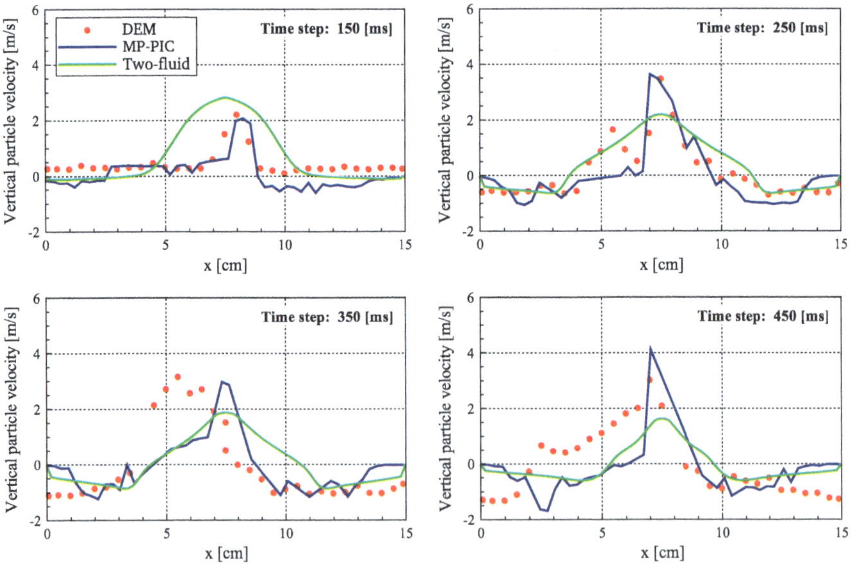

Fig. 4.118 Vertical particle velocity profiles (z = 15 cm) for Euler-Lagrange/DEM model, Euler-Lagrange/MP-PIC model and two-fluid model at gas mass flow rate 0.006 kg/s

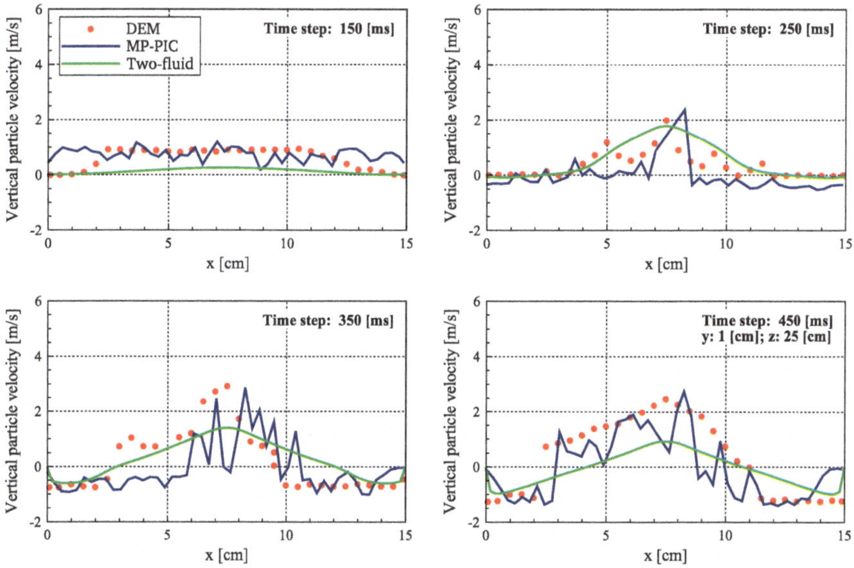

Fig. 4.119 Vertical particle velocity profiles (z = 25 cm) for Euler-Lagrange/DEM model, Euler-Lagrange/MP-PIC model and two-fluid model at gas mass flow rate 0.006 kg/s

the particle velocities in this backflow, while the Euler-Lagrange/MP-PIC model can correctly predicate the magnitude of the particle velocities in the annular region. However, the backflow is not directly found on the walls, but rather it is shifted from the wall with about 1 cm.

In Fig. 4.119, the quantitative comparison of vertical particle velocity profiles at z = 25 cm for the Euler-Lagrange/DEM, Euler-Lagrange/MP-PIC and two-fluid models for the gas mass flow rate of 0.006 kg/s is shown at different time steps (150, 250, 350 and 450 ms). After 150 ms from the start, the simulated particle velocities using the Euler-Lagrange/MP-PIC approach agree very well with the DEM model. The simulated particle velocities in the two-fluid model, by contrast, fall short of the DEM results. At t = 350 ms, the negative values of the vertical particle velocities refer to a strong downward movement of the particles in the direction of the jet region, i.e. circulation in the annulus. The particles in the fountain region move slowly upwards, leading to positive particle velocities in the positions between x = 4 cm and x = 11 cm. In the period of time between t = 350 ms and t = 500 ms, the particles in the fountain region are separated towards the bed wall, moving back to the jet zone by the reason of the gravity force. It can be seen that Euler-Lagrange/MP-PIC and two-fluid models show good agreement towards the Euler-Lagrange/DEM model. The spout and annular regions are clearly observed in all models. It should be mentioned here that the symmetry in the solid distribution in the two-fluid model (see Figs. 4.115 and 4.116) can also be noticed regarding the vertical particle velocity profiles.

Further comparison is carried out between the Euler-Lagrange/DEM, Euler-Lagrange/MP-PIC and two-fluid models for the mass flow rate of 0.005 kg/s. Only the quantitative comparison of the experimentally observed and simulated bed expansions as a function of time is presented (see Fig. 4.120). At the beginning, all simulation approaches have an initial bed height of nearly 0.17 m. However, slight deviations in the initial bed height are observed, which is correlated to the model used. The maximum packing limit for the Euler-Lagrange/MP-PIC model is slightly bigger than for the two-fluid model, so that the particles occupy more space in the same control volume. Hence, the initial bed height for the MP-PIC model is slightly smaller than for the two-fluid model. In the Euler-Lagrange/DEM model, the stiffness coefficient influences the allowed penetration between particles. In general, the higher the stiffness coefficient, the less penetration allowed between particles. The defined stiffness coefficient in the Euler-Lagrange/DEM model is set equal to 4.1×10^5 N/m, which seems to be a reasonable value in terms of allowed particle penetration depth and resulting accuracy due to particle-particle contact forces. All in all, the three methods describe the experimental initial bed height of 0.17 m quite accurately.

Within the first 100 ms, all numerical methods are capable of predicting the typical bubble formation that starts from the centre of the inlet nozzle. This can be observed in Fig. 4.120, where the variations of bed height and equivalent bubble diameter over time are shown for all approaches. From point of time t = 100 ms, the Euler-Lagrange/MP-PIC model overestimates the measured bed height, while the Euler-Lagrange/DEM and two-fluid models fall short the experimentally observed bed height. In the period of time between t = 150 ms and t = 300 ms, the bed height increases sharply in the Euler-Lagrange/MP-PIC model. The bed heights in the Euler-Lagrange/DEM and two-fluid models, by contrast, are close to the experiment. The measured bubble continues slowly growing and the top layer of the particles loses more of its kinetic energy till it reaches a standstill with a maximum measured bed height of about 25 cm. In the Euler-Lagrange/MP-PIC model, the bed height reaches its maximum value approximately 100 ms earlier than the measurement. Similar to the experiment, the Euler-Lagrange/DEM model reaches

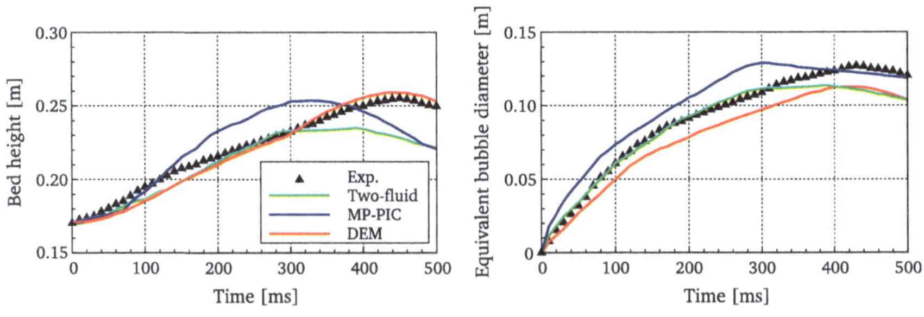

Fig. 4.120 Bed expansion behaviour for Euler-Lagrange/DEM model, Euler-Lagrange/MP-PIC model and two-fluid model at gas mass flow rate 0.005 kg/s, (left) bed height and (right) equivalent bubble diameter

its maximum bed height at t = 425 ms. The two-fluid model underestimates the measured bed height with a relative error of about 8%. After the bubble collapses at t = 500 ms, the DEM model exceeds the experimentally observed bed height with a relative error of 3%. The bed heights obtained from the Euler-Lagrange/MP-PIC and two-fluid models seem to be clearly below the average measured bed height with a relative error of 12%. The equivalent bubble diameter is basically matched by all three modelling approaches. However, the bubble size is over-predicted by the Euler-Lagrange/MP-PIC model with a maximum relative error of about 22%. One possible explanation is the high packing limit, which results in a larger resistance for the passing gas. As a result, the exerted drag force from the gas is over-predicted, which results in higher expansion rates at the beginning of the numerical simulation. In the period of time between t = 0 and t = 300 ms, the two-fluid model predicts the same experimental equivalent bubble diameter, while the Euler-Lagrange/DEM model underestimates the experiment with a maximum relative error of 18%. At approximately t = 350 ms in the experiment, the bubble starts to burst and the particles begin to fall into the spout region. Here, the Euler-Lagrange/DEM and two-fluid models under-predict the experimental bubble diameters. The equivalent bubble diameter simulated by the Euler-Lagrange/MP-PIC model shows good agreement, although the bed height (see Fig. 4.120, left) and the bubble shape is completely different from the measurements (the snapshots for the spatial distribution of solid are not illustrated here). This is probably caused by the non-accurate shear and collision forces between the particles, in addition to the over-predicted drag force.

In Figs. 4.121 and 4.122, the vertical particle velocity profiles on the midline at different locations of the bed heights (z = 2 cm and z = 12 cm) at different time steps are illustrated for the gas mass flow rate of 0.005 kg/s. The heights of the observation plans are selected according to the mass flow rate, so that the typical flow pattern of a spouted bed including the three main regions can be observed. Looking at Fig. 4.121, the vertical particle velocity profiles obtained from the two-fluid and Euler-Lagrange/MP-PIC approaches show a good agreement towards the DEM model. Here, the spout in the centre of the bed (between x = 6 cm and x = 9 cm) can be clearly seen. The particles accelerate in this region and move upwards with a moderate velocity (ca. 3 m/s) into the fountain. All models indicate zero particle velocity in the dead zone (between x = 0 to x = 6 cm and x = 9 cm to x = 15 cm).

Figure 4.122 shows the development of the spouted zone at the location z = 12 cm during the time. At t = 50 ms, the spouted zone is not formed yet. However, the particle velocities from the Euler-Lagrange/MP-PIC model are higher in comparison to the other modelling methods. The particle velocities at t = 150 ms and t = 250 ms confirm also the quantitative observations, where the Euler-Lagrange/DEM, Euler-Lagrange/MP-PIC and two-fluid models have positive velocities within the spout region and negative velocities in the annulus region. Because the particles are absent at the walls, the time-sampled statistical data results in zero particle velocities for the two-fluid model. The typical regions of the spouted bed at t = 350 ms and t = 450 ms are recognized, namely the spout in the bed

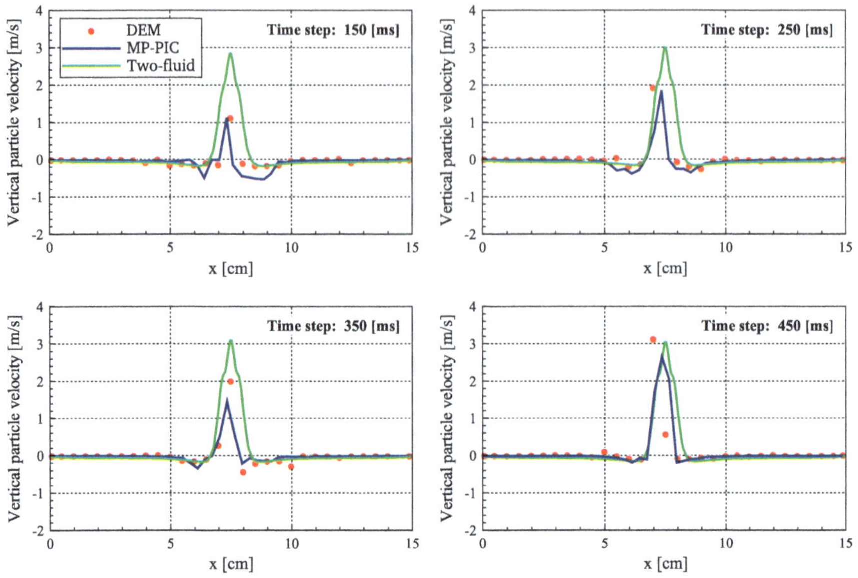

Fig. 4.121 Vertical particle velocity profiles (z = 2 cm) for Euler-Lagrange/DEM model, Euler-Lagrange/MP-PIC model and two-fluid model at gas mass flow rate 0.005 kg/s

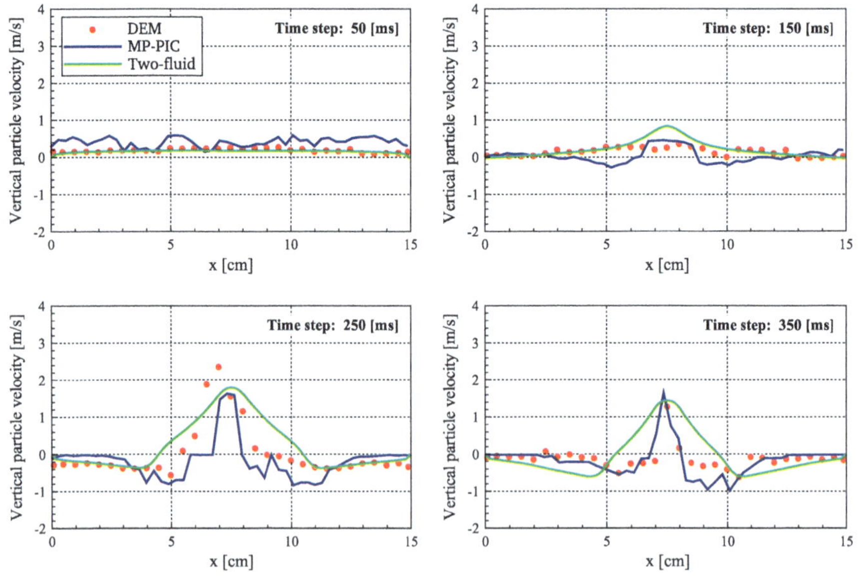

Fig. 4.122 Vertical particle velocity profiles (z = 12 cm) for Euler-Lagrange/DEM model, Euler-Lagrange/MP-PIC model and two-fluid model at gas mass flow rate 0.006 kg/s

centre, the fountain above the bed surface and the annular region. The dead zone width is here decreased to be between x = 0 to x = 3 cm and x = 12 cm to x = 15 cm at t = 350 ms as well as to be between x = 0 to x = 2 cm and x = 13 cm to x = 15 cm at t = 350 ms.

As previously demonstrated, the simulated results including the spatial distribution of solid, the bed height and the equivalent bubble diameter is analysed for different mass flow rates within 500 ms. It is clear that the numerical results show a good agreement with the measurements. However, deviations of bed height expansion and bubble shape developments between simulations and experiments, especially in the nozzle region, may be caused by several reasons:

- The inlet assumption of the centrally placed nozzle in case of the Euler-Lagrange/DEM and Euler-Lagrange/MP-PIC models as well as the set of nine squares in the two-fluid model instead of the nine holes in the test rig represents the most significant measure for these deviations (see Fig. 4.113).
- Other important factor is the long-range adhesive force, such as the electrostatic force. For different material pairings, the electrostatic force that occurs between the oppositely charged surfaces plays an important role in fluidization. While the particles with opposite charges attract each other, the particles with identical polarity are repulsive. During the fluidization, the particle-particle/wall collisions lead to generation of electric charges that result in particle accumulation on the walls (Sowinski et al. 2010). An examination of the test rig after the fluidization experiment with material paring of glass-Plexiglas indicates considerable electrostatic charges on the particles (Alobaid and Epple 2013). Due to the continued particle-particle/wall collisions, the electrostatic charges of particles are increased gradually. The models are not able to simulate the cluster build-up when the adhesive forces, particularly the electrostatic force, are neglected. Therefore, the particles are distributed more evenly in the computation domain. It can be concluded that for the material combination of glass and Plexiglas, the electrostatic force is a crucial parameter. As a result, the tensile forces become effective and lead to significant damping of particles collisions and disturbances. This stabilizing effect causes an early achievement of a quasi-steady flow state.
- Further reason regarding the deviations between simulations and experiments is related to the inaccurate solid parameters that are used in the calculation. The simulation of the solid phase requires a lot of material and geometry properties of particle-wall pairing that are not completely available. In the DEM model, the applied stiffness coefficient of 4.1×10^5 N/m gives a good compromise between accuracy of the results and the simulation duration. In fact, the real stiffness coefficient is within the range 10^4–10^8 N/m (Alobaid et al. 2014a; Di Renzo and Di Maio 2004). Increasing the stiffness coefficient combined with a smaller particle time step leads to improve the prediction of particle collisions, especially in the bottom region of the reactor. Regarding the Euler-Lagrange/MP-PIC model, a detailed analysis on the particle probability distribution function may reveal their influence, e.g. in dense zones, which could improve

the overall flow pattern prediction. However, the results of the probability distribution function depend on the cell size, which at the moment is required to be much larger than the particle size.
- Additional investigations into the drag models used are necessary to improve the simulation results for all approaches together with more developed empirical models that are applied to calculate the friction forces and particle stresses.
- The calculation of fluid phase with the aid of finite volume method involves several errors such as modelling, discretisation and iteration errors (Ferziger and Perić 2002), which is directly related to the grid resolution. As a result of fluid-solid interaction, these numerical errors of the fluid phase are forwarded to the particle phase. Apart from the calculation error of the fluid phase, the particle calculation itself is associated with numerical errors such as the penetration depth between the collision partners in the DEM model and the exact calculation of inter-phase values (e.g. volumetric void fraction and momentum transfer).

In conclusion, a qualitative statement and direct comparability of the computational performance for all approaches is not feasible. This is due the fact that the applied mesh quality differs considerably between the applied approaches. While the DEM and two-fluid models contain a numerical mesh of approximately 380,000 cells, the Euler-Lagrange/MP-PIC simulation is performed using a mesh of 25,000 cells by the reason of cell volume restriction (the cell volume should be much bigger than particle size).

Offset-Method

The modelling of the fluid phase in gas-solid flows requires the extension of the conservation equations for the single-phase flow. This is due to the fact that the control volumes of the computational domain consist of different fluid and solid volume fractions. The impact of solid on fluid phase (inter-phase values), i.e. volumetric void fraction and the rates of momentum and heat transfer should be known as volume average values in each control volume. These average values are functions of geometrical and physical properties of each particle (position, velocity and temperature) that is located in the investigated control volume. The exact calculation of the inter-phase values requires considering all particles that are completely located inside the control volume and other particles that are partly located in this CV (spherical sections). Due to the computational effort required, this can only be achieved in simple cases, where the number of particles is relatively low. Numerical approximations should therefore be introduced. Here, the particle centre method represents the simplest approach. In this method, the spherical sections for the particles are not considered and the entire volume of the particle is allocated to the fluid cell, where the centre of this particle is found. The efficiency gained in this case is, however, associated with high uncertainty for the determination of the inter-phase values. By applying the so-called offset-method (see Fig. 4.123), which results in several numbers of the spatial displacements of the grid cells the determination of the inter-phase values can be significantly improved.

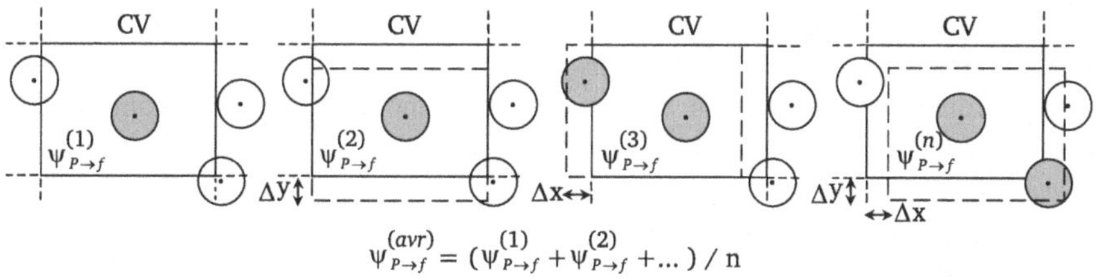

Fig. 4.123 Determination of the inter-phase values using the particle centre method in combination with offset-method

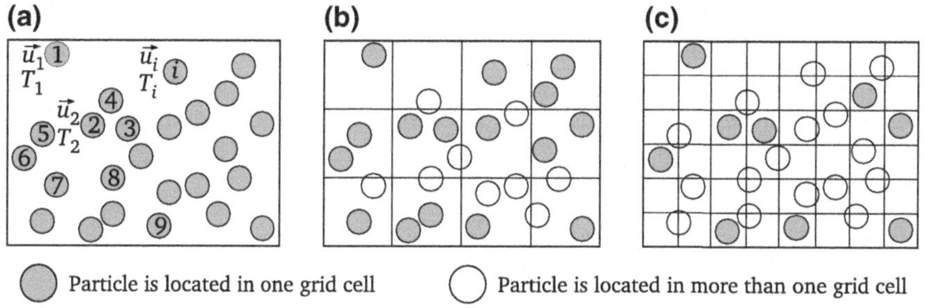

Fig. 4.124 Spatial resolution of local inhomogeneities and averaging accuracy in dependence on grid resolution: **a** coarse grid and (**b** and **c**) dense grids

In offset-method, the inter-phase values $\psi_{P\to f}$ in a grid cell are determined then by averaging all calculated values with regard to the number of displacements. For 2D simulations, 9 inter-phase values are obtained by 8 displacements in all directions with:

$$\Delta x,\ \Delta y \in [-r_P, 0, +r_P], \tag{4.30}$$

in addition to the first value without displacement. For 3D simulations, one obtains 27 inter-phase values with 26 displacements in all directions with

$$\Delta x,\ \Delta y,\ \Delta z \in [-r_P, 0, +r_P], \tag{4.31}$$

in addition to the first value without displacement.

The averaging accuracy depends strongly on the local inhomogeneities, i.e. the particle distribution in the computational domain and thus on the grid resolution. A grid refinement allows the spatial resolution of the local inhomogeneities, but on the other hand leads to an inaccuracy in the averaging process. By improving the grid resolution, the number of particles that belong to different control volumes is significantly increased (see Fig. 4.124). If the particle diameter is equal or greater than the grid size, the local averaging is not possible. In the literature, this problem is well-known and can be solved using different approaches such as Porous Media

Concept, Diffusion Based Concept, Particle Meshing Method and Two-Grid Concept (Particle-Grid Method). In these approaches, the grid resolution can be further refined independent of the particle size and the local averaging is still possible. In the Porous Media Concept, the particles are represented as porous cubes with a certain size (Link 2006), while in the Two-Grid Concept, a new grid so-called particle-grid is introduced (see Sect. Particle-Grid Method).

In order to investigate the accuracy of the particle centre method with and without the offset-method, k sample cells with known volumetric void fraction $\bar{\varepsilon}_f$ are generated. While the particles are uniformly distributed over the entire computational domain, the particles are randomly distributed in each sample cell. It is assumed that every sample cell consists of n fluid cells (see Fig. 4.125a). The volumetric void fraction is then computed in each fluid cell with the aid of both approaches. By determining the average volumetric void fraction for the n fluid cells, the volumetric void fraction of the corresponding sample cell $\varepsilon_{f,i}$ is calculated. In order to evaluate the accuracy, the standard deviation for the entire computational domain is applied:

$$\sigma_{\varepsilon_f} = \sqrt{\frac{1}{k}\sum_{i=1}^{k}\left(\varepsilon_{f,i} - \bar{\varepsilon}_f\right)^2}. \tag{4.32}$$

Furthermore, the ratio of the fluid cell length l to the particle diameter d_P is varied and the corresponding standard deviation is then determined. From Fig. 4.125b, it can be observed that the mean errors of the particle centre method with and without the offset-method increase with decreasing the ratio of the fluid cell length to the particle diameter.

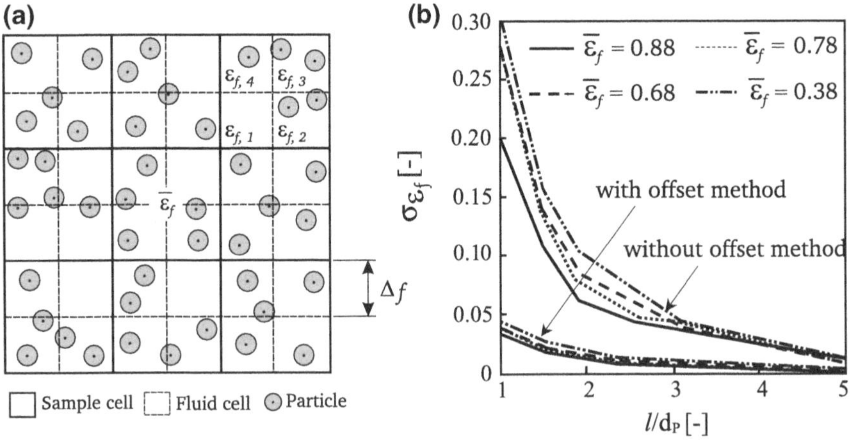

Fig. 4.125 Particle centre and offset-method in direct comparison: **a** sample cells with equal predefined volumetric void fraction but their fluid cells have different volumetric void fractions calculated by the particle centre approach with and without the offset-method, **b** standard deviation of the particle centre method with and without the offset-method plotted against the ratio Δ_f/d_P for different solid loadings

The application of the offset-method improves the calculation accuracy up to an order of magnitude. The accuracy of averaging using the offset-method depends highly on the grid resolution. Here, an optimal ratio between the grid size and the particle diameter $\Delta_f / d_P \approx [2 - 3]$ is derived. The solid loading has almost no influence on the accuracy of the calculation, whereas the accuracy of the approximation without the offset-method deteriorates with increasing solid loading.

In order to illustrate the benefits of the offset-method application to gas-solid flows, a set of simulations is performed. Here, two different operation conditions, namely air mass flow rates of 0.006 kg/s and 0.005 kg/s, are investigated. In Fig. 4.126, the simulated bed height and the equivalent bubble diameter with and without the offset-method are compared towards the measurement. At the mass flow rate of 0.006 kg/s and in the period of time between t = 0 and t = 350 ms, the bed height in both models agrees very well with the experiments. A relative small discrepancy can be detected in the model without the offset-method. From the time t = 350 ms, a large deviation in the simulated bed height occurs. At t = 500 ms, the numerically obtained bed height from the DEM model without the offset-method has a relative error of 40%. The error by calculation the equivalent bubble diameter increases over time till it reaches a maximum value of 32% during the gas bubble burst. At the mass flow rate of 0.005 kg/s, the simulated bed height using the DEM model without the offset-method agrees well with the experimentally observed one. During the last stage of the bubble formation, a small deviation can be observed. It seems that the DEM model without the offset-method can predict, on the one hand,

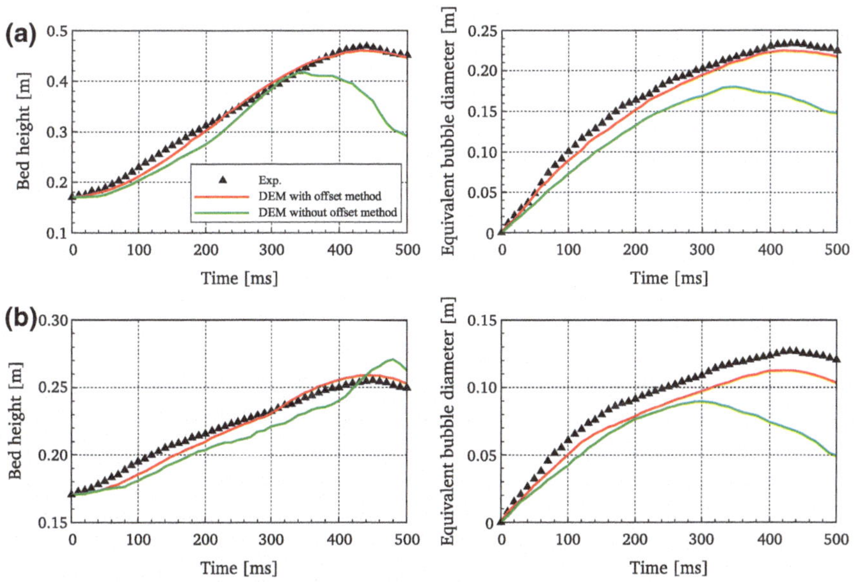

Fig. 4.126 Comparison of bed expansion with and without the offset-method: **a** mass flow rate of 0.006 kg/s and **b** mass flow rate of 0.005 kg/s

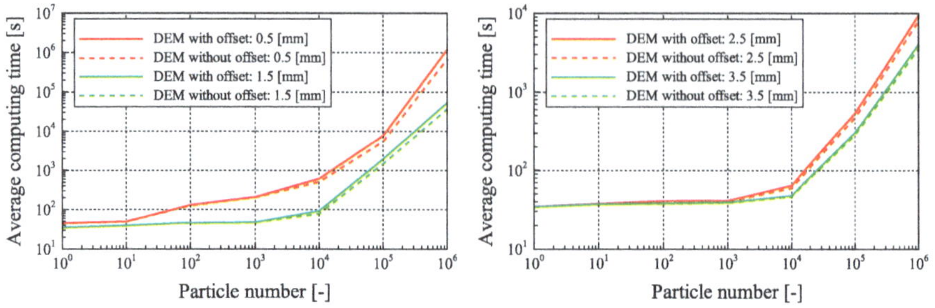

Fig. 4.127 Comparison of computational effort with and without the offset-method at different solid loadings and particle diameters

the correct behaviour of the bed at lower mass flow rates. The calculated equivalent bubble diameter reveals, on the other hand, the weakness of this model. From the t = 300 ms, the DEM model without the offset-method underestimates the bubble size with a large discrepancy of approximately 60%. According to the numerical result obtained and the quantitative comparisons, the DEM model with the offset-method matches very well with the measurements, while the DEM model without the offset-method deviates obviously. The discrepancy is attributed to the inaccurate determination of the inter-phase values.

One important point is the computational time that should always be discussed when Euler-Lagrange/DEM model is applied to simulate the hydrodynamic behaviour of dense granular flows. In this context, the additional computational effort that results from the using of the DEM model with the offset-method is calculated at different particle numbers and various particle diameters. In Fig. 4.127, the average computing time per time step is plotted against the particle number. The additional computational time rises with increasing the solid loading. Furthermore, the particle diameter has a considerable influence on the computational effort. A decrease in particle diameter leads to an increase in the average computing time per time step.

It should be mentioned here that the calculation of the momentum and heat transfers in each control volume can also be carried out using the offset-method.

Particle-Grid Method

The averaging accuracy by determination the inter-phase values depends highly on the solid distribution in the computational domain, i.e. on the grid resolution. An optimal ratio between the grid size and the particle diameter $\Delta_f / d_P \approx [2-3]$ is already suggested. A grid refinement beyond this ratio leads to an inaccuracy in the averaging process (see Fig. 4.125b). If the particle diameter is equal or falls short of the grid size, the local averaging is not possible; for example: the volumetric void fraction becomes zero (see Fig. 4.128a). By the reason of numerical simplifications

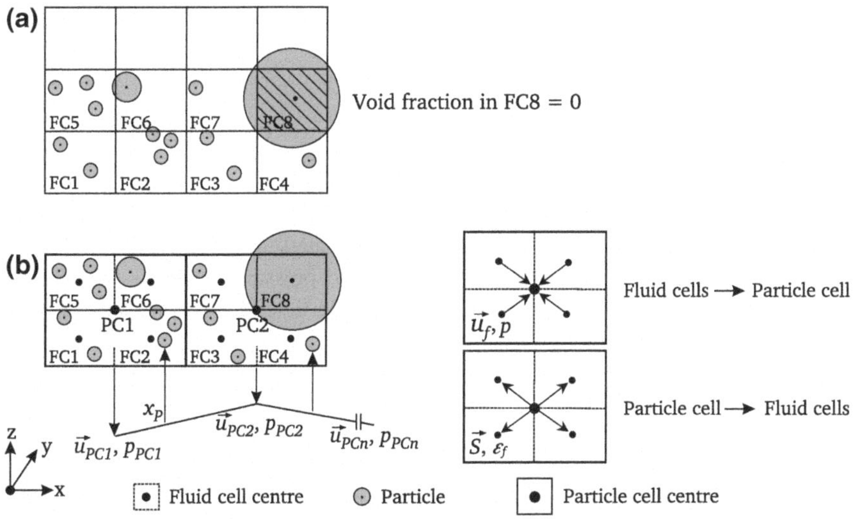

Fig. 4.128 Fluid and particle grids: **a** particle bigger than the fluid cell but smaller than the particle cell and **b** data mapping between fluid and particle cells

in the previous Euler-Lagrange/DEM works, the physical values of the fluid and solid phases are calculated in a common grid, namely the fluid grid. This assumes that the dimensions of the largest particle must be smaller than the smallest fluid grid cell. Relative big particles or polydisperse systems with a broad range of particle sizes result in a rough fluid grid resolution and thus an inaccuracy in the fluid calculation. In this study, an additional grid, so-called particle-grid, in which the physical values of the solid phase is computed, is introduced. A geometric allocation between the fluid and particle-grids is necessary to enable the data transfer from fluid to solid and vice versa. The refinement of the fluid grid resolution beyond the particle scale will not result in local extremes in the gas-fraction around the centre particle mass (Link 2006). Therefore, the proposed method is consistent with the basis of using the extended Navier-Stokes equations.

The physical values of the fluid and solid phases are determined in separated grids, namely the fluid and particle grids. The proposed procedure maintains the averaging accuracy during the determination of the interaction values, i.e. volumetric void fraction, momentum and heat transfers by selecting the optimal ratio between the particle-grid size and the particle diameter as:

$$\Delta_P/d_P \approx [2-3] \tag{4.33}$$

The particle-grid method allows the variation of the fluid grid resolution Δ_f independent of the particle size and consequently improves the calculation accuracy of the fluid phase. Here, the conservation equations of the gas phase are solved in a fine fluid grid using an iterative method. The obtained fluid physical variables are transferred from the fine fluid cells (FC) to the coarse particle cells (PC) using an arithmetic average as (e.g. uniform grid):

$$\phi_{PC} = \sum_{i=1}^{n} \phi_{FC,i} = \frac{\phi_{FC,1} + \phi_{FC,2} + \phi_{FC,3} + \ldots + \phi_{FC,n}}{n}, \tag{4.34}$$

where the parameter ϕ is either fluid velocity or fluid pressure and the symbol n denotes the number of fluid cells that are allocated in one particle cell (see Fig. 4.128b). In the particle cells, the velocity and pressure gradients of the fluid phase are built and accordingly transferred to the particle phase, for example, with the aid of a linear interpolation as follows:

$$\phi_P = \phi_{PC,i} + \frac{\phi_{PC,j} - \phi_{PC,i}}{x_{PC,j} - x_{PC,i}}(x_P - x_{PC,i}). \tag{4.35}$$

The parameter ϕ_P represents the fluid gradient values on the particle at the position x_P. The same procedures are repeated to the other axes (y and z). By transferring the fluid physical variables to the particles, the aerodynamic surface forces can be determined. The particle volume forces as well as the short-range forces between particles (e.g. contact force) are defined employing the DEM model. The translational and angular velocities of particles are then determined with integrating Newton's equations of motion. Finally, the volumetric void fraction, the momentum transfer and the heat transfer are calculated in the particle cells using the averaging procedures; for example, the particle centre method in combination with offset-method. Subsequently, the inter-phase values are forwarded to the allocated fluid cells using a weighting function:

$$\Phi_{FC,i} = fun(\Phi_{PC}) \tag{4.36}$$

Here, the parameter Φ denotes to one of the inter-phase values.

It should mentioned here that each fluid cell will participate in the fluid calculation even when is fully occupied with a particle (see fluid cell FC8 in Fig. 4.128b). In this cell, the volumetric void fraction is not equal to zero. This is due to the fact that the volumetric void fraction will be calculated in the coarse particle cell (PC2). Obviously, the volumetric void fraction in the PC2 has a value bigger than zero. The obtained volumetric void fraction in the PC2 is then transferred to FC3, FC4, FC7 and FC8 avoiding the zero porosity.

In order to investigate the influence of using the particle-grid method on the simulation accuracy, two different numerical models of the quasi-2D fluidized bed (see Sect. 4.2.2.1) are generated and then compared with the measurements. In both cases, the simulation parameters and the applied boundary conditions are identical. While in the first case the gas and solid phases are calculated by the same fluid grid, each phase is computed on its grid in the second case. For the first case as mentioned previously, the dimension of the smallest fluid grid cell must be much coarser than the largest particle in the system, but smaller than the characteristic scales in a fluidized bed such as bubbles (Anderson and Jackson 1967). Accordingly, the number of grid nodes in x direction, y direction and z direction must be less than (bed width/particle diameter: 15/0.25 = 60 nodes), (bed width/ particle diameter: 2/0.25 = 8 nodes) and (bed height/particle diameter: 100/ 0.25 = 400 nodes), respectively, i.e. the finest grid resolution is (60 × 8×400 = 192,000 cells). As a result, the fluid grid resolution for the first case is selected equal to (56 × 6×380 = 127,680 cells). The fluid grid in the second case has the resolution of 72 × 12 × 440 = 380,160 cells. The particle phase is computed on a particle-grid with a resolution of 36 × 6×220 = 47,520 cells. Two different operation conditions, namely air mass flow rates of 0.005 kg/s and 0.006 kg/s, are applied for the case 1 and the case 2.

In Fig. 4.129, the simulated bed heights are compared with the experiment over time. For the high mass flow rate of 0.006 kg/s, both models reproduce the measurement in the period of time between t = 0 and t = 100 ms. From time point t = 150 ms, the simulated bed height in the case 1 starts deviating from the test rig. In contrast to the case 1, the case 2 still shows a very good agreement. After 420 ms, the bed height in the test rig and in the case 2 reaches a maximum value of about 45 cm. On the other hand, the simulated bed height for the case 1 differs from the experiment with a relative error of about 30%. The numerically observed bubble size agrees very well with the measurement. From time point t = 300 ms, the simulated equivalent bubble diameter in the case 1 deviates from the case 2 as well as from experiments. Here, the bubble size exceeds the measured data with a

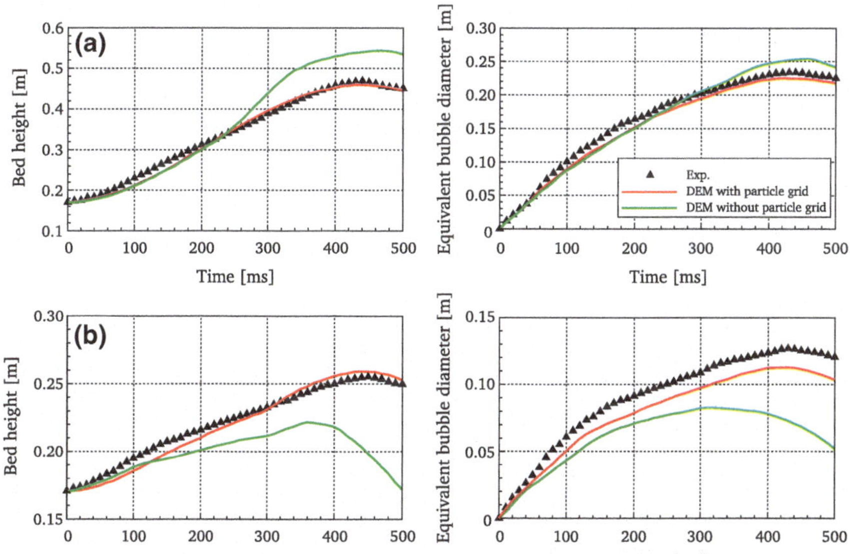

Fig. 4.129 Comparison of bed expansion with and without the particle grid method: **a** mass flow rate of 0.006 kg/s and **b** mass flow rate of 0.005 kg/s

relative error of 13%. At the mass flow rate of 0.005 kg/s, a considerable discrepancy in the simulated bed height can be seen in the period of time between t = 150 and t = 350 ms. During the bubble burst, a large deviation between the numerically observed bed height and experiments occurs, suggesting that the first model is not able to predict the bed expansion. At t = 500 ms, a relative error of 32% in the bed height for the case 1 is registered. The error in the simulated bubble size regarding the case 1 increases slightly between t = 50 ms and t = 300 ms and then increases sharply in the remaining time.

A grid dependence study shows that a further refinement of the fluid grid more than 380,160 cells improves the simulation accuracy very slightly. However, the average computing time increases extremely.

The computational efficiency of the DEM model with and without the particle-grid method is also investigated. In Fig. 4.130, the average computing time per time step is plotted against the particle number at different particle diameters. The additional computational time, which results from the different numbers of the fluid grid cells used in both cases, is considered in this comparison. By increasing the solid loadings, a considerable increase in the computing time per time step is observed. At constant fluid time step, the computational effort is substantially raised by increasing the solid loading. This can be easily explained by the reason of the DEM model. Here, each particle has a set of equations to be solved. Furthermore, the computational effort of collision detection is correlated with the number of investigated particles. The particle diameter has also an influence on the computational effort if the particle-grid method is employed.

At constant particle number, the average computational time rises considerably with decreasing the particle diameter. This significant increase in the computing time results from the decreasing in the particle time step that is a function of the stiffness coefficient and the particle diameter. Since the stiffness coefficient is set constant equal to (5×10^5 N/m) in these simulations, the particle time step is then related to the particle diameter. At lower particle numbers (10^0, 10^1, 10^2, 10^3 and 10^4), the flattening in curves is caused due to the fact that a part of the computing time is required for the calculation of the fluid phase. This part is negligibly small for higher numbers of particles (10^5 and 10^6), but it takes at low particle numbers

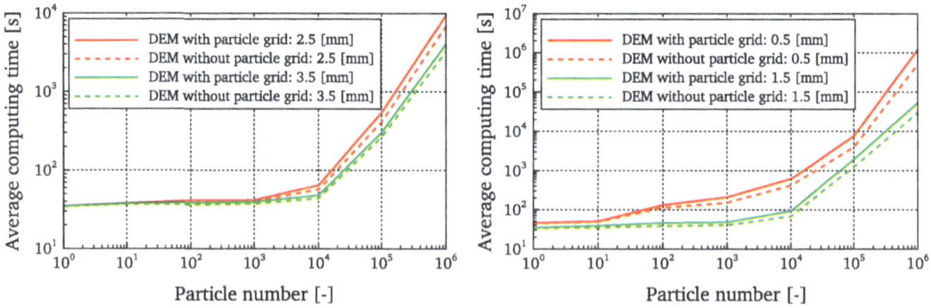

Fig. 4.130 Comparison of computational effort with and without the particle-grid method at different solid loadings and particle diameters

into account. In general, it can be concluded that the application of the particle-grid method for DEM simulations relates always to an additional computational time. This is due the fact that the data transfers between fluid and particle grids lead to additional numerical operations.

The ratio between the fluid and particle grids has an influence on the computational time as well as on accuracy of the calculation. However, the selection of the particle-grid size depends on the largest particle diameter in the system. In order to maintain the accuracy of the averaging procedure, the particle-grid size is selected around $\Delta_P \approx 2d_P^{\max}$. For polydisperse flows with a wide range of particle size distributions, the particle-grid size should be selected very coarse. In this case, not only the computational effort of the particle-grid method decreases, but also the accuracy of the calculation declines. The recommended ratio that shows a very good compromise between an efficient computing time and an acceptable accuracy is between 4 and 10.

Particle Search-Grid Method

The collision detection between particles is based on the particle search-grid method (see Sect. Deterministic Collision Detection Models). This method requires an additional grid, known as the particle search-grid, which is independent of the fluid and particle grids. The correct selection of the grid size is of crucial importance in order to obtain physically meaningful results. By skilful selection of search-grid size, the computational effort may behave proportional to the number of particles. Basically, the search-grid size should not be smaller than the particle diameter in case of monodisperse systems and the largest particle in case of polydisperse systems. The effect of using a search-grid with a smaller size than the particle diameter on the simulation accuracy is here investigated. A simulation case of the quasi-2D fluidized bed is carried out using a search-grid with a high resolution $\Delta_{SG} = 0.5d_P$. The deterministic detection of particle-particle/wall collisions is then performed on the particle search-grid with a number of nodes: 120 (W) $\times$ 16 (D) $\times$ 800 (H). The simulation parameters and the boundary conditions are same as in Sect. DEMEST Code Validation. It is clearly that the DEM model with a fine search-grid deviates from the measurements (see Fig. 4.131). The kinetic energy conservation is not guaranteed and thus the simulated bed height and the equivalent bubble diameter are underestimated.

Parameter Study

The efficiency and accuracy of the Euler-Lagrange/DEM model are analysed. Different important parameters, namely stiffness coefficient and fluid time step are varied under the consideration of various solid loadings and particle diameters. For the efficiency study, a 3D fluidized bed model is constructed (see Fig. 4.132).

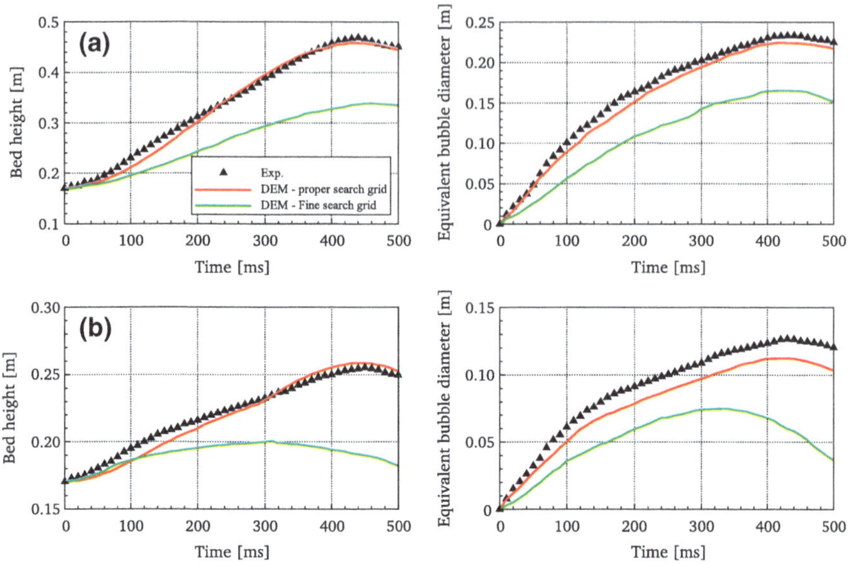

Fig. 4.131 Comparison of bed expansion using proper search-grid and fine search grid: (upper row) mass flow rate of 0.006 kg/s and (lower row) mass flow rate of 0.005 kg/s

The model has a height of 150 cm, a width of 15 cm, and a depth of 20 cm. The air is supplied through a centrally placed nozzle (1 cm (W) × 15 cm (D)), while the outlet is completely opened (15 cm (W) × 20 cm (D)). A block-structured fluid grid with number of nodes: 64 (W) × 80(D) × 640 (H) is generated. Besides the fine fluid grid, three increasingly coarser multi-grids are built. In the first level coarse grid, the number of cells is reduced to 409,600 (number of nodes: 32 (W) × 40 (D) × 320 (H)). For the second and third level coarse grids, the numbers of cells are further decreased to 51,200 with number of nodes: 16 (W) × 20 (D) × 160 (H) and 6400 with number of nodes: 8 (W) × 10 (D) × 80 (H), respectively. The calculation of solid is carried out on the particle grid with a resolution of 409,600 cells (number of nodes: 32 (W) × 40 (D) × 320 (H)). The deterministic detection of the particles collisions is performed on the particle search grid (number of nodes: 40 (W) × 40 (D) × 200 (H), a total of 256,000 cells. The monodisperse particles have the glass properties. The particle number is varied between 10^0, 10^1, 10^2, 10^3, 10^4, 10^5 and 10^6. Furthermore, the particle diameter is varied between 3.5, 2.5, 1.5 and 0.5 mm. Depending on the solid loading, the mass flow used in the efficiency study varies in the range between 0.01 kg/s and 0.5 kg/s, so that the bubble flow regime is observed. The forces acting on the particles are the gravitational, buoyancy, drag, pressure, Saffman, Magnus and contact forces. For the validation study, the previously explained quasi-2D fluidized bed model is used (see Sect. DEMEST Code Validation).

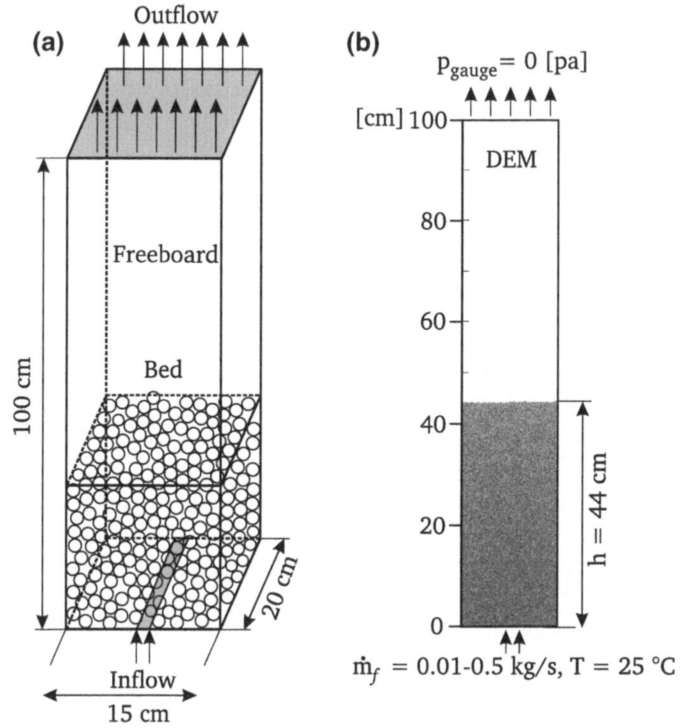

Fig. 4.132 Numeric model used for the efficiency study: **a** schematic diagram (only 100 cm presented) and **b** boundary and initial conditions, e.g. particle diameter (2.5 mm) and particle number (10^6)

Variation of stiffness coefficient

In Table 4.32, the normal and tangential stiffness coefficients are calculated for identical spherical particles with different diameters (0.5, 1.5, 2.5 and 3.5 mm) under the consideration of glass material properties. The determined stiffness coefficients have high-orders of magnitude and vary in the range between 10^8 N/m and 10^9 N/m depending on the diameter of collision partners. If the stiffness coefficients are defined according to Table 4.32 and the particle time step is calculated according to Eq. (3.288), the unphysical penetrations between the particles and the particles/walls are avoided. However, the use of realistic stiffness coefficients results in a very small

Table 4.32 Calculation of realistic normal and tangential stiffness coefficients

Stiffness coefficients (N/m)	Particle diameters (mm)			
	0.5	1.5	2.5	3.5
k^n (Eq. (3.157))	5.5×10^8	9.6×10^8	1.2×10^9	1.47×10^9
k^t (Eq. (3.291))	2.1×10^9	3.7×10^9	4.9×10^9	5.8×10^9

Table 4.33 Calculation of penetration depths for glass collision partners with different particle diameters (0.5, 1.5, 2.5 and 3.5 mm) and relative velocity of 10 m/s

Stiffness coefficients (N/m)	Penetration depths (mm)			
	$d_p = 0.5$ (mm)	$d_p = 1.5$ (mm)	$d_p = 2.5$ (mm)	$d_p = 3.5$ (mm)
$k^n = 10^2$	14	73	160	260
$k^n = 10^4$	1.4	7.3	16	26
$k^n = 10^6$	0.14	0.73	1.6	2.6
$k^n = 10^8$	0.014	0.07	0.16	0.26

particle time step and thus extremely computational time. Therefore, it is a fact that the stiffness coefficients are set lower than their real values. With decreasing the stiffness coefficients and hence increasing the particle time step, larger penetration depths can occur between the collision partners. At high relative velocity of collided particles, there is a risk to arise unrealistic penetrations. A complete penetration for the particle-wall collisions represents the worst case, since the particles leave the computational domain and accordingly are no longer available for the subsequent time steps. This leads to a continuous decline in the number of particles during the Euler-Lagrange/DEM simulations. In Table 4.33, the penetration depth is calculated depending on the stiffness coefficients for different diameters of the collision partner and a maximum relative velocity of 10 m/s.

By selecting smaller stiffness coefficients (for example 10^3 N/m), the normal penetration depth can reach up to 24 times the particle diameter. Previous numerical studies of the DEM model such as (Tsuji et al. 1992, 1993) showed that the variation of the stiffness coefficient over several-orders of magnitude has only a minor influence on the results. Therefore, a stiffness coefficient of 800 N/m is recommended by Tsuji, which was used by several other authors. Di Renzo and Di Maio(2004) have found that the proposed value of 800 N/m for the applied material properties and the simulation parameters causes a normal penetration depth up to 74 times the particle diameter. The underestimating of the stiffness coefficients produces not only unphysical penetrations, but also leads in to the shrinkage of the entire bed volume (i.e. the random-close packing limit increases). In Fig. 4.133, the bed height for 36,500 spherical glass particles (monodisperse) with a diameter of 2.5 mm and different stiffness coefficients is presented. Compared to the experiment, a reduction in the stiffness coefficient by four-orders of magnitude (from 10^2 N/m to 10^6 N/m) results in a reduction of the entire bed volume with about 10%.

In order to obtain the minimal normal stiffness coefficient that prevents the unphysical penetration, a condition should be defined. A penetration depth is unrealistic if the maximum penetration depth is greater than the particle radius ($\delta^n_{\max} > r_P$). In Table 4.34, the minimal normal stiffness coefficients are calculated using Eq. (3.292) at different particle diameters, diverse relative velocity of the collision partners and a maximum penetration depth ($\delta^n_{\max} = r_P$). Independent of the particle diameter, the resulted minimal stiffness coefficient values have a range

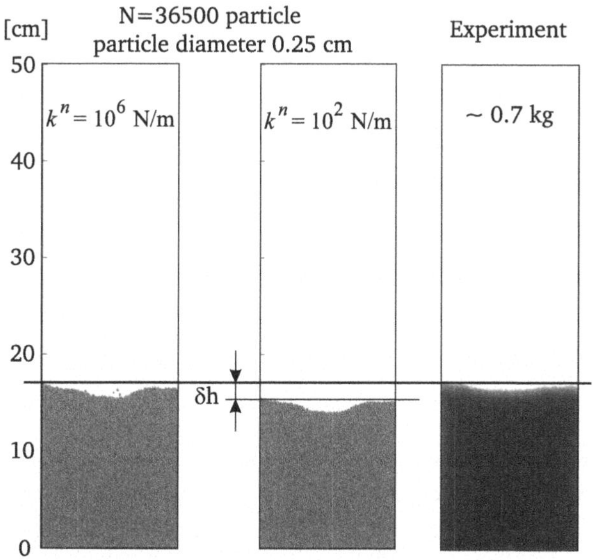

Fig. 4.133 Bed heights at different stiffness coefficients (simulation) and experiment

Table 4.34 Calculation of minimum normal stiffness coefficients at different particle diameters (0.5, 1.5, 2.5 and 3.5 mm), relative velocities (10, 5 and 2 m/s) and maximum penetration depth $\delta^n_{\max}=r_P$

Stiffness coefficients (N/m)	Particle diameters (mm)			
	0.5	1.5	2.5	3.5
k^n (Eq. (3.292)) $u^n_{\max}$ = 10 (m/s)	5.1×10^6	5.1×10^6	5.1×10^6	5.1×10^6
k^n (Eq. (3.292)) $u^n_{\max}$ = 5 (m/s)	6.3×10^5	6.3×10^5	6.3×10^5	6.3×10^5
k^n (Eq. (3.292)) $u^n_{\max}$ = 2 (m/s)	4.1×10^4	4.1×10^4	4.1×10^4	4.1×10^4

between 5×10^4 and 5×10^6 N/m and depend mainly on the relative velocity of collision partners.

The influence of the stiffness coefficient variation on the computation time at different particle numbers and diverse particle diameters is investigated (see Table 4.35). For the first simulation series, the number of particles is set constant equal to 10^0 at different particle diameters, while the stiffness coefficient is varied from 10^1 to 10^9 N/m. The number of particles is changed to 10^1 and the entire simulations are repeated again and so on. All other parameters and numerical methods are same in the simulations. The parallel simulation is performed by decomposing numerical domain in 9 sub-blocks using commercial computers with multi-core processors (4th Generation Intel i7®). For each case, ten steps are simulated and the average computing time per time step is then determined through the division of the entire simulation time by the number of calculation steps.

Table 4.35 Simulation parameters for the stiffness coefficient variation

Variable parameters	Stiffness coefficients (N/m)	10^1, 10^2, 10^3, 10^4, 10^5, 10^6, 10^7, 10^8, 10^9
	Particle numbers	10^0,10^1, 10^2, 10^3, 10^4, 10^5, 10^6
	Particle diameters (mm)	0.5, 1.5, 2.5, 3.5
Fixed parameters	Fluid time step (ms)	2
	Number of calculation steps	10
	Number of decompositions	9
Grids	Fluid multi-grid	Activated
	Particle grid	Activated

In Fig. 4.134, the effect of the stiffness coefficient variation on the average computing time per time step at different particle numbers (10^0, 10^1, 10^2, 10^3, 10^4, 10^5, 10^6) and different particle diameters (0.5, 1.5, 2.5 and 3.5 mm) are illustrated. It can be seen that an increase in the stiffness coefficient values leads to an exponential rise in the average computing time. This is evident from the order of magnitude 10^3 N/m, where a jump in the stiffness coefficient by two-orders of magnitude results in an increase in the simulation time by one-order of magnitude. The flattening of the curves at the stiffness coefficient values below 10^3 N/m is

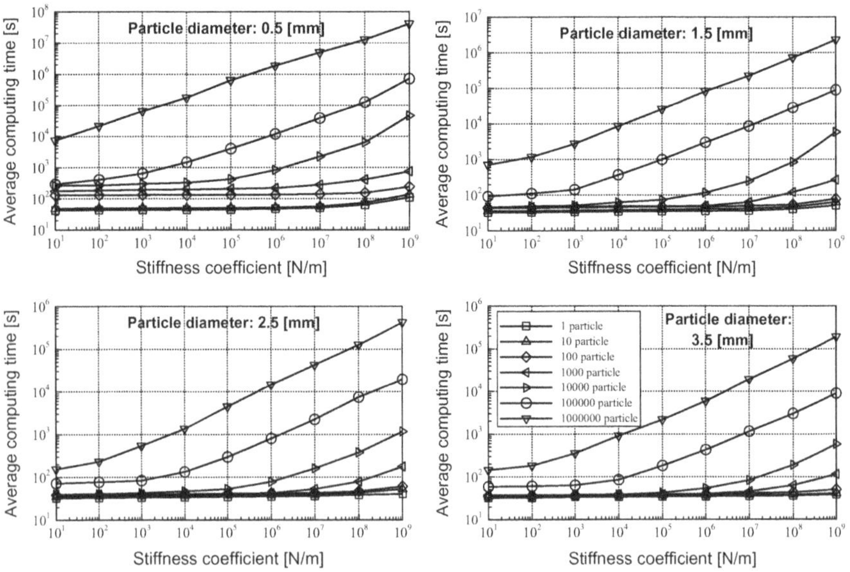

Fig. 4.134 Influence of stiffness coefficient variation on the average computing time per time step at different particle diameters and various particle numbers

caused due to the fact that a certain part of the computing time is required for the calculation of the fluid phase. For stiffness coefficients with high-orders of magnitude, this part is not significant in percentage terms; however, it takes at very low stiffness coefficients into account.

The particle diameter has a relevant influence on the average computation time. The reason of this is that the size of the particle time step is a function of the particle diameter and thus smaller particle diameters lead to finer particle time steps for an identical stiffness coefficient value. From the logarithmic plots, a power function can be obtained for the average computing time T and the stiffness coefficient as:

$$T = A\sqrt{k^n} \approx (k^n)^{1/2}, \quad (4.37)$$

where the symbol A represents the slope of the curve. In order to understand explicitly the influence of the particle diameter on the average computing time per time step, the previous obtained results are replotted again (see Fig. 4.135). For various particle numbers, the stiffness coefficient is varied from 10^1 N/m to 10^9 N/m at different particle diameters (0.5, 1.5, 2.5 and 3.5 mm). The obtained curves behave almost the same, whereas the average computation time rises exponentially with increasing the stiffness coefficient. As it is expected, the average computing time escalates considerably for higher solid loadings. This can be explained by the application of the DEM model, where the computational effort depends on the particle number and the particle time step. For an identical stiffness coefficient as well as a constant particle diameter, the particle time step remains unaltered. With increasing the solid loading, the number of equations to be solved is raised and consequently the simulation time increases. Contrary to expectations, the particle diameter plays a crucially important role concerning the computational effort. The computing time per time step increases sharply with decreasing the particle diameter. At constant particle diameter (for example 10^6), a reduction in the particle diameter from 3.5 mm to 0.5 mm results in an increase in the simulation time by two-order of magnitude.

In order to investigate the influence of the stiffness coefficient variation on the simulation accuracy, the numerical model of a quasi-2D fluidized bed is simulated and the obtained results are compared with the measured data. For the simulation, 36,500 identical glass particles with a diameter of 2.5 mm are used. The simulation parameters, the applied boundary conditions and the grid resolutions are identical for all simulation cases (see Sect. DEMEST Code Validation), while the stiffness coefficient is set to 10^2, 10^3, 10^4, 10^5 and 10^6 N/m. In Figs. 4.136 and 4.137, the simulated bed heights at different stiffness coefficients are compared with the experiments over time (500 ms). The application of stiffness coefficients with low-orders of magnitude has the benefit of a low computational effort, but the accuracy of the Euler-Lagrange/DEM model decreases sharply. This can be clearly detected in case of applying stiffness coefficients with values of 10^2 and 10^3 N/m. Here, the simulated bed heights at mass flow rates of 0.006 kg/s and 0.005 kg/s deviate from the measured data with maximum relative errors of 20% and 45%, respectively. On the other hand, stiffness coefficients with high-order of magnitude

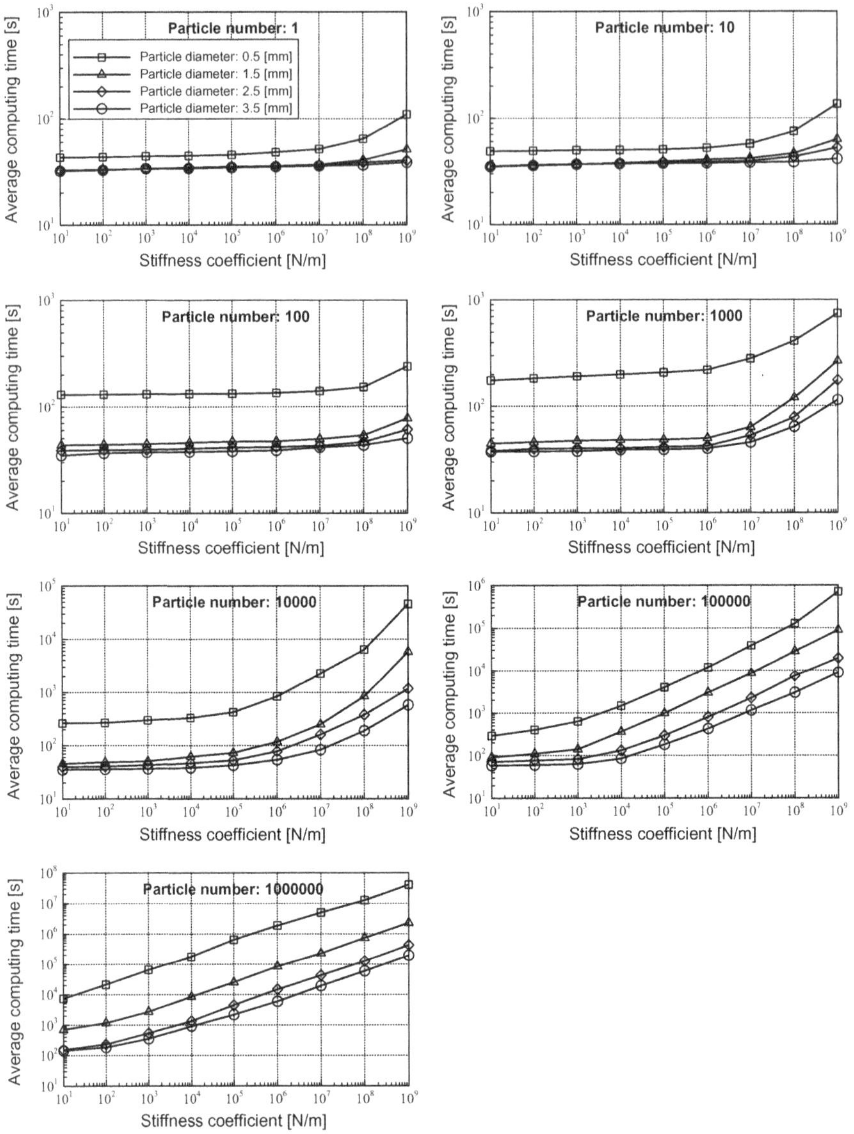

Fig. 4.135 Influence of stiffness coefficient variation on the average computing time per time step at different particle diameters and various particle numbers (replotted)

are related to an extreme computation time, but they result in a high simulation accuracy. Moderate stiffness coefficients in the range between 5×10^4 N/m and 5×10^6 N/m show a very good compromise between an acceptable computing time and a good accuracy. Although the application of stiffness coefficients with

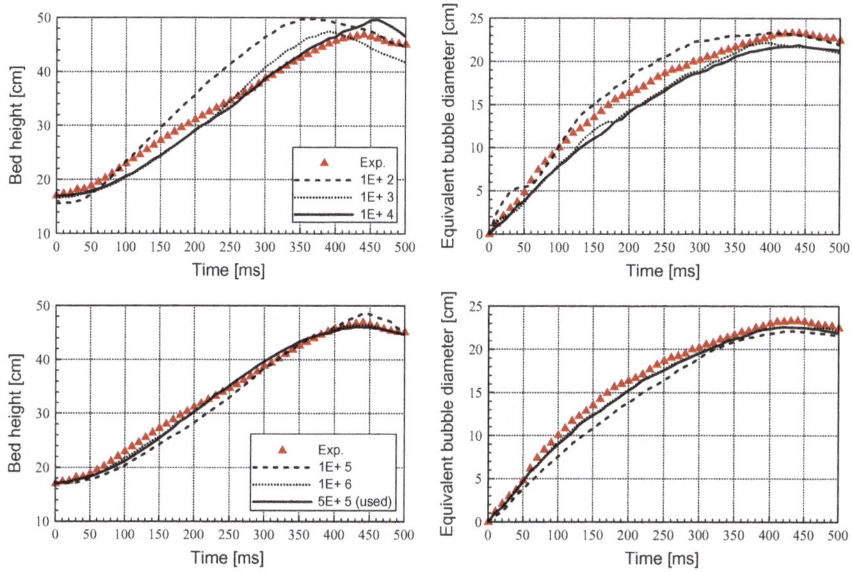

Fig. 4.136 Influence of stiffness coefficient variation on the simulation accuracy of Euler-Lagrange/DEM model (number of particles is 36,500 and the particle diameter is 2.5 mm for mass flow rate of 0.006 kg/s

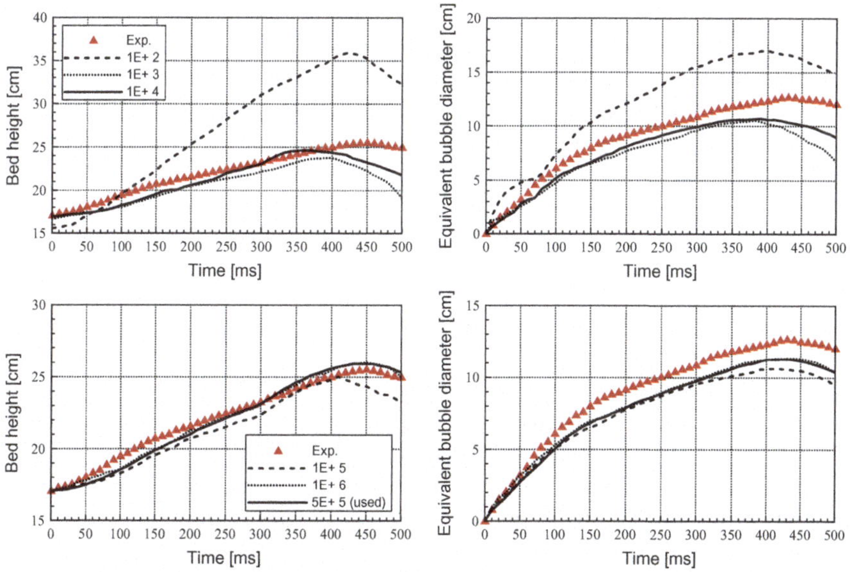

Fig. 4.137 Influence of stiffness coefficient variation on the simulation accuracy of Euler-Lagrange/DEM model (number of particles is 36,500 and the particle diameter is 2.5 mm) for mass flow rate of 0.005 kg/s

moderate values causes normal penetration depths up to the particle radius (see Table 4.33), they still show good final results. So they are recommended to apply for the simulation of the fluidized bed especially since the maximum relative velocities of collision partners in this system are smaller than 10 m/s.

Variation of fluid time step

As previously mentioned in the Sect. 3.2.3.4, the fluid time step can be selected arbitrarily and is set constant during the simulation. Generally, the fluid time step is set large and accordingly consists of several particle time steps. The size of the particle time step depends directly on the applied stiffness coefficient. In Table 4.36, the particle time steps are determined at different stiffness coefficients and particle diameters. The calculated particle time step decreases with increasing the stiffness coefficient and reducing the particle diameter. Independent of the particle diameter, an increase in the stiffness coefficient by four-orders of magnitude leads to a decrease in the particle time step of 100 times. At constant stiffness coefficient, a reduction in the particle diameter from 3.5 mm to 0.5 mm results in a decrease in the particle time step of 26 times.

The influence of the fluid time step variation on the computational time is investigated. Here, the stiffness coefficient is set constant equal to 5×10^5 N/m, which represents a compromise between a realistic simulation result and an acceptable simulation time. While the resulted particle time step is constant depending on the particle diameter and the stiffness coefficient used, the fluid time step size is varied in the range between 0.1 ms and 105 ms. The effect of the particle diameter and the particle number variation on the average computing time is also examined. The detailed simulation parameters for the following series of simulations can be found in Table 4.37. For each fluid time step, ten steps are simulated and the average computing time per time step is then determined through the division of total simulation time by number of calculation steps.

In Fig. 4.138, the effect of the fluid time step variation on the average computing time per time step at different particle numbers (10^0, 10^1, 10^2, 10^3, 10^4, 10^5, 10^6) and different particle diameters (0.5, 1.5, 2.5 and 3.5 mm) are displayed. It is assumed that for identical simulation duration, an increase in the size of the fluid

Table 4.36 Calculation of particle time step for glass collision partners at different particle diameters (0.5, 1.5, 2.5 and 3.5 mm)

Stiffness coefficients (N/m)	Particle time steps (s)			
	$d_p = 0.5$ (mm)	$d_p = 1.5$ (mm)	$d_p = 2.5$ (mm)	$d_p = 3.5$ (mm)
$k^n = 10^2$	0.1×10^{-2}	0.7×10^{-2}	1.5×10^{-2}	2.6×10^{-2}
$k^n = 10^4$	0.1×10^{-3}	0.7×10^{-3}	1.5×10^{-3}	2.6×10^{-3}
$k^n = 10^6$	0.1×10^{-4}	0.7×10^{-4}	1.5×10^{-4}	2.6×10^{-4}
$k^n = 10^8$	0.1×10^{-5}	0.7×10^{-5}	1.5×10^{-5}	2.6×10^{-5}

Table 4.37 Simulation parameters for the fluid time step variation

Variable parameters	Fluid time steps (ms)	0.1, 0.3, 0.5, 1, 2, 5, 10, 15, 30, 45, 60, 75, 90, 105
	Particle numbers	10^0,10^1, 10^2, 10^3, 10^4, 10^5, 10^6
	Particle diameters (mm)	0.5, 1.5, 2.5, 3.5
Fixed parameters	Number of calculation steps	10
	Stiffness coefficient (N/m)	5×10^5
	Number of decompositions	9
Grids	Fluid multi-grid	Activated
	Particle grid	Activated

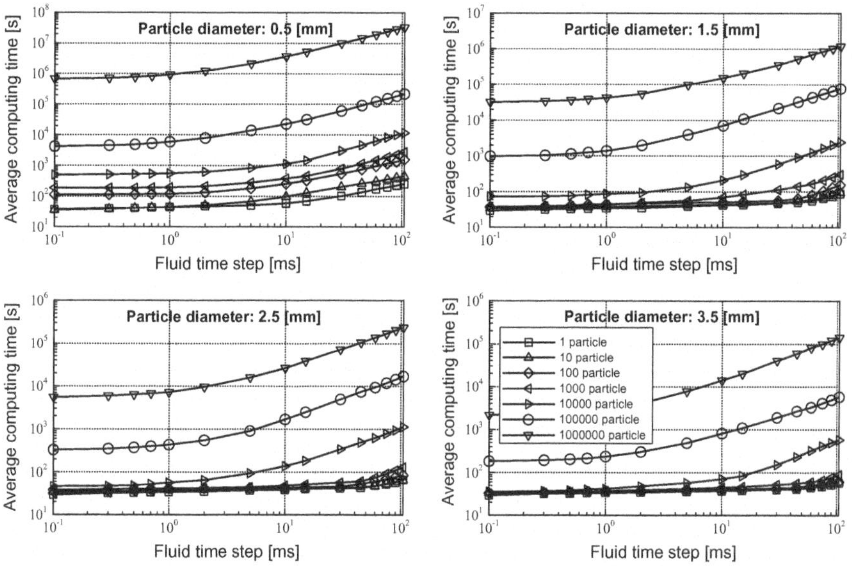

Fig. 4.138 Influence of fluid time step variation on the average computing time per time step at different particle diameters and various particle numbers

time step accelerates the simulation. Figure 4.138 shows that this relationship is not necessarily correct. An enlargement of the fluid time step in the range between 10 ms and 105 ms does not accelerate the Euler-Lagrange/DEM calculation, i.e. leads to increase in the average computing time per time step. This is due to the fact that the ratio of the fluid time step to the particle time step is bigger than one. This means, during one fluid time step several particle time steps must be carried out. The number of particle time steps depends on the stiffness coefficient and the particle diameter used as well as on the size of the fluid time step. Subsequently, the computation time to calculate the particle phase has a significantly larger proportion of the total simulation time than the computational effort of the fluid phase. From the curves illustrated in Fig. 4.138, it can be supposed that fluid time steps smaller than 10 ms could not cause any substantial increase in the computational time. Finer fluid time step sizes produce more accurate simulation results, which is aimed by practical applications. An excessive reduction of the fluid time step leads, however, to a considerable increase in the computational effort. It raises the question, at which ratio of fluid to particle time step, a reduction in the size of fluid time step will not cause a significant increase in computational time. By decreasing the fluid time step until the size of the particle time step, the independent correlation between the fluid time step and the computational time is not any more valid.

It can be observed in Fig. 4.138 that the particle numbers have a major impact on the computational time. At constant fluid time step, the computational effort is substantially raised by increasing the solid loading. This can be easily explained by the reason of the DEM model used. Here, each particle has a set of equations to be solved. Furthermore, the computational effort of collision detection is correlated with the number of investigated particles. At defined particle diameter, an increase in the particle numbers from 10^5 to 10^6 causes a jump in the simulation time by one-order of magnitude. The obtained results show that the average computing time depends highly on the particle diameter. This is due to the fact that the variation of the particle diameter has a relevant influence on the determined size of the particle time step.

In order to understand the effect of the particle diameter on the average computing time per time step, the previous obtained results are replotted again (see Fig. 4.139). For different particle numbers (10^0, 10^1, 10^2, 10^3, 10^4, 10^5, 10^6), the fluid time step is varied from 0.1 to 105 ms at various particle diameters (0.5, 1.5, 2.5 and 3.5 mm). At constant particle number, the average computational time rises considerably with decreasing the particle diameter. This significant increase in the computing time results from the decreasing in the particle time step that is a function of the stiffness coefficient and the particle diameter. Since the stiffness coefficient is set constant equal to (5×10^5 N/m) in these series of simulations, the particle time step is then related to the particle diameter (see Table 4.36). At lower particle numbers (10^0, 10^1, 10^2, 10^3 and 10^4), the flattening in curves is caused due to the fact that a part of the computing time is required for the calculation of the fluid phase. This part is negligibly small for higher numbers of particles (10^5 and 10^6), but it takes at low particle numbers into account.

The consideration of the average computing time as a function of the fluid time step is not informative in terms of the efficiency analyse. Therefore, the computation time per simulated millisecond is displayed in Figs. 4.140 and 4.141 at different solid loadings (10^0, 10^1, 10^2, 10^3, 10^4, 10^5, 10^6) and various particle diameters (0.5, 1.5, 2.5 and 3.5 mm). For large fluid time steps, the average computing time per one millisecond is remained almost constant depending on the particle number and the particle diameter. At particle diameter 1.5 mm for example, the average computing time per one millisecond is about 1000 s and 10,000 s for particle numbers 10^5 and 10^6, respectively. An exponential increase in the average

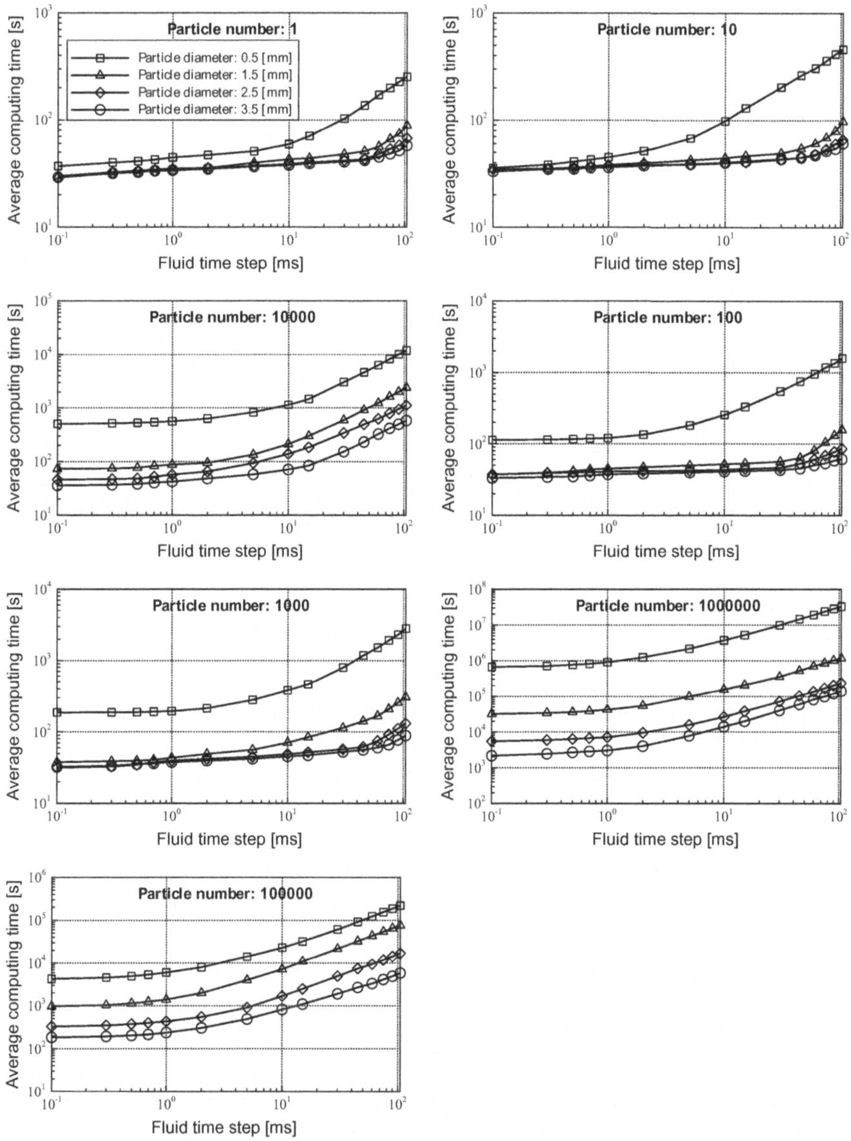

Fig. 4.139 Influence of fluid time step variation on the average computing time per time step at different particle diameters and various particle numbers (replotted)

computing time per simulated millisecond is appeared when finer fluid time steps are applied.

From Fig. 4.140, a fluid time step of about 5 ms ensuring the efficiency of simulations can be extracted. A further increase in the size of the fluid time step more than 5 ms is not correlated with any efficiency gain; but the fluid phase is

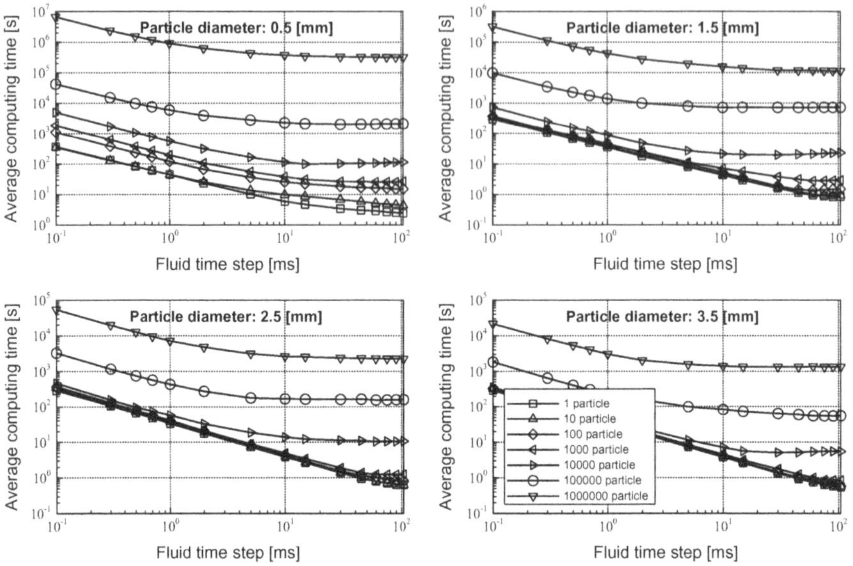

Fig. 4.140 Influence of fluid time step variation on the average computing time per one millisecond at different solid loadings and various particle diameters

resolved temporally from bad to worse. A reduction in the fluid time step less than 5 ms in order to achieve higher temporal resolutions leads to a significant loss in the efficiency, since the computing time per simulated millisecond increases exponentially. The variation of particle numbers has almost no influence on the efficient size of the fluid time step (Fig. 4.141).

In order to estimate the effect of the fluid time step variation on the Euler-Lagrange/DEM model accuracy, a numerical model of a quasi-2D fluidized bed with 36,500 identical glass particles (diameter: 2.5 mm) is simulated. While the fluid time step is varied, other simulation parameters remain unaltered (see Sect. DEMEST Code Validation). The fluid and particle grids resolutions are identical for all simulation cases. In Fig. 4.142, the simulated bed height and the bubble size are compared with the experiments over time (500 ms). At finer fluid time steps below 5 ms, a very good agreement between the simulated bed expansion and the experimental observed bed height and area can be detected regarding the mass flow rate of 0.006 kg/s. A close comparison between the simulated bed height as well as the equivalent bubble diameter at fluid time step 0.5 ms and 2 ms shows almost exact characteristic. This suggests that a further decrease in the fluid time step below 0.5 ms will not result in an improvement in the Euler-Lagrange/DEM model accuracy. At this point, the discrepancy with measurements is related to other factors, presented in Sect. DEMEST Code Validation. At fluid time step 10 ms, the bed starts expanding about 200 ms later compared to the model with 2 ms fluid time step. For coarser fluid time steps, the bed did not move and remains almost unchanged. The bed expansion at the mass flow rate of 0.005 kg/s indicates

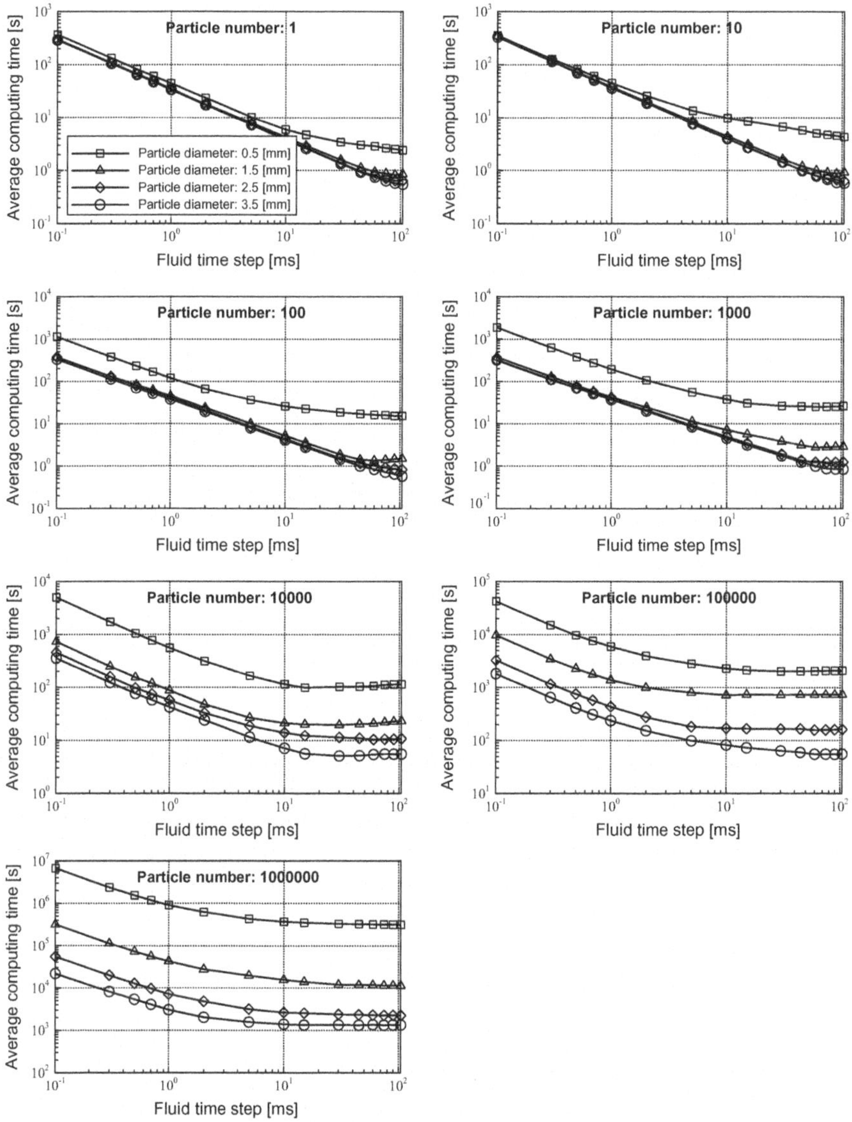

Fig. 4.141 Influence of fluid time step variation on the average computing time per one millisecond at different solid loadings and various particle diameters (replotted)

similar behaviour to the mass flow rate of 0.006 kg/s. At fluid time step of 5 ms, the bed height exceeds the measurement with a relative error of 13%. A large deviation in the bed height and the bubble size can be observed again for fluid time steps bigger than 5 ms.

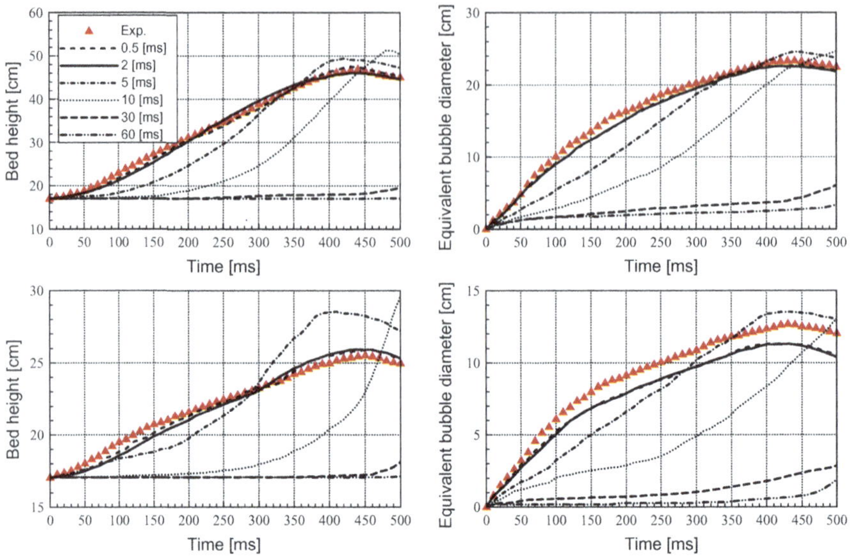

Fig. 4.142 Influence of fluid time step variation on the simulation accuracy of Euler-Lagrange/DEM model (number of the particles is 36,500 and the particle diameter is 2.5 mm): (upper row) mass flow rate of 0.006 kg/s and (lower row) mass flow rate of 0.005 kg/s

On the basis of above discussion together with the comparison illustrated in Fig. 4.142, it can be concluded that the application of larger fluid time steps (more than 10 ms) displays a negative impact on the Euler-Lagrange/DEM results, although the computational effort remains almost unaltered. On the other hand, finer fluid time steps (less than 1 ms) lead to a negligible improvement in the Euler-Lagrange/DEM accuracy, but to considerable increase in the computing time. Fluid time steps in range between 1 ms and 5 ms show a reasonable compromise between the required accuracy and simulation efficiency. So they are recommended to apply for the simulation of the fluidized bed independent of the solid loading and the particle diameter.

DEM with Thermochemical Reaction Model

In addition to the particle trajectory and the gas flow streamlines, the local concentrations of solid and gaseous substances, the local gas and particle temperatures, the local heat release and the heat transfer rates can be calculated using the DEM model. In order to evaluate the DEM model under the consideration of thermochemical reaction mechanism, the quasi-2D fluidized bed model, explained in the Sect. DEMEST Code Validation is used. However, the supplied air enters through the completely opened bottom (15 cm (W) × 2 cm (D)). The dispersed phase consists of 36,500 spherical particles with an average diameter of 2.5 mm. The

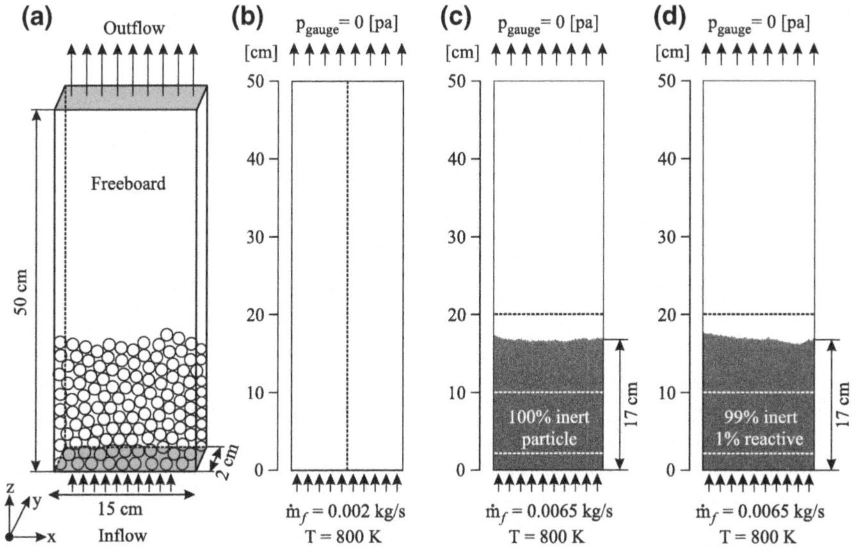

Fig. 4.143 Numeric model used for the simulation of reactive gas-solid flow: **a** schematic diagram, **b** combustion with fuel gas (no particle), **c** combustion with fuel gas with sand particles and **d** combustion with solid fuel (99% sand particles +1% pine wood particle)

forces on the particles are gravitational, buoyancy, drag, pressure, Saffman, Magnus, and contact forces. The bubbling flow regime (monodisperse particles of group B) is selected. Three series of warm simulations are performed (see Fig. 4.143). For the first case, the model has no particle and a mixture of CH_4 (0.04 kg/kg$_{mix}$), O_2 (0.23 kg/kg$_{mix}$) and N_2 (0.73 kg/kg$_{mix}$) is supplied through the inlet with a mass flow rate of 0.002 kg/s. For the second case, the reactor is filled with inert particles. The particles have the properties of sand and the fluidization medium has the same properties as case 1 with a mass flow rate of 0.0065 kg/s. Finally, the reactor is filled with a mixture of 99% inert sand particles and 1% pine wood particles. The fluidizing medium is air with a mass flow rate of 0.0065 kg/s. The same configurations used for the second simulation case are applied here.

Combustion with fuel gas

The combustion with the fuel gas is investigated for different operation pentameters. While the simulation is performed without the solid phase in the first case, inert particles are added to reactor in the second case. In both simulation cases, a mixture of methane and air is supplied as a fluidization medium.

For the first simulation case, the particle phase is not considered and thus no particles are generated (see Fig. 4.143b). At the reactor inlet, a mixture of CH_4 (0.04 kg/kg$_{mix}$), O_2 (0.23 kg/kg$_{mix}$) and N_2 (0.73 kg/kg$_{mix}$) is supplied. The mass flow rate amounts to 0.002 kg/s at temperature of 800 K. The temperature of walls is set constant equal to 1000 K. The initial temperature in all grid cells is 1000 K and the initial mass fractions are O_2 (0.23 kg/kg$_{mix}$) and N_2 (0.77 kg/kg$_{mix}$). At this stage, the reaction mechanism is limited to the homogeneous reactions (methane and carbon monoxide oxidations):

$$\begin{aligned} &R1: CH_4 + {}^3\!/_2O_2 \rightarrow CO + 2H_2O \\ &R2: CO + {}^1\!/_2O_2 \rightarrow CO_2 \end{aligned} \tag{4.38}$$

The local mass fraction of each chemical component i is calculated based on the gaseous substance transport equation. In order to balance the energy in the fluid phase, the transport equation for the specific enthalpy is solved in each control volume. The fluid temperature is then determined from the specific enthalpy and the mean specific heat capacity. The source terms in the species transport equations consider the formation/reduction of gaseous components by the reason of chemical reactions (R1 and R2). The heat source term takes into account the heat released by the homogeneous gas phase reactions. The radiation source term is omitted from the calculation. In Fig. 4.144, the species mass fractions and the fluid temperature

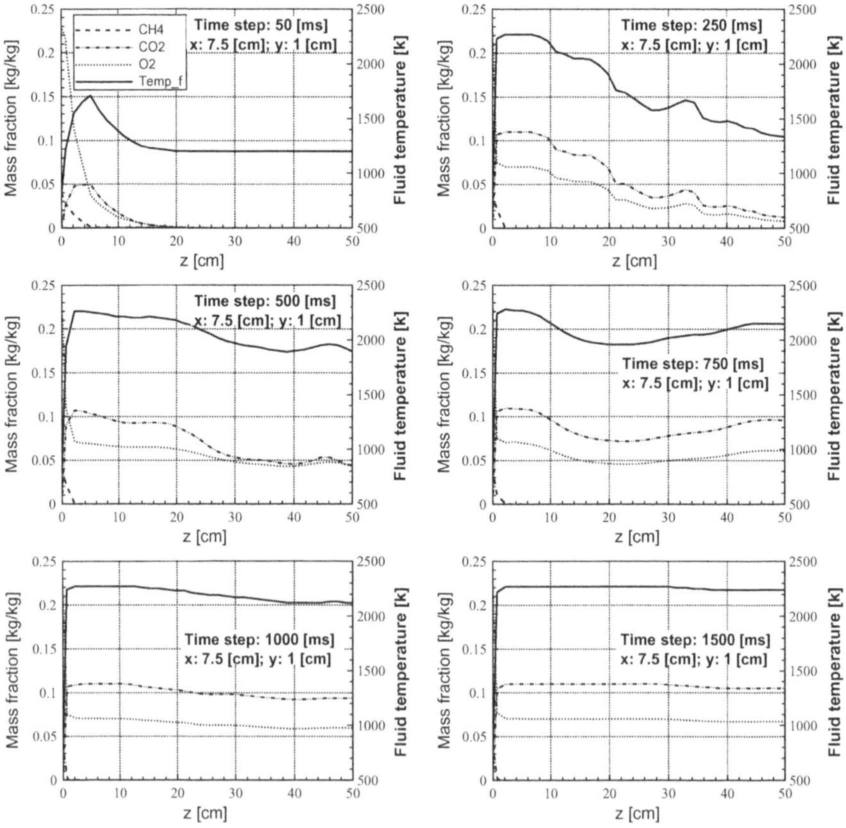

Fig. 4.144 Species mass fractions (first axis) and fluid temperature profiles (second axis) along z axis of the reactor at different time step (combustion with fuel gas)

profiles along the reactor (see Fig. 4.143b; dashed line) at different time steps (50, 150, 250, 350, 450 and 500 ms) are displayed. At t = 50 ms, the supplied methane reacts with the oxidation agent to CO_2. The methane is completely consumed about 5 cm far from the inlet. By the reason of methane combustion, the fluid temperature increases gradually and thus the combustion zone becomes closer to the inlet. After 1500 ms from the start, the model reaches the steady-state and the fluid temperature is about 2100 K. The mass fractions of CH_4, O_2 and CO_2 at the rector outlet are 0, 0.07 and 0.11 kg/kg_{mix}, respectively.

For the second simulation case, 36,500 inert sand particles with 2.5 mm diameter are generated and distributed evenly in the entire computational domain. During this stage, the mass flow rate is set equal to zero. The particles drop freely into the reactor due to the gravity. The static height of sphere particles is 17 cm and the bulk solid fraction is ca. 0.6 (see Fig. 4.143c). A mixture of CH_4 (0.04 kg/kg_{mix}), O_2 (0.23 kg/kg_{mix}) and N_2 (0.73 kg/kg_{mix}) is supplied at the reactor inlet. The mass flow rate is set equal to 0.0065 kg/s at temperature of 800 K. The mass flow rate is selected, so that the bubbling flow regime is obtained. The initial temperature of particles is 800 K. The temperature of walls and the initial temperature in all fluid grid cells are set equal to 1000 K. The initial mass fractions of species are O_2 (0.23 kg/kg_{mix}) and N_2 (0.77 kg/kg_{mix}).

Besides the previously explained reaction mechanism, the following energy balance equation for each particle is solved:

$$\underbrace{m_P c_{p,P} \frac{dT_P}{dt_P}}_{1} = \underbrace{\alpha_P A_P \left(T_f - T_P\right)}_{2} + \underbrace{A_P q_{rad}}_{3}. \tag{4.39}$$

Equation (4.39) describes the change of the particle temperature T_P (term 1) as a result of the heat transfer from fluid to particle due to convection (term 2) and radiation (term 3). Here, α_P is the heat transfer coefficient of particles, $c_{p,P}$ represents the solid specific heat capacity, A_P and q_{rad} denote to the particle surface and radiation heat flux, respectively. It should be mentioned here that the radiation term is not considered in the simulation. Similar to the first case, the source terms in the species transport equations take into account the formation/reduction of gaseous components by the reason of chemical reactions (R1 and R2). Besides the heat released by the homogeneous gas phase reactions, the heat source term in the energy equations of fluid phase consider the heat transfer from solid to fluid by convection.

In Fig. 4.145, the solid and fluid temperature profiles are presented on the midline at different locations of the bed height (z = 2 cm, z = 10 cm and z = 20 cm) at various time steps (0.1, 1, 5, 10, 15 and 25 s) (see Fig. 4.143c; dashed lines). The heights of the observation lines are chosen, so that the bubbling flow regime can be covered. In a little while after the start, the particle and fluid temperatures have a value of 800 K at all locations. Due to homogeneous reactions (R1 and R2), the fluid temperature increases in the computational domain. The temperature of the particles increases accordingly because of the heat transfer from

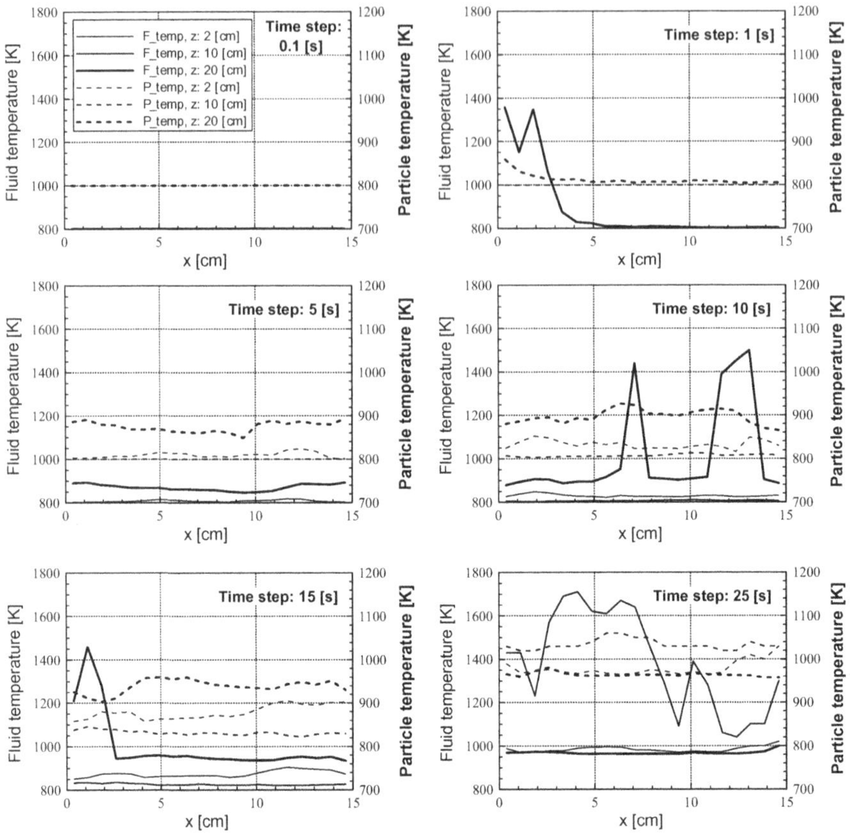

Fig. 4.145 Fluid temperature (first axis) and particle temperature profiles (second axis) at different bed height of the reactor at various time steps

fluid to solid by convection. At t = 10 s, the temperature of the particles is about 900 K at location z = 20 cm, while it is 850 K and 800 K at positions z = 10 cm and z = 2 cm, respectively. The reason of the solid temperature differences is that the homogeneous reactions occur effectively, where the solid phase is not existent. This means the gaseous species pass through the inert particles and react to each other in the upper region of the bed, resulting in higher fluid temperatures compared to the lower region. The peaks in the fluid temperature indicate the existence of gas bubbles, where the methane and oxygen can burn. After 25 s, the fluid and solid temperatures reach the values of 1700 K and 1050 K, respectively. The particles are well mixed and the difference in the solid temperature at various locations is gradually decreased. One important point is the heat transfer from solid to fluid, which it can be clearly detected at the location z = 2 cm. Here, the fluid temperature increases to 1000 K, although the CH_4 as well as CO cannot burn at this region due to the dense existence of the solid phase (lower volumetric void fractions).

Combustion with solid fuel

For combustion with the solid fuel, the same configurations used for the second simulation case (combustion with fuel gas and inert particles) are applied here. Air is supplied as a fluidizing medium at the bottom of the reactor. The solid phase consists of a mixture of sand and pine wood particles (36,135 + 365). It is assumed that the parameters of the particle-particle/wall collisions of both solid materials have the same values. Here, the collision parameters of sand are also applied to the pine wood particles. This is due to the fact that on the one hand the bed consists of 99% sand particles and on the other hand the experimental data for the heterogeneous material pairing of sand-pine wood particles are not yet available in the literature.

The multi-step reaction mechanism, which it is illustrated in Fig. 4.146, comprises of drying (R1), pyrolysis (R2), char-oxidation (R3) and combustion of volatile gases, namely methane and carbon monoxide (R4 and R5). The drying process takes place when the temperature of a particle exceeds 100 °C. The kinetics of the pyrolysis is calculated with the aid of the single-step global model that considers the pyrolysis as one reaction of first-order. After (Kaltschmitt et al. 2009), the char-oxidation starts after the complete pyrolysis of biomass particles. For the kinetic of the biochar-conversion, the particle shrinkage model is used. The homogeneous gas chemical reactions are described by CO and CH_4 oxidations.

$$\begin{aligned}
&\text{R1: Raw biomass} \rightarrow \text{biomass} + H_2O \\
&\text{R2: Biomass} \Rightarrow C + CH_4 + CO + CO_2 + H_2O + N_2 \\
&\text{R3: } C + {}^1\!/_2 O_2 \rightarrow CO \\
&\text{R4: } CH_4 + {}^3\!/_2 O_2 \rightarrow CO + 2H_2O \\
&\text{R5: } CO + {}^1\!/_2 O_2 \rightarrow CO_2.
\end{aligned} \tag{4.40}$$

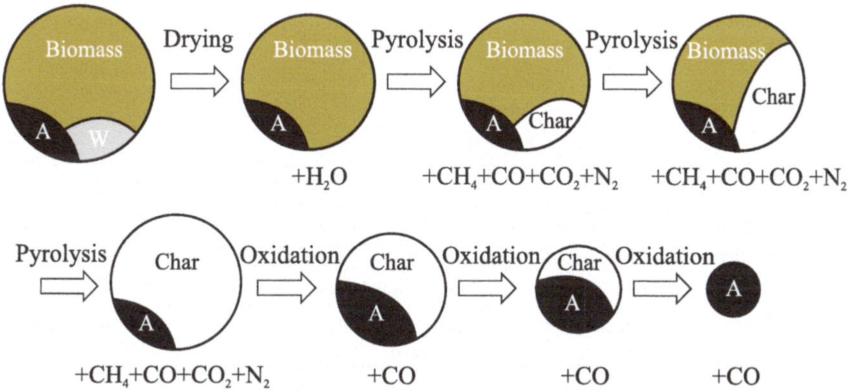

Fig. 4.146 Representation of the multi-step reaction model of a particle

If a complex pyrolysis model is selected, e.g. the single-step multi-reaction mechanism that consists of a primary and a secondary reaction, the following reactions, namely steam gasification, Boudouard reaction and hydrogenating gasification can be considered during the combustion process:

$$\begin{aligned} &R6: C + H_2O \rightarrow CO + H_2 \\ &R7: C + CO_2 \rightarrow 2CO \\ &R8: C + 2H_2 \rightarrow CH_4. \end{aligned} \tag{4.41}$$

By the application of complex pyrolysis models and heterogeneous reactions, the homogenous gas phase reaction can also be extended with additional reactions as:

$$\begin{aligned} &R9: CH_4 + H_2O \Leftrightarrow CO + 3H_2 \\ &R10: CO + H_2O \Leftrightarrow CO_2 + H_2 \\ &R11: CO + 3H_2 \Leftrightarrow CH_4 + H_2O \\ &R12: H_2 + {}^1\!/_2O_2 \Leftrightarrow H_2O. \end{aligned} \tag{4.42}$$

The rate of the devolatilisation is calculated according to the following relation:

$$\frac{dm_P^{bio}}{dt} = -k_{pyr}^{bio} m_P^{bio}, \tag{4.43}$$

where k_{pyr}^{bio} is the pyrolysis reaction rate constant. The released gas species during the pyrolysis process can be determined as:

$$\frac{dc_{i,pyr}}{dt} = k_{pyr}^{bio} v'_{i,pyr} c_{i,pyr}. \tag{4.44}$$

The symbols $c_{i,pyr}$ and $v'_{i,pyr}$ are the concentration and the stoichiometric coefficient of a species i in the pyrolysis reaction (R2). The temporal change in the molar concentration of a chemical component i due to homogenous gas reactions is deafened as:

$$\frac{dc_i}{dt} = \sum_{k=1}^{k} \left(v''_{i,k} - v'_{i,k} \right) r_k^{hom} \quad (i = 1, \ldots, N). \tag{4.45}$$

Here, the subscript k is the number of the homogenous gas reactions, N denotes to total number of the chemical components, $v'_{i,k}$ and $v''_{i,k}$ represent the stoichiometric coefficient of the species i as reactant and product, respectively. The reaction rate r^{hom} of a homogenous gas reaction k is determined from the reaction rate constant of forward $k_{for,k}^{hom}$ and backward $k_{bac,k}^{hom}$ reactions:

$$r_k^{hom} = r_{for,k}^{hom} - r_{bac,k}^{hom} = k_{for,k}^{hom} \prod_{i=1}^{N} c_i^{\nu'_{i,k}} - k_{bac,k}^{hom} \prod_{i=1}^{N} c_i^{\nu''_{i,k}}. \quad (4.46)$$

The reaction rate constants k_{kin} in Eqs. (4.43) and (4.46) is defined using Arrhenius expression including a pre-exponential factor A and the activation energy E:

$$k_{kin} = A \exp\left(\frac{-E}{RT_j}\right), \quad (4.47)$$

where j is either a fluid or a solid. The char-combustion (R3) to carbon monoxide is modelled according to the kinetics/diffusion limited reaction model (Epple et al. 2012). The model assumes that reaction rate of the char-combustion combines the kinetic reaction rate constant and the diffusion reaction rate constant of oxygen to particle surface:

$$\frac{dm_P^{cha}}{dt} = -k_{ox}^{cha} A_P^{cha}, \quad (4.48)$$

with:

$$k_{ox}^{cha} = \frac{p_{O_2}}{\left(\frac{1}{k_{kin}^{cha}} + \frac{1}{k_{dif}^{cha}}\right)}. \quad (4.49)$$

Here, A_P^{cha} represents the specific particle surface, p_{O_2} is the oxygen partial pressure and k_{ox}^{cha} denotes to the overall reaction rate of the char-combustion. The diffusion rate reaction constant is expressed by:

$$k_{dif}^{cha} = \frac{48 D_{dif}^{O_2,N_2}}{RT_0} 10^5 \frac{T_P^{0.75}}{d_P^{cha}}. \quad (4.50)$$

The binary diffusion coefficient $D_{dif}^{O_2,N_2}$ of oxygen and nitrogen has the value of 3.49 10^{-4} m^2/s at reference state temperature $T_0 = 1600$ K.

It should be mentioned here that an additional source term $S_{P \to f}^{m,P}$ on the right side of the mass conservation equation (see Eq. (3.87)) should appear as a result of the heterogeneous reactions. This term corresponds to the mass transfer due to the phase transformation from solid to fluid phase. The source term $S_f^{c,i}$ in the species transport equation (see Eq. (3.90)) considers the formation/reduction of the chemical components due to the homogenous and heterogeneous reactions (R3, R4 and R5). The heat source term $S_{P \to ff}^{h}$ in energy equation of the fluid phase (Eq. (3.89)) is expressed by the relation:

$$S^h_{P\to f,f} = \underbrace{S^h_f}_{S^{hom}_f + S^{rad}_f} + \underbrace{S^{h,P}_{P\to f}}_{S^{conv}_{P\to f} + S^{rad}_{P\to f} + S^{mas}_{P\to f}} . \tag{4.51}$$

It consists of three main parts, namely the heat released by homogeneous gas phase reactions (R4 and R5) as well as the radiation source term of the fluid phase and the heat transfer from solid to fluid by convection, radiation and mass transfer due to the phase transformation. At this stage, the heat released by homogenous and heterogeneous reactions as well as the heat transfer between both phases by convection are only considered in the DEMEST programme. The heat transfer from solid to fluid phase is determined for each control volume taking into account all particles that are located in the control volume during the time interval.

The change of the particle temperature (term 1) is calculated from the energy balance of the particle governed by the heat evaporation (term 2), the heat release due to the char-oxidation (term 3) and the heat transfer from fluid to solid phase associated with convection and radiation (term 4 and term 5):

$$\underbrace{m_P c_{p,P} \frac{dT_P}{dt_P}}_{1} = \underbrace{m_{P,wat} k_{wat} \Delta h_{wat}}_{2} + \underbrace{m_{P,cha} k_{cha} \Delta h^0_{cha}}_{3} + \underbrace{\alpha_P A_P \left(T_f - T_P\right)}_{4} + \underbrace{A_P q_{rad}}_{5}$$
$$+ \underbrace{\alpha_P \sum_{\substack{k=1 \\ k\neq i}}^{k=N} A_{ik}\left(T_{P,i} - T_{P,k}\right)}_{6} . \tag{4.52}$$

Here, Δh_{wat} is the enthalpy of vaporisation of water, Δh^0_{cha} represents the standard enthalpy of char-oxidation, k_{wat} and k_{cha} are to evaporation and reaction rates constants, respectively. The quantities $m_{P,wat}$ and $m_{P,cha}$ denote the mass fractions of water and char in the investigated particle.

The conduction heat transfer during the particle-particle and particle-wall collisions is of relevance to the simulation accuracy. The transient heat conduction between the collision partners can be determined by a non-steady heat conduction equation (Zhou et al. 2008). This procedure leads to considerable additional computational time. For reasons of simplicity, the heat transfer by conduction can explicitly be expressed by term 6 in the particle energy equation. The index N represents the number of contacts for the particle i. The area A_{ik} corresponds to the contact area between the particle and a collision partner (particle or wall) and can be defined from the known tangential penetration depth:

$$A_{ik} = \pi r^2_{ij} = \frac{\pi\left(\delta^t_{ij}\right)^2}{4} . \tag{4.53}$$

In case of 2D simulations, the contact line between the collision partners is equal to the tangential penetration depth. In Eq. (4.52), the radiation term is usually not considered on the one hand because of the high complexity of the problem. On the other hand, the fluidized bed is characterised by a high heat and mass transfer rates between both phases, resulting in uniform temperature distribution. It should be noted here that for all non-reactive particles (99% sand particles), the particle mass remains constant and the reaction terms 2 and 3 are therefore omitted.

Figure 4.147 shows the particle and fluid temperature profiles on the midline at different height positions (see Fig. 4.143d; dashed lines) and at various time steps. At t = 0.1 s, the particles have almost a temperature of 800 K. The released volatile components as well as the char are burned, resulting in an increase in solid and fluid temperatures. After 5 s, the solid temperature rises from 800 K to approximately 850 K at z = 10 cm and 950 K at z = 20 cm. The peaks in the fluid temperature profiles occur in areas with large fluid volume fractions. Here, large amounts of oxygen and methane accumulate, whose combustion causes these high temperature

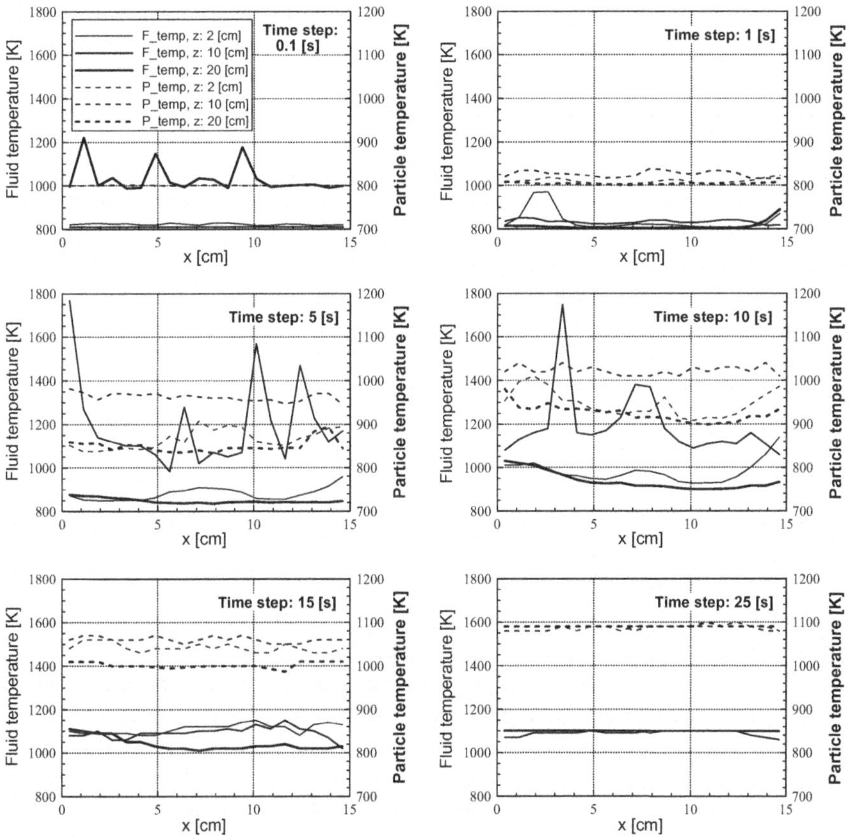

Fig. 4.147 Fluid temperature (first axis) and particle temperature profiles (second axis) at different bed height of the reactor at various time steps (combustion with solid fuel)

peaks. The pine wood particles are completely burned after 25 s. Due to intense mixing, the fluid and solid temperatures at all height positions show nearly constant temperature gradients of about 1100 K. From time point t = 25 s, the solid starts cooling down since the fluidization medium enters the bed with temperature of 800 K.

The Euler-Lagrange/DEM model under the consideration of the thermochemical reaction mechanism, implemented in the in-house programme (DEMEST), is applied to simulate the hydrodynamic behaviour of the gas-solid flow in a cold-flow and a warm-flow quasi-2D fluidized bed. The validity of the developed programme should be further examined first towards a cold-flow 3D fluidized bed model of Plexiglas and second to a warm-flow 3D fluidized bed test rig. The temporal change of pressure determined by pressure sensors at different positions and the species concentrations at the reactor outlet measured by a gas chromatography can be applied for the validation of the numerical models.

The coupling of the Euler-Lagrange/DEM model with the thermochemical reaction mechanism enables the modelling and simulation of reactive and dense gas-solid flows. The modelling of multiple energy systems can be covered by this concept for example the reactive fluidized bed, the ash deposition on boiler walls, the gasification or combustion of solid fuels as well as the CO_2 capture in the thermal power plants by chemical and carbonate-looping processes. The simulation of cyclone for particle separation and the milling process as found in coal pulveriser, in which thermal gradients are present, can be realised in the course of time when the performance of computers is further increased.

All in all, the Euler-Lagrange/DEM model in combination with the reaction and heat transfer models may become a standard tool for design and optimisation of process and energy systems. Nevertheless, significant research in terms of numerical and experimental investigations are required for making the DEM model as competitive as the present stand of other CFD models.

Parcel DEM Model

In the DEM model, the number of tracked particles is currently limited to the availability of computational resources. In order to simulate large-scale industrial systems, few real particles (micro scale particles) can be represented by a so-called parcel with same material properties as the particles in the system. The parcel approach (also known as the coarse grain particle model) is necessary to overcome exhaustive computational time that would occur in the case of tracking real sized particles. The amount of represented micro scale particles per one parcel is defined as the volume ratio between the parcel volume and the particle volume as:

$$\zeta_{str} = \frac{V_{par}}{V_P} = \left(\frac{d_{par}}{d_P}\right)^3. \tag{4.54}$$

Here, the parcel can represent up-to several thousands of real particles. The number of tracked parcels can be obtained according to the following formula:

$$N_{par} = \frac{N_P}{\zeta_{str}}. \tag{4.55}$$

This simplification is required for the simulation of granular systems containing billions of particles. As application examples of the parcel DEM model, implemented in the software ANSYS-FLUENT, a 3D cold-flow model, the carbonator and calciner reactors of the carbonate-looping process are presented.

A lab-scale 3D cold-flow fluidized bed test rig, illustrated in Fig. 4.148, is constructed and built using stainless steel. The test rig has a height of 3.1 m and diameter of 0.21 m. The fluidizing medium (air with humidity of about 70%) is injected with a superficial gas velocity of about 2.5 m/s into the bed through a porous plate at the bottom of the test rig. Calcium carbonate is applied as a bed material (bed inventory of about 11 kg) with a particle size distribution (median diameter d_{50} = 180 μm) and a density of about 2600 kg/m^3. The experiments are performed at room temperature and the freeboard pressure is atmospheric. At the outlet of the test rig, the particles are separated from the gas-solid mixture in a

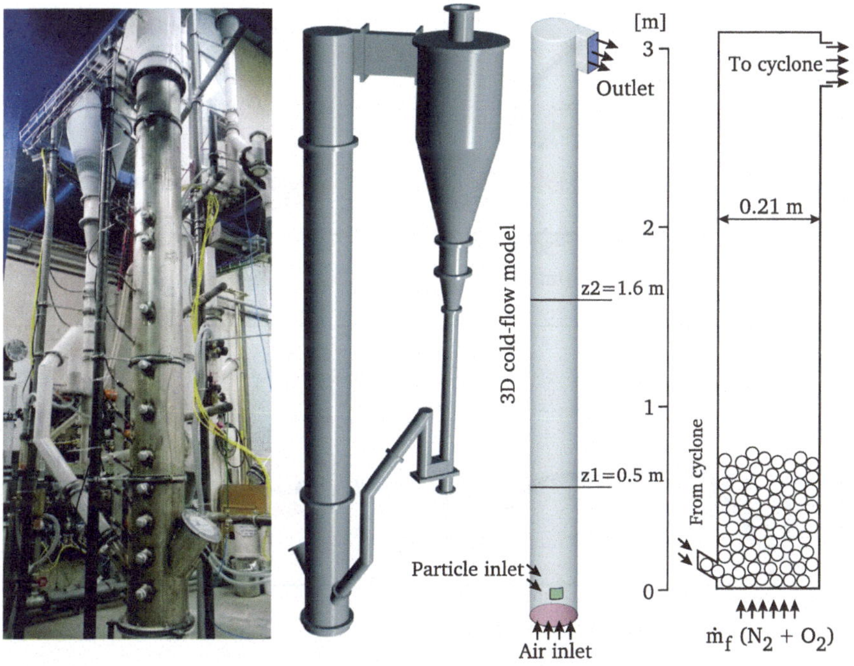

Fig. 4.148 Test rig of the 3D cold-flow fluidized bed at EST

cyclone and the particles circulate back into the bed. The particle flow and the bubble behaviour inside the test rig involve pressure drop fluctuations that are measured using 22 pressure sensors along the test rig height. Furthermore, 25 capacitive measurement devices are installed that enable the predications of the solid velocity and the solid volume fraction.

For the numerical simulation, only the riser of the 3D cold-flow fluidized bed test rig is considered. A structured mesh with about 75,000 hexahedron cells is generated. All boundary conditions are set similar to experiment and the parcel DEM model with the Wen and Yu drag model is applied. In order to keep the mass inventory constant, User-Defined-Function is used that injects solid particles through the inlet "Particle inlet".

In Fig. 4.149, the simulated solid velocity and the simulated solid volume fraction at z1 = 0.5 m and z2 = 1.6 m are compared with measurement data. Here, the typical flow pattern of a circulated fluidized bed can be seen. At z1 = 0.5 m, the solid velocity is smaller than at z2 = 1.6 m, but the solid volume fraction is clearly higher. Furthermore, and for both heights (z1 = 0.5 and z2 = 1.6 m), the solid velocity in the centre of the test rig is much higher than the solid velocity near the wall, where negative velocities are observed due to the back flow on sidewalls.

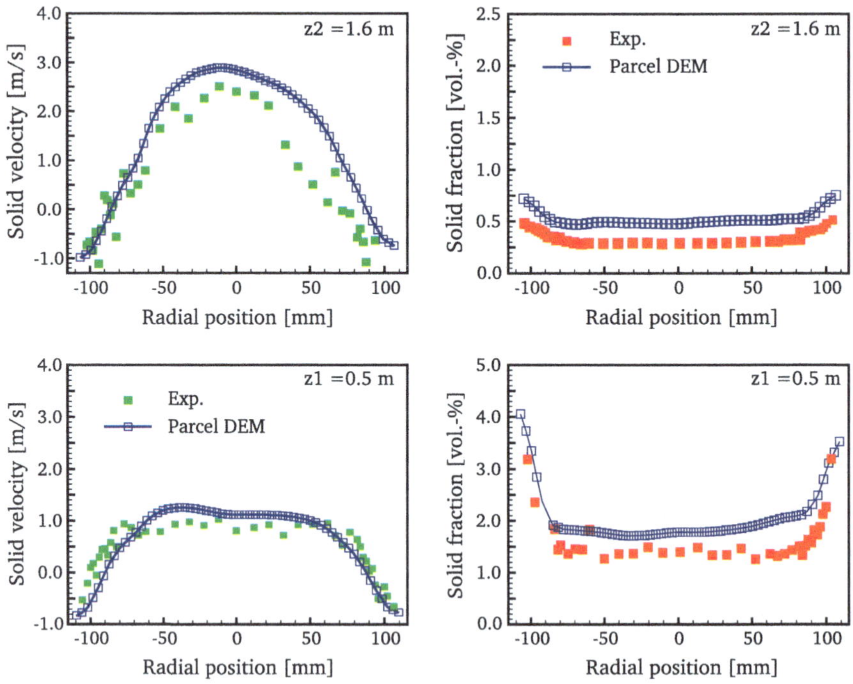

Fig. 4.149 Solid velocity and solid volume fraction at height z1 = 0.5 m and z2 = 1.6 m (simulation and measurement)

The calcium carbonate-looping (CCL) process, known also as carbonate-looping process, is a post-combustion CO_2 capture method using limestone as solid sorbent. Compared to other CO_2 capture methods (e.g. pre-combustion and oxyfuel processes), the CCL process has a high efficiency and low CO_2 avoidance, in particular as a retrofit of existing thermal power plants and energy systems, e.g. cement, steel, chemical industries. The 1 MW_{th} pilot plant for CCL process, erected at Technische Universität Darmstadt, consists of two interconnected circulating fluidized bed reactors (carbonator and calciner), fuel and limestone feeder as well as flue gas treatment.

In the carbonator reactor that is operated at around 650 °C, the calcium oxide (CaO) reacts with CO_2 from the flue gas, producing calcium carbonate ($CaCO_3$). Because of the exothermic CO_2 capture reaction, the released heat is extracted directly from the carbonator reactor by means of axial internal cooling tubes. The $CaCO_3$ is transferred to the calciner reactor, where CO_2 and CaO are released by an increase of the temperature to around 900 °C. Here, pulverised coal is fired with oxygen enriched air. Furthermore, a make-up stream of fresh $CaCO_3$ is constantly entering the calciner reactor.

For the numerical investigation of the carbonator reactor, a structured mesh with about 105,000 hexahedron cells is generated. The carbonator reactor has a height of 8.66 m with an inner diameter of 0.6 m and includes two loop-seals for a continuous solid feed during steady-state operation. The loop-seal 4.1 (LS 4.1) circulates the solids internally and the loop-seal 4.5 (LS 4.5) feds the entrained solids from the calciner into the carbonator. The cooling lances can be introduced to about 7 m into the carbonator. The boundaries are represented by the distributor inlet, the burner inlet, the loop seal (LS 4.5) inlet, the loop seal (LS 4.1) inlet and the pressure outlet at ambient pressure (see Fig. 4.150 and Table 4.38).

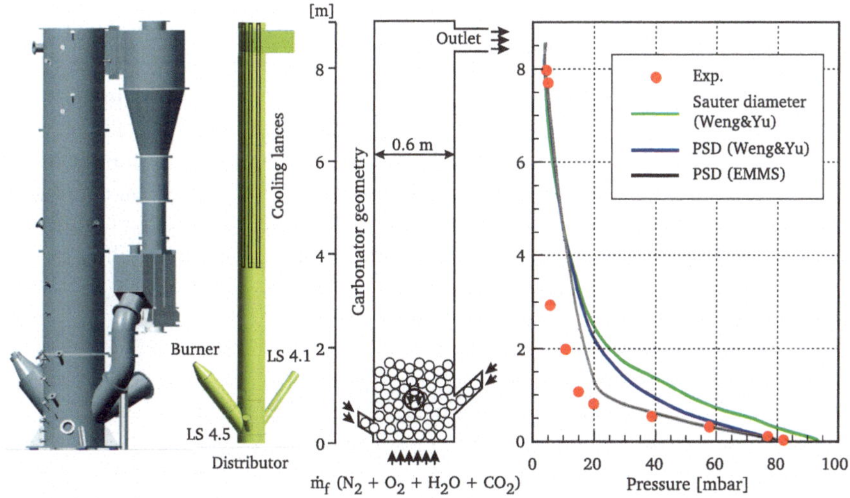

Fig. 4.150 Carbonator geometry of CCL process with associated inlets and time averaged pressure profiles along the reactor height (simulation and measurement)

Table 4.38 Boundary conditions of the carbonator model

Location	Temperature (°C)	Mass flow rate (kg/s)	Species mass fraction (wt%)	
Distributor inlet	110	0.21	CO_2	20.1
			O_2	6.2
			H_2O	5.2
			N_2	68.5
Burner inlet	35	0.029	O_2	23
			N_2	77
			Propane	0
L S4.5 inlet	35	0.0075	O_2	23
			N_2	77
	800	1.02	CaO	82.9
			$CaCO_3$	0.4
			$CaSO_4$ + Ash	16.7
L S4.1 inlet	35	0.009	O_2	23
			N_2	77
	635	User-Defined-Function	Solid	100
Pressure outlet	635	0		

The carbonation of calcium oxide to calcium carbonate is an exothermic heterogeneous reaction and is given by the following equation:

$$\mathrm{R1: CaO + CO_2 \rightarrow CaCO_3 + \Delta H_{25\,^{\circ}C}}\left(-178\,\frac{\mathrm{kJ}}{\mathrm{mol}}\right). \tag{4.56}$$

Thermogravimetric analysis (TGA) experiments were carried out at 650 °C with sample data from the pilot plant in order to measure the reaction rates close to the conditions in the 1 MW_{th} plant. Sorbent material with high number of carbonation-calcination cycles form a $CaCO_3$ product layer that limits the reaction rate by diffusional resistance. The heterogeneous reaction rate between CaO and CO_2 was modelled according to Bhatia and Perlmutter (1981, 1983), an expression derived from a random pore model that combines the kinetic controlled regime and the diffusional controlled regime using an overall rate expression as follows:

$$r_{carb}^{\mathrm{CaO}} = \frac{dX_{carb}^{\mathrm{CaO}}}{dt} = \frac{k_{carb}^{\mathrm{CaO}} a_{sorb}\left(X_{carb,\max}^{\mathrm{CaO}} - X_{carb}^{\mathrm{CaO}}\right)\left(C_{\mathrm{CO_2}} - C_{\mathrm{CO_2},eq}\right)\sqrt{1 - \psi * ln\left(1 - X_{carb}^{\mathrm{CaO}}\right)}}{\left(1 - \varepsilon_0^{\mathrm{CaO}}\right)\left(1 + \frac{\beta Z}{\psi}\left(\sqrt{1 - \psi * ln\left(1 - X_{carb}^{\mathrm{CaO}}\right)} - 1\right)\right)}, \tag{4.57}$$

where, X_{carb}^{CaO} represents the conversion of the calcium oxide, a_{sorb} denotes the initial surface area per unit volume of the sorbent (specific surface area for reaction), $\varepsilon_0^{\mathrm{CaO}}$ is the random-close packing limit of the calcium oxide (initial porosity) and Z is ratio of volume after reaction to that before reaction. The diffusion parameter β and the structural parameter ψ are expressed using the following equations:

$$\beta = \frac{2k_{carb}^{\mathrm{CaO}} a \rho_{\mathrm{CaO}} \left(1 - \varepsilon_0^{\mathrm{CaO}}\right)}{b M_{\mathrm{CaO}} D_{dif}^{\mathrm{CaO}} a_{sorb}}, \tag{4.58}$$

$$\psi = \frac{4\pi L_0 \left(1 - \varepsilon_0^{CaO}\right)}{a_{sorb}^2}, \tag{4.59}$$

with the stoichiometric coefficients a and b, the mass of the calcium oxide per unit volume of solid phase ρ_{CaO}, the molar mass of the calcium oxide M_{CaO}, the effective product layer diffusivity D_{dif}^{CaO} and the initial total length of the calcium oxide pore system per unit volume L_0. The kinetic rate constant for surface reaction k_{carb}^{CaO} is determined using the Arrhenius rate expression with the assumption of a zero activation energy:

$$k_{carb}^{\mathrm{CaO}} = A_{pre}^{\mathrm{CaO}} \underbrace{\mathrm{e}^{-\frac{E_{act}^{\mathrm{CaO}}}{RT}}}_{\approx 1} = A_{pre}^{\mathrm{CaO}} \tag{4.60}$$

The time dependent conversion degree of the sorbent (Eq. (4.57)) is limited by the maximum conversion degree $X_{carb,\max}^{\mathrm{CaO}}$. The sorbent specific constants were either determined experimentally or empirically and were adjusted in order to fit the measured overall reaction rate by TGA. The original expression was modified by the term $\left(X_{carb,\max}^{\mathrm{CaO}} - X_{carb}^{\mathrm{CaO}}\right)$ and the equilibrium CO_2 concentration in order to assure that the reaction rate is limited either by the maximum sorbent conversion degree or by the chemical equilibrium. It is worth to mention that the diffusion parameter β influences the transition time for the diffusion controlled regime. Generally, the bigger diffusion parameter, the more the reaction rate is dominated by diffusion through a product layer on the sorbent surface. Table 4.39 summarizes the sorbent related constants that were used in order to represent a realistic reaction rate of the sorbent.

The inventory in the carbonator is approximately 200 kg. Almost 16% of the particles are modelled as inert particles, representing the ash and calcium sulphate inside the carbonator. The mass flow rates are set equal to 0.21 kg/s ($O_2 + N_2 + H_2O + CO_2$) at distributor plate, 0.009 kg/s (air) as well as circulated solid at LS 4.1, 0.0075 kg/s (air) as well as 1.02 kg/s (solid) at LS 4.5 and 0.029 kg/s (air) at burner. User-Defined-Functions are used to keep the mass inventory constant during the transient run by injecting solid particles through the

Table 4.39 Sorbent specific parameters

Parameter	Description	Value	Unit
A_{pre}^{CaO}	Pre-exponential factor	6×10^{-10}	(m^4/mol s)
a_{sorb}	Total specific surface area for reaction	2.5×10^6	(1/m)
$X_{carb,\max}^{CaO}$	Maximum sorbent conversion level	0.32	[–]
ψ	Structural parameter	0.5	[–]
β	Diffusional resistance term	28	[–]
Z	Volume ratio after reaction to that before reaction	2.16	[–]
ε_0^{CaO}	Porosity initial porosity	0.5	[–]

inlet of LS 4.1. A pressure outlet boundary condition with atmospheric pressure is applied at the outlet. An isothermal carbonator temperature of 635 °C is defined as it was measured in the experiment. The simulation is performed for approximately 30 s until no notable differences in the time average pressure profile are observed.

The simulated pressure profile (solid line black) is in good agreement with experimental measurements (red in colour) obtained from the 1 MW_{th} pilot plant for CCL process (see Fig. 4.150). The Sauter mean diameter that is often used in numerical studies leads to an over-prediction of the pressure profile along the whole reactor height, while the use of the PSD improve the simulated pressure. In comparison to Wen and Yu drag model, the EMMS drag model shows better agreement to the measured pressure. The concentration of the solid phase, the solid z-velocity, the gas z-velocity, the mass fraction of steam, oxygen and carbon dioxide in the carbonator are illustrated in Fig. 4.151, while the comparison between the mass fraction of steam, oxygen and carbon dioxide at the carbonator outlet is presented in Table 4.40. Further results can be found in (Nikolopoulos et al. 2017; Stroh et al. 2017).

The calciner reactor has a height of 11.3 m with an inner diameter of 0.4 m in a cylindrical top section and a cone section at the bottom with a diameter of 0.28 m. The cylindrical reactor section starts at a height of 1.95 m. The generated mesh contains 105,000 hexahedron cells. The boundaries are represented by the distributor inlet, the burner inlet, the solid inlet and the pressure outlet at ambient pressure (see Fig. 4.152 and Table 4.41).

The numerical model of the calciner reactor consists of two phases. The primary phase is mixture phase, containing the gas species together with raw coal, char and ash species. The secondary phase includes the discrete element phase, considering the calcium oxide and the calcium carbonate. The primary phase consists of 10 species, where for each one transport equation has to be solved. The specific heat capacity is assumed to be constant for the solid species, while for the gas species a function of the temperature is defined. A global four-step reaction mechanism is employed for the pulverised coal combustion, which consists of two heterogeneous reactions and two homogeneous gas phase reactions as:

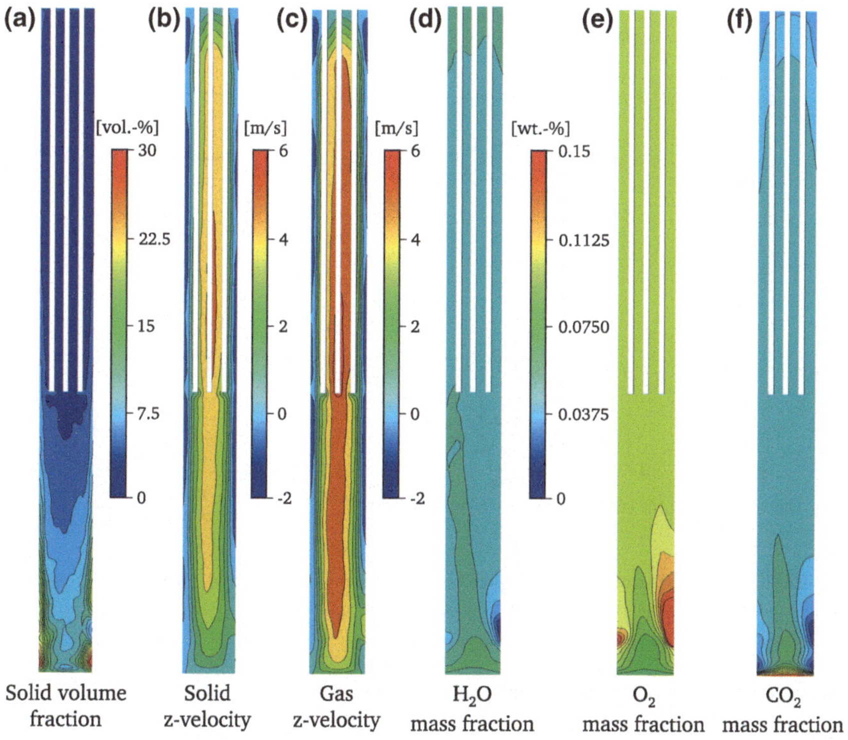

Fig. 4.151 Snapshots for the spatial distributions of gaseous species and solid: **a** Concentration of solid phase, **b** solid z-velocity, **c** gas z-velocity, **d** mass fraction of steam, **e** mass fraction of oxygen and **f** mass fraction of carbon dioxide

Table 4.40 Mass fraction at carbonator outlet (simulation and measurement)

Parameter	Unit	Experiment	Simulation	Absolute deviation
CO_2 (outlet)	(vol%)	2.2	2.34	+0.14
O_2 (outlet)	(vol%)	9.32	9.89	+0.57
N_2 (outlet)	(vol%)	80.93	80.68	−0.25
H_2O (outlet)	(vol%)	7.22	7.42	+0.20

$$\begin{aligned}
&R1: \text{Coal} \Rightarrow C + CH_4 + CO + CO_2 + H_2 + H_2O + N_2\\
&R2: C + {}^1\!/_2 O_2 \rightarrow CO\\
&R3: CH_4 + {}^3\!/_2 O_2 \rightarrow CO + 2H_2O\\
&R4: CO + {}^1\!/_2 O_2 \rightarrow CO_2
\end{aligned} \tag{4.61}$$

Taldinsky hard coal (lower heating value of 26.6 MJ/kg) is used. The ultimate and proximate analyses as well as products from devolatilisation are displayed in

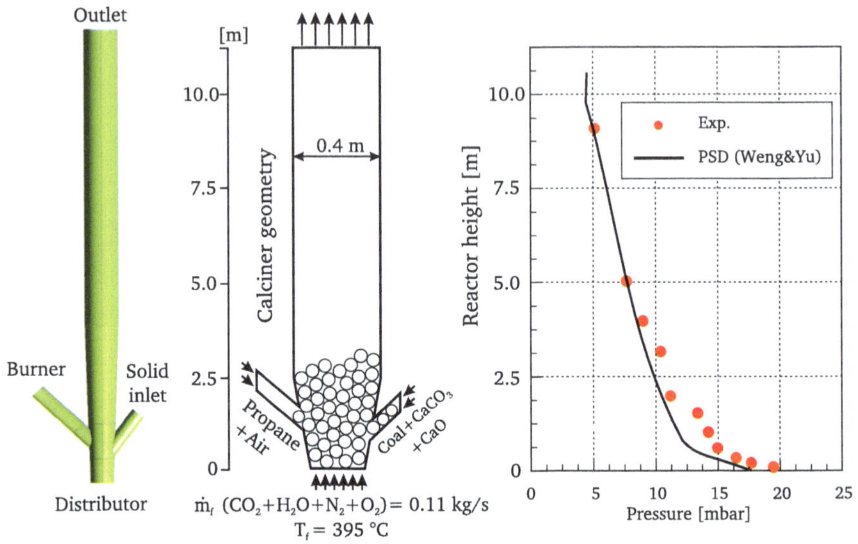

Fig. 4.152 Calciner geometry of CCL process with associated inlets and time averaged pressure profiles along the reactor height (simulation and measurement)

Table 4.41 Boundary conditions of the calciner model

Location	Temperature (°C)	Mass flow rate (kg/s)	Species mass fraction (wt%)	
Distributor inlet	395	0.11	CO_2	26.2
			O_2	48
			H_2O	4.2
			N_2	21.6
Burner inlet	35	0.0075	O_2	23
			N_2	77
			Propane	0
Solid inlet	35	0.025	O_2	5
			N_2	95
	615	1.05	CaO	76
			$CaCO_3$	8
			$CaSO_4$ + Ash	16
	35	0.034	Raw coal	100
Pressure outlet	860	0		

Tables 4.24 and 4.25. The complex conversion of char to carbon monoxide is computed by a global char-oxidation reaction. This reaction takes place both on the surface and inside the pores of the char particles, so that the transport mechanisms such as the absorption and desorption of chemical species on the particle surface,

pore diffusion and the diffusion process within the fluid phase have a crucial influence on the char reaction rate (see Sect. DEM with Thermochemical Reaction Model).

The calcination of calcium carbonate to calcium oxide is an endothermic heterogeneous reaction, given by:

$$\mathrm{R1}: \mathrm{CaCO_3} \rightarrow \mathrm{CaO} + \mathrm{CO_2} + \Delta\mathrm{H}_{25\,^\circ\mathrm{C}}\left(+ 178\frac{\mathrm{kJ}}{\mathrm{mol}}\right). \tag{4.62}$$

The calcination reaction is modelled under the consideration of the shrinking core model (SCM) in association with an additional term considering the adsorption/desorption of CO_2 at active surface sites. The model was developed by Garcia et al. (2002) and tested for different sorbents and operating conditions. The overall reaction rate is written as:

$$r_{calc}^{\mathrm{CaCO_3}} = \frac{dX_{calc}^{\mathrm{CaCO_3}}}{dt} = k_{calc}^{\mathrm{CaCO_3}} a_{sorb}(1-\Theta)\left(1-\frac{p_{\mathrm{CO_2}}}{p_{\mathrm{CO_2},eq}}\right), \tag{4.63}$$

with the chemical rate constant $k_{calc}^{\mathrm{CaCO_3}}$ for surface reaction according to the shrinking core model, the partial and equilibrium pressures of the carbon dioxide $p_{\mathrm{CO_2}}$ and $p_{\mathrm{CO_2},eq}$ as well as the adsorption mechanism of carbon dioxide Θ, given as:

$$\Theta = k_{ads}^{\mathrm{CaCO_3}} p_{\mathrm{CO_2}}^{1/n}. \tag{4.64}$$

The chemical and adsorption reaction rate constant are modelled according Arrhenius type expression with the kinetic data obtained from Table 4.42.

Figure 4.152 shows the mean static pressure profile along the reactor central axis. The pressure profile is time averaged for 15 s after obtaining a mixed particle structure within the reactor. The pressure profile is qualitatively in good agreement with the measurements. However, slight deviations can be observed in the bottom part of the reactor, where the pressure value is underestimated in the numerical simulation. The main reason for that is the measuring location of the pressure signal. Here, the pressure signal might be affected by the dynamic portion of the pressure that is large due to high velocities of the gas streams coming from the distributor, burner and coal inlet. It is also possible that the pressure signal has a

Table 4.42 Kinetic parameters of the calcination reaction

	Description	Value	Unit
Chemical reaction mechanism	Activation energy	131	(kJ/mol)
	Pre-exponential factor	254	(mol/m² s)
Adsorption mechanism	Activation energy	−90	(kJ/mol)
	Pre-exponential factor	3.7×10^{-10}	(mol/m² s)
	Reaction order	2	[–]

systematic offset due to blockage. Another explanation is the low inventory of the calciner (here is assumed to be approximately 25 kg) that will be increased in the next parametric investigation. Furthermore, the conventional Weng and Yu drag model was applied, and therefore advanced drag models such as three-zone, four-zone or energy minimization multi-scale should be used in order to accurately describe the momentum transfer between gas and solid phases.

4.3 Conclusion

Modern thermal power plants must be designed for high efficiency, low emissions and high flexibility with regard to load changes, start-ups and shutdowns. An important complement to experiments is the mathematical modelling that contributes to a better understanding of the process, capabilities and limitations. In general, it is distinguished between one-dimensional process simulation and three-dimensional computational fluid dynamics. While 3D simulations are often used for individual components to visualise flow patterns, 1D process simulations can be applied to model the entire thermal power plant or energy system.

Optimisations and new designs of thermal power plants begin with steady-state modelling. Mathematically, the steady-state models are based on mass, momentum, energy and species balances. Although the thermodynamic properties of the working fluid as well as process efficiency can be conducted for a series of operating points, the steady-state simulation tools do not allow any information about transient operations. The process analysis during load changes, start-ups and malfunction cases is achieved using dynamic models that can also be applied to the proposal stage of the power plant project. However, investigation into the dynamic performance of thermal power plants requires detailed information of the process and its control structures, making the dynamic simulation codes sophisticated and complex software.

Relevant studies on dynamic process simulations were presented in this chapter for different technologies of combined-cycle power plant, pulverised coal-fired power plant, concentrated solar power plant and waste incineration as follows:

Combined-cycle power Combined-cycle power plants are widely recognized in the field of power generation for their high efficiency, fast start-up capability and comparatively small environmental impact. The technology also supports increasing shares of renewable feed-in due to flexible unit dispatch. Modern combined-cycle power plant consists of a 1 + 1 arrangement of GT and ST units in combination with a triple-pressure, drum-type, reheat heat recovery steam generator.

Detailed dynamic simulation models of Prai Power Station, located in Malaysia were generated using the process simulation software tools (APROS and ASPEN PLUS DYNAMICS). The real power plant consists of a gas turbine and a vertical heat recovery steam generator, which is installed downstream in the flue gas path. Waste heat from the passing flue gas in the HRSG is absorbed by the fin tubes heat exchangers, generating live steam for turbine sections. The HRSG is divided in three pressure levels (low, intermediate and high-pressure) with forced circulation evaporators. The individual heat exchangers are arranged to match the temperature curves of flue gas path and water/steam side. The capability of APROS and ASPEN PLUS DYNAMICS models to predict the real behaviour of Prai Power Station during part loads, off-design operation, start-ups (hot and warm) as well as shutdown procedure were investigated, resulting in the following conclusions according to the comparison between the numerical results and the operation data:

- The numerical models can represent quantitatively the real power plant at different steady-state base loads. The error percentage for mass flow rate, temperature and pressure are all within 5%. Several parameters show, however, a relative error of less than 1%. Furthermore, the steam mass flow rate and the temperature are reproduced in both programmes with higher accuracy compared to the pressure.
- During off-design operation, the high accuracy obtained at base loads is decreased. The dynamics models can qualitatively follow the measurements. A maximum relative error of about 12% is reported. Nevertheless, several parameters show a relative error of less than 5% or even the accurate characteristic of the real power plant can be predicted. In addition, it is found that the simulated feedwater mass flow rate follows the measurement with a higher accuracy than the steam mass flow rate and pressure. By contrast, the temperature displays over some period of time disagreements with the real power plant.
- Qualitative behaviour of the real power plant parameters can be reproduced by the dynamic models during the warm start-up procedure. Deviations between simulations and measurements are observed. The early steam production in the numerical models has been a major problem, which is mainly caused by lower thermal inertia of the models compared to the real system. This suggests the importance of implementing the auxiliary HRSG systems such as vent and draining circuits. Furthermore, considerable oscillations in the measured parameters such as the feedwater mass flow rate and the drum level could not be predicted by the simulation models. The reason of smooth characteristic of these parameters is the rigid setting of the three-term controller (proportional, integral and derivative values). The drum levels and the feedwater mass flow rates are captured in both models with higher accuracy compared to the steam mass flow rate, the temperature and the pressure.
- During hot start-up and shutdown procedure, the simulated steam mass flows, pressures and temperatures are in good agreement with operation data. However, discrepancy between simulations and measurements is observed, which can be attributed to incomplete information on components and control circuits. On the one hand, the thermal inertia of the water/steam side is underestimated since auxiliary systems and supporting structures in contact with the primary flow paths are not available and thus neglected. This is reflected in the calculations by premature generation of first steam (−6, −7, −9 min for HP, IP and LP circuits, respectively) as well as higher temperature gradients during

start-up and shutdown procedures. On the other hand, the observed oscillations of the feedwater mass flow rates and drum's levels in the power plant are not reproduced in the simulations due to more suitable setting of the valve controllers.

- APROS and ASPEN PLUS DYNAMICS models demonstrate similar development of the simulated parameters at different steady-state base loads, during the off-design operation as well as during different start-up procedures. Compared to the ASPEN PLUS DYNAMICS model, the APROS model can generally reproduce the real behaviour of the power plant with higher accuracy. This is due to the fact that the combined-cycle power plant model in APROS is based on the six-equation flow model, while the combined-cycle power plant model is built in ASPEN PLUS DYNAMICS using the mixture flow model that provides less accurate results form a physical point of view.
- A complete set of operation data for the response of the most important parameters from an existing real combined-cycle power plant during the off-design operation, hot and warm start-up procedures were provided. These measurements are of high relevance for other researchers since publication of detailed experimental data of the real power plant is often not possible due to proprietary reasons.

Modern combined-cycle power plant with short start-up time is effective plant for saving fuel, lower CO_2 emissions and attached great importance for the grid stability. Supplementary-fired large combined-cycle power plant allows higher flexibility in operation, since this plant can produce more power output during periods of high electricity prices by switching on the duct burner of the HRSG and can decrease sharply its power output during periods of low electricity prices by switching off the duct burner as well as by reducing the GT load. Despite this important advantage, the process efficiency of the supplementary-fired combined-cycle power plant is lower than an unfired plant, which in turn makes the interest of this technology relatively low. Therefore, the benefit of the supplementary-fired combined-cycle power plant is a trade-off between number of hours with high and low electricity prices and depends on several other factors such as the cost of fuel, the cost of plant and the demand for electricity.

A dynamic simulation model of a 300 MW_{el} combined-cycle power plant with horizontally-based, natural circulation heat recovery steam generator was developed using the advanced process simulation software (APROS). The water/steam side of the HRSG consists of three pressure circuits with a natural circulation in the evaporator paths for HP, IP, LP and a reheater section after the high-pressure turbine. The model predictions were compared with design data at different operation loads. The comparison shows good agreement with an average relative error of 5%. Hot, warm and cold start-up procedures were simulated with the validated HRSG model and then the start-up times of the gas turbine were shortened by factors of 0.3, 0.8 and 0.93 in hot, warm and cold start-ups, respectively. This study reveals that the HRSG can reach faster its nominal steam power output by 38%, 31% and 20% in hot, warm and cold start-ups respectively, compared to baseline case. The gas turbine shows higher power generation rate since it runs to its full load within 15 min in case (a), 24 min in case (b) and 38 min in case (c). At full load, the supplementary firing of the HRSG can be switched on. The supplementary firing adds approximately 50 MW_{el} to the 100 MW_{el} steam turbine power output, resulting in a total power output of the combined-cycle power plant of about 350 MW_{el} (ST = 150 MW_{el} and GT = 200 MW_{el}).

Horizontally-based HRSGs, by contrast to vertically-based HRSGs, are designed with natural circulation evaporator systems. Here, the density difference between water in the downcomers and water/steam mixture in the risers creates an uplift force, ensuring stable and uniform natural circulation of fluid. A circulation pump is therefore not required in the horizontally-based HRSG, while it is an essential part of the evaporator systems of the vertically-based HRSG in order to avoid two-phase flow instabilities.

Based on the 450 MW_{el} Deir Aamar Power Station II, located in Lebanon (3 HRSGs +1 steam turbine), the design of the vertically-based HRSGs has been modified by replacing the forced circulation with natural circulation systems in the HP and LP evaporators. In order to analysis the flow stability in the vertical heat recovery steam generator with natural circulation evaporators, a dynamic simulation model, based on six-equation flow model, was developed. The static and dynamic instabilities such Ledinegg, density wave and pressure-drop oscillations were analysed during part loads, shutdown and start-up procedures. The obtained data of the mass flow, pressure, temperature and other parameters were investigated using fast Fourier transform algorithm (FFT) and histogram. The design of vertical HRSG with nature circulation (thermo-hydraulic and geometry) was modified in order to avoid flow instabilities in the HP and LP evaporators.

According to the previous evaluations, it can be conducted that the dynamic simulation models are capable to reliably predict the real system response to failure malfunctions, modifications in design and control structures. The major interest for designers in this field is to develop new strategies for the combined-cycle power plants in order to get higher flexibility and efficiency. The most promising feature, as the gas turbines become larger and the exhaust temperatures rise, is the once-through technology with sub-critical or super-critical properties. The once-through HRSG can follow with better dynamic characteristic the gas turbine transients, e.g. fast start-up and short load variation time. This is due to the absence of the high-pressure drum that greatly limits the allowable temperature transients with its large wall thickness.

A dynamic model of a triple pressure sub-critical, once-through HRSG was developed to understand the influence of fast gas turbine start-up on the dynamic behaviour of the investigated power plant. The model includes a new feedwater control circuit and a set of enhanced steam feedwater bypass systems. The developed model and its advanced control circuits were evaluated towards the design data at different part loads. The received results demonstrate good agreement with a relative error less than 3%. The model analyses during the fast start-up processes (hot, warm and cold within 20 min) provide evidence that the once through HRSG is able to meet the gas turbine fast start-up requirements. The results show that the sub-critical, once-through HRSG reaches its nominal load 50, 90 or 150 min after the GT start, depending on whether a hot, warm, or cold start-up procedure is performed. An important result is also obtained by the application and the verification of a new high-pressure feedwater control circuit that considers a large number of plant parameters and derivative functions, e.g. the degree of sub-cooling at the evaporator inlet, the degree of superheating at the evaporator outlet, the derivative elements for considering the fluid side mass storage processes in the economiser and evaporator heat exchange surfaces etc. The developed high-pressure feedwater control circuit offers the possibility to operate the HRSG in different modes, namely level or once-through.

In order to investigate the influence of the HRSG design on its dynamic behaviour, the previous triple pressure, sub-critical, once-through HRSG model was modified by replacing the once-through evaporator in the high-pressure circuit with a natural circulation evaporator, including the associated control circuits. Both models were designed to supply equal steam mass flows with equal steam parameters (temperature and pressure) at full load, which enables a balanced assessment of the two technologies. For the natural circulation HRSG, the required heat exchanger surface of the high-pressure circuit decreases by approximately 8%. The responses of the models to the gas turbine start-up from various initial conditions (hot, warm and cold) were performed, resulting in the following conclusions:

- Subjected to the same hot, warm and cold start-up procedures, both models meet the design requirements for fast start-up and reach full CCPP load after 50, 90 and 170 min, respectively.
- The calculations show similar dynamic behaviour of HP steam conditions in general. However, pressure build-up is slightly accelerated for the once-through evaporator due to lower thermal inertia of the system. This effect amplifies with extended gas turbine standstill before the start-up.
- The increased thickness of the drum walls causes significantly higher temperature deviations and thus higher thermal stress in comparison to the HP separator. Additional temperatures oscillations at the inner wall are observed, which are explained by water level fluctuations. While the temperature deviations in the separator wall decline less than 60 min after GT start, up to 257 min are required for uniform temperature distribution in the drum wall. Consequently, the drum-type HRSG will only sustain a reduced amount of start-up procedures, if full load is to be reached after a given time in particular cold start-up.
- The results indicate that installing a sub-critical once-through HRSG downstream the gas turbine is a favourable choice for plants with enhanced flexibility requirements, where frequent load changes, start-ups and shutdowns are expected in operation. However, this comes at the cost of increasing heat exchanger surface for the HP evaporator.

Several sub-critical once-through HRSGs are in operation, but are not as frequently applied as the drum-type HRSGs. The pressure in the once-through high-pressure circuit is ranged between 130 bar and 180 bar. A once-through HRSG with a super-critical steam properties was not constructed yet. In this work, a dynamic simulation model of a super-critical once-through HRSG with 250 bar in the HP circuit was developed based on a validated sub-critical drum-type HRSG model. The verification and evaluation of the developed model at full load and at different part loads were performed, showing good agreement towards the design data. The study displays a potential in the thermal steam power with a net gain in the steam power of 5% compared to the drum-type HRSG. The super-critical once-through HRSG not only offers a gain in the steam power, but also provides a reliable response with the dynamic transient of the gas turbine. However, an increase of the heat exchanger's surfaces of about 30% compared to the drum-type HRSG is expected. The combined-cycle power plant with the super-critical once-through HRSG is now under investigation whether the investment and operating costs are comparable to once-through sub-critical HRSGs.

Pulverised coal-fired power plant With regard to installed capacity, coal-fired power plants in many countries are the most important generating units in order to cover the base load demand. Although all coal-fired power plants have the same working principle, each power plant is individually engineered, leading to different dynamic behaviour. Here, the load change and start-up behaviour by increasing flexibility requirements are in the focus of interest.

A single-pass pulverised coal-fired power plant model of an existing unit (800 MW_{el} sub-critical, once-through Heilbronn Power Station, located in Germany), including all power plant components and its associated control schemas was generated to investigate the operation flexibility. The only boundary conditions are the hard coal composition, the ambient temperature and the temperature and amount of cooling water in condensers. The developed model was validated at a load point of 105% and during part load transients (for a load change from 100% to 27.5% and 27.5% to 100%). According to the evaluations together with comparisons of the simulation results with the operational data obtained from Heilbronn Power Station, it can be concluded that the developed model can simulate the dynamic behaviour of the real power plant with high accuracy, even at lower loads. Deviations between the simulations and measurements are, however, observed (a maximum relative error of about 5%). Uncertainties, like the exact composition of the burned coal during the operation of the real power plant, the fouling level within the furnace and the flue gas system are a major problem for the simulation. Furthermore, the estimation of the difference between parallel paths of the flue gas flow represents an additional challenge for the numerical model. Due to the high degree of detail regarding the process and the implemented automation, the developed model can be applied for further flexibility investigations, e.g. faster load changes, new control concepts during low loads, economiser bypass and heat storage system.

The convective heat exchangers in the single-pass boiler are arranged one above the other, which in turn results in smaller differences in the temperature and the velocity of flue gas path and water/steam side compared the top. Furthermore, the single-pass boiler is characterised with several additional advantages such as rectilinear flow paths with few deflections, low deposition problem, low thermal stresses and thermal expansion only in one direction, but the construction height and construction time are much higher than the two-pass boilers.

A pulverised coal-fired power plant model of an existing unit (550 MW_{el}, located in South Korea) was developed. The 1190 MW_{th} pulverised hard coal-fired power plant is a super-critical, once-through and two-pass steam generator with a nominal superheated steam mass flow rate to the HP turbine of 445 kg/s at 260 bar and 602 °C as well as a reheated steam mass flow rate to the IP turbine of 359 kg/s at 46 bar and 612 °C. The model describes the air/flue gas path and the water steam side in detail. On the air/flue gas path, all real components of the plant are considered, including the primary air supply with its pulverisers, the secondary air supply system, the air preheater, the furnace chamber, the vestibule chamber and the parallel pass heat recovery area (HRA) chamber, in addition to the flue gas cleaning devices (e.g. electrostatic precipitator and selective catalytic reduction unit). The circuit of the water/steam side starts from the boiler feedwater pumps that supply the feedwater to the evaporator system via the high-pressure preheater system. The evaporator system consists of five parts: the economiser, the lower furnace walls, the mixing bottle, the vestibule sidewall and the upper furnace wall. At the evaporator outlet, the steam enters the separator, so that the water droplets within the steam flow is removed during part loads and start-up procedures. A circulation pump forces the collected water to re-enters the steam generator at the economiser inlet. At nominal load (super-critical pressure), no phase separation can take place. The steam leaves the separator and flows into the primary superheater, the platen superheater and the final superheater. An inter-stage and final attemperators are installed to control the steam temperature. The superheated steam is directed to the high-pressure turbine, where it is expanded and re-enters the steam generator via the reheater system that consists of the primary reheater and the final reheater. The reheated steam is then directed to the intermediate and low-pressure turbines and finally condenses in the condenser. The condensed water is fed through the feedwater preheaters back to the feedwater storage tank of the steam generator. Employing the developed model, several dynamic simulations were carried out, including load change from 100% to 75% to 100%, from 100% to 50% to 100% and from 100% to 30% to 100%. Furthermore, the dynamic behaviour of the coal-fired power plant during restart, hot start-up, warm start-up and cold start-up were investigated.

Recently, the focus of dynamic simulation of coal-fired power plants is shifted towards carbon capture and storage, especially to oxyfuel coal combustion. Here, the combustion of coal is taken place with a nitrogen-free oxidant. The water content in the flue gas is condensed and then the pure CO_2 flow is prepared for storage or further utilisation. In this field of research, the dynamic simulation tools are mostly used to evaluate the switch over from air to oxygen combustion. Since no large scale oxyfuel coal-fired power plants exist, measured data is not available either.

A dynamic simulation model of an oxygen pulverised coal-fired power plant (Jaenschwalde Power Station, located in Germany) was developed in order to investigate the maximal pressure difference along the gas path between the nominal load and the transient operation during malfunctions, e.g. during fuel trip and blackout. The results show that the minimal furnace pressure is dependent on the fire extinguish time, the behaviour of the dampers and the fans. The numerical results confirm that the fuel trip and the blackout scenarios do not have any negative influence on the investigated gas path. This is due to the fact that the observed lowest negative gauge furnace pressure of −36 mbar during malfunctions can be handled with current state of the art technology used in conventional pulverised coal-fired power plant. Further investigations such as switch over from air to oxygen or oxygen to air, load changes, start-up and shutdown procedures can also be carried out with the developed model.

Waste incineration Due to the advantages it offers, the waste incineration becomes a quite reasonable option for the large-scale disposal of municipal solid waste. Advantages of this technology include the decrease of original waste volume as well as producing stable and odourless residue. However, the investing cost is relatively high, in addition to the continuous maintenance over the lifetime. Generally, waste incinerators operate on off-design load conditions in view of the fluctuations of the waste heating value. The changes in operating conditions can influence the plant efficiency and consequently the plant emissions. Although the waste incinerators have been applied commercially for many decades in industrialised countries, the dynamic behaviour has not yet been investigated in the literature.

A dynamic simulation model of the 60 MW_{th} waste incineration built in Finland was developed in order to investigate the impact of waste heating value variation on the water/steam circuit and accordingly on the plant operation and efficiency. The numerical results show good agreement with the design data at 100% nominal load and at different part loads. The relative errors of steam parameters are less than 1%. For the flue gas path, a slight deviation in the post-combustion zone can be seen, which might be attributed to the secondary air. Here, the simulation model under-predicts the design temperature with a relative error of about 5%. In addition to the part loads, the dynamic behaviour during a start-up and shutdown procedure was evaluated with the validated model.

Concentrated solar power plant The concentrated solar power plant is a renewable energy technology that concentrates solar rays on a receiver to heat a fluid in order to generate electricity. There are different technologies available, including parabolic dish, Fresnel reflector, solar tower and parabolic trough. Both latter technologies are showing increasing interest in the field of research due to the simple integration with a thermal energy storage system and with a fossil fuel combustion facility, increasing the power plant availability.

A dynamic simulation model of an existing 50 MW_{el} parabolic trough solar thermal power plant (Andasol II) built in Spain was developed. For the model validation, sets of operation data obtained from the real power plant during clearly days with little scattered clouds and with strongly cloudy periods were used. The comparisons show a quantitative agreement between the numerical results and the measurements for clearly days with little scattered clouds (a maximum relative error of about 10%). For days with strongly cloudy periods, discrepancies between the model predictions and the real power plant are observed. The main reason of these differences is the unknown decisions from the operator during power plant operation. In some periods for example, the operator redirected the mass flow rate of the heat transfer fluid to the thermal storage system instead of the power block. Furthermore, the operator in a certain period stopped the circulation of the heat transfer fluid within the solar field, despite the fact that solar radiation is again increased after a strongly cloudy period.

In addition to dynamic process simulations, the computational fluid dynamics (CFD) provides a promising tool to investigate numerically the thermochemical conversion systems, describing flow turbulence, heat and mass transfer, gasification or combustion process of solid fuels (including drying, pyrolysis and char burn-out) and gas phase reactions. For the simulation of gas-solid flows under the consideration of thermochemical reaction model, there are different numerical methods available: quasi-single-phase, two-fluid and single-particle method. The quasi-single-phase model offers an efficient approach and is therefore frequently used to simulate the power plants components such as heat exchangers and combustion chambers. For a precise description of the solid phase, the classic application of quasi-single-phase model is not sufficient. Here, the two-fluid model is related to a better accuracy and a wide application range. The basic idea behind the two-fluid model is that both phases are treated as fully interpenetrating continua. Representing the solid phase as a continuum results in a challenge of how to model the kinetic and collisional transports, i.e. gas-particle interactions, particle-particle and particle-wall collisions. Hereby, assumptions have been proposed by several authors regarding the gas-particle drag correlations and the rheology of solid phase. The single-particle model, by contrast, combines the continuum description of the gas phase with the Lagrange representation of the solid phase on the basis of Newton's transport equations. This combined method for the modelling of granular flows provides detailed information regarding the microscopic behaviour of the particle phase. The collision detection between particle-particle and particle-wall can be detected either with deterministic collision models or with stochastic collision models. In the latter, the collision treatment between particles is only performed with the hard sphere model, where the particles' properties after the collision are a function of the particles' properties before the collision. If the collision is detected with a deterministic collision model, the subsequent collision treatment can be achieved either with the hard sphere model or the soft sphere model that enables particles to overlap each other or penetrate into the wall.

Relevant simulation studies of entrained-flow reactor, 1 MW_{th} combustion chamber, 1 MW_{th} fluidized bed pilot plant, 60 MW_{th} waste incineration, large-scale lignite and hard coal-fired power plants were presented in this chapter using different CFD models. The obtained results are as follows:

Quasi-single-phase model Different combustion models for coal and biomass, which are based on the quasi-single-phase method, were developed and validated towards measurements obtained from an entrained-flow reactor, erected at Technische Universität Darmstadt. The entrained-flow reactor that is a ceramic tube with an inner diameter of 70 mm and a length of 2.2 m consists of a preheating step for the reaction gas, a reaction zone as well as a water-cooled quenching section. The operation parameters can be varied between 900 °C to 1600 °C and 1 bar to 20 bar, respectively. For models validations, concentrations of main species such as CO, CO_2, CH_4, SO_2, and O_2 as well as corrosive substances as for example HCI, H_2S and COS are applied. The results show that the coal combustion model can reproduce the entrained-flow reactor with higher accuracy than the coal/biomass combustion model. This is due to the fact that the modelling of biomass is complicated by the inhomogeneous composition of raw material. The biomass consists of hemicellulose, cellulose, lignin, inorganic ash and small amounts of other organic substances. The heterogeneous structure of biomass and its anisotropic physical properties have the consequence that different reaction rates and diverse reaction mechanisms take place during the pyrolysis process and the char-combustion.

In the next step, the developed combustion models were validated towards the experimental data obtained from the 1 MW_{th} test facility, erected at Technische Universität Darmstadt. Gas concentrations such as CO_2, CH_4, CO and O_2, in addition to the solid samples at several positions on the sidewall of the furnace are used for validation. Furthermore, the numerical models can be further verified with the continuous measurements of slagging and corrosion processes. The combustion behaviour of three fuel blends, including pure coal, coal with 9.14% and 17.04% torrefied sawdust (thermal basis) were investigated. The numerical predictions show good agreement with the experimental data obtained from the 1 MW_{th} test facility. While the temperature distribution of the three cases did not differ significantly, more differences for the gas species distribution inside the flame zone are observed. This can be explained by the different calorific value and particle size distribution of the torrefied sawdust. The combustion temperature is decreased with increasing the biomass share, which can be linked to the lower calorific value of the biomass. For the 17.04% torrefied sawdust fuel blend, the lignin component was not completely decomposed in the simulation due to lower pyrolysis reaction rate and coarser biomass particles. According to the numerical and experimental results from the 1 MW_{th} combustion chamber, the possible used of torrefied sawdust as a co-firing fuel in pulverised coal-fired power plant it is confirmed.

Using the validated combustion models, the 60 MW_{th} waste incinerator (located in Finland) was simulated using different CFD programmes: ANSYS-CFX and ANSYS-FLUENT. The waste incineration burns 5.7 kg/s of waste with an ideal lower heating value of 10.5 MJ/kg, generating about 16 MW_{el} with a gross efficiency of 26.5%. The distributions of velocity, temperature and species concentrations are almost similar in both programmes and show a good agreement with the design data. Furthermore, large-scale lignite and hard coal-fired power plants were simulated using the quasi-single-phase method in combination with the validated combustion models, showing a good agreement with the operation data, obtained from real power plants.

Two-fluid model A cold-flow, quasi-2D spouted fluidized bed was simulated using the two-fluid model, implemented in ANSYS-FLUENT programme. Two different fluidization mass flow rates (0.005 kg/s and 0.006 kg/s) were considered, while maintaining the same computational parameters that were identified through a parameter study. The investigated input parameters were gas-particle drag models, specularity coefficient, granular temperature approach and the particle-particle restitution coefficient. The results show that the two-fluid model is able to capture the flow patterns obtained in the experiment at a mass flow rate of 0.005 kg/s, but it fails in reproducing the experiment when increasing the mass flow rate to 0.006 kg/s, especially during the final stage of bubble formation. According to comparison between the numerical results and the measurements, conclusions with regard to the present quasi-2D spouted fluidized bed are obtained as follows:

- The drag model has a significant effect on the flow behaviour of the spouted fluidized bed characterised by the bed height and the bubble size. Compared with the drag models of Gidaspow, Wen and Yu, the drag model of Syamlal and O'Brien offers good agreement with the measurements. However, the model accuracy decreases with increasing the mass flow rate from 0.005 kg/s to 0.006 kg/s. This can be explained by the fact that the selection of an appropriate drag model might probably not improve the simulation results for all cases. Other parameters that are not appropriately modelled or neglected in the framework of the two-fluid model e.g. frictional viscosity model, particle-particle interactions and the particle rotation must be investigated, too.
- The solid wall boundary conditions in terms of the specularity coefficient have a significant impact on the flow behaviour of the spouted beds. In the range of values considered between 0 and 1, a value of 0.5 shows the best predicted flow patterns.
- Compared with the partial differential approach, the algebraic granular temperature approach is more efficient and can reproduce the flow structure more reasonably.
- In the range of values considered for the particle-particle restitution coefficient, a value of 0.6 provides the best predictions.

In the next step, the two-fluid model was combined with in-house codes for the consideration of the thermochemical reaction mechanism. The extended two-fluid/thermochemical reaction model was validated with six hours of worldwide first auto-thermal tests of the 1 MW_{th} chemical-looping pilot plant, erected at Technische Universität Darmstadt.

For the simulation of the air reactor of the CLC, the geometry of the air reactor with a height of 8.66 m and a diameter of 0.59 m was built and a block-structured grid with a resolution of approximately 150,000 cells was generated. The comparison between the simulation results and the pilot plant data results in the following conclusions:

- The simulated temperature and the outgoing concentration of carbon dioxide agree well with the experiment. This shows that the unwanted char coming from the fuel reactor was reasonably estimated.
- Discrepancies were observed regarding the outgoing concentration of oxygen and the pressure profile in the air reactor.
- The mass flow rate of ilmenite leaving the air reactor was slightly higher in the simulation than in the experiment.

For the simulation of the fuel reactor of the CLC, the geometry of the fuel reactor with a height of 11 m and a diameter of 0.4 m was built and a block-structured grid with a resolution of approximately 100,000 cells was generated. The comparison between the simulation results and the pilot plant data results in the following conclusions:

- The simulated pressure distribution shows good agreement with measured data with a maximum deviation value of approximately 8 mbar.
- Deviations were observed regarding the temperature and the outgoing gas species concentrations.
- The solid mass flow rate at the outlet of the fuel reactor shows a good agreement in the weight fractions of species. Nevertheless, in comparison to an ASPEN PLUS model of the process, the outgoing mass flow seems to be high.

According to the qualitative and quantitative comparisons between measurements obtained from 1 MW_{th} chemical-looping pilot plant and the numerical results of the air and fuel reactors, suggestions regarding the improvements of the simulation are summarised as follows:

- Drag models represent the most important factor regarding the characteristic of gas-solid flows. Conventional drag models (e.g. Syamlal and O'Brien, Gidaspow, Wen and Yu) usually over-predict drag forces and under-predict slip velocities. Newly developed drag models that are based on flow structures (e.g. three-zone, four-zone and energy minimization multi-scale) can improve the results considerably. Furthermore, a parametric study of different restitution coefficients and specularity coefficients must be carried out in order to analysis the influences of these values on the outgoing mass flow rate of solid.
- The modelling of turbulence in dispersed flows is usually neglected. However in practical applications (e.g. circulating fluidized beds or pulverised coal-fired power plants), the gas velocities are high and there are zones with low volumetric void fractions, which in turn make the turbulence modelling of relvance. In these simulations, the statistical mixture realisable $k-\varepsilon$ turbulence model was used to consider the turbulent fluctuations. Complex turbulence models such as Reynolds stress model or large eddy simulation may affect the result positively.

- In the fuel reactor, the gasification reactions rates were calculated by neglecting the actual conversion of char. This simplification leads to a slightly increased reaction rate. In further simulations, a mean conversion over the whole fuel reactor should be considered.
- The distribution of species in the fuel reactor showed that the char does not fall into the dense bed of the fuel reactor. Reason is the definition of char as part of the gaseous phase. Since the mean diameter of coal was approximately 3 mm during the experiments, different behaviour of char is expected. The char particles should fall into the dense bed forming carbon monoxide, carbon dioxide and hydrogen. After the gasification, products react with ilmenite also mostly in the dense bed. In order to avoid this behaviour of char, the char should be defined as component of the second phase or as a third phase.

Single-particle model A numerical simulation programme (DEMEST), based on the deterministic Euler-Lagrange/DEM model under the consideration of thermo-chemical reaction mechanism was developed. Validation studies were performed to assess the results obtained from the DEM model and the quasi-2D spouted fluidized bed of Plexiglas. Here, the spouted flow regime with particle of group D was investigated. The dispersed phase consists of 36,500 spherical particles with an average diameter of 2.5 mm and a density of 2500 kg/m^3. The particles have the properties of glass and the fluidizing medium is air at room temperature. The spatial distribution of solid, the bed height and the equivalent bubble diameter are analysed at different mass flow rates. The results suggest that the Euler-Lagrange/DEM model can accurately predict the typical flow pattern of a spouted bed including the spout, the fountain and the annulus between the spout and the walls. However, a relatively small deviation of simulated solid distribution from the experiments in the nozzle region may be caused by the inlet assumption of the centrally placed nozzle instead of nine holes.

The numerical results obtained from the DEM model implemented in the DEMEST programme were also compared with the numerical results of the two-fluid model implemented in the CFD package ANSYS-FLUENT and the Euler-Lagrange/ MP-PIC model implemented in the programme CPFD-BARRACUDA. According to the qualitative and quantitative comparisons, the results show that the two-fluid model tends to predict smaller bubble and bed expansion than the experimentally observed bed behaviour, especially during the final stage of the bubble formation. In addition, the solid distribution in the two-fluid model shows an exact symmetrical contour of particle volume fraction with respect to the $y-z$ midplane. The Euler-Lagrange/MP-PIC model, although it reaches earlier the maximum bed expansion, is able to predict the bed expansion. These deviations from experiments may have several reasons. The inlet assumption of the centrally placed nozzle instead of the nine holes in case of the Euler-Lagrange/MP-PIC model as well as the set of nine squares in two-fluid model instead of the nine holes in the test rig represents the most significant measure for these deviations. Other important factor is the electrostatic force. For different material pairings, the electrostatic force that occurs between the oppositely charged surfaces plays an important role in fluidized beds.

During the fluidization, the particle-particle/wall collisions lead to generation of electric charges that result in particle accumulation on the walls. The models are not able to simulate the cluster build-up when the adhesive forces, particularly the electrostatic force, are neglected. Therefore, the particles are distributed more evenly in the computation domain. Further reason is related to the inaccurate solid parameters that are used in the calculation. The simulation of the solid phase requires many material and geometry properties of particle-wall pairing that are not completely available. The stiffness coefficient, the damping coefficient and the friction coefficient of the particle-wall are estimated from the particle-particle coefficients. The reason of this assumption is that experimental data for the heterogeneous material pairing of Glass-Plexiglas is presently not existent in the literature. Considering the deviations in the bed behaviour, the two-fluid and Euler-Lagrange/MP-PIC models still show potential to simulate the complex behaviour of dense gas-solid flows in a fluidized bed with an acceptable accuracy.

The gas-solid interaction was performed in the the Euler-Lagrange/DEM model using a new procedure, the so-called offset method. Here, the fluid grid is shifted in space several times and the interaction values are calculated for each displacement. The inter-phase values in a grid cell are then determined by averaging all calculated values with regard to the number of displacements. The results show that the application of the offset-method improves the calculation accuracy up to an order of magnitude. The accuracy of the procedure depends highly on the resolution of the fluid grid. Accordingly, an optimal ratio between the grid size and the particle diameter in the range of 2–3 is obtained. Furthermore, the results display that the offset-method can maintain the high accuracy independent of the particle numbers. The additional computational effort that results from the using of the offset-method is negligibly small for granular flow with a low solid content. However, the computational effort of the offset-method increases at higher particle numbers. The particle diameter has a remarkable impact on the additional computational effort. By constant number of particles, smaller particles cause more additional computing time compared to bigger ones.

In the DEM literature, the physical values of the gas and solid phases are calculated in a common fluid grid with a size of at least two times larger than the particle diameter. If the particle diameter is equal or bigger than the grid size, the determination of the inter-phase values, i.e. volumetric void fraction and the rate of momentum and heat transfer is not possible (for example: volumetric void fraction becomes zero), which in turn lead to numerical errors. The physical values of the fluid and particle phases were determined by means of separate grids for fluid phase and particle phase. The so-called particle-grid method allows the refinement of the fluid grid resolution beyond the particle size. The results show that the application of the particle-grid method improves significantly the calculation accuracy. The variation of the fluidization mass flow rate has no influence on the procedure accuracy. The ratio between the fluid and particle grids has an influence on the computational time as well as on the quality of the obtained result. A ratio in the range between 4 and 10 is suggested that presents a compromise between an acceptable computing time and good accuracy. The resolution of the fluid grid should be fine enough to resolve the scales of motion accurately, while the resolution of the particle-grid depends on the particle diameter. In order to maintain the accuracy of the averaging procedure (determination of inter-phase values), the particle-grid size should be selected around two to three times bigger than the particle diameter. The additional computational effort, resulting from the use of the particle-grid method, rises with increasing the solid loading. The particle diameter has a considerable influence on the computing time. When the particle diameter decrease, the computational time increases accordingly.

The collision detection between particle-particle/wall is based on the assumption of the DEM model and carried out with the aid of the particle search-grid method. The use of the particle search-grid method enhances the efficiency calculation of collision detection. However, it requires an additional grid, the so-called particle search-grid that is independent of the fluid and particle grids. The size of the particle search-grid depends highly on the particle diameter. A decrease in the size of the search-grid leads to a gain in the efficiency up to limit of the particle diameter. The results show that the size of the search-grid should not fall short of the particle diameter in order to avoid incorrect results from a physical point of view.

Parameters studies were performed in order to understand the influence of the stiffness coefficient and the fluid time step variations on the efficiency and accuracy of the Euler-Lagrange/DEM model. The obtained results were compared with the measurements in order to derive the optimum parameters for the DEM simulations. The results suggest that the application of higher stiffness coefficients (more than 10^6 N/m) improves the simulation accuracy slightly, however, the average computing time increases exponentially. For time intervals larger than five milliseconds, it has been found that the average computation time is independent of the applied fluid time step, while the simulation accuracy decreases extremely by increasing the size of the fluid time step. Fluid time steps smaller than five milliseconds lead on the one hand to a negligible improvement in the simulation accuracy, but on the other hand to exponential rise in the average computing time.

In order to evaluate the DEM model under the consideration of thermochemical reaction mechanism, three series of warm simulations were performed. In order to achieve this, the kinetic models for drying, pyrolysis, char-oxidation, homogenous gas phase reactions and the associated heat and mass transfer processes were implemented in the DEMEST programme. The bubbling flow regime (particle of group B) was selected for the warm fluidized bed simulations. For the first series of simulations, a mixture of methane and air was delivered at the reactor inlet. While the solid phase in this investigation was disregarded, the reactor was filled with inert sand particles in the second simulation case. The 36,500 particles used are spherical with an average diameter of 2.5 mm and a density of 2200 kg/m^3. In the last series of warm simulations, the reactor was filled with a mixture of 99% inert sand particles and 1% pine wood particles. Here, pure air was applied as a fluidizing medium. On the basis of these warm simulations, the coupling of the Euler-Lagrange/DEM model and the chemical reaction model is realised.

For the simulation of real systems containing billions of particles, the DEM model cannot be currently used by the reason of limited computational resources. Here, the parcel DEM model offers an efficient approach, where each tracked particle (parcel) represents up-to several thousands of real particles. As an example, the parcel DEM model, implemented in ANSYS-FLUENT programme, is applied to simulate the hydrodynamic behaviour of a lab-scale 3D cold-flow fluidized bed test rig as well as the reactive gas-solid flow behaviour in the carbonator and calciner reactors of the calcium carbonate-looping process. The simulated solid velocity and solid volume fraction, in addition to the pressure profiles at different height positions are in good accordance with the measured data. The good agreement between simulations and experiments supports the reliable application of the parcel DEM model to large-scale and reactive granular systems.

References

Abad A, Adánez J, Cuadrat A, García-Labiano F, Gayán P, Luis F (2011) Kinetics of redox reactions of ilmenite for chemical-looping combustion. Chem Eng Sci 66:689–702

Al-Maliki WAK, Alobaid F, Kez V, Epple B (2016a) Modelling and dynamic simulation of a parabolic trough power plant. J Process Control 39:123–138

Al-Maliki WAK, Alobaid F, Starkloff R, Kez V, Epple B (2016b) Investigation on the dynamic behaviour of a parabolic trough power plant during strongly cloudy days. Appl Therm Eng

Almohammed N, Alobaid F, Breuer M, Epple B (2014) A comparative study on the influence of the gas flow rate on the hydrodynamics of a gas-solid spouted fluidized bed using Euler-Euler and Euler–Lagrange/DEM models. Powder Technol 264:343–364

Alobaid F (2013) 3D Modelling and simulation of reactive fluidized beds for conversion of biomass with discrete element method. tuprints, Germany

Alobaid F (2018a) Dynamic simulation and flow stability analysis of a vertical natural circulation heat recovery steam generator. Appl Energy, Submitted

Alobaid F (2018b) Start-up improvement of a supplementary-fired large combined-cycle power plant. J Process Control 64:71–88

Alobaid F, Epple B (2013) Improvement, validation and application of CFD/DEM model to dense gas-solid flow in a fluidized bed. Particuology 11:514–526

Alobaid F, Postler R, Ströhle J, Epple B, Kim H-G (2008) Modeling and investigation start-up procedures of a combined cycle power plant. Appl Energy 85:1173–1189

Alobaid F, Ströhle J, Epple B, Kim H-G (2009) Dynamic simulation of a supercritical once-through heat recovery steam generator during load changes and start-up procedures. Appl Energy 86:1274–1282

Alobaid F, Busch J-P, Ströhle J, Epple B (2012a) Investigations on torrefied biomass for co-combustion in pulverized coal-fired furnaces. VGB PowerTech

Alobaid F, Pfeiffer S, Epple B, Seon C-Y, Kim H-G (2012b) Fast start-up analyses for Benson heat recovery steam generator. Energy 46:295–309

Alobaid F, Baraki N, Epple B (2014a) Investigation into improving the efficiency and accuracy of CFD/DEM simulations. Particuology 16:41–53

Alobaid F, Karner K, Belz J, Epple B, Kim H-G (2014b) Numerical and experimental study of a heat recovery steam generator during start-up procedure. Energy 64:1057–1070

Alobaid F, Ohlemüller P, Ströhle J, Epple B (2015a) Extended Euler-Euler model for the simulation of a 1 MW_{th} chemical–looping pilot plant. Energy 93:2395–2405

Alobaid F, Starkloff R, Pfeiffer S, Karner K, Epple B, Kim H-G (2015b) A comparative study of different dynamic process simulation codes for combined cycle power plants–Part A: part loads and off-design operation. Fuel 153:692–706

Alobaid F, Starkloff R, Pfeiffer S, Karner K, Epple B, Kim H-G (2015c) A comparative study of different dynamic process simulation codes for combined cycle power plants–Part B: start-up procedure. Fuel 153:707–716

Alobaid F, Abed Kattea Al-Malikia W, Lanz T, Haaf M, Brachthäuser A, Epple B, Zorbach I (2018) Dynamic simulation of a municipal solid waste incinerator. Energy 149:230–249

Anderson TB, Jackson R (1967) Fluid mechanical description of fluidized beds. Equations of motion. Ind Eng Chem Fundam 6:527–539

Badzioch S, Hawksley PGW (1970) Kinetics of thermal decomposition of pulverized coal particles. Ind Eng Chem Process Des Dev 9:521–530

Baum M, Street P (1971) Predicting the combustion behaviour of coal particles. Combust Sci Technol 3:231–243

Benim A, Epple B, Krohmer B (2005) Modelling of pulverised coal combustion by a Eulerian-Eulerian two-phase flow formulation. Prog Comput Fluid Dyn Int J 5:345–361

Bhatia S, Perlmutter D (1981) A random pore model for fluid-solid reactions: II. Diffusion and transport effects. AIChE J 27:247–254

Bhatia S, Perlmutter D (1983) Effect of the product layer on the kinetics of the CO2-lime reaction. AIChE J 29:79–86

British Coal Utilisation Research, A., Field MA (1967) Combustion of pulverised coal. B.C.U.R. A., Leatherhead (Sy.)

Di Renzo A, Di Maio FP (2004) Comparison of contact-force models for the simulation of collisions in DEM-based granular flow codes. Chem Eng Sci 59:525–541

Epple B, Leithner R, Linzer W, Walter H (2012) Simulation von Kraftwerken und Feuerungen. Springer, Berlin

Epple B, Ströhle J, Alobaid F, Busch J-P (2014) Qualifizierung torrefizierter biogener Reststoffe für den Einsatz in Kraftwerksfeuerungen

Ferziger JH, Perić M (2002) Computational methods for fluid dynamics, 3rd edn. Springer, Berlin

Garcıa-Labiano F, Abad A, De Diego L, Gayan P, Adanez J (2002) Calcination of calcium-based sorbents at pressure in a broad range of CO_2 concentrations. Chem Eng Sci 57:2381–2393

García IL, Álvarez JL, Blanco D (2011) Performance model for parabolic trough solar thermal power plants with thermal storage: comparison to operating plant data. Sol Energy 85:2443–2460

Golovina E, Klimov A (1999) On the true kinetic constant of the heterogeneous C+CO_2 gasification reaction. Combust Explosion Shock Waves 35:393–396

Götz S (2006) Gekoppelte CFD-DEM-simulation blasenbildender Wirbelschichten. Shaker

Kaltschmitt M, Hartmann H, Hofbauer H (2009) Energie aus Biomasse: Grundlagen, Techniken und Verfahren. Springer, Berlin

Ketterer JC (2014) The impact of wind power generation on the electricity price in Germany. Energy Econ 44:270–280

Klutz H-J, Maser C, Block D (2006) WTA-Feinkorntrocknung, Baustein für die Braunkohlekraftwerke der Zukunft: Entwicklung und Betriebsergebnisse der Versuchsanlage. VGB PowerTech 86:57–61

Lee MS (1986) Mathematical model and control of a coal pulverizer. University of Austin

Link JM (2006) Development and validation of a discrete particle model of a spout-fluid bed granulator. PrintPartners Ipskamp

Magnussen BF, Hjertager BH (1977) On mathematical modeling of turbulent combustion with special emphasis on soot formation and combustion. In: Symposium (International) on combustion. Elsevier, Amsterdam, pp 719–729

May J, Alobaid F, Ohlemüller P, Stroh A, Ströhle J, Epple B (2017a) Reactive two-fluid model for chemical-looping combustion—Part A: simulation of fuel reactor. Int J Greenhouse Gas Control. Submitted

May J, Alobaid F, Ohlemüller P, Stroh A, Ströhle J, Epple B (2017b) Reactive two-fluid model for chemical-looping combustion—Part B: simulation of air reactor. Int J Greenhouse Gas Control. Submitted

Mertens N, Alobaid F, Starkloff R, Epple B, Kim H-G (2015) Comparative investigation of drum-type and once-through heat recovery steam generator during start-up. Appl Energy 144:250–260

Mertens N, Alobaid F, Lanz T, Epple B, Kim H-G (2016a) Dynamic simulation of a triple-pressure combined-cycle plant: hot start-up and shutdown. Fuel 167:135–148

Mertens NJ, Alobaid F, Epple B, Kim H-G (2016b) Combined-cycle start-up procedures: dynamic simulation and measurement. In: ASME 2016 Power Conference collocated with the ASME 2016 10th International Conference on Energy Sustainability and the ASME 2016 14th International Conference on Fuel Cell Science, Engineering and Technology. American Society of Mechanical Engineers, pp V001T011A006–V001T011A006

Nikolopoulos A, Stroh A, Zeneli M, Alobaid F, Nikolopoulos N, Ströhle J, Karellas S, Epple B, Grammelis P (2017) Numerical investigation and comparison of coarse grain CFD–DEM and TFM in the case of a 1 MW_{th} fluidized bed carbonator simulation. Chem Eng Sci 163:189–205

Ohlemüller P, Alobaid F, Gunnarsson A, Ströhle J, Epple B (2015) Development of a process model for coal chemical looping combustion and validation against 100 kW th tests. Appl Energy 157:433–448

Ohlemüller P, Busch J-P, Reitz M, Ströhle J, Epple B (2016) Chemical-looping combustion of hard coal: autothermal operation of a 1 MWth pilot plant. J Energy Res Technol 138:042203

Ohlemüller P, Alobaid F, Abad A, Adanez J, Ströhle J, Epple B (2017) Development and validation of 1D process model with autothermal operation of a 1 MW_{th} chemical looping pilot plant. Int J Greenhouse Gas Control. Submitted

Postler R (2012) Modellbildung und simulation instationärer Prozesse in einer Oxyfuel-Kraftwerksanlage. Cuvillier

Sharma R, May J, Alobaid F, Ohlemüller P, Ströhle J, Epple B (2017) Euler-Euler CFD simulation of the fuel reactor of a 1 MW_{th} chemical-looping pilot plant: Influence of the drag models and specularity coefficient. Fuel 200:435–446

Sowinski A, Miller L, Mehrani P (2010) Investigation of electrostatic charge distribution in gas-solid fluidized beds. Chem Eng Sci 65:2771–2781

Spalding DB (1982) The "Shadow" method of particle-size calculation in two-phase combustion. In: Symposium (International) on combustion. Elsevier, Amsterdam, pp 941–951

Spliethoff H (2010) Power generation from solid fuels. Springer, Berlin

Starkloff R, Alobaid F, Karner K, Epple B, Schmitz M, Boehm F (2015) Development and validation of a dynamic simulation model for a large coal-fired power plant. Appl Therm Eng 91:496–506

Starkloff R, Postler R, Al-Maliki WAK, Alobaid F, Epple B (2016) Investigation into gas dynamics in an oxyfuel coal fired boiler during master fuel trip and blackout. J Process Control 41:67–75

Stroh A, Alobaid F, Busch J-P, Ströhle J, Epple B (2015) 3-D numerical simulation for co-firing of torrefied biomass in a pulverized-fired 1 MW_{th} combustion chamber. Energy 85:105–116

Stroh A, Alobaid F, von Bohnstein M, Ströhle J, Epple B (2017) Numerical CFD simulation of 1 MWth circulating fluidised bed using the coarse grain discrete element method with homogenous drag models and particle size distribution. Fuel Process Technol, pp 1173–1189

Tsuji Y, Tanaka T, Ishida T (1992) Lagrangian numerical simulation of plug flow of cohesionless particles in a horizontal pipe. Powder Technol 71:239–250

Tsuji Y, Kawaguchi T, Tanaka T (1993) Discrete particle simulation of two-dimensional fluidized bed. Powder Technol 77:79–87

Wolf C (2005) Erstellung eines Modells der Verbrennung von Abfall auf Rostsystemen unter besonderer Berücksichtigung der Vermischung. Universität Duisburg-Essen, Fakultät für Ingenieurwissenschaften

Zhou J, Yu A, Horio M (2008) Finite element modeling of the transient heat conduction between colliding particles. Chem Eng J 139:510–516

Chapter 5
Conclusion

Thermal power generation remains indispensable in the near future in order to cover the base load demand. In many countries however, the desirable expansion of intermittent energy sources is drastically altering the traditional landscape of thermal power plants. New operating conditions are therefore required to mitigate these constraints such as the higher load gradients for both positive and negative load changes, the high efficiency at off-design operations, while maintaining low emissions and the fast start-up procedures. A thermal power plant that meets these new requirements has a competitive advantage in the electricity market. Experiments can contribute to this field of research, providing fundamental insight into the hydrodynamic behaviour of the process. However, the availability of detailed operation data from real systems is difficult because of the lack of accessibility, the harsh environment and the costs of measuring devices. Mathematical models, including one-dimensional dynamic process simulation and three-dimensional computational fluid dynamics, offer significant contributions to the direct measurements.

5.1 General Review

Dynamic process simulation becomes an integral part of design, operation and optimisation of energy systems. The basic task for dynamic simulation is to calculate the response of the thermal power plant and its control circuits to malfunction or to a change in load demand as well as to assess the capabilities and limitations of the system with regard to materials, emissions or economics. This implies strong requirements on both model accuracy and efficiency of the numerical solver. Relevant process components required for modelling the energy systems such as tubes, flow valves, heat exchangers, turbomachines etc. are discussed, complemented by automation and electrical modules vital for power plant control. The latter include analogue and binary modules, signal sources, controllers, generator,

F. Alobaid, *Numerical Simulation for Next Generation Thermal Power Plants*,
Springer Tracts in Mechanical Engineering,
https://doi.org/10.1007/978-3-319-76234-0_5

transformer, inverter and others. Generally, the process simulation codes are based on the governing conservation equations of mass, momentum, species and energy. The specific mathematical formulation of the balance equations depends on the underlying flow model. Many approaches that can be found in the literature are reviewed such as mixture flow model and two-fluid models, which in turn can be divided into four-equation, five-equation, six-equation and seven-equation flow models. Due to its relative simplicity and suitability for a wide range of practical applications, the mixture flow model is of considerable relevance since the calculation of average mixture properties is often sufficiently accurate for system-level analysis. The two-fluid model offers the possibility to consider thermodynamic non-equilibrium phenomena. Thus, it is more suitable for detailed analysis of specific components and application cases characterised by intense mass and heat transfer between the phases. The resulting partial differential equation system is discretised and typically closed with empirical correlations, which are selected according to the prevailing flow regime.

In this work, dynamic process simulation models for different technologies of combined-cycle power plant, pulverised coal-fired power plant, concentrated solar power plant and municipal waste incineration were developed. According to the comparison between the numerical results and the operation data obtained from real thermal power plants (e.g. Prai Power Station, Heilbronn Power Station and Andasol II Solar Power Station), the following conclusions are obtained:

- At different part loads, the dynamic simulation models can reproduce the behaviour of the investigated thermal power plant with high accuracy. Here, the error percentage for mass flow rate, temperature, pressure and other parameters lies in the range between 1 and 5%.
- At off-design operations, the dynamic simulation models can follow the measurements with a maximum relative error of about 12%. Several parameters show, however, a relative error of less than 5% or even a quantitative behaviour to the real thermal power plant is simulated.
- During warm start-up, hot start-up and shutdown procedures, the dynamic simulation models can reproduce the qualitative behaviour of the real power plant. Discrepancies between simulations and measurements are observed. Here, the early steam generation and the smooth development of feedwater mass flow rates in the numerical models are major problems. These are attributed to incomplete information on real operation, control circuits as well as the lower thermal inertia, suggesting the importance of knowing the correct setting of the valve controllers and the relevance for modelling of auxiliary systems such as vent and draining circuits.
- Compared to the mixture flow model, the two-fluid model simulates more accurately the behaviour of the thermal power plant during the start-up procedure. The correct time prediction of the steam mass flow rate and temperature jumps as well as the better representation of the measured parameter oscillations are advantages for the two-fluid flow model.

Validated dynamic simulation models can reliably predict the thermal power plant response to malfunctions (e.g. gas turbine load rejection, steam turbine trip and blackout), to changing in operative parameters (e.g. high load gradients and fast start-up), to modifications in design and control structures (e.g. once-through super-critical heat recovery steam generator and oxyfuel pulverised coal-fired power plant). These dynamic simulation models, although they can describe the entire process with relatively low computational effort, are not able to reveal a detailed insight into the flow pattern. By contrast, the computational fluid dynamics that is more appreciated for the simulation of individual components of thermal power plants provides valuable information on the hydrodynamic behaviour of reactors. For gas-solid flows, there are different numerical methods available: quasi-single-phase, two-fluid and single-particle model. In the quasi-single-phase model, the gas-solid flow is modelled with the conservation equations of the single-phase flow, while the gas and solid phases are modelled separately in the two-fluid model. The single-particle model combines Eulerian description of fluid phase (conservation equations of the single-phase flow is solved) with Lagrangian representation of dispersed phase (for each particle a set of transport equations is solved). The particle-particle and particle-wall collisions in the two-fluid model are considered with the aid of additional assumptions, while these can be stochastically or deterministically detected in the single-particle model.

In this work, several combustion models for coal, biomass and municipal waste that are based on quasi-single-phase, two-fluid and Euler-Lagrange/MP-PIC models were developed. Comparison studies were carried out between the numerical models and the measurements obtained from entrained-flow reactor, quasi-2D fluidized bed model, 1 MW_{th} combustion chamber, 1 MW_{th} fluidized bed, 60 MW_{th} waste incineration, large-scale lignite and hard coal-fired power plants. The results show the following points:

- The distribution of velocity, temperature and pressure in the reactor, the flue gas composition as well as the solid fuel conversion can be predicted with good accuracy.
- Quantitative deviations are observed, suggesting that the validation against the experiments should not be waived.
- Detailed reaction mechanisms for the gas species that include nitrogen, sulphur and chlorine chemistry can be included in the simulation in order to estimate the corrosion process in thermal power plants.
- The coal combustion model can reproduce the behaviour of the gas-solid flow in the power plant with higher accuracy than the coal/biomass or municipal waste combustion model. In addition to the long residence time, the heterogeneous structure of biomass or waste has the consequence that diverse reaction rates and different reaction mechanisms occur during the pyrolysis process and the char-combustion, which in turn make the problem of high complexity.

Furthermore, a numerical code, so-called DEMEST was developed, in which the particle-particle/wall and gas-solid interactions are modelled by tracking all individual particles. For this purpose, the Euler-Lagrange/discrete element method is applied and further extended as follows:

- The gas-solid interaction is achieved using a new procedure, known as the offset-method, which results in several numbers of spatial displacements of the fluid grid. The proposed method is highly precise in determining the inter-phase values (volumetric void fraction, rates of momentum and heat transfer). The results show an improvement in the calculation accuracy up to an order of magnitude. The accuracy of the procedure depends on the fluid grid resolution, but independent of the solid loading. An optimal ratio between the grid size and the particle diameter in the range of 2–3 is suggested. The additional computational effort rises with increasing the number of particles. At same solid loading, smaller particles result in more additional computing time compared to particles with bigger diameters.
- An additional grid, the so-called particle-grid, in which the physical values of the solid phase are computed, is introduced. The proposed procedure allows the refinement of the fluid grid resolution beyond the particle size. The results show a significant improvement in the calculation accuracy. The quality of the result and the additional computational time are a function of the fluid grid to the particle-grid ratio. A ratio in the range between 4 and 10 is suggested, showing a very good compromise between an acceptable computing time and high simulation accuracy. The particle-grid size should be selected around two to three times bigger than the particle diameter in order to maintain the averaging procedure accuracy regarding the determination of inter-phase values. At constant grid ratio, the additional computational time rises with increasing the particle numbers. A reduction in particle diameter results in an increase of the computing time.
- The collision detection between particle-particle and particle-wall is performed using the particle search-grid method. This method requires an additional grid, the so-called particle search-grid that is independent of fluid and particle grids. The results show an efficiency dependence on the size of the search-grid. A refinement in the search-grid size leads to a gain in the calculation efficiency. However, the size of the search-grid should not be smaller than the particle diameter in order to maintain the energy conservation.
- A validation study is carried out to assess the results delivered from the extended Euler-Lagrange/DEM model and a quasi-2D spouted fluidized bed of Plexiglas. Air at different mass flow rates (0.006 kg/s and 0.005 kg/s) is supplied from the bottom of the model, i.e. through nine holes. The results show that the Euler-Lagrange/DEM model can accurately predict the hydrodynamic behaviour of the gas-solid flow in the fluidized bed. The simulated spatial distribution of solid, the bed height and the equivalent bubble diameter agree very well with the measurements. A relatively small deviation is noticed, especially in the nozzle region, caused by the inlet assumption of the centrally placed nozzle instead of the nine holes. Furthermore, the numerical results of the Euler-Lagrange/DEM model are compared with the numerical results of the two-fluid and

Euler-Lagrange/MP-PIC models. The results show that the two-fluid model tends to predict smaller bubble sizes and bed expansions than the measurements, while the Euler-Lagrange/MP-PIC model reaches faster the maximum bed expansions.

- The results show that the stiffness coefficient and the fluid time step need to be set up very precisely in order to achieve efficient and accurate DEM calculations. The application of the realistic stiffness coefficients is not possible due to the exponential increase in the computational time. Depending on the relative velocity of collision partners, stiffness coefficient values between 10^4 N/m and 10^6 N/m are recommended, since they offer a good compromise between the efficiency and accuracy. The obtained results show that fluid time steps with interval size larger than five milliseconds cannot accelerate the simulation, but lead to worse temporal resolution of the fluid phase. The reduction of the fluid time step size smaller than five milliseconds in order to achieve higher resolutions causes an efficiency loss together with a negligible improvement in the simulation accuracy. Independent of solid loading and particle diameter, a fluid time step in the range between 1 ms and 5 ms is suggested according to the simulations and the experiments.
- Three series of warm simulations in a quasi-2D fluidized bed model are performed: combustion with fuel gas without and with inert sand particles as well as combustion with solid fuel (a mixture of inert sand and pine wood particles). The results confirm the possible coupling of the Euler-Lagrange/DEM model and the chemical reaction mechanism.
- For industrial applications including reactive gas-solid flows, the use of the Euler-Lagrange/DEM model is computationally very expensive and is difficult to achieve with the performance of currently available computers. The two-fluid and Euler-Lagrange/MP-PIC models offer here alternative approaches with comparatively lower computational effort. Recently, the parcel DEM model, in which each tracked parcel represents up-to several thousands of real particles, presents a promising approach. The parcel DEM model is applied to simulate a 3D cold-flow fluidized bed test rig as well as the carbonator and calciner reactors of a 1 MW_{th} calcium carbonate-looping pilot plant, showing good agreement.

5.2 Future Prospects

Detailed dynamic simulation modellings were presented in this book for different thermal power plants during off-design operation, start-up procedure and malfunction. However, the following points should be considered for future works:

- The dynamic of the combined-cycle power plant is well demonstrated. In contrast, the integrated gasification combined-cycle (IGCC) process and the interaction of syngas path components are not well understood and have to be the subject of further study.

- Although the technology of the coal-fired power plant is well-known, there is still potential for further improvement regarding load gradients, minimal load limit, start-up procedure and oxyfuel combustion, which can be explored using dynamic simulation.
- The dynamic of the municipal waste incineration is not reported in literature. Additional studies to that presented in this work are recommended, e.g. for investigating the influence of the waste heating value variation on the water/steam circuit and accordingly on the plant operation and efficiency. Furthermore, operation data from an existing waste incinerator have to be provided to the literature. These measurements are of high importance for the validation of dynamic models.
- Most studies of the concentrated solar power plant focus on solar field dynamics and use a simplified steady-state model rather than a detailed dynamic model of the power block. Additional studies to that presented in this work are required in order to analyse the dynamic interaction of sub-systems (solar field, thermal energy storage and power block).
- Simple dynamic models are applied to flue gas cleaning devices (e.g. dust removal, NO_x removal, SO_x removal and CO_2 capture) by the reason of insufficient library components of process simulation programmes. The flue gas cleaning devices are modelled as pressure drops as well as thermal masses and thus detailed dynamic models should be developed.
- Novel thermal power plants with energy storage systems or hybrid power plants that integrate renewables in conventional power plants have to be the subject of further research. Furthermore, the supercritical CO_2 (S–CO_2) Brayton cycle is a promising power conversion system for application to next generation thermal power plants. Advantages of this technology include the possible use of several heat sources (e.g. fossil fuels, waste heat and renewable heat sources), the high thermal efficiency at intermediate turbine inlet temperature in the range between 400 and 600 °C, the simple cycle layout and the compactness of components (e.g. turbomachinery and heat exchangers).
- The dynamic simulation and optimisation of energy systems should shift away from commercial codes towards more openly accessible models and software tools.

Detailed CFD simulation results of various energy systems were presented in this book. However, the following points should be kept in mind for future works:

- Two-fluid and Euler-Lagrange/MP-PIC models, by contrast to quasi-single-phase model, are rarely applied to simulate the hydrodynamic behaviour of reactive gas-solid flows. Additional studies are required in order to gain a better understanding of the heat transfer between fluid and solid phases as well as the heat transfer within the particle-particle and particle-wall collisions.
- Commercial CFD programmes used for the simulation of energy systems are considered as black boxes with minimum knowledge of their internal processes. Some programmes give flexibility to implement new algorithms and physical or chemical models. For example, the ANSYS-FLUENT programme enables the users to write their own sub-models using the programming language "C" and to

hook them as so-called User-Defined-Functions to the programme. However, the use of the commercial CFD programmes is related to license cost. Recent studies are dedicated to shift the field of CFD simulation away from commercial software towards open source codes such as OpenFOAM, a promising approach that the author would encourage other researchers to follow.

- The calculation of the adhesive forces without physical contact (van der Waals and electrostatic forces) and the adhesive forces with physical contact (due to liquid or solid bridges) are still an open issue that needs further investigations.
- The Sauter mean diameter that often used in the CFD studies over-predicts the pressure profile along the whole reactor height. It is strongly recommended to use a particle size distribution for representation of the solid phase.
- Drag models have a crucial effect on the characteristic of gas-solid flows. The conventional drag models (e.g. Syamlal and O'Brien, Wen and Yu as well as Gidaspow) assume that in each computational cell homogeneous conditions exist, neglecting any heterogeneities induced by the complex gas-solid flow. These drag models tend to over-predict the induced drag force and under-predict the slip velocity. Many advanced drag models (e.g. three-zone, four-zone and energy minimization multi-scale), which can describe in a more precise manner the momentum transfer between gas and solid phases, have been developed recently. However, the research should be active on improving these drag models and on proposing new theories and computational formulations.
- The modelling of turbulence in dispersed flows is usually neglected. The reason of this is that on the one hand the effect of the turbulence on the velocity field in dense gas-solid flows is low and on the other hand low Reynolds numbers are considered. For practical engineering applications, a strong influence of intensive turbulence on particle motions is detected. Accordingly, the statistical two-equation turbulent model $k - \varepsilon$ recently is applied to model the motion of the fluid phase. Great efforts in terms of meaningful experimental and numerical investigations must be performed in order to have a better understanding of turbulence-particle interactions using complex turbulence models such as Reynolds stress model or large eddy simulation.
- The modelling of heat transfer through radiation in granular flows is generally not considered due to the problem complexity. Detailed models that describe the radiative heat transfer between gas and solid phases are actual state of research and further experimental and theoretical investigations are required.
- The modelling of particles with complex geometries in the DEM model is limited by the unknown mechanical behaviour of these particles. The conventional models that are used to calculate the volume and surface forces as well as the moments of force in case of spherical particles should be extended based on the particle shapes. There remains also a need for research into the parcel DEM model for particles with complex geometries under the consideration of the thermochemical reaction mechanism.

The manufacturer's authorised representative in the EU is Springer Nature Customer Service Centre GmbH, Europaplatz 3, 69115 Heidelberg, Germany. If you have any concerns regarding our products, please contact ProductSafety@springernature.com

Printed and bound by CPI Group (UK) Ltd, Croydon, CR0 4YY
05/10/2026
02228147-0003